Hunting for Sustainability in Tropical Forests

Biology and Resource Management Series

Biology and Resource Management Series

EDITED BY MICHAEL J. BALICK, ANTHONY B. ANDERSON,
AND KENT H. REDFORD

Hunting for Sustainability in Tropical Forests

EDITED BY

JOHN G. ROBINSON

ELIZABETH L. BENNETT

Columbia University Press

New York

Columbia University Press
Publishers Since 1893
New York   Chichester   West Sussex

Copyright © 2000 Columbia University Press
All rights reserved

Library of Congress Cataloging-in-Publication Data

Hunting for sustainablity in tropical forests / John G. Robinson and
    Elizabeth L. Bennett, editors
        p.   cm. — (Biology and resource management series)
    Includes bibliographic references.
    ISBN 978-0-231-10976-5 (alk. paper)— 978-0-231-10977-2 (pbk.: alk. paper)
    1. Hunting   2. Hunting and gathering societies.   3. Wildlife
    management   I. Robinson, John G.   II. Bennett, Elizabeth L.
    III. Series: Biology and resource management series.
    SK35.H774   1999
    333.95'416'0913—dc21   99–20178

Casebound editions of Columbia University Press books are printed on permanent and
durable acid-free paper.

Printed in the United States of America

Contributing Authors

Michael Alvard
State University of New York
380 MFAC
Buffalo, NY 14261
U.S.A.

Philippe Auzel
Faculté des Sciences Agronomiques de Gembloux
Unité de Sylviculture
Passage des Déportes 2
B 5030
Gembloux, Belgique

Elizabeth L. Bennett
Wildlife Conservation Society
7 Jalan Ridgeway
Kuching, Sarawak
Malaysia

Richard E. Bodmer
Program for Studies in Tropical Conservation
Department of Wildlife Conservation
University of Florida
Gainesville, FL 32611
U.S.A.

Lynn Clayton
Ecosystems Analysis and Management Group
Department of Biological Sciences
University of Warwick
Coventry CV4 7AL
United Kingdom

Bryan Curran
Wildlife Conservation Society
B.P.3055
Messa, Yaounde
Cameroon

Ruben Cueva L.
Fundación Ecuatoriana de Estudios Ecológicos (EcoCiencia)
Isla Cristobal 1523 e Isla Seymour
P.O. Box 17-12-257
Quito, Ecuador

Heather Eves
School of Forestry and Environmental Studies
Yale University
New Haven, CT 06511
U.S.A.

John H. Fanshawe
Wildlife and Conservation Research Unit
Department of Zoology
South Parks Road
Oxford, OX1 3PS
United Kingdom

John E. Fa
Jersey Wildlife Preservation Trust
Les Augrès Manor
Trinity, Jersey JE3 5BP
Channel Islands
United Kingdom

Cheryl Fimbel
Wildlife Conservation Society
185th St. and Southern Blvd.
Bronx, NY 10460
U.S.A.

Clare D. FitzGibbon
Large Animal Research Group
Department of Zoology
Cambridge CB2 3EJ
United Kingdom
and
Estación Biológica de Doñana (CSIC)
Apdo. 1056
41080 Sevilla
Spain

P. Bion Griffin
Department of Anthropology
University of Hawaii at Manoa
Honolulu, HI 96822
U.S.A.

Marcus B. Griffin
Department of Anthropology
University of Illinois at Urbana-Champaign
Urbana, IL 61801
U.S.A.

John Hart
Wildlife Conservation Society
185th St. and Southern Blvd.
Bronx, NY 10460
U.S.A.

Kim Hill
Department of Anthropology
University of New Mexico
Albuquerque, NM 87131
U.S.A.

Jeffrey P. Jorgenson
Departamento de Biología
Pontificia Universidad Javeriana
Apartado Aéreo 56710
Santafé de Bogotá, DC
Colombia

Ullas Karanth
Wildlife Conservation Society
185th St. and Southern Blvd.
Bronx, NY 10460
U.S.A.

Margaret F. Kinnaird
Wildlife Conservation Society
P.O. Box 311
Jl. Ciremai No. 8
Bogor 16003
Indonesia

Rob J. Lee
Wildlife Conservation Society
185th St. and Southern Blvd.
Bronx, NY 10460
U.S.A.

Frans T. Leeuwenberg
Centro Pesquisa Indígena
Caixa Postal 25.945
Sao Paulo SP 05599-970
Brazil

M.D. Madhusudan
Centre for Ecological Research and Conservation
Mysore 570 002
India

Patricio Mena V.
Fundación Ecuatoriana de Estudios Ecológicos (EcoCiencia)
Isla Cristobal 1523 e Isla Seymour
P.O. Box 17-12-257
Quito, Ecuador

E.J. Milner-Gulland
Ecosystems Analysis and Management Group
Department of Biological
Sciences
University of Warwick
Coventry CV4 7AL
United Kingdom

Hezron Mogaka
Kenya Indigenous Forest Conservation Project
Kenya Forestry Research Institute
P.O. Box 20412
Nairobi, Kenya

Andrew Noss
Wildlife Conservation Society
Casilla 2417
Santa Cruz, Bolivia

Adrian J. Nyaoi
Wildlife Conservation Society
7 Jalan Ridgeway
Kuching, Sarawak
Malaysia

Timothy G. O'Brien
Wildlife Conservation Society
P.O. Box 311
Jl. Ciremai No. 8
Bogor 16003
Indonesia

Jonathan Padwe
Mbaracayu Forest Reserve
Canindeyu, Paraguay

Carlos Peres
School of Environmental Sciences
University of East Anglia
Norwich NR4 7TJ
United Kingdom

Pablo E. Puertas
Centro de Investigaciones Veterinarias Tropicales y de Altura
Universidad Nacional Mayor de San Marcos
Apdo. 575
Iquitos, Perú

Jhanira Regalado B.
Fundación Ecuatoriana de Estudios Ecológicos (EcoCiencia)
Isla Cristobal 1523 e Isla Seymour
P.O. Box 17-12-257
Quito, Ecuador

John G. Robinson
Wildlife Conservation Society
185th St. and Southern Blvd.
Bronx, NY 10460
U.S.A.

Richard Ruggiero
School of Forestry and Environmental Studies
Yale University
New Haven, CT 06511
U.S.A.

George B. Schaller
Wildlife Conservation Society
185th St. and Southern Blvd.
Bronx, NY 10460
U.S.A.

Jephte Sompud
Wildlife Conservation Society
7 Jalan Ridgeway
Kuching, Sarawak
Malaysia

Jody R. Stallings
CARE Ecuador
151 Ellis St.
Atlanta, GA 30303
U.S.A.

Allyn MacLean Stearman
Department of Sociology and Anthropology
University of Central Florida
Orlando, FL 32816
U.S.A.

Wendy Townsend
Casilla 6266
Santa Cruz, Bolivia

Leonard Usongo
WWF Cameroon
B.P. 2417
Douala, Cameroon

David S. Wilkie
Associates in Forest Research and Development
18 Clark Lane
Waltham, MA 02154-1823
U.S.A.

Contents

Foreword

Societies thrive on myths, of ancient heroes and divine origins, that unite and give identity. However, in an age when population growth, climate change, pollution, habitat destruction, and extinction of species endanger the future of many organisms, including ourselves, new myths have appeared that threaten all societies by hampering worldwide conservation efforts. As promulgated by developed nations, the new myths include the idea that natural resources are infinite, that technological fixes will solve all problems, that poverty can be eliminated by producing and consuming more, and that sustainability in use of resources can be achieved without a radical transformation of economics, culture, society, and politics. Conservation is in crisis. Conventional approaches have not worked. Development and conservation are on an accelerating collision course, and proposed solutions are no more than hopeful improvisations. Myths, even surreal ones, provide comfort and validate the denial of problems. Psychological agility, sentimental optimism, and a tendency to favor clichés undiluted by insight enable people to ignore fact or fiction. Instead of confronting uncertainty honestly, all too much of the conservation agenda consists of hollow and confused verbiage that promotes dogma rather than dialogue. Blindness to reality is dangerous.

These and other thoughts come to mind as I perused this volume on the impact of hunting wildlife by local peoples in tropical forests. The chapters are resonant with authority and illuminating in perceptions on an important subject that has until now been almost devoid of meticulous and intellectually rigorous data. It is noteworthy that the information is based on prolonged, strenuous field work by a core of dedicated biologists and anthropologists, not on opinions or on calculations derived from remote sensing, computer modeling, and other technological tools. Although considered archaic by some, natural history remains the cornerstone of knowledge about species (including the human species) and their role in ecosystems, and it provides the basis for elucidating the biological patterns and principles upon which conservation must depend. By studying the demographics of duikers and measuring the grams of meat eaten per day by a Wana villager, the way we think about and treat the world is affected. This volume focuses on a specific problem, but at the same time it addresses two basic global issues, both of which are in need of hard-eyed scrutiny: the sustainable use of resources and the attitudes toward nature by local people.

Sustainability is the most commonly invoked concept in the arsenal of solutions to a resource problem. But it is a vague and slippery concept that has commonly been designated by three different terms. One is "sustainable growth," an oxymoron because nothing can continue to grow indefinitely.

The second term is "sustainable development," enshrined by the 1992 Earth Summit in Rio de Janeiro as gospel. Most countries find the idea congenial because it represents, in effect, legalized environmental degradation. No matter what the rhetoric, development strives for economic growth, a way to increase consumption—not protect biodiversity. The creation of more wealth inevitably means greater environmental degradation. Development for whom? A cynic could point out that sustainable development is based on a Western perspective whose principal aim is to promote access to resources so that the developed nations can continue their profligate life-styles. Development has created the environmental crisis. The proposed solution is more of the same. Free markets will not preserve biodiversity and maintain ecosystems. One concern is that wealthy international corporations are becoming like nation states, beyond the reach of law of any one country, with massive power to develop and destroy almost at will. Half of the world's hundred largest economies are not countries but corporations (Kaplan 1997).

The third term, "sustainable use," is possible, as this volume shows, in that it aims to regulate consumption in such a way that a resource is not used up. Unfortunately, far too much of the earth's natural heritage is still carelessly compromised. We cannot avert a large disaster by permitting many small ones without making a major effort at sustainability. I am concerned that in recent years the value system via which nature is perceived has shifted dramatically away from ethics, esthetics, and religious sentiment toward wealth, toward dollars. This is perhaps a predictable outcome of the Western world's trajectory from the Age of Reason (seventeenth century) and Age of Enlightenment (eighteenth century) to what could be called the Age of Greed (twentieth century). An ancient Greek philosopher noted that if horses had gods, these gods would resemble horses. Similarly, in societies that idolize wealth and elevate billionaires almost to the status of gods, nature's virtue will be measured in dollar values.

A goal of many studies in this volume was to measure whether wildlife was being killed sustainably, and a main conclusion was that the "hunting of many species important to local inhabitants of tropical forests is not presently sustainable" (Bennett and Robinson this volume). Having for decades observed local peoples snare, trap, and shoot wildlife with skill and persistence from the Arctic to the Amazon and elsewhere, this conclusion did not startle me. A positive aspect of these reports is that although wildlife has been decimated, the habitat itself remains. I have for weeks wandered through fine rain forests in Laos with few signs or sightings of large birds and mammals. The best place to encounter barking deer, wild pig, squirrel, and even small birds such as bulbuls and barbets was not in the seemingly pristine forests, but in villages and markets. Virtually all vertebrate species in Laos become recipes. However, several studies in this volume offered me a new and disconcerting insight: hunting of a species may actually be sustainable, but the population is maintained at such a low level that its function in the ecosystem has almost been eliminated. If species have a critical function, such as fruit-eating bats and hornbills dispersing seeds, their reduction in number may ultimately result in many extinctions. Tornados, floods, and other natural disasters strike with sudden violence, whereas extinctions usually unfold imperceptibly through relentless attrition, sometimes over decades. How many species can an ecosystem lose before it is unable to compensate and collapses?

To determine ecologically sustainable methods of harvesting any resource is obviously complicated, as these studies make clear. Years of detailed observations are required, and even then solutions may remain elusive. Optimum harvest levels are more often than not based on trial and error. And limits must be imposed— limits on number of people, on number of trees cut, on medicinal herbs collected, on monkeys shot. Demand will inevitably exceed supply. Without enforced limits there can be no sustainability. Distrust facile claims of long-term sustainability! The motto of the Royal Society of London is *Nullius in verba* ("No one's word shall be final"). Look at the facts—and retain a measure of doubt. The facts will vary from place to place and from society to society. In North America, successful wildlife management is based on denying market value to dead animals (Geist 1988); in Kenya, game ranching and other commercial harvests may lead to better conservation of species and habitats (Kock 1995). Whatever the method, wildlife will survive only if local people want to retain it.

Native peoples in the Americas, subdued and decimated by colonists, have in recent years been burdened with the patronizing myth of the "ecologically noble savage" (Redford 1990), and this idea has been extended to Africa and Asia, where hundreds of millions of people are native and rural. Such people, it is said, tend to live in harmony with nature, harvesting resources sustainably; we should learn from them, emulate their ancient wisdom. Simplicity is romanticized, a balm to our complicated lives. Besides, it is easier to ascribe virtue to the poor and distant than to strive for it oneself. Guilt over misdeeds of the past contributes to the myth as well, so much so that even notorious resource abuse by local people, such as the massive clear-cutting of forests for quick profit by some Alaskan native corporations (Wuerthner 1988; Cornelius and Creed 1996), receives at most a gentle and politically correct admonition. It is not necessary to grant local people noble attributes in order to sympathize with rightful claims to land and desire for cultural continuity.

People have never lived wholly in balance with the environment. True, many propitiated the spirits of the animals they killed, but they were still pragmatists, not conservationists with an ecological conscience as defined in today's terms. They used what they needed, even if this depleted the resource. Early hunters with simple weapons helped to exterminate the mammoth and giant ground sloth, and various island birds, such as the moas of New Zealand, were eaten into extinction. Where people are few, their weapons primitive, and the market economy small, their impact has usually been modest. All too often this is no longer true. Half of the forest destruction in the world is by local people, not by logging companies. Wholly traditional people have virtually vanished.

If you buy or barter an aluminum pot, no matter how remote your home, you have become part of the global market economy. Subsistence now means not just food but purchasing power. By extracting resources for profit—whether gorilla meat or walrus ivory—and by succumbing to the need for radios, shotguns, and store food, people irrevocably exchange their old life-style for one based on a modern material existence. It is, of course, everyone's own choice, even if it means traditional dislocation. Some peoples also claim that hunting is necessary for their mental health and spiritual well-being (Kancewick and Smith 1991).

Ecological and ethical standards of conservation must apply to everyone—urban and rural, rich and poor, a Yuquí in the Amazon forest and the CEO in a corporate headquarters—if the earth's biodiversity is to persist in all its marvelous unity, complexity, and variety. Fortunately, no culture remains static. Traditions evolve as people accept new attitudes and habits and discard old ones, as they adapt to changing circumstances. It should be remembered that tradition does not necessarily represent the best in a culture but merely what survived from the past. Relentless hunting may well become less important economically and socially in various societies. Certainly some traditions have no place in today's world, among them the killing of rare or endangered species, even if ancient custom is invoked.

Local control over resources will not alone assure sustainability. This is an important insight of this volume, implicit in almost all chapters, and a most timely one because such control is uncritically promoted by many organizations. Communities need technical assistance to determine the biological limits of harvesting resources and of monitoring and managing them. Given the social and economic realities—and limits to human perfectibility—this is such a complex and expensive task that the chances of failure are greater than those of success, as shown by various programs in Africa (Gibson and Marks 1995). Communities need to work in partnership with government, universities, conservation organizations, and others toward a sustainable future.

We are fortunate that large expanses of forest and other habitats still persist, even if most have to some extent been modified by humans. The option to protect, conserve, and manage remains. But as populations grow, industry obtains large tracts of land to exploit, and habitats become increasingly fragmented, placing many species at risk, options will become ever more circumscribed. There must be limits on the amount of natural habitat that is destroyed through neglect and negligence. A part of every landscape must become a reservoir of biodiversity. Unless such is done, the idea of sustainability will remain a vision of ecological utopia, a search for a chimera forever beyond our grasp. People must accept the need for wild places where nature can function unimpeded and the land is treated with love, respect, and responsibility and where pillaging is prohibited. Leopold (1994) provided a guiding principle when he wrote: "A thing is right when it tends to preserve the integrity, stability and beauty of the biotic community. It is wrong when it tends otherwise."

GEORGE B. SCHALLER

Preface

Throughout the world, people are concerned about the decline of tropical forests and their wildlife. Trying to arrest such decline is now integral to the policies and programs of governments, aid agencies, and conservation organizations. Yet we are becoming increasingly aware that there is a basic dilemma in knowing how to achieve this. Recent field studies show that hunting has a dramatic impact on wildlife in many tropical forests throughout the world, frequently diminishing wildlife populations and sometimes driving species to local extinction. This has wider repercussions on the diversity and health of the forests themselves. Studies also have documented the importance of wildlife as a source of animal protein and cash income for rural peoples. All of these studies converge on the following questions: Are hunting rates as practiced by rural peoples sustainable? If not, what are the biological, social, and cultural implications of such hunting?

Answering these questions is ever more important as national and international agencies seek to integrate the development of local peoples with the conservation of tropical forest systems and species. One frequently used rationale for this approach is that the best way to conserve natural systems in the tropics is to give control of management to the local people and allow them to use the area's natural resources. However, this approach has been questioned because much local resource use is unsustainable, especially in the context of expanding development and participation in market economies. What ecological and socioeconomic factors affect the sustainability of resource use? There are answers to this question, but much of the information is unavailable to policy makers and project managers. The aim of this volume is to compile as many of the relevant studies as possible to examine the sustainability of hunting as practiced by rural peoples. If hunting is determined not to be sustainable, then we must address what can be done to meet the needs of people living in and around tropical forests and to achieve the long-term goal of conserving tropical forest systems.

This volume includes chapters by biological and social scientists to try to achieve a balance of viewpoints. The introductory chapter outlines some of the fundamental dilemmas of hunting covered by the book: how hunting is defined, what sustainability of hunting means, and how it can be assessed. The rest of the book is divided into five main parts. The first looks at the biological context of hunting, examining the impact of hunting on wildlife in tropical forests throughout the world. This starts with a detailed analysis of the carrying capacity of tropical forests for human hunters, the biological limits to the system, and implications for policy. We then have three case studies from the neotropics: a wide-ranging comparative study of hunting throughout the Brazilian Amazon, a consideration of the effects of changing technology on the sustainability of hunting in Ecuador, and an examination of "sources and sinks" in Paraguay and their importance to management. The theme of how the impact of hunting changes across the landscape continues with a long-term, detailed study of the effects of hunting on duiker and other wildlife populations in Congo-Zaire, consideration of the effects of hunting on mammal communities in northern Kenya at different distances from the forest edge, and examination of the effects of the wild meat trade in Equatorial Guinea on wildlife populations at different distances from the markets. The section finishes with two chapters examining the impacts of hunting on the unique faunal communities of Sulawesi in Indonesia.

Part II examines the sociocultural context of hunting: the importance of hunting to local communities and how that affects sustainability and management options. Again, chapters in this part cover a wide geographical and ethnic span, from the Sirionó in Bolivia, to a range of indigenous hunters in the Central African Republic and Borneo, to the Agta on the Philippine island of Luzon.

If hunting is to be sustainable in the modern world, management capacity is essential, be it in the local communities, local or central government, or other agency. Thus part III examines the institutional capacity for management and how that affects sustainability. It looks at the effect of different levels and types of anti-hunting enforcement on wildlife populations in India and considers the involvement of local communities in management: the Bantu and Baka in Cameroon, and the Xavante in Brazil. The final chapter of this section presents a case example of translating research into management by local communities in the Peruvian Amazon and examines how successful this has been in increasing the sustainability of hunting.

Part IV looks at hunting from an economic perspective. It explores how economic mechanisms shape the context in which hunting occurs and how they can be used to develop mechanisms to promote sustainability. This includes two chapters examining the interrelationships of logging, wild meat trade, and other economic changes on hunting practices in Congo, as well as two chapters looking at the economics of hunting in Sulawesi. The first investigates the importance of wildlife to the economies of local communities, and the second examines the role of wildlife trade in Sulawesi and how the balance between benefits and costs in the market system affects the sustainability of hunting in the forests.

Part V of the book is a synthesis, drawing together all of the chapters into an overview of hunting, summarizing the factors that influence sustainability both positively and negatively, and the implications for management.

As ever in an enterprise such as this, a large number of people have contributed in bringing the book to fruition. We are grateful to all those who have written chapters and for their patience and good nature throughout. We also thank those who took the time to review chapters: some of the authors themselves, as well as Katrina Brandon, Julian Caldecott, Alejandro Grajal, Mikaail Kavanagh, Dale

McCullough, Mike Fay, Ricardo Godoy, John Oates, Alan Rabinowitz, Kent Redford, Andrew Taber, William Vickers, and Tony Whitten. Particular thanks also go to Tracy Van Holt for her sterling work in compiling reference lists, formatting, and editing the book.

Finally, it is our hope that this book will bring the fundamental dilemma of hunting sustainability in tropical forests to the public arena, move forward the debate, and thereby bring us one step further toward knowing how best to conserve such forests throughout the world.

JOHN G. ROBINSON AND ELIZABETH L. BENNETT

1

Hunting for the Snark

ELIZABETH L. BENNETT AND JOHN G. ROBINSON

> "'They sought it with thimbles, they sought it with care;
> They pursued it with forks and hope;
> They threatened its life with a railway-share;
> They charmed it with smiles and soap—'"
>
> ('That's exactly the method,' the Bellman bold
> In a hasty parenthesis cried,
> 'That's exactly the way I have always been told
> That the capture of Snarks should be tried!')
> *From* The Hunting of the Snark, *by Lewis Carroll*

The hunt of the Snark, a search undertaken by the Beaver, the Barrister ("brought to arrange their disputes"), the Butcher ("He could only kill beavers"), the Banker (with "the whole of their cash in his care"), and the rest of the motley crew could almost be Lewis Carroll's allegory for the quest for sustainability, but if so he was before his time. The search for sustainability, in such an intractable area as hunting, is as elusive as the Snark in this classic tale. One reason for this is the wide spectrum of activities involving hunting. Hunting can be for subsistence or for commerce. Commercial hunting can be small scale or heavily capitalized. People may say that they hunt for necessity, pleasure, or out of cultural imperatives. The meat from hunted animals may be distributed within the community or may enter the marketplace. Meat can be consumed, stored, bartered, or sold. Yet in the study of hunting, rarely are these distinctions clear. The distinctions between subsistence and commercial, between traditional and modern, between sport and necessity are often fuzzy at best. Any case study of hunting, therefore, is challenged to specify carefully the socioeconomic context of the activity.

Once the context is defined, the next challenge is to quantify hunting patterns. Collecting meaningful data on hunting levels (Redford and Robinson 1990) is

1

difficult, and describing overall harvest rates problematic. Data on hunting in tropical forests are often scanty, collected over a relatively short period, under rapidly changing local conditions in the face of swift economic development. Interpretation of results often requires making assumptions on scanty grounds, and applying conclusions to policy is usually done with considerable hubris.

Yet it is imperative to understand the cultural and socioeconomic contexts, and to collect accurate data on hunting and its effects, if we are to base wildlife management on solid foundations. Only then will we be able to answer the question of whether hunting is sustainable and know what measures are needed to ensure that it is. Only if hunting is sustainable will wildlife populations be conserved and will people be able to continue to use wildlife resources.

In this chapter we explore certain key issues with the aim of providing context to the rest of the book. In particular, we have encouraged the authors to be consistent in their definitions and use of terms. Here we define hunting as it is used in the book. We describe the types of hunting with which the book is concerned, and we explain how we define sustainability in the context both of hunting and of the conservation of biodiversity in tropical forests.

DEFINITIONS

Hunting

Hunting has had different definitions over the millennia. The ancient Greeks and medieval writers distinguished different kinds of animal harvest based on the methods used by people. Hunting involved the active pursuit of large mammals and was distinguished from trapping, fishing, and other means of catching animals such as falconry (Cartmill 1993). In this book, hunting is used to include all capture by humans of wild mammals, birds, and reptiles, whether dead or alive, irrespective of the techniques used to capture them. This usually involves killing animals for human use, especially meat for eating, for traditional medicines or trophies, but it also can include taking of live animals as pets, or for the biomedical or zoo trades. It also encompasses taking animals both for personal consumption and for commercial sale. This definition of hunting is an ecological one, based as it is on the extraction of the wild animal from its environment. That extraction affects the species population being harvested and, to some extent, the entire biological community. From this perspective, it does not matter how an animal is killed: whether through shotguns, blowpipes, dogs, spears, arrows, machetes, traps, snares or nets.

Under this overall definition of hunting, it is still useful to distinguish active hunting, which involves active pursuit of animals, from trapping or passive capture of animals. This is largely because the two activities are often regarded very differently by the hunters; they are frequently conducted by different members of society, for different reasons and under different circumstances. So although the effect of both activities is to remove animals from the ecosystem, the study and management of them often needs to be handled in different ways.

In this book with its ecological perspective, we regard fishing as a separate activity from hunting. This is partly because the two are traditionally distinguished, but more important in this context is that fishing affects a different ecosystem to hunting (rivers, lakes, and seas rather than forests), so the effects will be different and management procedures often dissimilar. Although there is some ecological overlap, especially in the case of reptiles (e.g., crocodilians and terrapins), it is useful to keep hunting and fishing separate when considering management of tropical forests.

Hunting is generally done for one or more of the following reasons:

1. *Nutrition.* This can be highly significant: in at least 62 countries worldwide, wildlife and wild fish contribute a minimum of 20% of the animal protein in rural diets (Prescott-Allen and Prescott-Allen 1982). In the Amazon, wildlife provides significant calories to rural communities, as well as essential nutrients such as protein and fats (Dufour 1983; Flowers 1983; Yost and Kelley 1983; Stearman 1994; Townsend this volume). In Nicaragua, wildlife provides 98% of the meat and fish consumed by Miskito Indians (Nietschmann 1973). Wild meat is eaten in 67% of all meals of the Kelabits in parts of Sarawak and is their main source of animal protein (Bennett et al. this volume). The effects of nutritional hunting on wildlife populations can be considerable. More than 13,600 animals are harvested from the Arabuko-Sokoke Forest in Kenya each year (FitzGibbon et al. this volume). In Amazonas State, Brazil, the rural population annually kill about 3.5 million vertebrates for food (Robinson and Redford 1991a), and in Sarawak, subsistence hunters eat a minimum of 23,513 tons of wild meat per year (Bennett et al. this volume).

2. *Economy.* This falls into two categories. First, to a rural hunter, wildlife is an essential source of animal protein that otherwise would have to be raised or bought. The economic value of this wild meat if replaced with domestic meat can be very significant. In Sarawak, for example, the economic value of wild meat consumed by rural people if replaced by domestic meat is about $75 million per year (Wildlife Conservation Society and Sarawak Forest Department 1996). Second, hunting is done to obtain income from selling the animals for their meat and pelts, or as pets or trophies. In Bioko, Equatorial Guinea, 63 kg of wild meat per square kilometer of forest per year was extracted for commercial sale (Fa et al. 1995). In Tahuayo, a closely monitored site in the lowland forests of Peru, 44 kg of meat per square kilometer per year was extracted, representing 1,278 mammals from a 500 km² area (Bodmer et al. 1994). About 14% of this was consumed by people in the thirty-four extended families that hunted in the area, the rest was sold in local markets. The total value of that commercial sale was over $17,000 per year. Wildlife sales by rural peoples can be an important source of cash; each gun owner in Bomassa, Congo, sells wild meat worth about $395 per year (Eves and Ruggiero this volume), and the total estimated value of the wild meat sold in Sarawak is about $3.75 million per year (Wildlife Conservation Society and Sarawak Forest Department 1996). For many tropical forest peoples, the distinction between subsistence and commercial is not clear, with wild meat from the forest supplementing both diets and incomes. Attempts to quantify the combined

subsistence and commercial value of wild meat per unit area produced in a year have been made (e.g., Godoy et al. 1993). In Korup National Park, Cameroon, for example, each square kilometer of forest produces wild meat with an annual value of about $106 (Infield 1988). The figure for the Arabuko-Sokoke Forest in Kenya is about $94 (FitzGibbon et al. 1995b) and for the Ituri Forest, Democratic Republic of Congo (Congo-Zaire), is about $318 (Wilkie 1989). The sustainability of this economic production is questionable at both the Kenya and Cameroon sites. In Tahuayo, Peru, Bodmer et al. (1994) estimated that the forest is producing $42 of wild meat per square kilometer. If harvests were required to be sustainable, then this figure would decrease to $31 per square kilometer. The total production of wild meat at a regional or national level can be startling. If one considers the whole of Sarawak, the figure is about $83 million (Wildlife Conservation Society and Sarawak Forest Department 1996), and for the whole of the Amazon Basin, the figure exceeds $175 million per year (TCA 1995).

3. *Culture.* Again, there are two aspects. First is the acquisition of animal trophies as cultural artifacts or for personal adornment, e.g., feathers, skins and teeth. This is widespread throughout tropical forest regions. Second is that animals and hunting are inextricably woven into the world view of many cultures; to be a hunter is essential in gaining respect, achieving manhood, or winning a bride (Kwapena 1984; Brosius 1986; Caldecott 1988).

4. *Recreation.* Many hunters hunt because they enjoy it. This includes a wide range of hunters, from children with catapults, to townsfolk who spend large amounts of time, and often money, hunting in either their own country or overseas. The tropical forest hunters with whom this book is concerned, however, seldom make a clear distinction between recreational and other forms of hunting. Most tropical forest hunters would say that they like to hunt, yet they also hunt to fulfill their nutritional and economic needs, to obtain trophies, to obtain wild meat because they prefer it to other forms of protein, and because it is part of their culture.

Wild Meat, Bushmeat, and Game Animals

In most tropical forests, a major portion of the meat harvested by forest hunters comes from a relatively small number of large-bodied species (Redford and Robinson 1987). These species are normally the larger ungulates and primates (Robinson and Bennett this volume). For example, 44% of the biomass harvested by the Huaorani in Ecuador comprises just two species (*Lagothrix lagothricha* and *Tayassu tajacu*) (Mena et al. this volume); three species of ungulates comprise 80% of the biomass harvested by rural hunters in Sarawak (*Sus barbatus, Muntiacus muntjac,* and *M. atheroides*) (Bennett et al. this volume). These animals, important for human nutrition, are often described as *wild game* or *bushmeat,* and the species commonly hunted, or even raised to be hunted, as *game species.*

This terminology implies that hunters rely on relatively few species, the game animals, and that other species are infrequently or not hunted. This implication is erroneous, however, for hunting in tropical forests is much more catholic. If one

examines the number of animals harvested of different species, the relative importance of these large-bodied "game" species declines. For example, *Lagothrix
lagothricha* and *Tayassu tajacu* only make up 27% of the animals killed by the
Huaorani (Mena et al. this volume), and the three ungulates only 22% of the
animals hunted in Sarawak (Bennett et al. this volume). The number of species
taken by forest hunters is often large; the Maracá Indians of Colombia kill at least
51 species of birds, including 10 species of hummingbirds (Ruddle 1970);
Amazonian Indians of Ecuador hunt about 26 species of mammals, 18 of birds,
and four of reptiles (Vickers 1991); the Huaorani of Ecuador take about 25 species
of mammals, 11 of birds, and five of reptiles (Mena et al. this volume); and the
Sirionó Indians of Bolivia hunt 23 species of mammals, 33 of birds, and nine of
reptiles (Townsend this volume). A similar picture is seen in Africa: in southwest
Central African Republic, hunters using snares capture 33 species of mammals,
seven of reptiles, and three of birds (Noss this volume), and in the Lobéké region
of Cameroon, hunters kill at least 36 animal species (Fimbel et al. this volume). In
Asia, the picture is comparable, with at least 26 species of mammals, 12 of birds,
and five of reptiles being eaten by rural hunters in Sarawak (E. L. Bennett and
A. J. Nyaoi, unpublished data), and in Laos, hunters kill and eat any species of
mammal, bird, and reptile, from small birds to large ungulates (A. R. Rabinowitz
and G. B. Schaller personal communication).

The few species that provide most of the diet are thus not the only ones significantly affected by hunting. Hunting affects populations of a far wider array of
species. Indeed, to most tropical forest hunters, *game* is any animal ranging from
small rodent and hummingbird size, to large ungulates, and almost everything
else in between. In this volume, therefore, for convenience we use the term *game*
to mean any animal hunted for food, but it has no further implications about its
importance in the human diet, nor the ecological importance of the species.

WHEN IS HUNTING SUSTAINABLE?

Defining Sustainability Simply

No matter how important wild species are to people, if hunting is not sustainable,
then the resource will be depleted. People are then forced to spend more time and
energy seeking to locate an ever diminishing source of animal protein. Wildlife
populations can dwindle to such low levels that they provide little to human
communities and can become locally extirpated.

When is hunting sustainable? In the original formulation on sustainability, that
outlined in the World Conservation Strategy (IUCN/UNEP/WWF 1980), the
answer was that natural resource use is sustainable when it does not significantly
affect the wild population. In other words, the original thinking on sustainability
assumed that there are conditions in which the use of a resource has little or no
impact on the resource itself. Natural populations could be understood as being

natural capital, and the harvest as the interest. Resource users could harvest the interest without touching the capital. However, biological populations do not work like economic systems because density-dependent effects mean that the interest is not always proportional to the capital (Anonymous 1996). A smaller population might generate greater production. In addition, any harvest reduces a population, and lowered densities alone do not mean that hunting is unsustainable.

A more complete answer is that hunting is sustainable when the use or harvest of the resource does not exceed the production. The harvest rate of a resource is driven by the demands of consumers on one hand, and is controlled by taboos, rules, regulations, incentives and enforcement on the other. The production rate of the resource is determined by the density or numbers of organisms and per capita reproductive rates. When the two sides of the equation are balanced, when production balances harvest, then sustainability is possible.

This answer remains incomplete, however, for a number of reasons. First, harvest can equal production at many different population densities. This condition can be met at populations reduced by hunting to well below carrying capacity (K); indeed, it could be met at population densities ranging from K to $0.1 K$ or less. But if a population is lowered to such a small level that is in imminent danger of extirpation, or if a reduced population no longer meets the social and economic needs of the people who are harvesting the resource, many people would not consider this harvest sustainable. Second, wildlife populations are composed of living animals that are part of a dynamic and complex system. They live in social groups that are disrupted by the loss of individual members. Loss of significant numbers of members will have wider repercussions throughout the ecosystem: of browsers on their plant foods, of prey species on their predators, and vice versa. If a harvest so disrupts the biological community as a whole, many people would not consider the harvest sustainable, even if the production of the target species balanced the harvest. In other words, levels of harvests, in addition to being strictly sustainable, must also meet wider management goals for the resource and the users of the resource, as well as the habitat in which it occurs.

Management Goals

The answer to the question "When is hunting sustainable?" therefore must incorporate these wider management goals. Robinson (1993) defined the sustainable use of a resource as "when the rights of different user groups are specified, when human needs are met, and when the losses in biodiversity and environmental degradation are acceptable." The challenge is to define "human needs" and "acceptable losses." Defining human economic needs and aspirations is a Sisyphean task—basic needs are plastic, and expectations and aspirations change with the social and economic landscape. Yet every specific discussion of sustainability must state what needs can be met. Defining acceptable biological losses is also intractable. A first approximation might be for population levels not to be reduced by hunting to levels at which species are vulnerable to local extinction, or where ecosystem function is affected (Robinson and Redford 1991b).

The importance of specifying these management goals can be illustrated in the following example. If the goal is to achieve the maximum productivity of wild meat in a forest, the management option of choice might be to kill all of the nonhuman predators in the forest, thereby allowing all productivity of those prey species to be channeled to human users (Robinson and Bennett this volume). Similarly, we might increase the productivity of preferred game species by altering the habitat to increase available food or other limiting resources for those animals. Both of those strategies might promote sustainability. However, if the aim is to conserve the full biodiversity and ecological functioning of a tropical forest, while allowing human hunting, this management option would be unacceptable.

In the discussion of hunting sustainability, is there any general agreement on the management goals? One goal would certainly be to avoid extreme situations. For example, the management goal might be not to allow a population to be reduced to such a low level that it no longer contributes a significant resource. For example, a model of market hunting of babirusa (*Babyrousa babyrussa*) in North Sulawesi indicates that populations will equilibrate when they reach about 4% of K (Clayton 1996). At that level, the population might still survive, but hunting would be unprofitable, so would only be done occasionally. Harvest might equal production, but offtake would be minimal. Hence, one management goal might be not to allow a population to be reduced to such a low level that it no longer supports significant use. Similarly, another management goal might be not to allow populations of animals to be knocked down to levels from which they cannot recover. That might be because of some intrinsic biological feature of the species (e.g., only one sex remaining; animals so rare that they cannot find mates; population so small that they become inbred), or to extrinsic factors (e.g., their niche in the environment has been taken over by another species).

However, avoiding such extreme situations should not be the management goal; by the time extremes are imminent, wider ecosystem changes are inevitable and recovery might be impossible. A more positive goal for management is to ensure that densities of the hunted wildlife population allow the species to retain its ecological role. However, defining *ecological role* is extremely difficult and is almost impossible to determine in practice. A more practical goal is to maintain densities so that even when there is significant human use of wildlife, the populations of hunted species are not declining. In the Ituri Forest, Democratic Republic of Congo, for example, blue duikers are hunted intensively, yet populations are stable (Hart this volume).

Conditions Enhancing Hunting Sustainability

What biological conditions maximize the probability of achieving the management goal of maintaining hunted populations while allowing human hunting? Maintaining a hunted population at a level that allows maximum productivity—a level that is generally considered to be below carrying capacity (Caughley 1977)—is a goal of traditional wildlife management. That is generally thought to occur at population densities below carrying capacity (K). For most species for which there are good data, as population densities decrease, fecundity and infant survival increase, and mortality declines; these are all presumably due to a lowering of intraspecific competition. Thus, at some level below K, productivity is maximized. The actual level at which this occurs varies between species. For species such as temperate deer, it can be as low as 50% of K (Caughley 1977), although for slower breeding, longer lived rain forest species, it is likely to be much closer to K; figures in the range of $0.65\ K$ to $0.90\ K$ have been suggested (Robinson and Redford 1991b; Peres this volume). However, it is unlikely that the traditional approach of managing wildlife populations at specific densities can be applied to many of the subsistence and commercial systems discussed in this volume. Most situations, especially in tropical forests, lack the monitoring and regulatory capacity necessary to manage population numbers directly. A number of authors in this book suggest that the only practical approach in tropical forests is to establish totally protected zones, which can act as refuges for wild populations and allow dispersal back into overhunted areas.

What social and political conditions, and what use regimens and management approaches, maximize the probability of achieving the management goal of maintaining hunted populations while allowing human hunting? This goal might be achieved under a variety of biological, social and political conditions. It might be attained if management of hunted populations is undertaken by local communities or by the agencies of central government. It might be realized through economic incentive or legal enforcement. It might be achieved whether hunting is for subsistence or commerce, whether for meat or for animal parts (e.g., for traditional Chinese medicines). Different management approaches will most likely be necessary for different use regimens, and it might be easier to achieve this goal under some socioeconomic conditions compared with others.

For all use regimens, effective approaches inevitably involve incentives to protect the resource and also regulation of harvest. The chapters in this book show that regulation of access to the resource is almost always necessary to prevent access by outsiders to the resource, and also to preclude overharvesting by local people allowed to hunt.

Achieving Hunting Sustainability

For hunting to be sustainable, (a) harvest must not exceed production, (b) the management goals must be clearly specified, and (c) the biological, social, and political conditions must be in place that allow an appropriate use and an effective management. By themselves, none of these demonstrate sustainability. Production might balance harvest, but the harvested population might be in danger of extirpation. Management goals might be in the national interest but might be incompatible with the particular government-based management approach being applied. A community-based management approach might be ethically justifiable and socially acceptable, but it might not be politically realistic, and the resource might be overharvested. Sustainability requires that all of these expressions be solved simultaneously—in effect solving a complex multinomial equation.

The chapters of this book use case examples to explore whether hunting is sustainable and to discuss the conditions—biological, social, political, cultural—that favor sustainability and those that oppose it. Identifying these conditions, applying them, and achieving sustainability, no matter how elusive, is critical if hunted wildlife populations are to remain viable, and if human societies that depend on them do not suffer by being tied into a declining resource base.

PART I

Biological Limits to Sustainability

2

Carrying Capacity Limits to Sustainable Hunting in Tropical Forests

JOHN G. ROBINSON AND ELIZABETH L. BENNETT

A fundamental tenet of modern conservation thinking is that wild areas will be conserved if specified groups of people—frequently indigenous peoples or rural populations—retain or are given the right to exploit resources in these areas. In its simplest form, the argument is that if people have the right to exploit natural resources, then they will take the responsibility for conserving them. For instance, *Caring for the Earth* (IUCN/UNEP/WWF 1991) urges as a priority action (7.1) that communities and individuals be provided with secure access to resources and an equitable share in managing them.

The link between effective resource management and community control is frequently stated categorically: "a community's rights to ownership and tenure of wildlife resources *must* be secured for sustainable wildlife management" (IIED, 1994, p. ix, emphasis added). "Sustained management of the forest requires the presence of forest-living people who know the forests and understand how to use their full range of resources" (Brookfield 1988). This conservation strategy is often supported primarily by political or ethical considerations. For instance, the *Global Biodiversity Assessment* (UNEP 1995) notes (p. 1024) that "it is increasingly recognized as neither politically feasible nor ethically justifiable to attempt to deny poor people the use of natural resources [unless they are provided] with some alternative means of livelihood." In other words, the argument goes that the ethically appropriate, politically realistic, and pragmatically effective way to manage natural resources is to allow or even encourage local communities to use them (Western and Wright 1994; IIED 1994).

This approach to conservation, however, makes two assumptions. First, it requires that local communities have the cultural traditions (Kleymeyer 1994), the political, legal, and economic power (Lynch and Alcorn 1994; Feldmann 1994; Bromley 1994), and the support of local, regional and national institutions (Murphree 1994) to manage the resource sustainably. In tropical forests, there are numerous examples of this concatenation of capacities and circumstances, especially when people are exploiting resources of plant origin (e.g., Langub 1988; Posey 1988). Equally, there are many cases, some reported in this volume, in which local communities do not possess the capacities or are not allowed to manage resources. Wildlife in tropical forests, in particular, is notoriously difficult to manage, primarily because of the fugitive nature of the resource (Naughton-Treves and Sanderson 1995).

The second requirement, assuming the local management capacity exists, is that the natural resources can sustain the harvest needs of the human population. The term *sustain* means that the resource extraction must meet the socioeconomic needs and aspirations of the people harvesting the resource while limiting the losses in biological diversity and environmental degradation to acceptable levels (Robinson 1993; Bennett and Robinson this volume). If the human population needs to harvest more than the natural system can produce, then natural populations will be systematically depleted. Depleted natural populations have a limited production, and the people will be trapped into a declining resource base. When the goal is also biodiversity conservation (as it is by definition in community-based conservation) then the necessary overharvest of resources will result in the extirpation of wild species and the deterioration of ecosystem functioning. Under these circumstances, no level of political empowerment, institutional strengthening, or wishful thinking would allow communities to manage and conserve their natural resources. Securing a community's right to resources would be a sham, and although it might be politically expedient, at least in the short term, it would not be ethically justifiable.

Therefore, if human communities in tropical forests are to obtain their animal protein by harvesting wild species, there is a limit to the number of people that a tropical forest can support if biodiversity is to be conserved. That limit is the carrying capacity of the forest for such people. The amount that people need to harvest is the product of per capita needs and the number of people that depend on the resource.

The aim of this chapter is to attempt to determine the carrying capacity for human communities living in tropical forests, given the assumption that the management goal is to limit the losses in biological diversity and environmental degradation. However, specifying the carrying capacity for people is difficult and contentious (Moffat 1996) because the human species is highly adaptable and flexible, depending on many types of resources, and our ability to switch to alternate resources and find substitutes has been established (Simons 1980). Rural peoples living in tropical forests, however, have fewer alternatives than do urban dwellers relying on technologically sophisticated production systems, and here we consider an important resource whose production and substitutability is limited.

The limiting resource that we consider are wild animals, which are ubiquitously harvested by rural peoples living in and around tropical forests. Meat from wild mammal, bird, and reptile species provides much of the animal protein needs for rural peoples around the world. Harvested animals are also an important economic commodity. The high value of meat relative to its bulk means that meat is frequently transported to local and urban markets (e.g., Castro et al. 1976; Caldecott 1988; Lee et al. 1988; Bodmer et al. 1990a; Falconer 1992; Fa et al. 1995) and is a major source of cash for many rural peoples. Tropical forest peoples can find substitutes for wild meat. They can catch fish, both freshwater and marine (Redford and Robinson 1987), and either raise domestic animals (frequently pigs and chickens) or buy them from the national markets (Ayres et al. 1991). But as we will argue below, cost considerations and peoples' preferences frequently do not favor this substitution.

This chapter examines the carrying capacity of the forest for traditional rural peoples who depend significantly on forest-dwelling animal species for their protein needs, in a context where the goals are the long-term maintenance of that resource base, and the long-term conservation of hunted species in particular and biodiversity in general. It is of concern that many of the studies in this volume have indicated that in these circumstances, much hunting is no longer sustainable. If this is a general trend, it raises the question of whether rural people, possibly through increased population density or through higher harvest rates, have exceeded the carrying capacity of tropical forests.

HUMANS AND TROPICAL FORESTS

Human beings have lived in tropical forests for thousands of years. They have been in the tropical forests of southeast Asia for at least 40,000 years (Zuraina 1982; Hutterer 1988). A similar age has been suggested for rain forest inhabitants in central Africa (Bahuchet 1993). In central Amazonia, palaeoindian peoples go back at least 10,000 years (Roosevelt et al. 1996). These humans have shaped the natural landscape. The lowland forests of Mexico and Guatemala bear the unmistakable imprint of Maya civilization some 400 years ago (Gómez-Pompa and Kaus 1992). Present-day *terra firme* forests in Amazonia have been structured by hundreds if not thousands of years of active indigenous management (Balée 1992). People have cultivated throughout Borneo's forests and extracted forest products for subsistence and sale for millennia (Padoch and Peluso 1996).

Throughout history, indigenous peoples have undoubtedly depended on wild meat and fish to meet their animal protein requirements. The long-term presence of humans living in tropical forests, and their dependence on wild meat and fish throughout the period, indicate that extant wild species must have been able to sustain human hunting. Those species that were unable to sustain human hunting have gone locally or globally extinct (Olson and James 1982; Lewin 1983).

Recent studies have indicated that today—although few people living in the forest secure all their protein needs from wild game—wild meat is still an important source of animal protein and is a commodity sold for cash needed to buy other subsistence requirements (e.g., Caldecott 1988; Bodmer et al. 1990a; Redford and Robinson 1991; Lahm 1993). Meat of wild origin remains a limiting resource for some rural peoples. The loss of wildlife species in rural areas is followed by decreases in animal protein consumption in many rural communities (Stearman 1990; Bennett et al. this volume). Wild meat is an important commodity that is sold by rural peoples in local markets (Castro et al. 1976; Hart 1978; Caldecott 1986; Bodmer et al. 1990b; Fa et al. 1995). And the well-being of many indigenous peoples depends on their continued access to wild game (Vickers 1984; Redford and Robinson 1987).

The most important wild species hunted by humans living in and around tropical forests are the large mammals. These tend to occur at relatively high densities, and to have large body masses. The single most important group are the ungulates. In the neotropics, peccary, deer, and tapir are preferred game (e.g., Wetterberg et al. 1976; Bodmer et al. 1988b). Where these animals are abundant, they constitute the most important part of the harvest by weight (e.g., Smith 1976; Vickers 1984). In Africa, forest duikers and pigs are preferred prey, and contribute a significant fraction of animal biomass harvested (e.g., Colyn et al. 1987b; Infield 1988; Fa et al. 1995). In Asia, it is pigs and deer (e.g., Caldecott 1986), which between them can comprise more than 80% of the total weight of wild meat hunted (Bennett et al. this volume).

GAME BIOMASS IN TROPICAL FORESTS

The harvest of wild species by human hunters depends in part on the standing biomass of the species, and compared with other habitats, mammalian biomass in tropical forests is low. In general, the degree of forest cover is inversely related to the overall standing biomass of mammals—forests have a much lower overall standing biomass than do grasslands. For instance, in pantropical comparisons, evergreen, closed canopy, moist forests have lower standing biomasses than more open, seasonally deciduous forests. The total biomass of large mammals (over 1 kg adult body mass, a category that includes most species important for human hunters) in evergreen forests rarely exceeds 3000 kg/km^2 (table 2-1 and figure 2-1). In more open forests, usually where trees are seasonally deciduous, and where the habitat is a mosaic of forest and grassland, standing biomasses can reach 15,000 kg/km^2. Open grasslands can support overall standing biomasses of mammals in excess of 20,000 kg/km^2.

This variation in mammalian biomass is largely accounted for by the variation in ungulate biomass (see table 2-1 and figure 2-2).

Forests, in comparison with grasslands, have a much lower and less available productivity for ungulates (Yalden 1996). The number of ungulates per square kilometer in tropical forests is low because of the scarcity of grasses and browse, and most forest ungulates are quite frugivorous. The densities of frugivores (figure

Table 2-1. Comparison of Large Mammal Biomasses at Tropical Sites

Area	Habitat	Biomass Ungulates (kg/km²)	Biomass Primates (kg/km²)	Biomass Rodents (kg/km²)	Total Biomass (kg/km²)	References
Barro Colorado Island, Panama	Evergreen, closed, upland forest	542	482	300	2264	Eisenberg 1980
Manu National Park, Peru	Evergreen, closed, alluvial forest	403	655	129	1400	Janson & Emmons 1990
Teiú, Amazonia, Brazil	Evergreen, closed, flooded forest				1087	Ayres 1986
Urucu, Amazonia, Brazil	Evergreen, closed, terra firme forest	341	391	70	891	Peres (in press)
Guatopo National Park, Venezuela	Evergreen, closed, late secondary forest	270	139	280	946	Eisenberg 1980
Ogooué-Maritime, Gabon	Evergreen, closed, late secondary forest	765	247	2	1050	Prins & Reitsma 1989
Lopé Reserve, Gabon	Semi-evergreen, closed forests	2776	319	5	3101	White 1994
Hato Masaguaral, Venezuela	Deciduous, open, forest mosaic	275 (+ 7600)	175	445	1084 (+7600) = 8684	Eisenberg 1980
Acurizal Ranch, Brazil	Deciduous, open, forest mosaic	224 (+3750)	20	50	380 (+3750) = 4130	Schaller 1983
Nagarahole, India	Deciduous, open, forest mosaic	14,510 (+350)	236	0	14,744 (+350) = 15,094	Karanth & Sunquist 1992
Simanjiro, Tanzania	Grassland savanna	2273 (+5936)	0	0	2273 (+5936) = 8,209	Kahurananga 1981
Hato El Frio, Venezuela	Grassland savanna	320 (+18,504)	0	2564	3730 (+18,504) = 22,405	Eisenberg 1980
Mara, Kenya	Grassland savanna	19,200 (+4400)	0	0	19,200 (+4400) = 23,600	Stelfox et al. 1986

Biomasses in brackets are those of domestic ungulates. Numerical values for Lopé Reserve are averages from five sites. Figures from Simanjiro are averages from wet and dry season.

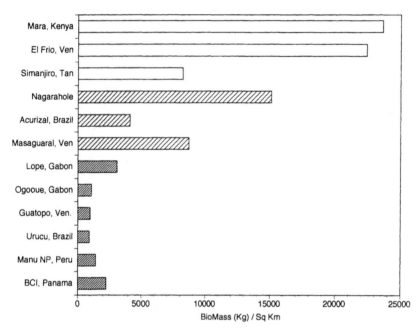

Figure 2-1. Total biomass (kg) of rodents, primates, and ungulates per square kilometer at 12 tropical sites, distinguishing grassland savannas (*light shading*), open, seasonally deciduous forests (*light hatching*), and closed, evergreen forests (*strong hatching*).

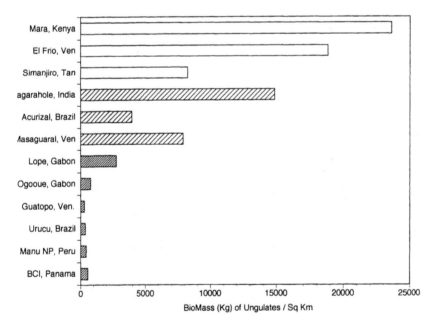

Figure 2-2. Total biomass (kg) of ungulates per km² at 12 tropical sites, distinguishing grassland savannas (*light shading*), open, seasonally deciduous forests (*light hatching*), and closed, evergreen forests (*strong hatching*).

2-3), at a specified body mass, are generally lower than those of animals that eat grass or browse (Robinson and Redford 1986a). Moreover, the body mass of forest ungulates is generally less than the body mass of open grassland ungulates (Jarman 1974), so the total standing biomass of ungulates in tropical forests is low. From the perspective of the human hunter in a tropical forest, the low biomass of ungulates is not significantly offset by the generally higher biomasses of primates and rodents in forests (figure 2-4).

The highest ever recorded biomass of forest primates, in Kibale Forest, Uganda, only reaches 2364 kg/km^2 (Struhsaker and Leland 1979), and total primate biomasses rarely exceed 500 kg/km^2. The same generalization holds for large-bodied rodents, whose biomasses also rarely reach 500 kg/km^2 (see table 2-1).

These generally low biomasses for ungulates in forested habitats are even more pronounced when one considers that not all of these species are available for human hunters. In some African and Asian forests, elephants make up a large proportion of the mammalian biomass. Because they are large and dangerous, elephants are rarely taken by subsistence hunters. If they are excluded from the totals, biomass of game species is lower. In the Lopé reserve in Gabon, for instance, elephants make up an average of 89% of the ungulate biomass (White 1994). In Nagarahole National Park, elephants make up 47% of the ungulate biomass (Karanth and Sunquist 1992).

Overall, then, the standing biomass of mammals in tropical forests can be an order of magnitude lower than in more open habitats. This significantly affects the

Figure 2-3. Wedge-capped capuchin monkey (*Cebus olivaceus*). Many forest primates, rodents, and ungulates are quite frugivorous. *Photo courtesy of Haven Wiley.*

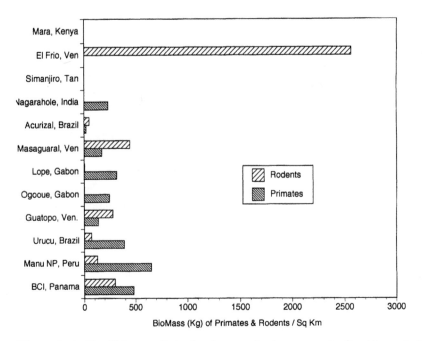

Figure 2-4. Total biomass (kg) of rodents and primates per km² at 12 tropical sites, distinguishing grassland savannas (*light shading*), open, seasonally deciduous forests (*light hatching*), and closed, evergreen forests (*strong hatching*).

amount of meat that is produced in a forest, and thus the maximum harvest that can be secured by human hunters.

GAME PRODUCTION FROM TROPICAL FORESTS

The maximum harvest of wild species by human hunters in tropical forests ultimately depends on the proportion of the overall biomass that can be harvested. That proportion can be estimated by first calculating the annual production (P) of each species, where production is defined as the numbers of animals that are added to a population over the year, and then estimating the proportion of that production that can be harvested.

Production is determined by two parameters: the standing population number and the intrinsic rate of natural increase of that population. Within an area of a given size, the number of animals in a population is affected by the body size of the species (with larger animals occurring at lower densities) and the diet of the species (with carnivorous species at a given body mass occurring at lower densities than frugivorous species, which occur at lower densities than herbivorous species) (Robinson and Redford 1986a; Fa and Purvis 1997). The intrinsic rate of natural increase in a population is affected principally by the body size of the species (with larger animals having lower rates) and by phylogeny (with primates and carnivores having lower rates than expected from body size, and ungulates and lagomorphs having higher rates) (Robinson and Redford 1986b).

The total meat that is produced from a forest and available for human hunters can be estimated, provided that densities of different species at a site and the rate of population increase of different species are known. Robinson and Redford (1991b) provide a model that estimates the potential production for human harvest for the different game species, assuming maximal conditions. Here we use this model to provide an estimate of total game meat production from a tropical forest, using observed densities of game species in the undisturbed, biologically diverse Manu National Park in southern Peru. Table 2-2 lists all mammalian species in Manu with body masses of at least 1 kg, together with observed densities at Cocha Cashu (Janson and Emmons 1990), a site in the Manu National Park, Peru. For each of these species, we took calculated maximum rates of population increase (r_{max}), converted these to finite rates of increase ($e^r = \lambda$), and then calculated maximum production at observed densities:

Table 2-2. Calculation of Maximum Potential Harvest from a Mammal Community Like That Found in Manu National Park, Peru

Species	Body Mass (kg)	Density (#/km²)	r_{max}	Production (kg/km²)	Life Expectancy	Maximum Potential Harvest (kg/km²)
Marsupials						
Didelphis marsupialis	1.0	55	2.92	577.5	very short	346.5
Primates						
Alouatta seniculus	6.0	30	0.17	20.5	long	4.1
Cebus apella	2.6	40	0.14	9.4	long	1.9
Cebus albifrons	2.4	35	0.17	9.6	long	1.9
Ateles paniscus	7.0	25	0.07	7.4	long	1.5
Edentates						
Dasypus novemcinctus	3.5	21.9	0.69	45.5	short	27.3
Rodents						
Hydrochaeris hydrochaeris	45.0	1.6	0.69	42.8	short	25.7
Agouti paca	8.0	3.5	0.67	16.0	long	3.2
Dasyprocta variegata	4.0	5.2	1.10	25.0	short	10.0
Myoprocta pratti	1.5	5.3	1.42	11.6	short	4.6
Perrisodactyls						
Tapirus terrestris	160.0	0.5	0.20	10.6	long	2.1
Artiodactyls						
Tayassu pecari	35.0	3.0	0.84	83.2	long	16.6
Tayassu tajacu	25.0	5.6	1.25	209.2	long	41.8
Mazama americana	30.0	2.6	0.40	22.9	short	9.2
Carnivores						
Nasua nasua	4.5	0.2	0.23	0.1	short	0.04
Potos flavus	2.2	25.3	0.30	11.7	long	2.3
Felis concolor	29.0	0.02	0.30	0.15	long	0.03
Felis pardalis	9.3	0.8	0.46	2.59	long	0.5
Panthera onca	35.0	0.04	0.23	0.19	long	0.04

Body mass from Janson and Emmons (1990), with the exception of *Dasypus novemcinctus*, whose mass was from Robinson and Redford (1986a). Densities from Janson and Emmons (1990), with the exception of *Dasypus novemcinctus*, whose density was from Robinson and Redford (1986a). Intrinsic rate of natural increase (r_{max}) values from Robinson and Redford (1986b), with the exception of *Didelphis marsupialis*, which is from Tyndale-Biscoe and Makenzie (1976), and *Myoprocta pratti*, which is from Kleiman (1970). Life expectancy data are from Robinson and Redford (1986b).

$$P_{max(D)} = M[(0.6D \times \lambda) - 0.6D]$$

where $P_{max(D)}$ is the maximum possible production at the observed densities, M is the average body mass of the species, and D is the observed density of each species at Manu. Because Manu is an unhunted and undisturbed site, densities are presumably at carrying capacity (K). Only a certain proportion of this production can be harvested annually by humans, and that proportion is somewhat arbitrarily taken as 20% of production for long-lived species, 40% for short-lived species, and 60% for very short-lived species (for rationale, see Robinson and Redford 1991b, and Robinson, this volume). Addition of the maximum possible harvest figures for all species hunted commonly for game in neotropical forests, converted to biomass, yields a total biomass harvest of 152 kg/km^2 (from table 2-2; this excludes possums and the large felids, which were not recorded as preferred game species by neotropical hunters) (Redford and Robinson 1987). Ungulates, edentates, and rodents contribute the most to this potential biomass harvest (figure 2-5). It is unlikely that wildlife harvest will greatly exceed this figure, at least in neotropical forests, because the wildlife community at Manu is diverse, and this calculation assumes maximal production rates for all species.

The potential biomass harvest of wild meat from tropical forests also can be estimated from actual harvest rates, with the caveat that actual harvests might or might not be sustainable. In a case study of a West African tropical forest, Infield

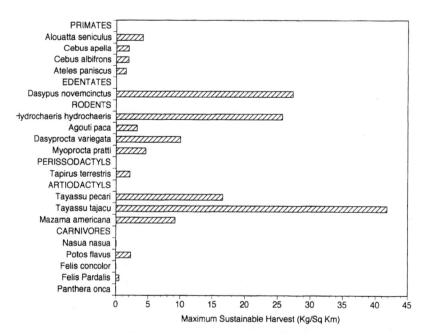

Figure 2-5. Maximum possible sustainable harvest (kg/km^2) of a diverse tropical forest mammal community (using densities from Manu National Park, Peru), including all species with a body mass over 1 kg.

reported on hunting in and around the Korup National Park (covering 1250 km²) in 1987 and 1988. Hunters from the six villages actually within the park annually harvested some 271,000 kg of wild meat, which indicates that the annual harvest from each square kilometer was approximately 217 kg. These rates were apparently not sustainable because wildlife population densities in the park in 1990 were clearly depleted (Edwards 1992; Payne 1992). A study of Efe pygmies and Lese horticulturalists in the northern Ituri forest revealed that hunting in this relatively undisturbed forest provided 50 kg/km² of wild meat per year. Almost all captures were of forest ungulates, with duikers predominating. A comparable study in East Africa is that of FitzGibbon et al. (1996), who studied subsistence hunting in the Arabuko-Sokoke nature reserve, which covered 372 km². The total biomass of game meat taken during each year was estimated at 130,000 kg, yielding a harvest rate of 350 kg/km². This harvest, however, was not sustainable for all species. At the time of the study in 1991 and 1992, large ungulates had already been overharvested and reduced to very low densities, and primates were being depleted.

Finally, potential biomass harvest from a forest can be estimated using natural predation offtake rates. Hart's (this volume) study of leopard and their duiker prey provides a unique data set of predator harvests in the Ituri forest of the Democratic Republic of Congo. Duikers are also the primary game species of the Mbuti hunters in the forest. At two sites, both protected from human hunting, leopards harvested 201 kg/km² and 42.5 kg/km² of duikers over the 4-year study. Although a 4-year period might not accurately map predator harvest rates over evolutionary time, these offtakes may be interpreted broadly as sustainable.

These potential harvest estimates are remarkably consistent. These studies indicate that actual sustainable harvest of game meat from most forests is generally under 200 kg/km². This fits well with the theoretically derived figure of 152 kg/km².

CARRYING CAPACITY OF FORESTS FOR PEOPLE

Given these estimates of the maximum potential harvest rate for game species in tropical forests, what is the carrying capacity of the forest? In other words, what is the maximum number of people depending on meat from wild species who can live in a forest while still conserving adequate populations of these game species? Note that the definition of carrying capacity used throughout this chapter accepts the constraint that where the goal is conservation, populations of game species and biodiversity in general must be conserved, even if people are harvesting the wildlife resources.

Any discussion of carrying capacity of people must specify the conditions under which human populations live. In this discussion, we first consider the case where people live in the tropical forest with little or no dependence on agriculture and domestic animals. This is the extreme case, although one still followed by a number of indigenous groups. We then consider the case in which people clear the forest for swidden agriculture. This allows people much greater access to food, including animal protein. However, it also raises the specter that associated human activities (e.g., extensive clearing of land for agriculture, hunting for sale in local markets) will be inconsistent with the conservation objectives.

Calculation of human carrying capacity is relatively intuitive for the extreme case in which people depend on wild meat for all of their animal protein needs. As argued above, maximum sustainable production of wild meat in tropical forests probably rarely exceeds 200 kg/km^2/yr and is likely to be around 150 kg/km^2 in most forests. Dividing this figure by human per capita animal protein needs provides an estimate of carrying capacity.

What are the per capita animal protein needs of people living traditionally in tropical forests? Recommendations from expert committees on protein and energy intakes do not take account of the activity levels and phenotypic plasticity of many rural populations. Nevertheless, the U.S. recommended daily intake of protein for a 70-kg man is 50 g. Assuming that boneless meat contains about 20% protein, this translates into a minimum recommended daily amount (RDA) of 0.28 kg of meat per day. This RDA is comparable with actual meat consumptions recorded from a range of different groups in Africa, Asia, and Latin America (table 2-3).

These figures are per capita consumption rates, which include men, women, and children. Clearly, there is great variation in the amount of meat that different groups actually consume, but a modal value for individual consumption is about 0.25 kg/meat/day.

If production of game meat from a forest is about 150 kg/km^2/yr, and if 65% of live weight is edible meat (Hill and Hawkes 1983), then each square kilometer of forest will produce 97 kg of edible meat per year. If one assumes that a healthy consumption of meat is 0.25 kg/person/day, then each person will consume about 91 kg of meat each year. These figures indicate that in tropical forests the carrying capacity for people depending exclusively on game meat will not greatly exceed one person per km^2, even under the most productive circumstances. Actual densities of people living in tropical forests are consistently lower than this figure,

Table 2-3. Per Capita Consumption of Wild Meat (kg/person/day) of Selected Human Groups

Human Group	Location	Per Capita Consumption of Meat (kg/person/day)	Reference
Yanomamö	Venezuela–Brazil	0.25 (+ 0.13 fish)	Chagnon and Hames 1979
Machiguenga	Peru	0.23	Alvard 1993
Piro	Peru	0.23	Alvard 1993
Waorani	Ecuador	0.28	Yost & Kelley 1983
Mbuti	Congo-Zaire	0.33–0.54	Harako 1981
Bakola	Cameroon	0.22 (+ 0.02 fish)	Koppert et al. 1993
Mvae	Cameroon	0.19 (+ 0.04 fish)	Koppert et al. 1993
Ogooué-Ivindo	Gabon	0.10–0.17	Lahm 1993
Efe	Congo-Zaire	0.16	Bailey & Peacock 1988
Hagahai	Papua New Guinea	0.39	Jenkins & Milton 1993
Iban	Sarawak	0.10 (+ fish)	Bennett (unpublished data)
Kelabit	Sarawak	0.21 (+ fish)	Bennett (unpublished data)
Wana	Sulawesi	0.03	Alvard (this volume)

supporting our conclusion. For example, Denevan (1992) estimated that traditionally most neotropical forest populations lived at densities of about 0.1 person/km². The Ache in Paraguay live at 0.2 person/km² (Hill this volume), whereas the Huaoroni (Mena et al. this volume) live at roughly 0.5 person/km². Even nowadays, few indigenous groups in the neotropics live at densities above 1 person/km², although the Jivaroans of Peru, at 2.6 persons/km² (Ross 1978), and the Yanoama of Venezuela, at 1.3 persons/km² (Smole 1976), are exceptions.

For people involved in swidden agriculture, however, carrying capacity might be greater. Agriculture allows the production of much greater amounts of carbohydrates. Tropical forests are generally depauperate in carbohydrates (e.g., Hart and Hart 1986). Fruit abundance is seasonal in most forests, and patchily distributed in space. Leaves, stems, and other vegetative matter are generally steeped in secondary compounds and are not commonly available to humans, although manioc leaves and ferns are consumed throughout the tropics. Tropical forests do contain starchy food sources—sago palms, wild yams, wild cassava, or manioc—but preparation for consumption is time consuming and energetically expensive (Ulijaszek and Poraituk 1993; Pasquet and Koppert 1993). However, with swidden agriculture, the availability of plant carbohydrates is much greater. For human groups that practice swidden agriculture, then the much greater reliance on cultivated plantains and bananas, tubers (manioc, yams, and taro), and grains (rice, millet, and maize) allows dramatic increases in the population density of human beings.

Despite this increase in carbohydrate production, swidden agriculture does not necessarily increase total animal protein available to people. Production of animal protein is increased through the raising of domestic animals (especially pigs and chickens). In addition, cultivated fields are exploited by wildlife species for food, augmenting wild meat production (e.g., Jorgenson 1995a). Animal protein production is further increased because abandoned fields often revert to secondary forests, which in addition to providing many forest products that can be harvested (Padoch and de Jong 1992), frequently are themselves productive for some wild game species, particularly ungulates. For instance, in Wilkie's (1989) study of subsistence hunting in the Ituri forest of Congo-Zaire, primary forests provided 50 kg of wild meat per km²/yr, whereas adjoining secondary forests provided 318 kg/km²/yr, mostly duiker.

This potential increase in the availability of animal protein is offset by the fact that people involved in swidden agriculture are also almost invariably incorporated into market economies. For many, an important commodity to be sold for cash is meat from wild species (Castro et al. 1976; Caldecott 1988; Bodmer et al. 1990a). People sell that which they might have eaten. Studies have not quantified the extent to which the increased production of both domestic and wild meat is offset by the increased need to sell meat in markets. What is clear is that it is sometimes the case that rural peoples often sell the preferred meat and themselves consume the less preferred. For instance, in Tahuayo-Tamshiyacu, Peru, rural peoples sell peccary meat in Iquitos, while themselves eating the less preferred primates and rodents (Bodmer et al. 1990a). What this means is that the absolute amount of meat per km² available to people is probably not greatly enhanced in swidden agricultural systems, and with the higher human population densities characteristic of these systems, the per capita availability of meat is significantly reduced in most circumstances.

For people of the swidden, therefore, plant carbohydrates replace animal protein in the diet, and these mixed agriculture–forest systems generally support much higher densities of human beings. However, these people continue to evince a strong interest in the consumption of meat (Redford 1993). People continue to hunt. Yet because they live at much higher densities than hunter-gatherer societies, and because they are involved in market economies and thus frequently sell large amounts of meat, agriculturalists have the potential to have much greater impacts on the populations of wild animal species. If through their desire to eat meat many agriculturalist groups exceed the production of meat from the forest, then populations of wild species will be depleted or extirpated. If we require the conservation of adequate populations of game species, as we do with community-based conservation, this means that the carrying capacity of the forest for these people would be exceeded. If swidden agriculture does not significantly increase the animal protein available to people, as suggested above, then the carrying capacity of agriculturalists in the forest might not be significantly different from one person/km^2.

This conclusion is supported by a review of human population densities in extractive reserves in Brazil. Extractivists in this area are principally rubber tappers, but they also clear the forest for crops, raise livestock, and hunt wildlife (Browder 1992). Game densities within reserves are often low (Schwartzmann 1989; Bodmer et al. 1990a; Peres this volume). Table 2-4 presents a list of reserves ranked by human population density. In Mamiraua, where human densities are the lowest, high densities of game species are still retained (Ayres et al.1999). But in Rio Yaco, which has a human density of only one person/km^2, Martins (1992) found that populations of large game species were at very low levels (table 2-5).

Even at such low human densities, people who depend both on agriculture and resource extraction from the forest have a significant effect on wildlife populations.

Table 2-4. Human Population Density in Extractive Reserves in the Brazilian Amazon

Reserve (state)	Inhabitants/km^2	Reference
Mamiraua (Amazonas)	0.52	Ayres et al. (in press)
Tejo (Acre)	0.67	Fearnside 1992
Macaua (Acre)	0.72	Fearnside 1992
Chico Mendes (Acre)	0.77	Fearnside 1992
Rio Yaco (Acre)	1.00	Martins 1992
Rio Jurua (Acre)	1.04	Fearnside 1992
Floresta (Acre)	1.39	Fearnside 1992
Terrua (Amazonas)	1.53	Fearnside 1992
Rio Cajari (Acre)	1.58	Fearnside 1992
Cachoeira (Acre)	1.60	Fearnside 1992
Maraca (Amapa)	1.61	Fearnside 1992
Antimary (Acre)	1.67	Fearnside 1992
Ouro Preto (Rondonia)	1.67	Fearnside 1992
Riozinho (Acre)	1.67	Fearnside 1992
Santa Quiteria (Acre)	1.70	Fearnside 1992
Porto Dias (Acre)	1.88	Fearnside 1992

Table 2-5. Comparison of Densities of Game Species in the Manu National Park, Peru, and an Extractive Reserve, the Rio Yaco Basin, Brazil

Species	Manu NP density (ind./km²)	Rio Yaco density (ind./km²)
Primates		
Alouatta seniculus	30	2.9
Cebus apella	40	3.2
Cebus albifrons	35	1.6
Ateles paniscus	25	0.1
Edentates		
Dasypus sp.	21.9	2.1
Rodents		
Agouti paca	3.5	2.4
Dasyprocta sp.	5.2	2.3
Myoprocta pratti	5.3	1.4
Hydrochaeris hydrochaeris	1.6	0.7
Perissodactyls		
Tapirus terrestris	0.5	0.01
Artiodactyls		
Tayassu pecari	3.0	0.01
Tayassu tajacu	5.6	0.3
Mazama spp.	2.6	0.3
Aves		
Tinamus spp.	34.2	9.5
Psophia leucoptera	5.4	1.0
Penelope jacquacu	1.2	2.0
Ara spp.	3.6	0.3
Amazona spp.	3.6	1.2

Manu National Park data from Terborgh (1983) and Janson and Emmons (1990). Rio Yaco data from Martins (1992).

INCREASING HUMAN CARRYING CAPACITY IN TROPICAL FORESTS

The production capacity of wildlife in tropical forests therefore defines a limit to the level of hunting that can be sustainable without decreasing biodiversity, and to the general applicability of community-based efforts in forests. But are there ways or perspectives that would increase the carrying capacity of these systems for human beings?

It might be expected that a tropical forest habitat could support more people if forest-dwelling people ate less meat. Although the amount of meat that people indigenous to the tropical forest consume is rarely much more than the minimum suggested by nutritionists, many human groups eat significantly less meat than the suggested minimum (see table 2-3). The long-term health consequences of this consumption level are unclear. However, these low per capita consumption rates are probably largely the consequence of low wildlife densities (see Alvard this volume; Bennett et al. this volume; Hill et al. this volume) and not the result of people choosing a low meat diet. People will continue to hunt if prey is available.

Expecting people to hunt less and eat less meat so as to manage their resources more effectively is probably an unrealistic strategy.

Another approach is for people to harvest a greater breadth of game species, thus taking many small-bodied, less preferred species (such as marsupials and small rodents). Using the hypothetical case from Manu National Park as an example, the potential meat harvest could be greatly increased if the possum *Didelphis marsupialis* and other species with body weights under 1 kg were taken. In practice, this tends not to happen because hunters will not switch to smaller bodied species until the larger bodied species are depleted. When they encounter large-bodied species they will hunt them. There are good economic reasons for this. Optimal foraging theory (MacArthur and Pianka 1966; Charnov 1976; Pyke et al. 1978) predicts that hunters will only take a given species when the ratio of benefits (measured for instance in kilograms or calories) to costs (measured for instance by the time and energy spent hunting and processing meat) equals or exceeds the average returns for all hunted species. A focus on smaller bodied game species tends to happen only when large bodied, preferred species are overexploited (Hill and Hawkes 1983; Stearman and Redford 1995; Peres this volume). This means that it would be difficult to ask or require human hunters to include smaller bodied species in their hunt if larger bodied species are still available. The range of species taken, and the body sizes of those species are determined by the hunter's assessment of profitability, not a need for sustainability. Accordingly, hunters will not switch to smaller bodied prey until they have depleted or extirpated the preferred larger bodied prey, a consequence that would not be acceptable if the primary goal was conservation.

Another approach that would increase the productivity of game species would be the extirpation of the large and medium-sized predators from an area, resulting in higher densities of herbivorous and frugivorous animals (see Terborgh 1992). Barro Colorado Island in Panama illustrates this phenomenon (Eisenberg 1980; Glanz 1991). Many preferred game species fall into this dietary category (Redford 1992, 1993; Robinson 1996), so predator extirpation would have the effect of potentially channeling more of the mammalian productivity to humans. However, this approach would be unacceptable if the primary goal were conservation.

Ultimately then, attempting to improve the carrying capacity of a forest by manipulating harvest patterns or by altering the biological community is a limited approach. Overall, tropical forests cannot support more than a very low human population density without becoming biologically depauperate.

THE SUSTAINABLE LANDSCAPE

Where the intent is the conservation of adequate populations of game species, this discussion of carrying capacity indicates that the density of people who can sustainably harvest meat from wild species is limited. Our discussion has assumed, however, that people and wildlife are distributed homogeneously across the tropical forest landscape. They are not. People are most frequently found in the lowlands and valleys, where soils are more likely to support agriculture. They are often found along water courses, where they have access to fish and where rivers provide transportation routes. And although human presence can augment wild-life densities in certain circumstances (e.g., Linares 1962; Greenberg 1992; Jorgenson 1995a), high-density wildlife populations generally are found where people are not. Wildlife populations tend to be extirpated near human habitations.

That human hunting acts to create heterogeneity in the spatial distribution of wildife abundances is well documented. Hunters take the most accessible prey first, so the impact of hunting varies with the distance from human settlements. An example is Hames's (1979) study of the Ye'kwana in the small village of Toki in Venezuela: with some 100 inhabitants, the impact of hunting on many species is quite localized. For instance, most of the pacas (*Agouti paca*) were killed quite close to the village. Perhaps the most complete study of this effect is the comparison of two villages in the Manú River drainage of southern Peru. Yomiwato is a small village inhabited by approximately 100 Machiguenga Indians who engage largely in subsistence hunting using bow and arrow, whereas Diamante is about twice the size and inhabited by Piro Indians, who depend largely on shotguns for their hunting and are actively involved the market economy of the region. The impact of hunting on the larger bodied, most preferred species was greater around Diamante, and indeed, most such species were extirpated from areas within 3 km of the village (Mitchell and Raez Luna 1991; Alvard 1993, 1995a). The impact was also more spatially heterogeneous, with distance from the village being a stronger determinant of wildlife densities around Diamante (Mitchell and Raez Luna 1991).

This is more than a local effect around villages. The Siona-Secoya of the northwest Amazon traditionally have hunted over an area of 2500 km² (Vickers 1991, 1994), and they report that wildlife densities are higher in those parts of their catchment areas not recently hunted. Hill and Padwe (this volume) have quantified this effect in their study of hunting impacts in the Mbaracayu reserve in eastern Paraguay. Wildlife densities increase with distances from settlements across the 600-km² reserve. This pattern is reported in a number of studies in this volume (e.g., Bennett et al.; Fimbel et al.; FitzGibbon et al.; Hart), and these results are consistent with those from other ecological systems. In the Serengeti National Park in Tanzania, densities of ungulates, dwelling in a relatively homogeneous savanna, vary with distance from villages along the western boundary of the park as a result of hunting pressure (Campbell and Hofer 1995; Hofer et al. 1996).

That wildlife abundances vary across the landscape means that consideration of the carrying capacity for humans (by definition requiring sustainable harvest rates) must take into account that spatial heterogeneity. In addition to depending on the productivity of the harvested species, the annual harvest that is sustainable will be affected by the replenishment rate from unhunted areas where wildlife is more abundant. Although the factors that might affect replenishment rates have been examined theoretically (e.g., Pulliam and Danielson 1991; McCullough 1991, 1996a; Joshi and Gadgil 1991), there is little empirical quantification of dispersal rates into hunted areas. Based on the evidence that harvest rates of the Aché in

Paraguay do not appear to be sustainable but wildlife populations show no evidence of decline, Hill and Padwe (this volume) suggest the existence of unhunted "source" wildlife populations within the Mbaracayu reserve. Similarly, Fimbel et al. (this volume) report that in the Lobeké area of southeastern Cameroon, harvests of red and blue duiker exceeded calculated sustainable rates, but both species still retained high densities in hunted areas. The authors posit that hunted populations were being replenished from high-density populations further removed from the road. Replenishment has been documented in a study of harvests of primates in northern Peru. Populations of tamarins recovered following cropping within 2 to 5 years, with immigration from surrounding source areas of considerable importance where cropping was extensive (Glander et al. 1984; Ramirez 1984). Perhaps the best information on replenishment of hunted populations comes from studies of species subject to commercial harvest for the fur trade (e.g., Knowlton 1972; Novarro 1995). Detailed demographic studies, which indicated that game populations could not persist in heavily hunted areas, contradicted observations that in fact they did persist, often for many years. Researchers have had to posit a significant immigration rate and the existence of unhunted source populations.

Carrying capacity for people, where the requirement is the continued conservation of the area's biological diversity, therefore can sometimes exceed calculated levels if game can be replenished from nearby unhunted areas. The establishment of protected areas where there is no hunting will frequently be the management option of choice if hunting in adjoining areas is to continue sustainably. Indeed, the intent of the juxtaposition of parks with other multiple-use areas is frequently to provide both a refuge for natural populations from harvest and a source for later replenishment for harvest. The very notion of the "sustainable landscape" (Robinson 1994) presupposes a mosaic of different land uses—protected areas, multiple use areas, agriculture areas—that taken together serve both to enhance the conservation of biological diversity and to augment the ability of people to meet their needs and aspirations. The carrying capacity of tropical forests for people can be augmented in a heterogeneous landscape that includes both extractive zones and protected areas. Except at very low human population densities, community-based conservation initiatives in tropical forests will probably always require the establishment and protection of adjacent "source" areas, be they national parks, wildlife reserves, or wildlife sanctuaries, if hunting for meat is to be sustained.

3

Evaluating the Impact and Sustainability of Subsistence Hunting at Multiple Amazonian Forest Sites

CARLOS A. PERES

L arge-bodied vertebrates are perhaps the most ecologically sensitive extracts of tropical forests, and their sustained management will become increasingly critical to the long-term integrity of these ecosystems. Throughout human history, however, the relentless harvest of wild meat by subsistence hunters around the tropics has resulted in conspicuous population declines and extinctions from local to global scales for many species of birds and mammals (e.g., Diamond 1984; Milberg and Tyrberg 1993). Growing concern for this widely acknowledged conservation problem is reflected in a flurry of research by conservation ecologists and applied anthropologists on the impact of hunting (see Redford and Robinson 1987, Robinson and Redford 1991a, and references therein).

Large vertebrates often have the greatest market value of any forest products extracted by tropical forest dwellers (e.g., Anstey 1991; Melnik and Bell 1996). Consequently, game species continue to be eliminated from otherwise undisturbed tropical forests, which may be left strangely quiet, reminiscent of silent cathedrals, once their most prominent megafauna has been seriously reduced in numbers if not extirpated (e.g., Terborgh et al. 1986; Emmons 1989; Peres 1991; Redford 1992). In areas where relatively high densities of humans rely on nonaquatic sources of animal protein, subsistence hunters have been forced to become increasingly unselective, to the point of targeting even small prey of low return value per unit of hunting effort. From the semi-arid deciduous caatinga of northeastern Brazil to the West African rain forests of Mali, Burkina Faso, Ghana, and Côte d'Ivoire, to mention but a few examples, many a dinner plate is now supplemented by even small morsels (Johansson 1995; personal observation). In sharp contrast, relatively abundant stocks of animal protein—mainly from highly productive aquatic sources—are often available in western Amazonia, allowing for relatively selective subsistence hunting practices, which largely ignore small-bodied forest vertebrates (Peres 1990; Bodmer 1995a).

Despite the importance of subsistence hunting, the methodology and field data required to form sound management policies are still in their infancy. Ideally, evaluation of the sustainability of a harvest system should consider its effect on age- or size-dependent fecundity, growth and survival rates of individuals, and the growth rate and age structure of populations (Lefkovitch 1967; Usher 1976). Conservation biology texts may even assume that long-term monitoring programs should be in place before a population decline caused by overharvesting can be properly diagnosed (e.g., Caughley and Gunn 1996). Yet the effects of different cull regimes on population dynamics cannot be explored because detailed demographic information is not available for any large-bodied tropical forest vertebrate. Unlike classic studies in the temperate zone on game hunting (e.g., McCullough 1979; Milner-Gulland 1994) and other resource harvest systems (Getz and Haight 1989), most studies on tropical subsistence hunting are based on harvest data obtained from household interviews (e.g., Ayres et al. 1991), counts of animal carcasses consumed at forest dwellings (e.g., Mena et al., this volume), and surveys of wildlife meat sold at urban markets (e.g., Fa et al. 1995). These data, however, cannot be easily translated into actual cull rates because the collateral mortality caused by hunters (e.g., nonretrieval of fatally wounded animals; deaths of unweaned infants) can be very significant. Although some of these studies derive average figures for game offtake per unit area using an estimated hunting range, they fail to consider the enormous variability in habitat productivity that results in natural differences in large vertebrate abundance independent of hunting (e.g., Arita et al. 1990; Robinson and Redford 1994b). This severely limits the predictive power of sustainable yield models as applied to a given forest site. In an attempt to model sustainable harvest rates for neotropical mammals, Robinson and Redford (1991b) acknowledge this problem, stating that the "population density estimate is usually the largest numerical contributor to the model, so an error in this estimate will generate the largest error in the final harvest value." Yet mammal population densities can vary widely across different spatial scales, as documented for entire continents (e.g., Chapman and Feldhamer 1982 and references therein), large tropical macrohabitats (August 1983; Emmons 1984; Oates et al. 1990; Peres 1997a), and even different parts of relatively small study areas (Peres 1993a; Chapman and Onderdont unpublished data). This is understandable given that the schedules of births, survival, deaths, and migration can be regulated by features of a habitat such as rainfall seasonality, soil fertility, availability of food and shelter, and vegetation structure and composition, which then define its species-specific carrying capacity (K). Yet data on population responses to varying degrees of hunting pressure under different environmental conditions remain scarce or nonexistent for most tropical game species.

From this perspective, comparing faunal densities at different sites to examine the effects of hunting pressure on game abundance (e.g., Bodmer 1995a; Glanz 1991; Peres 1990, 1996, 1997b; Silva and Strahl 1991) would seem pointless. However, in the absence of long-term studies in one area using data on both cull rates and stock recovery, cross-site comparisons using control sites can be a powerful tool to assess the impact of subsistence hunting. In this chapter, I consider a long-term series of line-transect censuses of tropical forest vertebrates conducted at multiple sites under varying degrees of subsistence hunting pressure. The aim is to examine the effects of hunting on vertebrate community structure. I begin by describing the overall impact of subsistence hunting on vertebrate densities and biomass at the population and community level. I then demonstrate the importance of considering habitat type, and its effects on game productivity, in quantifying whether a given harvest program is sustainable. In addition, I use a literature review and my own field data to assess the variability in population density of two of the most ubiquitous neotropical primates under different harvest regimens. Finally, I apply the results of these line-transect censuses to Robinson and Redford's (1991b) model of sustainable harvest rates to evaluate its predictive power under the scenario, usually faced by researchers, in which local densities of game populations are not known.

METHODS

This study is primarily based on a long-term series of standardized line-transect censuses of diurnal forest reptiles, birds, and mammals undertaken over 10 years (1987–1996) at 24 Amazonian forest sites (table 3-1). This is the most comprehensive set of quantitative inventories of medium to large-bodied tropical forest vertebrates based on this technique. The surveys have focused on a limited number of diurnal primary consumers, including primates, ungulates, squirrels, caviomorph rodents, cracids, tinamous, trumpeters, wood-quails, and tortoises. These comprise the bulk of the total vertebrate community biomass in neotropical forests (Eisenberg and Thorington 1973; Terborgh 1983; Terborgh et al. 1990; Peres in press). Surveyed species are placed into 32 discrete functional groups corresponding to a single species, or a few ecologically similar taxa (see appendix 3-1 for species names and groupings) and are collectively referred to in this chapter as *vertebrates*. Because of their size and abundance, these taxa are also the most amenable to direct observations, allowing relatively accurate estimates of population density using census data. The diurnal line-transect censuses conducted here thus include the vast majority of game species consumed by subsistence hunters in Amazonian forests. The main exceptions are a few species of nocturnal mammals, such as pacas (*Agouti*

Table 3-1. Key Habitat Features of Forest Sites Censused in This Study

Forest Types and Site Localities	Latitude (S), Longitude (W)	Sources of Natural Disturbance[a]	River Type[b]	Flood[c] Level (m), Duration (mo)	Hunting Pressure[d]	Transect Length (km)	Cumulative Census Distance (km)	Time of Census
Terra firme forests								
1. Urucu	4°50', 65°16'	BW (l)	B (hw)	—	N	4.5	359	Mar 88–Sep 89
2. Igarapé Açú	4°35', 64°29'	—	B (c)	—	N	4.0	51	May 87
3. SUC-1	4°50', 65°26'	—	B (hw)	—	N	4.0	47	Feb 88
4. Oleoduto	4°42', 65°23'	—	B (c)	—	N	4.0 + 4.0	124	May 94
5. Igarapé Curimatá	4°26', 65°39'	BW (m)	B (c)	—	N	3 × (4.0)	305	Oct–Nov 96
6. Igarapé Jaraquí	4°21', 65°31'	G (l)	B (l)	—	M	4.2	50	Feb–Mar 88
7. Riozinho	4°28', 67°06'	G (m)	B (c)	—	L	4.0	44	Dec 88
8. Vira Volta	3°17', 66°14'	G (l)	W (l)	—	L	5.0	110	May–Jun 92
9. Vai Quem Quer	3°19', 66°01'	BW (l)	W (l)	—	L	4.2	107	May 92
10. Fortuna	5°05', 67°10'	—	W (c)	—	H	4.5	56	May–Jun 87
11. Barro Vermelho I	6°28', 68°46'	—	W (c)	—	M	4.2	117	Oct–Nov 91
12. Altamira	6°35', 68°54'	T (l)	W (c)	—	L	4.0 + 0.8	113	Nov–Dec 91
13. Condor	6°45', 70°51'	T (l)	W (c)	—	L	4.2	119	Sep–Oct 91
14. Penedo	6°50', 70°45'	B (l), G (m)	W (c)	—	H	4.0	102	Aug–Sep 91
15. Sobral	8°22', 72°49'	B (l)	W (hw)	—	H	5.0	109	Feb–Mar 92
16. Porongaba	8°40', 72°47'	B (m), G (l)	W (hw)	—	H	4.6	115	Feb 92
17. São Domingos	8°55', 68°20'	T (l)	W (c)	—	H	4.0	52	Jun 87
18. Kayapó Reserve	7°46', 51°57'	G (m)	C (hw)	—	L	3.5 + 2.3 + 5.0	224	Sep–Nov 94, 95
Alluvial forests								
19. Kaxinawá Reserve	9°23', 71°54'	T (l)	W (hw)	0.2 (every 7 yr)	H	4.2 + 3.5	91.5	Aug–Sep 93
25. Cocha Cashu[e]	11°54', 71°22'	F (m)	W (hw)	0.4 (every 5 yr)	N	?	?	—[e]
Várzea forests								
20. Lago da Fortuna	5°05', 67°10'	F (m)	W (c)	1.0–2.0, 4–5	M	4.0	48.0	May–Jun 87
21. Barro Vermelho II	6°28', 68°46'	F (l)	W (c)	0.5–0.8, 4–5	M	4.3	91.4	Oct–Nov 91

Table 3-1. (*continued*)

Forest Types and Site Localities	Latitude (S), Longitude (W)	Sources of Natural Disturbance[a]	River Type[b]	Flood Level (m), Duration (mo)	Hunting Pressure[d]	Transect Length[b] (km)	Cumulative Census Distance (km)	Time of Census
22. Boa Esperança	6°32'; 68°55'	F (m)	W (c)	0.8–1.2, 4–5	L	3.8	101.8	Nov–Dec 91
23. Nova Empresa	6°48'; 70°44'	F (l)	W (c)	0.5–0.8, 3–4	M	3.6	96.0	Aug–Sep 91
24. Sacado	6°45'; 70°51'	F (l)	W (c)	0.7–1.0, 3–4	M	4.2	92.4	Sep–Oct 91

Site numbers correspond to those shown in figure 3-1.

[a]Indicates most important sources of natural disturbance: B, bamboo regeneration cycles; G, small canopy gaps; B, larger gaps generated by wind blowdowns; BW, backwater palm swamps; F, prolonged seasonal flooding. Classes of intensity are: none (—), light (l), moderate (m), and heavy (h).

[b]Indicates geochemical characteristics of nearest river—W, white-water river; B, black-water river; and C, clear-water river—and geographic position of survey sites along a given watershed—hw (headwaters) and c (central) or l (lower) sections of the rivers.

[c]For várzea sites represented by the high-water mark in typical years above the soil level, as indicated by a band of discoloration on tree trunks. Flooding periodicity at alluvial sites is supra-annual and represented in years.

[d]Hunting pressure: N, none; L, light; M, moderate; H, heavy.

[e]Data from Terborgh (1983), Terborgh et al. (1990), and Janson & Emmons (1990).

paca), armadillos (e.g., *Dasypus* spp., *Priodontes maximus*), and night monkeys (*Aotus* spp.), and large-bodied canopy birds, such as macaws (*Ara* spp.) and large toucans (*Ramphastos* spp.). The latter tend to be highly mobile and notoriously difficult to census, but are less frequently killed.

Sampling Sites

Sampling sites comprised 17 unflooded (hereafter, *terra firme*) forests, five season-ally flooded (hereafter, *várzea*) forests, one alluvial forest inundated at irregular supra-annual intervals, and one transitional *terra firme*/palm forest punctuated by small enclaves of edaphic savannas (*cerrados*). These are located within the water-sheds of several major rivers of eastern (upper Xingú river basin, one site) and western Brazilian Amazonia (Juruá, upper Tarauacá, Jutaí, upper Purús, upper Tefé, and upper Urucu rivers, 23 sites; see figure 3-1 and table 3-1). Forest type

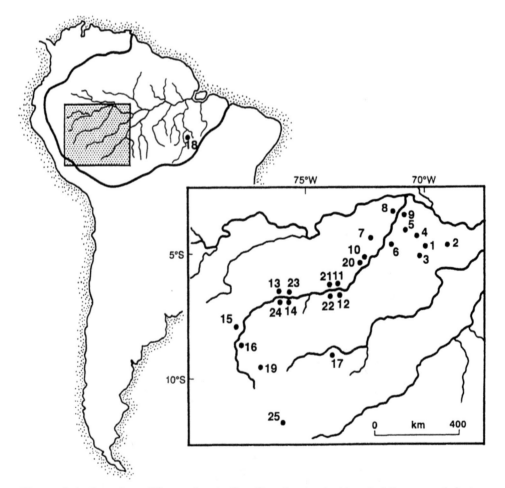

Figure 3-1. Location of forest sites in Brazilian Amazonia (sites 1–24) censused during this study. Data for Cocha Cashu, Peru (site 25) are from Terborgh et al. (1990) and Janson and Emmons (1990). Site numbers correspond to those listed in table 3-1.

and level of hunting pressure were the main criteria used in selecting sampling sites within a river basin. Wildlife populations sampled at these sites are assumed to be spatially independent of one another because any two sites within the same forest type were at least 45 km from one another or separated by a major river. Data from one additional alluvial Amazonian forest site were also incorporated into the analysis: Cocha Cashu, southern Peru, for which reliable density estimates are available for all large-bodied bird (Terborgh et al. 1990) and mammal species (Janson and Emmons 1990).

Unhunted sites are defined as those entirely uninhabited by Amerindians, nontribal extractivists (known in Brazil as *caboclos*), or rubber-tappers of European origin (*seringueiros*), and which had no evidence of any hunting this century (e.g., core heartwoods severed by axes; bark scars on *Couma*, *Brosimum*, and *Hevea* tree boles). These sites could not be economically reached on foot by hunters, and access to them was largely restricted to helicopters made available to us by the Brazilian Oil Company (Petrobrás, S.A.). The term *unhunted* as used in this chapter is thus reserved for truly pristine forests of remote interfluvial basins and headwater regions of Amazonia, rather than to refer to areas only rarely visited by hunters.

With one exception, hunting in all hunted sites was purely for subsistence. At Igarapé Curimatá (site 5 in figure 3-1) wild meat was harvested for both local consumption and sale, but the scale of the latter was small. At hunted sites, reliable data on game offtake were rarely available, and it was difficult to quantify accurately the hunting history over the past several decades. I therefore assigned them to one of three broad categories of wildlife harvest: light, moderate, or heavy hunting pressure. This was based on (a) semi-structured interviews with hunters who either lived or had lived at a given site for at least 2 years; (b) present and past human population densities quantified on the basis of the number of active households in each area, as determined using a high-resolution map (1:250,000) (Radam 1973–1981); and (c) the number of gunshots heard during each line-transect census. Interviews with hunters were unbiased with respect to fear of disclosing any illegal hunting activity because interviewees in such remote parts of Amazonia did not seem aware of any hunting restrictions or laws on protected species. The abundance of species most susceptible to hunting was not used to infer the previous history of hunting at a given site, for this relationship can be confounded by environmental heterogeneity. I therefore assume that the crude scale of hunting pressure used here was the most refined this methodology could afford in the absence of more accurate hunting records.

Sites also were assigned to one of three broad levels of topsoil fertility according to differences in seasonal influx of alluvial sediments. In lowland Amazonia, these are almost the only sources of key soil macronutrients, including carbon, nitrogen, potassium, calcium, magnesium, and phosphorus (C. Peres unpublished data). Hence, I classified all sites into *terra firme*, alluvial, or *várzea* according to their positions within a watershed, and to concentrations of macronutrients found in 20 soil samples collected along census transects at each site (C. Peres, unpublished data). The only outlier is represented by Pinkaiti, Kayapó Reserve (site 18 in figure 3-1), an eastern Amazonian forest sustained by exceptionally rich soil nutrients compared with *terra firme* forests of western Amazonia.

Population Censuses

Each survey comprised a cumulative census distance of 113.5 ± 78.0 km (range = 44–359 km), giving a total of 2724 km of census walks between the 24 sites. At each site, censuses were conducted from early morning to mid-day (0600–1130 hr) by at least two independent observers walking on separate transects each 4–5 km long. Transects were cut and marked every 50 m, and then left to "rest" for 1 or 2 days. This was immediately followed by 10 to 15 days of census walks. At previously hunted sites, we avoided using established hunting paths because this might introduce a bias, resulting in density underestimates. This is because animals could (a) shy away from such areas; (b) be excessively shy of observers using them; or (c) present more elusive escape responses, so were less likely to be detected by observers. No censuses were conducted on rainy days because that could also introduce bias. Censuses were conducted by one observer per transect (CAP and other observers with at least 5 years previous experience) with walking velocities of approximately 1250 m hr^{-1}, and were usually completed within 30 days.

An index of the numbers of animals (or groups in the case of social species) encountered per 10 km walked (hereafter, *sighting rates*) was calculated to control for overall differences in sampling effort. Density estimates were derived from either the Hazard rate or Uniform models with a cosine adjustment (Buckland et al. 1993) using ungrouped perpendicular distances from the sampling transect to the first animal sighted. Data truncation was used for a few highly vocal species (e.g., *Alouatta* spp., *Callicebus* spp.) by excluding vocally mediated outlier records far from the transect because those appeared to be detrimental to the performance of model estimators. For social species, detection is biased toward species in large, uncohesive groups, so a correction factor based on the group spread of a given species was also incorporated into the model (Peres 1997a). These models provided a better fit for species-specific data sets than did the Fourier series and half-normal models, as determined by the minimum Akaike Information Criterion (Buckland et al. 1993). To avoid unrealistic distortions of detection curves due to small sample sizes of relatively rare species, a pooled analysis was conducted for all surveys yielding fewer than 40 independent sightings—the minimum recommended by Burnham et al. (1980). This was justified because between-site variances in perpendicular distances were no greater than those within sites (analysis of variance [ANOVA], $p > 0.05$ in all cases).

Population densities (D) were then calculated using mean group sizes at each site from the group counts obtained during censuses that were considered to be accurate. To calculate crude population biomass, I used the mean body mass of an individual of a given species, defined as 80% of the average body mass of adults in Amazonian populations. Body mass data were compiled from Terborgh et al. (1990), Bodmer (1993), Peres (1993b), and C. Peres and H. Nascimento (unpublished data). Densities were assumed to be greater than zero if a species had been recorded in a site, either during or outside a census. A species was assumed to be extinct at a site if local hunters interviewed during surveys agreed unanimously that the species had once been common, but had not been sighted for at least 5 years prior to surveys. Further details on sampling sites, census methodology, and data analysis can be found elsewhere (Peres 1993a, 1997a, 1997b; Peres et al. 1997).

Data from Other Sources

To examine further the variation in game populations, density estimates were compiled from a comprehensive literature review of two widely distributed mammal taxa of neotropical forests: brown capuchin monkeys (*Cebus apella*) and howler monkeys (*Alouatta* spp.). These represent some of the most extensively censused neotropical mammals. Including my own census data from Amazonia, I obtained density estimates for 72 South American populations of brown capuchins (unpublished report), as well as 113 howler monkey populations (Peres 1997b) spanning from southern Mexico to northern Argentina and covering a wide range of neotropical forest types.

Density estimates of mammals also were incorporated from five hunted sites in western Amazonia, obtained using comparable line-transect censuses (Martins 1992; Bodmer 1993; Calouro 1995). These studies, however, were not used to derive estimates of total vertebrate biomass because they did not include birds in their censuses.

Statistical Analysis

Factorial ANOVAs were used to test for the effects of independent environmental variables on community-wide parameters. An indirect gradient analysis (CANOCO; ter Braak 1988) also was used to examine the overall site-by-species abundance matrix. Detrended correspondence analysis (DCA), which reduces the multiple dimensions of n species across m forest sites to a few ordination axes, proved to be the most appropriate ordination technique for this matrix. The performance of other techniques applied to the same matrix was consistently poor and showed systematic distortions of the first axis (arch effect). Species occurrences were weighed by their untransformed density estimates at each site. Points in the joint plots corresponding to forest sites are thus located at the center of gravity of the species occurring there, particularly those that are most abundant and widespread. Species were entered in the ordination according to discrete functional groups (see appendix 3-1), thus allowing for geographic replacements of ecologically equivalent congeners across different interfluvial regions of Amazonia. Two species—piping guans (*Aburria pipile*) and bearded saki monkeys (*Chiropotes satanas*)—were excluded from this analysis because they only occurred at a few censused sites.

VERTEBRATE DENSITIES AT UNHUNTED AND HUNTED SITES

Considering all sample sites, density and biomass estimates were calculated for two species of reptiles (one genus), 17 birds (eight genera), and 32 mammals (18 genera). These comprised assemblages of diurnal vertebrates dominated by small (<1 kg) and medium-sized species (1–3 kg; see appendix 3-1). Density estimates at unhunted and hunted sites are presented for 23 *terra firme* forests (table 3-2); this partially controls for differences in abundance due to forest type. Population densities of preferred game species at unhunted and lightly hunted sites were significantly higher than those at moderately to heavily hunted sites. The differences were particularly pronounced for tortoises (*Geochelone* spp.), curassows

Table 3-2. Population Densities (Mean ± SD) of Vertebrate Species Censused at 23 Amazonian *Terra Firme* Forests Subject to Different Levels of Hunting Pressure

Species	Hunter[a] Preference	Nonhunted and Lightly Hunted Ind./km^2	N	Moderately and Heavily Hunted Ind./km^2	N	F-value	p[b]
Geochelone carbonaria/denticulata	+++	3.1 ± 7.8	8	0.0 ± 0.0	5	0.332	+
Crypturellus spp. (*variegatus/ cinereus/undulatus/soui*)	+	13.2 ± 5.6	11	8.9 ± 3.5	7	0.083	NS
Tinamus guttatus/tao/major	++	8.3 ± 3.9	11	4.9 ± 1.4	7	0.049	+
Odontophorus gujanensis/stellatus	+	11.1 ± 5.7	11	7.4 ± 3.5	7	0.151	NS
Penelope jacquacu/pileata	++	8.5 ± 4.4	12	5.2 ± 2.9	9	0.073	NS
Mitu mitu/Crax fasciolata/globulosa	+++	2.43 ± 3.29	12	0.07 ± 0.21	9	0.046	+
Psophia leucoptera/viridis	++	7.8 ± 4.7	12	2.9 ± 2.3	9	0.011	+
Microsciurus flaviventer	–	2.1 ± 1.0	11	1.7 ± 1.2	7	0.472	NS
Sciurus ignitus	–	2.5 ± 1.5	11	3.5 ± 1.6	7	0.211	NS
Sciurus spadiceus	+	3.7 ± 3.2	10	3.5 ± 4.2	9	0.897	NS
Myoprocta pratti/acouchy	–	3.5 ± 2.0	10	4.3 ± 2.6	10	0.418	NS
Dasyprocta leporina/fuliginosa	+	7.6 ± 10.5	13	2.9 ± 1.7	10	0.176	NS
Tayassu tajacu	+++	6.5 ± 3.3	13	1.6 ± 1.9	10	0.001	+
Tayassu pecari	+++	1.70 ± 1.36	13	0.13 ± 0.41	10	0.002	+
Mazama gouazoubira	+++	1.44 ± 0.81	13	0.49 ± 0.29	10	0.002	+
Mazama americana	+++	1.25 ± 0.69	13	0.38 ± 0.53	10	0.004	+
Tapirus terrestris	+++	0.52 ± 0.23	13	0.11 ± 0.18	10	0.000	+
Cebuella pygmaea	–	2.9 ± 1.0	4	5.9 ± 6.8	4	0.403	NS
Saguinus fuscicollis	–	19.9 ± 11.9	10	41.2 ± 18.7	8	0.010	–
Saguinus mystax/imperator	–	31.5 ± 21.1	11	45.4 ± 22.4	7	0.201	NS
Saimiri boliviensis/sciureus	–	18.8 ± 15.9	12	17.9 ± 17.1	10	0.906	NS
Callicebus cupreus/moloch	–	2.7 ± 1.9	13	8.9 ± 9.9	10	0.037	–
Callicebus torquatus	–	5.2 ± 3.1	7	5.1 ± 7.0	3	0.964	NS
Pithecia albicans/irrorata/monachus	+	9.6 ± 5.8	11	7.6 ± 3.4	8	0.390	NS
Cacajao calvus	–	13.5 ± 2.3	2	18.7 ± 18.2	2	0.727	NS
Cebus apella	++	19.6 ± 12.0	13	11.9 ± 11.9	10	0.145	NS
Cebus albifrons	+	8.9 ± 5.6	12	9.9 ± 9.6	10	0.767	NS
Alouatta seniculus	+++	4.3 ± 3.7	13	1.6 ± 1.3	10	0.042	+
Lagothrix lagothricha	+++	15.1 ± 9.4	11	1.7 ± 3.0	7	0.002	+
Ateles paniscus	+++	4.2 ± 2.9	8	0.7 ± 1.3	10	0.003	+

N indicates the number of sites at which a given species was censused. This includes data from two lightly hunted (Fazenda Unio and Yavari Miri) and three heavily hunted sites (Rio Iaco, Tahuayo-Blanco and Antimari) of western Amazonia censused by Martins (1992), Bodmer (1993), and Calouro (1995).

[a]Indicates to what extent hunters were likely to ignore different species during a hunting foray: –, always ignored; +, often ignored; ++, rarely ignored; +++, never ignored.

[b]Plus or minus signs denote species for which densities at nonhunted and lightly hunted sites were either significantly higher (+) or significantly lower (–) than those at moderately and heavily hunted sites (one-way ANOVAs); comparisons of all other species were nonsignificant (ns).

(*Mitu mitu* and *Crax* spp.), large tinamous (*Tinamus* spp.), trumpeters (*Psophia* spp.), the three genera of large-bodied prehensile-tailed primates (*Ateles, Lagothrix,* and *Alouatta*), and all five species of ungulates (two species of peccaries, two species of brocket deer, and tapir), all of which appeared to be particularly susceptible to hunting. In contrast, small-bodied species (e.g., small squirrels, acouchys, tamarins, squirrel monkeys, and titi monkeys), which were usually ignored by hunters, maintained comparable or even greater densities at persistently hunted sites. Although medium-bodied taxa generally failed to show significantly higher densities in unhunted areas, species such as common guans (*Penelope* spp.), agoutis (*Dasyprocta* spp.), and brown capuchins (*Cebus apella*) occurred at persistently hunted areas in noticeably small numbers.

COMMUNITY-WIDE DIFFERENCES IN VERTEBRATE DENSITIES AND BIOMASS

There was a six-fold difference in the biomass of surveyed vertebrates across all sites, from less than 200 kg/km^2 at one of the most intensively hunted sites (São Domingos) to nearly 1200 kg/km^2 at one of the least hunted sites (Boa Esperança, figure 3-2). Assuming that the categories of hunting pressure used in this study can be fitted to a four-point scale of previous game offtake (ranging from 0 to 3), hunting levels alone explain 55% of the variation in total vertebrate biomass ($r = -0.744$, $p < 0.001$, $n = 25$). Whether hunting levels scale as an ordinal or continuous variable, vertebrate biomass was inversely correlated with hunting intensity at all sites ($r_s = -0.820$, $p < 0.001$, $n = 25$).

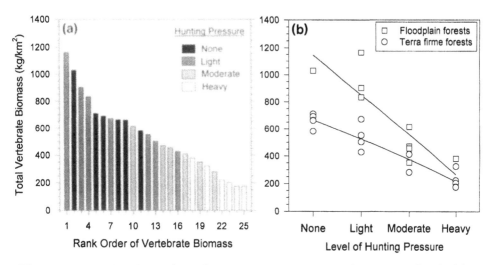

Figure 3-2. (a) Total vertebrate biomass at 25 Amazonian forest sites that had been subjected to varying degrees of hunting pressure **(b)** and the relationship between hunting pressure and vertebrate biomass (kg/km^2) for sites of two broad classes of forest types (see text for definitions).

The decrease in vertebrate biomass from unhunted to gradually more hunted sites is not mirrored by a concomitant decline in the density of all species combined (r_s = 0.206, p = 0.322, N = 25), which ranged between 167 and 409 individuals per km^2 (table 3-3). Indeed, the variable hunting pressure failed to explain a significant proportion of the variation in overall vertebrate density (table 3-4). In other words, the total number of animals detected at different sites was not affected significantly by hunting pressure. However, the size structure of the vertebrate assemblage shifted downward with greater hunting pressure, which significantly affected the total community biomass (see table 3-3). This suggests that hunters select strongly for species in large size classes (>3 kg), which contributed most to the overall vertebrate biomass at unhunted sites (see table 3-2). Species weighing at least 3 kg accounted for over three-fourths of the biomass at unhunted sites, but only one-fourth or less at heavily hunted sites. Large-bodied species thus contributed a major proportion of the variation in total vertebrate biomass (r^2 = 0.625, p < 0.001, n = 25). The reduction of large-vertebrate abundance at hunted sites thus resulted in a gradual decline in mean body mass. For example, the average weight of individual vertebrates was more than 3.5 kg at unhunted sites, but only 1 kg or less at heavily hunted sites (tables 3-3 and 3-5).

These patterns remain consistent whether one looks at all sites, or at *terra firme* sites only (see table 3-5). The vertebrate biomass of *terra firme* areas under increasing hunting pressure declined from nearly 700 to less than 200 kg/km^2. This was largely a result of marked declines in densities of large terrestrial birds, large primates, tapir, both species of brocket deer, and both species of peccaries, which are the main contributors to the overall community biomass, rather than to vertebrate numbers of all vertebrate species combined (ANCOVA, $F_{2,22}$ = 3.5, p = 0.047). Several lightly hunted sites have a contained biomass greater than that of some unhunted sites. This almost certainly reflects the large between-site differences in habitat quality, which also accounted for a significant proportion of the variation in vertebrate biomass. Alluvial and *várzea* forest sites sustained by nutrient-rich soils tended to have a greater wildlife biomass than those in nutrient-poor *terra firme* forests (see figure 3-2). *Terra firme* forest sites at best contained a

Table 3-3. Average Density, Biomass, Percentage of Biomass Contributed by Large-Bodied Species, and Mean Body Mass of Vertebrates Censused at 25 Amazonian Forest Sites Under Varying Degrees of Hunting Pressure

Hunting Regimen	Total Density (ind./km²)		Total Biomass (kg/km²)		Biomass of Species >3 kg (%)		Mean Vertebrate Body Mass (kg)		
	Mean	SD	Mean	SD	Mean	SD	Mean	SD	N
None	227.6	92.4	723.7	156.4	77	6	3.34	0.53	6
Light	264.9	76.0	723.1	260.0	70	6	2.72	0.59	7
Moderate	234.0	79.0	431.1	113.9	46	12	1.87	0.23	6
Heavy	237.4	23.2	247.2	85.1	25	10	1.03	0.10	6
All sites	241.9	69.7	538.9	262.3	55	22	2.26	0.97	25

Table 3-4. Analysis of Variance of Total Vertebrate Density and Total Vertebrate Biomass for 25 Amazonian Forest Sites Entering Levels of Hunting Pressure and Soil Fertility as the Main Factors

Factors	Overall Game Density (ind./km²)				Overall Game Biomass (kg/km²)			
	SS	df	MS	F	SS	df	MS	F
Hunting	13,126	3	4375	1.09	1,114,516	3	371,505	22.46[a]
Soil fertility	34,790	2	17,395	4.32[b]	314,582	2	157,291	9.51[b]
Explained	40,217	5	8043	2.00	1,337,120	5	267,424	16.16[a]
Residual	76,429	19	4023		314,323	19	16,543	
Total	116,646	24	4860		1,651,444	24	68,810	

[a]$p < 0.001$.

[b]$p < 0.05$.

Table 3-5. Average Density, Biomass, Percentage of Biomass Contributed by Large-Bodied Species, and Mean Body Mass of Vertebrate Species at 18 *Terra Firme* Forest Sites Under Varying Degrees of Hunting Pressure

Hunting Regimen	Total Density (ind./km²)		Total Biomass (kg/km²)		Biomass of Species >3 kg (%)		Mean Vertebrate Body Mass (kg)		
	Mean	SD	Mean	SD	Mean	SD	Mean	SD	N
None	191.2	27.7	662.5	49.6	78	5	3.50	0.38	5
Light	241.7	49.2	649.7	189.3	70	7	2.69	0.64	6
Moderate	194.6	27.6	346.7	92.3	51	8	1.77	0.22	2
Heavy	232.6	22.4	231.8	85.2	22	7	0.99	0.29	5
All *terra firme* sites	219.9	39.9	503.5	230.0	57	24	2.34	1.09	18

biomass of 904 kg/km², but this was only attained at one eastern Amazonian site that had atypically high soil fertility and canopy heterogeneity. Yet this figure is considerably lower than that of a *várzea* forest of western Amazonia (Boa Esperança, 1164 kg/km²), even though it had been subject to light hunting pressure for more than 100 years. Under entirely unhunted conditions, linear regressions predict a vertebrate biomass of 1140 kg/km² ($r^2 = 0.761$, $p = 0.002$, $n = 9$) for nutrient-rich *várzea*, alluvial, and one eutrophic *terra firme* forest, but only 699 kg/km² for oligotrophic *terra firme* forests of central-western Amazonia ($r^2 = 0.791$, $p < 0.001$, $n = 16$; see figure 3-2). Although these linear functions presented different negative slopes ($b_1 = -285.5$ and -162.9 for floodplain and *terra firme* forests, respectively), floodplain forests almost always supported a greater wildlife biomass than did *terra firme* forests, particularly under little or no hunting.

Overall vertebrate biomass of *terra firme* forests was therefore lower than that of floodplain forests, particularly when hunting pressure was controlled for (ANCOVA, $F_{2,22} = 31.6$, $p < 0.001$). For instance, vertebrate biomass of unhunted remote *terra firme* sites was lower than that of abutting floodplain forests that had been lightly hunted.

Several *várzea* forests subject to persistent hunting had a relatively low verte-brate biomass (see figure 3-2). This can be explained by the fact that typical flooded forests contained few strictly terrestrial vertebrate species (e.g., tortoises, ungulates, caviomorph rodents), which otherwise account for a large fraction of the biomass. Large-bodied arboreal taxa such as ateline primates, on the other hand, have high biomass at unhunted *várzea* sites (Peres 1997b), but appear to respond to hunting with sharp population declines. This is because of intrinsic life-history traits that make them particularly susceptible to hunting. Also they are one of the few large-bodied targets available within the more impoverished pool of prey species of *várzea* forests. Species such as howlers and spider monkeys are thus expected to take the brunt of the hunting pressure in this forest type, partic-ularly during the high-water season when subsistence fishing is far less feasible.

SPECIES BODY SIZE AND SHIFTS IN COMMUNITY STRUCTURE

The relationship between body mass and population density indicates that inter-mediate-sized species at unhunted sites could be more abundant than both their smaller and larger counterparts (figure 3-3). At hunted sites, however, body mass was a good inverse predictor of population densities ($r = -0.742$, $p < 0.001$, $n = 17$) because most large-bodied species showed sharp population declines with hunt-ing. In contrast, the densities of small and medium-sized taxa were comparable in hunted and unhunted sites.

Shifts in the relationship between population biomass and species body mass were largely due to declines of large vertebrates, as well as a greater variance in the density and biomass of most species, at intensively hunted sites (see figure 3-3). This can be understood given the massive contribution of large-bodied species to the vertebrate biomass at unhunted sites (see table 3-3). Body mass thus explained 71% of the variation in population biomass in unhunted sites ($r = 0.841$, $p < 0.001$, $n = 13$), but only 37% at hunted sites.

ORDINATION OF SAMPLING SITES

In the DCA assessing the overall matrix of 25 sampling sites by 30 vertebrate species, 53% of the total variance was explained by the first two axes, and 62% by the first four axes (axis I, 35.3%; II, 18.2%; III, 6.1%; IV, 2.6%). Because the first two axes alone uncovered such a large proportion of the total variance, other dimensions were disregarded.

On the basis of species densities, *terra firme*, alluvial, and *várzea* forest sites formed fairly distinctive natural groups diverging primarily along the first DCA dimension (figure 3-4). Loadings on the second DCA axis, on the other hand, corresponded primarily to levels of hunting pressure. The larger number of *terra firme* sites, which were distributed along a wider range of environmental gradients,

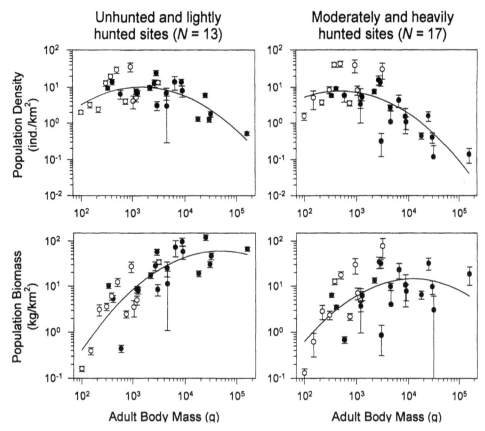

Figure 3-3. Relationship between body mass and mean (SE) population density (*top*) and biomass (*bottom*) for vertebrate species censused at 30 Amazonian forest sites under two broad classes of hunting pressure. Regression lines represent best-fitting functions based on all data points. Solid circles represent larger-bodied species, which were preferred by hunters; open circles represent smaller bodied species, which were usually ignored by hunters. Data from five additional sites censused by Martins (1992), Bodmer (1993), and Calouro (1995) were also incorporated.

allowed the model to detect a greater variability in the structure of their vertebrate assemblages. Remote interfluvial sites that had never been hunted formed a relatively tight cluster that nevertheless fell close to lightly hunted *terra firme* sites under the influence of white-water rivers. Not surprisingly, alluvial sites such as Cocha Cashu and the Kaxinawá Reserve, fell precisely between *terra firme* and typical *várzea* sites, reflecting their intermediate characteristics in hydrology, soil fertility, and patterns of vertebrate abundance.

The smaller number of *várzea* sites form a highly cohesive cluster. This reflects their peculiar species composition marked by the conspicuous absence of terrestrial species, an impoverished subset of understory species, and the numeric dominance of a few primate species (Peres 1997a). The outlier position of the single eastern Amazonian site (Pinkaiti, Kayapó Reserve) might be due to the fact that this atypical *terra firme* forest safeguards terrestrial vertebrate populations,

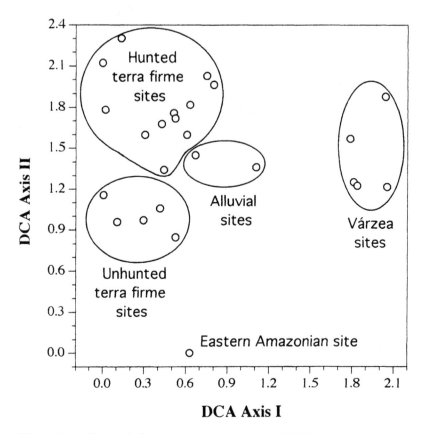

Figure 3-4. Detrended correspondence analysis (DCA) plot showing the centers of gravity of 25 forest sites, weighed by their respective vertebrate population densities, along the first two ordination axes. Loadings on the first and second DCA axes correspond primarily to forest types and levels of hunting pressure, respectively.

including ungulates and dasyproctids, in far greater numbers than those of western Amazonia (Peres 1996). This is probably because of the greater soil fertility and structural heterogeneity of this site (Peres et al. 1997), as well as underlying differences in the original fauna. Centers of gravity of forest sites were thus primarily influenced by both forest type and degree of hunting; these uncover most of the variation captured by the first and second axes, respectively.

GAME DENSITY, PRODUCTION, AND YIELD

Field estimates of population densities of hunted species can be applied to a maximum sustainable production model (Robinson and Redford 1991b, 1994b) to evaluate how differences in population densities affect estimates of maximum sustainable harvest levels. Robinson and Redford's model calculates the total annual sustainable production (population increments through births and immigrations), assuming that maximum production (P_{max}) occurs at about 60% of carrying capacity (K). This figure accounts for any density-dependent effects on population growth and is intermediate between the maximum sustainable yield (MSY) at 50% of K, as used in the classic logistic model of population growth (Caughley 1977; Sinclair 1989), and models where MSY is reached at about 70% of K (McCullough 1982). The model is possibly more realistic for tropical forest fauna and is expressed as follows:

$$P_{max} = (0.6\, D_2 \times \lambda_{max}) - 0.6\, D_2$$

D_2 is the "predicted density," as predicted from a linear regression of population density against body mass for subsets of species broken into dietary categories (Robinson and Redford 1986a). λ_{max} is the maximum finite rate of increase for a given species from time t to time $t + 1$ (measured in years); this depends primarily on the number of adult females reproducing and the average number of offspring produced per female per year. In addition, Robinson and Redford (1991b) calculate maximum sustainable harvest quotas for a species by assuming that the proportion of the potential production that can be harvested is 60% for very short-lived species, 40% for short-lived species, and 20% for long-lived species.

Assuming no prior knowledge of the population status of any species surveyed here in relation to K, the field data from unhunted and lightly hunted sites (in which populations are probably near K) were used to generate estimates of maximum production. As predicted by the model, these increase linearly as a function of density. Figure 3-5 shows this relationship for 10 of the 14 game species of large neotropical mammals considered by Robinson and Redford (1991b).

Densities for all but one of these 10 species were usually lower than these authors' predicted densities. For brown capuchins (*Cebus apella*), on the other hand, the predicted density underestimated maximum production at all Amazonian sites. This indicates that production estimates based on average densities at K, which ultimately rely on data from suitable study areas for any given species, can signify overestimate or underestimate potential game production in typical Amazonian forests. Robinson and Redford's predicted densities were greater than actual densities for tapir, both species of brocket deer, and white-lipped peccaries, and one of the highest values for agoutis, collared peccaries, howler monkeys, and spider monkeys (see figure 3-5). In contrast, densities of Amazonian populations of brown capuchins were actually higher than in those studies compiled in the Robinson and Redford (1986a) review.

Of particular concern is the fact that for seven of the ten species of large mammals considered here, production figures derived from predicted densities were substantially higher than those derived from actual densities for all five unhunted *terra firme* forests in remote interfluvial regions. Brown capuchins, woolly monkeys, and collared peccaries were the only species for which the predicted density fell below that of at least one actual density estimate for this forest type.

This is further illustrated by calculations of potential sustainable harvests (indi-

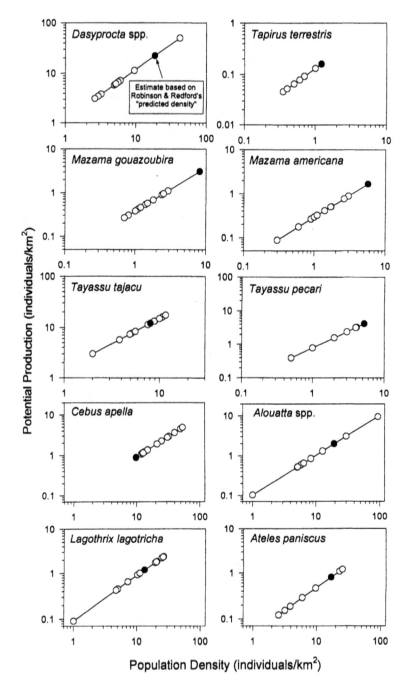

Figure 3-5. Relationship between population density and potential production for ten mammal species at unhunted and lightly hunted sites according to Robinson and Redford's (1991b) sustainable harvest model. Open circles show the range of production values based on actual density estimates for each species at each forest site. Solid circles indicate the predicted density for each species based on a linear regression of population density against body mass for samples of neotropical forest mammals divided into dietary classes (see Robinson and Redford 1986a, 1991b).

viduals/km²/yr) for all ten species, based on (a) density estimates from unhunted and lightly hunted sites in which populations are assumed to be at or near K; and (b) the longevity classes assumed by Robinson and Redford (1991b). At the five unhunted remote *terra firme* forests, the entire range of yield values for seven of the ten mammal species considered fell well below the average yields calculated by these authors (figure 3-6).

At *várzea*, alluvial, and *terra firme* forests adjacent to floodplains, however, average densities were closer to the predicted values; agoutis, collared peccaries, and both species of brocket deer were the only species in those forest types for which theoretical (predicted) yields continued to overestimate sustainable offtakes based on actual densities. Theoretical yields for howler monkeys, spider monkeys, woolly monkeys, and white-lipped peccaries fell within the 25th and 75th percentiles of density-derived yields, but underestimated those for brown capuchins and tapirs. Hence, wildlife densities in *terra firme* forests in remote interfluvial regions tend to be intrinsically low, and harvest estimates based on average densities can exceed sustainable levels. This could result in population declines, even when harvest systems are presumed to be sustainable in the absence of local density estimates.

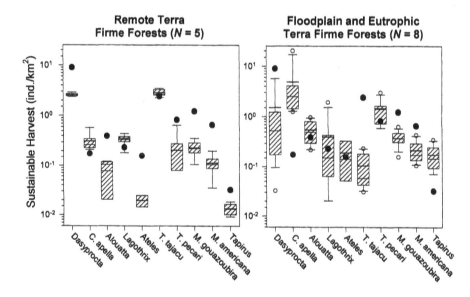

Figure 3-6. Tukey box plots of sustainable harvests for ten taxa of forest mammals within two broad classes of Amazonian forest types calculated from data obtained at unhunted and lightly hunted sites. Hatched boxes are delimited by thin lines representing the 25th, 50th, and 75th percentiles. Heavy lines represent mean harvest values. Capped bars indicate the 10th and 90th percentiles, and small open circles indicate points outside this range. Solid circles represent the maximum sustainable harvest for each species as calculated by Robinson and Redford (1991b) based on predicted densities from a wide range of neotropical forest sites.

CHANGES IN COMMUNITY STRUCTURE

Subsistence hunting in Amazonia is primarily targeted at large-bodied frugivores, granivores, and browsers. It can result in profound changes in the structure of tropical forest vertebrate communities through (a) shifts in the relative abundance of different size classes, (b) significant reductions in the overall community biomass, and (c) changes in guild structure.

Comparisons between hunted and unhunted sites suggest that population densities of most small- and medium-sized species are not significantly depressed by past and present hunting. In contrast, densities of large-bodied species, which tend to have long generation time and low fecundity, are adversely affected by hunting. Population declines also have been shown elsewhere in the neotropics for several groups of large-bodied birds (Thiollay 1986; Silva and Strahl 1991) and mammals (Peres 1990, 1996; Glanz 1991; Bodmer 1995a). Vertebrate species particularly sensitive to hunting include large tinamous (*Tinamus* spp.), large cracids (*Crax* spp. and *Mitu mitu*), trumpeters (*Psophia* spp.), all four species of artiodactyls (*Tayassu pecari, T. tajacu, Mazama americana,* and *M. gouazoubira*), tapirs (*Tapirus terrestris*), and all three genera of Amazonian ateline primates (*Alouatta, Ateles,* and *Lagothrix*). This relatively small subset of species comprises a large proportion of the overall biomass of diurnal primary consumers in unhunted Amazonian forests (cf. Terborgh et al. 1990; Peres in press), and parts of the Guianan shields (Eisenberg and Thorington 1973; Thiollay 1986). These taxa also comprise the best candidates for indicators of hunting in neotropical forests because they are particularly prone to depletion in persistently hunted areas, although they clearly vary in the extent to which they are susceptible to overharvesting (e.g. Bodmer 1995b; Peres 1996).

Although hunting affects the densities of large-bodied species, it does not result in significant crashes in the total vertebrate density. This could be due to a mechanism of density compensation because several small and some medium-sized species were more common at intensively hunted sites. However, the fact that a few species, (e.g., saddle-back tamarins and dusky titi monkeys) occur at higher densities in hunted areas could be explained by differences in forest structure rather than competitive release following the selective removal or depletion of large-bodied taxa. The density-compensation hypothesis is also weakened by the fact that several obviously small-bodied candidates for within-guild compensation (e.g., small tinamous and wood-quails at sites with low post-hunting densities of large tinamous and curassows) also showed density declines from unhunted to hunted areas.

Our census data partly confirm widely recognized relationships between body size and abundance; large-bodied species are on average less abundant than small-bodied species, and intermediate-sized species are on average more abundant than either small or large-bodied species (Peters and Wassenberg 1983; Cotgreave 1993; Blackburn and Lawton 1994). The negative slope of this relationship (see figure 3-3) becomes more obvious at moderately and heavily hunted sites following population declines of species weighing at least 1.2 kg. Differences in the shape of these curves cannot be predicted from the harvest alone; native

Amazonians and caboclos often hunt more intermediate-sized than large animals (e.g., Stearman and Redford 1995; C. Peres and H. Nascimento unpublished data). It is more likely that the curve reflects differences in the recovery potential and reproductive productivity of the species rather than their baseline densities under conditions of low or no hunting pressure (May 1981; Robinson and Redford 1986b; Bodmer 1995a).

Between-site variation in the biomass of forest vertebrates, and particularly that of large-bodied species, can to a large extent be explained by levels of hunting pressure. This is remarkable considering the enormous differences in hydrological regimes, soil fertility, forest structure and composition, and source faunas, each of which can account for part of the variation in vertebrate community structure in Amazonia (Emmons 1984; Terborgh et al. 1990; Janson and Emmons 1990; Peres 1997a, in press). The effects of soil fertility on the abundance and guild structure of vertebrates appear to be particularly important. Once hunting pressure has been controlled for, the vertebrate biomass in floodplain forests is greater than that in *terra firme* forests, and folivores are much more abundant in floodplain forests (Peres 1997b). Indeed, differences in vertebrate biomass between unhunted and lightly hunted areas are more a function of forest types than the degree of hunting. The availability of carbon, nitrogen, potassium, calcium, and magnesium in young sedimentary soils of *várzea* and alluvial forests is far greater than in *terra firme* forests, particularly those far removed from large white-water rivers (Duivenvoorden and Lips 1995; C. Peres, unpublished data). Yet nearly 90% of lowland Amazonia has received no sediment deposits since the Triassic or lower Cretaceous (Putzer 1984). Proximity to major white-water rivers is therefore a decisive factor in the local nutrient cycle in a region where only 6% of soils have no major nutrient limitations to plant growth (Sanchez 1981).

Although the mechanisms whereby which nutrient-poor soils curb tropical forest productivity remain unclear, it seems that large-scale differences in Amazonian vertebrate abundance are ultimately related to the region's geochemical gradients. Other studies are required to address how demographic parameters affecting wildlife recovery respond to forest productivity under a similar harvest regimen. Data presented here, however, suggest that forest productivity can compensate for the effects of hunting, in that game densities in hunted, high-quality habitats may be greater than those of unhunted, low-quality habitats. From a wildlife management perspective, populations can be encouraged to grow at their fastest rate within the limits set by resources available to them, but sustainable offtakes in *terra firme* forests would have to be adjusted in proportion to the low game productivity of this habitat. Major differences in habitat that affect yields of game populations should thus be explicitly considered in determining management restrictions that promote a sustainable harvest.

VARIANCE IN POPULATION DENSITIES

The substantial differences in densities of Amazonian vertebrates in different forest types can be further illustrated by compiling published population estimates for two of the most extensively censused and widely hunted primate taxa in the neotropics, *Cebus apella* and *Alouatta* spp. (figure 3-7). Even when hunting pressure is controlled for, these taxa show enormous variations in population abundance, as indicated by the large error bars about their mean densities (figure 3-8). Most sites at which the density of these widespread taxa was determined were described as unhunted (brown capuchins, 35%, *n* = 72; howler monkeys, 50%, *n* = 113). This clearly reflects an investigator bias in deliberately selecting high-density animal populations, which tend to occur in either protected or inaccessible study areas.

In unhunted areas, the mean densities of brown capuchins and howler monkeys at 15 and 45 sites, respectively, show enormous standard deviations. This presumably indicates the wide variance in forest quality that these ubiquitous primates can tolerate, quite independently of hunting pressure (Peres 1997b). Densities of the smaller bodied brown capuchin did not necessarily decline with increasing hunting, although they were significantly lower at heavily hunted sites where even the smaller females became fair game. However, this species was not a preferred target at lightly hunted sites where larger quarries had not been markedly depleted (Peres 1990; but see Hill 1996). The larger-bodied howler monkey, on the other hand, succumbed to a nonlinear decline in densities at all levels of hunting pressure, which confirms their being choice prey items even in lightly hunted areas.

GAME PRODUCTION AND SUSTAINABLE YIELDS

Estimates of production within a species can be directly proportional to estimates in population, so density errors in the latter can result in proportionally large errors in the former. Population densities are likely to be the most habitat-depen-

Figure 3-7. Red howler monkey (*Alouatta seniculus*). *Photo courtesy of John G. Robinson.*

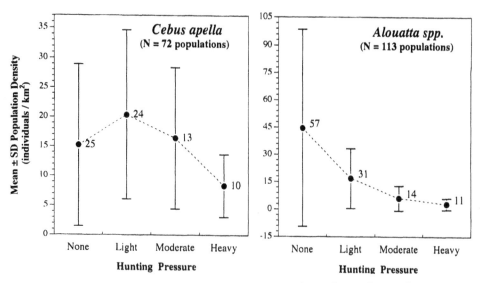

Figure 3-8. Relationship between hunting pressure and population density for two extensively surveyed taxa of neotropical primates, based on an extensive literature review (Peres 1997b, unpublished data). The number of density estimates considered at each level of hunting pressure are indicated near the means.

dent parameter entered in Robinson and Redford's (1991b) yield model and make the greatest contribution to possible errors in production and yield estimates. The other equation parameter, the finite rate of increase (λ), is derived directly from the intrinsic rate of population increase (r_{max}), which is likely to be more stable within a species at different sites because of its dependence on more phylogenetically inert characters such as body size and diet (e.g., Robinson and Redford 1986b). Ideally then, evaluations of the sustainability of harvests should include information on the harvest itself as well as actual density estimates of unhunted wildlife populations at nearby control sites (Martins 1992; Bodmer 1993; FitzGibbon et al. 1995b; Calouro 1995; Mena et al. this volume) rather than average densities obtained from literature reviews. Reproductive and survivorship data of harvested populations, particularly females, would also be highly appropriate to refine these evaluations, but such data are difficult to obtain and require longer term monitoring programs (but see Bodmer et al. 1994).

Use of predicted estimates of maximum sustainable yields (Robinson and Redford 1991b) and also the results obtained here suggest that harvest rates in most of Amazonia need not be very high before they begin to drive game populations into precipitous declines. This calls for concerns about the implementation of harvest programs and the persistence of game populations in vast expanses of *terra firme* forest in Amazonia, particularly within the central and eastern parts of the basin, and throughout the Guianan shields. This can be illustrated by the rapid depletion and local extinction of game populations across several taxonomic groups, brought about by agricultural settlements subsidized by wild meat wherever roads have intersected these forests (Smith 1976; Saffirio and Hames 1983; Peres 1990, 1993c, 1996).

This study highlights the importance of environmental heterogeneity when considering the potential production of wildlife, even in structurally uniform expanses of undisturbed tropical forests such as much of Amazonia. Subtle environmental gradients in this region, such as differences in soil fertility generated by fluvial dynamics, are also shared by other areas of tropical forest dominated by poor soils such as Borneo (Rajanathan 1992) and the African Congo basin (Maisels et al. 1994). In assessing the general application of this study, wildlife surveys across different forest types will have to be conducted in other poorly known parts of the tropics. It is high time for research to integrate the effects of species life-history traits and environmental variables to predict maximum sustainable yields of game populations in a given forest type. Understanding how environmental heterogeneity limits game population growth will be of crucial importance as advancing colonists are increasingly forced to extend their march into previously "unclaimed" *terra firme* forests supporting small numbers of large vertebrates. Otherwise, our images of tropical forest wildlife will increasingly rest on nostalgic memories of the "good old days" when ephemeral meat bonanzas were once enjoyed in many frontier settlements.

ACKNOWLEDGMENTS

Data presented here were obtained during a long-term series of line-transect censuses funded by the Bay Foundation (1996–1997), Wildlife Conservation Society (1991–1995), the National Geographic Society (1991–1992), and the World Wildlife Fund-US (1987–1989). I am grateful to the Brazilian Oil Company (Petrobrás, S.A.) for providing critical logistical support and helicopter transportation to some of the survey sites. João Pompilho, Raimundo Nonato, Luís Lopes, Paul Honess, "Toshiba," Josimar Pereira, Edvar Corràa, Pedro Develey, and Hilton Nascimento helped in conducting censuses. I wish to thank Jay Malcolm, Jim Patton, Maria N.F. da Silva, and Claude Gascon for sharing the pains and joys of the Projeto Juruá with me. J. Márcio Ayres provided crucial political support in the early preparation of this embattled expedition. John Robinson and Liz Bennett provided insightful comments on the manuscript. I am grateful to the Brazilian Science and Research Council (CNPq) for funding a postdoctoral fellowship that made part of the fieldwork possible.

APPENDIX 3-1. Nomenclature and body mass of vertebrate species referred to in this paper. Species sharing the same ecological functional groups were assigned to a common mnemonic.

Taxonomic Group	Mnem.	Species Linnaean Name	Vernacular Name	Adult[a] Body Mass (gm)
Reptiles				
Testudines	GEOCH	Geochelone carbonaria	Red-footed tortoise	4580
	GEOCH	Geochelone denticulata	Yellow-footed tortoise	4580
Birds				
Tinamids	CRYPT	Crypturellus variegatus	Variegated tinamou	350
	CRYPT	Crypturellus cinereus	Cinereous tinamou	450
	CRYPT	Crypturellus undulatus	Undulated tinamou	540
	CRYPT	Crypturellus soui	Little tinamou	205
	TINAM	Tinamus guttatus	White-throated tinamou	800
	TINAM	Tinamus tao	Gray tinamou	2000
	TINAM	Tinamus major	Great tinamou	1170
Wood-quails	ODONT	Odontophorus gujanensis	Marbled wood-quail	310
	ODONT	Odontophorus stellatus	Starred wood-quail	310
Cracids	PENEL	Penelope jacquacu	Spix's guan	1280
	PENEL	Penelope pileata	White-crested guan	1250
	ABURR	Aburria pipile	Common piping guan	1200
	CRAX	Mitu mitu	Razor-bill curassow	3060
	CRAX	Crax globulosa	Wattled curassow	3000
	CRAX	Crax fasciolata	Bare-faced curassow	2515
Trumpeters	PSOPH	Psophia leucoptera	Pale-winged trumpeter	990
	PSOPH	Psophia viridis	Dark-winged trumpeter	990
Mammals				
Sciurids	MIFLA	Microsciurus flaviventer	Amazon dwarf squirrel	96
	SCIGN	Sciurus ignitus	Bolivian squirrel	220
	SCSPA	Sciurus spadiceus	Southern Amazon red squirrel	600
Rodents	MYOPR	Myoprocta pratti	Green acouchy	950
	MYOPR	Myoprocta acouchy	Red acouchy	950
	DASYP	Dasyprocta fuliginosa	Black agouti	4500
	DASYP	Dasyprocta leporina	Red agouti	3200
Ungulates	TATAJ	Tayassu tajacu	Collared peccary	25000
	TAPEC	Tayassu pecari	White-lipped peccary	32000
	MAGOU	Mazama gouazoubira	Gray brocket deer	18000
	MAAME	Mazama americana	Red brocket deer	30000
	TAPIR	Tapirus terrestris	Lowland tapir	160000
Primates	CEPYG	Cebuella pygmaea	Pygmy marmoset	150
	SAFUS	Saguinus fuscicollis	Saddle-back tamarin	390
	SAMYS	Saguinus mystax	Moustached tamarin	510
	SAMYS	Saguinus imperator	Emperor tamarin	510
	SAIMI	Saimiri boliviensis	Black-capped squirrel monkey	940
	SAIMI	Saimiri sciureus	Gray-capped squirrel monkey	940
	CACUP	Callicebus cupreus	Red titi monkey	1050
	CACUP	Callicebus moloch	Dusky titi monkey	1050
	CATOR	Callicebus torquatus	Collared titi monkey	1200
	PITHE	Pithecia albicans	White saki monkey	2200
	PITHE	Pithecia irrorata	Bald face saki monkey	2200
	PITHE	Pithecia monachus	Monk saki monkey	2200
	CHIRO	Chiropotes satanas	Bearded saki monkey	2650

Taxonomic Group	Mnem.	Species Linnaean Name	Vernacular Name	Adult[a] Body Mass (gm)
	CACAL	Cacajao calvus	White uakari	3175
	CEAPE	Cebus apella	Brown capuchin	2910
	CEALB	Cebus albifrons	White-face capuchin	2700
	ALOUA	Alouatta seniculus	Red howler monkey	6500
	ALOUA	Alouatta belzebul	Red-handed howler monkey	6500
	LALAG	Lagothrix lagothricha	Lowland woolly monkey	8710
	ATPAN	Ateles paniscus	Black spider monkey	9020

[a]Body mass data from Terborgh et al. 1990; Janson and Emmons 1990; Bodmer 1993; Peres 1993b; C. Peres and H. Nascimento unpubl. data.

4

The Sustainability of Current Hunting Practices by the Huaorani

PATRICIO MENA V., JODY R. STALLINGS,
JHANIRA REGALADO B., AND RUBEN CUEVA L.

The notion of sustainable harvests of tropical wildlife derives in part from Leopold's (1933) classic treatise on game management, which argues for the use of natural resources without destroying the possibility of future generations using these resources in a similar fashion. This sustainable model of resource use was based firmly upon the management of upland game populations, habitat management, and control of hunting, principally by sport hunters. A tropical model of sustainable hunting must involve the same variables, but at a time in history when habitat destruction is seen as the main threat to wildlife and wildlife management is nascent. A fourth, and critical, variable involves the welfare of forest-dwelling people who, in many cases, depend almost entirely upon wildlife resources for their animal protein, in comparison with northern sport hunters. Finally, tropical game species exhibit distinct life history strategies (Bodmer et al. 1997a) and might react differently to habitat manipulation in comparison with northern game species (Robinson and Redford 1991b).

The concept of sustainable harvests of tropical wildlife by forest-dwelling people has been the subject of considerable discussion by conservation and development professionals. Recent studies show that the long-held popular myth of the ecologically noble savage has been partially debunked (Redford 1990; Robinson 1993; but see Vickers 1994) and that traditional people, in many cases, extract a living in any way that they can without regard to the prevailing ecological balance.

The cultures of forest-dwelling people are fast changing. New and modern technologies have affected the lives and livelihoods of these people. The associated changes, especially in the area of hunting techniques, could affect the availability of natural resources for these peoples over the short and long terms. These changes in hunting techniques have produced a rich academic debate regarding the impact on local resources (Yost and Kelley 1983; Hames 1979; Alvard 1995a).

Changes in hunting technology can have a significant impact on the way people affect local alpha diversity and natural resources. Shifts in the use of traditional weapons to modern technology might lead to depletion of game in some tropical areas. In Ecuador, the Huaorani have experienced the shift from traditional hunting weapons, including blowguns and spears, to using firearms in less than three decades. Comparisons with past Huaorani studies indicate that this shift is nearly complete. The associated impacts on the local resource base and its implications for future sustainability are the focus of this chapter.

THE HUAORANI: FROM BLOWGUNS TO SHOTGUNS IN 30 YEARS

Traditionally, the Huaorani were a semi-nomadic indigenous group occupying the present day Napo province in Ecuador's south-central Amazon region (Kaplan et al. 1984). The Huaorani formed nuclear family groups dedicated to hunting, fishing, and collection of forest products, and practiced rotating agroforestry, cultivating yucca, corn, banana, palms, and other species of lesser importance (Yost 1981). Missionary groups first contacted the Huaorani in 1958. In 1990, the Ecuadorian government granted 612.5 km² to the Huaorani, forming the Huaorani Ethnic Territory Reserve, which includes part of the western Amazonia "hot spot" for tropical forests (Myers 1988) and abuts Yasuni National Park (figure 4-1). Under the conditions of the land grant, the government prohibits the sale of Reserve land and requests that the Huaorani develop and follow the guidelines of a land use management plan, which includes the rational use of renewable resources. Presently, there are approximately 1282 Huaorani scattered among 18 communities throughout the Huaorani Reserve (Smith 1993), resulting in roughly 0.5 persons per km².

Traditionally, hunting was the most important livelihood activity because it provided the majority of the animal protein for the family. Prior to western contact, the Huaorani feared rivers, especially large rivers, and perhaps this fear led to the taboo on taking larger fish and game from waterways (Yost and Kelley 1983). Fishing did not provide much protein compared with hunting (figure 4-2). According to Yost and Kelley (1983), without the use of dogs for hunting, tapir, brocket deer, and the collared peccary were difficult to corner and kill, and the Huaorani had a taboo on hunting these species. However, once the Huaorani obtained dogs and learned to use them in hunts, these species were in the range of the spear. In recent decades, these taxa also have been hunted.

The blowgun was the traditional weapon used to hunt arboreal prey, whereas

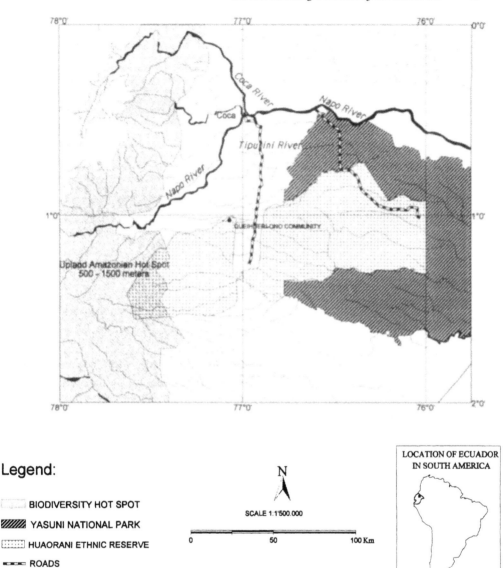

Figure 4-1. The Huaorani Ethnic Territory Reserve in relation to the Yasuni National Park and a proposed biodiversity hot spot in Ecuador.

the spear was reserved for peccaries and other large terrestrial animals. Yost and Kelley (1983) described the radical sociocultural changes in the Huaorani and their change from dependence on the blowgun and spear to firearms for hunting.

The Ecuadorian Amazon basin is the focus of transnational development of numerous oil fields. Huaorani men are occasionally employed by the oil companies to work in the oil field camps, which forces them to leave the community for extended periods of time. This change in the socioeconomic situation has numerous impacts on a Huaorani community. Men who do not leave the community are

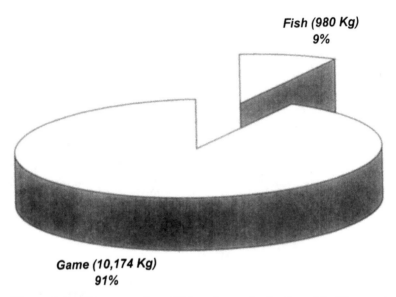

Fish (980 Kg)
9%

Game (10,174 Kg)
91%

Figure 4-2. The proportion of fish and game in the diet of the Huaorani of Quehuerini-ono.

frequently required to hunt on behalf of their neighbors and extended families of men who left for the oil fields. The newly acquired income generated by the employment has been used, in part, to purchase modern technologies such as firearms and ammunition. This helps to explain why many of the Huaorani men have firearms, and could help to explain why, 15 years ago, the blowgun was the more abundant and frequently employed technology (Yost and Kelley 1983).

The Huaorani of Quehueiri-ono

Quehueiri-ono, located along the Shiripuno River headwaters, was formed in 1989 by Huaorani families migrating from the Nushino River settlements. The community is located at 1° 01' 12" south and 77°09' 53" west (figure 4-3), 250 m above sea level, in northwestern Napo Province, about 25 km west of the Via Auca road and 120 km southeast of the town of Coca on the Napo River.

The predominant vegetation in the area is primary upland humid tropical forest, with alluvial forests along the Shiripuno River. According to the Holdridge system, the area belongs to the humid tropical forest life zone, with a mean annual temperature of 24°C and an annual mean precipitation between 2000 and 3000 mm (Cañadas 1983).

Quehueiri-ono, located along the southeastern limit of the Huaorani territory and composed largely of families from the Dayuno community, hosts a total of 167 Huaorani: 63 adults (30 men and 33 women) and 104 children under 18 years of age in 29 families, with 0.91 inhabitants per km^2. The Huaorani homes are distributed on both sides of the Shiripuno River along a distance of 800 m. Three of the families are formed by unions with Quichua Indian women or men. There were eight births and no deaths during a 12-month period from 1994 to 1995, representing a 2.5% increase in population.

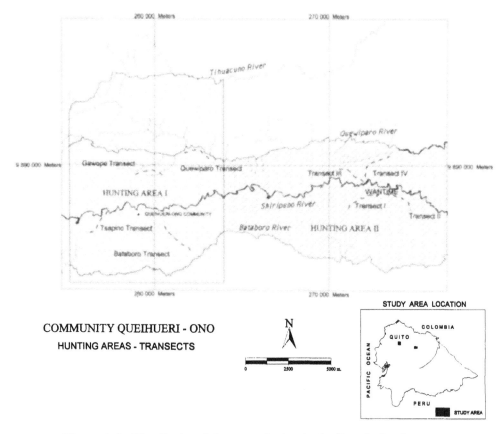

Figure 4-3. The Quehuerini-ono study sites in the Ecuadorian Amazon.

Quehueiri-ono is an isolated community. Access from the main road Via Auca is along difficult passage of the Shiripuno River in motorized or manually propelled dugout canoes, taking at least 6 hours or 1.5 days, respectively. The Huaorani also use a footpath parallel to the river, which takes 6 hours to reach the Via Auca. Imported goods are rare and expensive due to the isolation. The population relies on small family agricultural plots composed primarily of plantains, manioc, and corn. The extraction of natural resources, such as hunting wildlife, gathering fibers, roots, and resins, and fishing in streams and small rivers, is an extremely important subsistence activity for all members of the community.

The Huaorani of Quehueiri-ono are dedicated hunters, using ancestral and modern technologies to obtain animal protein. The repertoire of hunting techniques for terrestrial vertebrates includes blowguns, spears, shotguns, machetes, dogs, sticks or clubs, and harpoons. In small rivers and streams, the Huaorani use nets, harpoons, dynamite, hooks, and natural poisons to obtain fish.

THE IMPORTANCE OF WILD SPECIES FOR THE HUAORANI

In order to document the importance of hunting to the Huaorani, and to examine the long-term impact of hunting on wildlife, we collected data on the presence of vertebrates in the area, on the species recognized as important by the Huaorani, and on the hunting success of Huaorani hunters. These data were collected in the community for 322 days during an 11-month period between July 1994 and May 1995.

The recognition of wild species by the Huaorani was evaluated using interview techniques for three principal uses among mammalian, avian, and reptilian taxa: food, mascots, and adornments. All species were identified by the Huaorani using visual guides such as Emmons and Feer (1990) and Hilty and Brown (1986). Interviews were conducted with 92% of the adult population to document preferences and uses of wildlife, taboos, rituals, and methods for preparing meat for consumption (Bernard 1988). Hunting yield surveys at the household level were conducted on a daily basis.

Table 4-1 demonstrates the knowledge and use of terrestrial vertebrate taxa by the Huaorani. The community adults reported that food was the most important use of the terrestrial vertebrates identified, with 45%, 34%, and 82% of the mammalian, avian, and reptilian taxa being used for food, respectively. These results are in line with the patterns recognized by Redford and Robinson (1987) for several neotropical indigenous and colonist hunters.

Of the 72 mammal species recorded for the area and recognized by the adult population, 45%, 30%, and 23% of the species were used for food, mascots, and adornments, respectively. There was no use of marsupials, bats, and canids. In contrast, edentates, procyonids, mustelids, felids, cervids, peccaries, primates, rodents, and lagomorphs constituted important taxa for the community.

Only 11 reptilian species were recognized by the community during the study, with nine, or 82%, having a use registered by the community. Reptile use was heavily slanted toward food consumption (82%), with mascots representing only 18%, and no registered use for adornments.

There were 341 avian species recognized in the study area, of which 34%, 8%, and 9% were used for food, mascots, and adornments, respectively. The hawks, eagles, cracids, trumpeters, pigeons, doves, macaws, parrots, parakeets, owls, trogons, motmots, tinamous, toucans, piculets, woodpeckers, woodcreepers, antbirds, manakins, crows, jays, oropendulas, caciques, orioles, and tanagers were evaluated as useful by the community. Ducks, vultures, rails, coots, lapwings, sungrebes, sunbitterns, sandpipers, hoatzins, potoos, nighthawks, swifts, hummingbirds, jacamars, puffbirds, gnatwrens, seedeaters, seedfinches, and sparrows were not used by the Huaorani.

Table 4-1. Uses of Mammal, Bird, and Reptile Species Documented in Quehueiri-ono by the Huaorani

Taxa	Number of Species	Uses		
		Food	Mascots	Adornments
Mammals	72	45%	30%	23%
Birds	341	34%	8%	9%
Reptiles	11	82%	18%	0%

In addition, the interview data demonstrated that birds and mammals, not reptiles, were the most preferred game, with *Mitu salvini* (60%) being the most preferred for consumption. Other preferred species of interviewees expressing a preference were *Tinamus major* (41%), *Tayassu tajacu* (40%), *Agouti paca* (33%), *Ramphastos cuviere* (24%), *Pipile cumanensis* (22%), *Penelope jacquacu* (21%), *Mazama americana* (16%), *Sciurus spadiceus* (14%), *Amazona farinosa* (14%), and *Alouatta seniculus* (14%). No taboos against hunting any species were reported, lending credence to the belief that prohibitions are breaking down among indigenous groups (Redford and Robinson 1987).

WILDLIFE HARVESTED DURING THE STUDY

Hunting success was evaluated by visiting each household in the morning and in the early afternoon to determine the day's success. Terrestrial vertebrate wildlife was identified, weighed (kg), sexed, measured (cm), and aged. Relative age was estimated using physical characteristics such as size, dentition, and pelage coloration, and distributed among adult, subadult, and juvenile categories. Mean weights of whole animals were used to estimate weights of incomplete animals. The results corroborated the interview data (table 4-2) and showed the amount of game in the diet of the Huaorani. Fifty-five community members (30 men, 7 women, and 18 children) were registered as hunters during this study. An additional 8 Huaorani men from other communities hunted while visiting the community. Game was taken from two forested areas. Area I comprised 49 km^2 and abutted the community, whereas area II covered 81 km^2 and was located 16 km from the community (see figure 4-3). During the study, 7202 kg of wild meat was taken from area I, whereas 2972 kg was taken from area II.

The Huaorani used two hunting techniques to obtain terrestrial vertebrates: the common 1-day hunter forays close to the community and the rarer multiple-day hunts in areas farther from the village. On the 1-day hunts, a Huaorani hunter, occasionally accompanied by an additional hunter, would leave the community between 0600 hr and 0800 hr and return before dark. These hunts would occur on average at least 2 days a week per hunter. On the rarer multiple-day hunts, a Huaorani hunter took his family to their distant agroforestry plots and remained

Table 4-2. Number of Species, Individuals, and Biomass (kg) of Mammals, Birds, and Reptiles Taken in Quehueiri-ono by 63 Hunters During an 11-Month Period, 1994–1995

Taxa	Number of Species	Number of Individuals (%)	Biomass (%)
Mammals	39	1160 (49.3)	8658 (85.1)
Birds	113	1108 (47)	783 (7.7)
Reptiles	8	87 (3.7)	733 (7.2)
Totals	160	2355 (100)	10,174 (100)

in this area for a few days. These multiple-day hunts typically occurred during the children's vacation period from school, during holidays, and prior to a community celebration.

The hunters took 39 mammal species, 113 bird species, and 8 reptile species, with mammals representing 49% of the individuals taken from a total of 2355 individuals and 85% of the total biomass of 10,174 kg. There were 574 hunter-days, or 3595.5 hunter-hours, recorded for this study. On average, the Huaorani harvested 2.8 kg of meat per hour of hunting. Men hunters provided 98% of the biomass obtained, whereas women and children hunters represented 45% of the 55 community hunters, but only provided 2% of the biomass. Women and children did not use shotguns or blowguns, but hunted opportunistically with spears, sticks, machetes, stones, and hands to capture smaller terrestrial animals.

The more important species hunted during the study, those accounting for a minimum of 10 kg per species, are presented in table 4-3. Mammals comprised the bulk of all wildlife biomass taken, with *Lagothrix lagothricha* and *Tayassu tajacu* (figure 4-4) accounting for 51.5% and 43.2% of the mammalian biomass and proportion of the total animals taken, respectively. Three primate, three ungulate, three rodent, and one carnivore species comprised the most important mammals in terms of biomass, whereas five primate, two ungulate, and five rodent species represented the most important mammals in terms of animals harvested. *Mitu salvini*, *Ramphastos cuvieri*, *Penelope jacquacu*, and *Tinamus major* represented 77% and 44.5% of the avian biomass and proportion of animals, respectively. *Melanosuchus niger* and *Geochelone denticulata* were the most important reptiles

Figure 4-4. Collared peccary (*Tayassu tajacu*). *Photo courtesy of John G. Robinson.*

Table 4-3. Most Commonly Taken Game Species in Quehueiri-ono for a Consecutive 11-Month Period, 1994–1995

Species	Biomass (kg)	Individuals	Mean Weight (kg)
Mammals			
Lagothrix lagothricha	2289	395	5.79
Tayassu tajacu	2166	106	20.43
Mazama americana	971	48	20.23
Agouti paca	658	98	6.71
Tapirus terrestris	629	5	125.8
Alouatta seniculus	597	85	7.02
Dasyprocta fuliginosa	264	48	5.5
Hydrochaeris hydrochaeris	236	7	33.71
Cebus albifrons	104	36	2.88
Felis concolor	70	1	70.0
Pithecia monachus	69	28	2.46
Ateles belzebuth	68	10	6.8
Felis pardalis	62	8	7.75
Nasua nasua	58	19	3.05
Sciurus spadiceus	51	80	0.64
Myoprocta acouchy	35	40	0.88
Mazama gouazoubira	34	2	17.0
Dasypus novemcinctus	31	7	4.43
Myrmecophaga tridactyla	31	1	31.0
Priodontes maximus	30	1	30.0
Potos flavus	29	12	2.42
Aotus vosciferans	26	29	0.89
Tamandua tetradactyla	20	3	6.66
Sciurus igniventris	19	28	0.68
Callicebus moloch	18	16	1.13
Birds			
Mitu salvini	207	66	3.14
Ramphastos cuvieri	168	209	0.80
Penelope jacquacu	134	125	1.07
Tinamus major	95	93	1.02
Psophia crepitans	55	53	1.04
Pipile cumanensis	41	30	1.37
Amazona farinosa	16	24	0.67
Nothocrax urumutum	15	10	1.5
Ramphastos culminatus	14	26	0.54
Pteroglossus flavirostris	11	29	0.38
Tinamus guttatus	10	17	0.59
Reptiles			
Melanosuchus niger	427	22	19.41
Geochelone denticulata	175	34	5.15
Caiman crocodilus	54	12	4.50
Podocnemis expansa	21	6	3.5
Podocnemis unifilis	13	3	4.33

Important game species were those contributing more than a total of 10 kg during the study. There were 41 such species with 61%, 27%, and 12% being mammals, birds, and reptiles, respectively.

Table 4-4. Relative Age and Sex of Animals Hunted by Huaorani Hunters in Quehueiriono in an Eleven-Month Period

Taxa	Relative Age			Sex	
	Adults	Subadults	Juveniles	Males	Females
Reptiles	38	31	16	29	20
Mammals	836	173	115	499	423
Birds	765	79	139	287	211

hunted by the Huaorani, comprising 82% and 64% of the biomass and individuals, respectively.

The Huaorani hunters reported that they did not select the terrestrial vertebrate prey with any predetermined management scheme, but indiscriminately took individuals of all age classes and both sexes for mammals, birds, and reptiles. These observations are in accord with those of Vickers (1994). In general, more adults were hunted than subadults and juveniles combined, and more males were taken than females (table 4-4). These observations hold for the most commonly taken species of the three taxa (table 4-5).

Most species of mammals and birds that the Huaorani hunted were large

Table 4-5. Sex and Relative Age of Animals Frequently Hunted by the Huaorani in Quehueiri-ono

Species	Males	Females	Unknown	Adults	Subadult	Juvenile	Unknown
Mammals							
Lagothrix lagothricha	166	152	76	304	57	21	13
Tayassu tajacu	59	30	19	74	21	10	1
Agouti paca	29	47	20	72	14	10	0
Alouatta seniculus	38	25	21	70	6	4	4
Sciurus spadiceus	39	19	19	65	6	2	4
Mazama americana	18	23	6	24	19	4	1
Dasyprocta fuliginosa	18	15	10	29	5	5	1
Myoprocta acouchy	20	13	7	28	7	6	0
Cebus albifrons	18	15	3	26	6	4	0
Aotus vosciferans	5	12	12	22	1	2	4
Birds							
Ramphastos cuvieri	39	28	136	144	21	13	26
Penelope jacquacu	33	20	73	97	7	16	6
Tinamus major	17	21	56	63	4	7	20
Mitu salvini	22	12	29	55	3	1	4
Psophia crepitans	11	16	26	39	5	3	5
Cacicus cela	2	1	33	27	0	7	2
Pipile cumanensis	9	8	11	20	3	0	5
Pteroglossus flavirostris	3	1	20	17	1	4	2
Odontophorus gujanensis	4	5	15	17	2	1	4
Amazona farinosa	4	5	14	11	3	4	5
Reptiles							
Geochelone denticulata	14	12	8	17	13	4	0
Melanosuchus niger	9	1	11	13	3	4	1
Caiman crocodilus	2	2	8	2	6	4	0

bodied. The tapirs, deer, peccaries, primates, and some rodents constitute the large-bodied mammals, whereas the chicken-sized cracids, toucans, marbled wood-quail, gray-winged trumpeter, and tinamous represent the large-bodied birds. These results corroborate the findings of other researchers (Vickers 1984, 1988; Alvard 1993a; Bodmer et al. 1997a) and support the notion that neotropical hunters prefer to hunt larger animals and do so if provided the opportunity.

IMPACT OF HUNTING: COMPARISONS OF WILDLIFE ABUNDANCE IN TWO HUNTED AREAS

Wildlife censuses were conducted in infrequently hunted areas and persistently hunted areas to determine the impact of hunting of these taxa in the community catchment area. To evaluate the impact of hunting on the wildlife in the study area, we compared the diversity and abundance of mammals and birds in area I (persistently hunted) and area II (infrequently hunted). Four transects each 2.0 km long were established in each area. Each transect was visited by experienced biologists for a 2-day period each month. Mammalian and bird census walks were conducted at 0600 hr, 1630 hr, and 1830 hr at a speed of 1.0 km/hr. Mist nets were used to capture birds in both areas. The species, number of individuals, perpendicular distance from the transect, and distance along the transect were noted upon the observation of mammals and birds. The density of each species per km^2 was estimated by calculating the mean perpendicular detection distance of species from the transects, determining the detection area by multiplying the length of the transects by twice the width of the detection distances, and then dividing the numbers of individuals observed for each species per km^2 by the detection area.

More than 117 km of repeat censuses for birds and mammals and 228 mist net hours for birds were conducted along the transects in each area (table 4-6). More species of mammals and birds were observed in area II than in area I (table 4-7), and more species and numbers of birds were captured in mist nets in area II than in area I (table 4-8).

In addition, density estimates for 15 mammal and 10 bird species indicated that, in 19 of 25 cases, there were more individuals of these important hunted species in area II as compared with area I (table 4-9). Those exceptions were *Aotus vosciferans*, *Agouti paca*, *Nothocrax urumutum*, *Tinamus major*, and *Tinamus guttatus*.

Table 4-6. Sampling Effort for Recording the Presence and Density of Mammals and Birds in Two Hunted Areas in Quehueiri-ono

Effort	*Area I (49 km^2)*	*Area II (81 km^2)*
Mammals		
Distance covered in transects (km)	117.2	118.6
Birds		
Distance covered in transects (km)	119.7	118.1
Mist Net Hours	228	228

Table 4-7. Number of Families and Species of Mammals and Birds Observed Along Transects in Two Hunted Areas in Quehueiri-ono

	Area I	Area II
Mammals		
Number of families	11	11
Number of species	16	23
Birds		
Number of families	35	37
Number of species	193	218

These data demonstrate that some species had been extirpated, or at least become rare, in area I. A quick examination of table 4-9 indicates that five species of primates were observed in area I, whereas eight species were seen in area II. Absent were *Lagothrix lagothricha*, *Saimiri sciureus*, and *Ateles belzebuth*. *Lagothrix*, the woolly monkey, was one of the most abundant mammal species in area II and the species most frequently hunted during the study. This species is not the most preferred game taken by the Huaorani, but due to its size, abundance, and ease of capture, it is the species most frequently taken.

The absence of *Saimiri* and *Ateles* from area I could be due to the biology of the species, such as habitat preferences and social structure, and not due to extirpation by hunting. According to some researchers, *Ateles* and *Lagothrix* tend to occur in mutually exclusive areas (Hernandez-Camacho and Cooper 1976). However, our data in this region demonstrate that *Ateles* and *Lagothrix* species occurred in the same area (area II) and even on the same transect (transect 3) and do not support the idea of separation between these species.

Other mammal species not extirpated, but substantially reduced in area I, were *Cebus albifrons*, *Pithecia monachus*, *Alouatta seniculus*, and *Myoprocta acouchy*. All are large-bodied species.

Tayassu pecari, the white-lipped peccary, is a species commonly taken by neotropical hunters (Yost and Kelley 1983; Redford and Robinson 1987; Vickers 1988; Alvard 1995b). This species was curiously absent from both areas in Quehueiri-ono during the study. The Huaorani informed us that since the construction of the Via Auca in the 1970s, this species disappeared from the catchment area of Quehueiri-ono. The hunters' interpretation of this disappearance

Table 4-8. Number of Families, Species, and Individuals of Birds Captured in Mist Nets in Two Hunted Areas in Quehueiri-ono

	Area I	Area II
Number of families	15	15
Number of species	53	76
Number of individuals	181	260

Table 4-9. Density Estimates (Individuals per km²) for
Some Hunted Mammals and Birds in Two Hunted Areas
in Quehueiri-ono

Species	Area I	Area II
Mammals		
Tayassu tajacu	26.4	34.6
Mazama americana	12.5	18.8
Lagothrix lagothricha	—	36.8
Cebus albifrons	15.9	22.2
Alouatta seniculus	6.3	17.2
Ateles belzebuth	—	9.4
Pithecia monachus	20.0	36.9
Callicebus moloch	14.3	19.4
Saimiri sciureus	—	22.2
Aotus vociferans	40.0	37.5
Potos flavus	12.3	12.5
Dasyprocta fuliginosa	16.7	16.7
Myoprocta acouchy	6.7	19.3
Agouti paca	30.0	14.3
Sciurus spadiceus	16.7	18.8
Birds		
Mitu salvini	—	12.5
Penelope jacquacu	21.4	31.7
Pipile cumanensis	—	24.1
Nothocrax urumutum	6.8	5.2
Tinamous major	26.6	23.7
Tinamous guttatus	12.8	7.4
Phosia crepitans	28.5	80.0
Odontophorus gujamensis	31.2	35.3
Rhamphastos cuvieri	39.4	52.0
Amazona farinosa	39.1	58.8

was that the peccaries changed their movement patterns to avoid crossing the highway and no longer entered the Huaorani territory. *Tayassu pecari* has been reported from other Amazonian sites in Ecuador as a commonly hunted species (Yost and Kelley 1983; Vickers 1994).

Nine of the ten bird species most frequently hunted during this study are large bodied. The most hunted bird species was the toucan, *Ramphastos cuvieri*, with the cracids, *Penelope jacquacu*, *Mitu salivini*, and the tinamou, *Tinamus major*, representing other important hunted species. *Mitu salvini* and *Pipile cumanensis* were absent from area I. Both species of tinamous, *Tinamus guttatus* and *Tinamus major*, as well as the cracid *Nothocrax urumutum*, occurred at higher densities in area I.

FACTORS AFFECTING HUNTING RATES

Effectiveness of Resource Acquisition Technology

Yost and Kelley (1983) described in detail the resource acquisition technology of the Huaorani, including construction materials, dimensions, and uses. In addition to traditional technology, during the present study the Huaorani of Quehueiri-ono used two types of shotguns: the *escopeta* and the *carabina*. For the Huaorani, the *escopeta* is a muzzle-loading firearm loaded with multiple shot, whereas the *carabina* is a 16-gauge breech-loading conventional shotgun. The *escopeta* was more common (63%) than the *carabina*, and was used to take 88% of the animals harvested between both types of firearms.

Shotguns were the most numerous and most used hunting weapon in the community during the study (figure 4-5). There were 38 shotguns, 35 blowguns, and 33 spears registered in the community among 30 adult men classified as

Figure 4-5. Huaorani hunter, armed with a shotgun, returning with red brocket deer (*Mazama americana*). *Photo courtesy of Jody Stallings.*

hunters, with an average of 1.3 shotguns, 1.2 blowguns, and 1.1 spears per person. Not all male hunters possessed all weapons, with 87%, 80%, and 77% of the men possessing shotguns, blowguns, and spears, respectively. Three hunters did not own shotguns, six did not own blowguns, and seven did not own spears. However, one hunter owned four shotguns, another owned three blowguns, and another owned five spears.

Shotguns were used to take 87% of the biomass and 85% of the total individuals, respectively, when considering the animals hunted with shotgun, blowgun, and spear. This is in sharp contrast to data obtained in other Huaorani communities (figure 4-6) more than 15 years ago (Yost and Kelley 1983). The Yost and Kelley study presented hunting data of 867 day hunts over a 5-year period in four communities. The comparison in hunting techniques shows a substantial change in hunting technology. The major difference is that 15 years ago the Huaorani used the blowgun to take the majority of the biomass, and the shotgun to take the majority of the total individuals (Yost and Kelley 1983).

It is clear that the Huaorani of Quehueiri-ono are shifting rapidly from traditional to modern hunting technologies, practically abandoning the use of the spear, and significantly reducing the dependence of the blowgun in favor of the more efficient shotgun. This tendency was apparent in the late 1970s (Yost and Kelley 1983), but not to the extent observed in our study. Table 4-10 presents comparisons of hunting weapons used to hunt five game species recorded from both studies, and illustrates the dramatic shift in weaponry.

Table 4-11 compares the hunting results between the two studies per unit effort. For this analysis, our data only include those animals that were taken with blowgun, spear, and shotgun in order to compare results.

Our data demonstrate a 15% increase in mean kilograms per unit effort, a 15% increase in number of animals killed per hour, but almost no change (2%) in mean kilograms per animal hunted when compared with Yost and Kelley's (1983) data. These trends suggest that the Huaorani of Quehueiri-ono are taking more individuals and more weight of game than the Huaorani in the four other communities per unit time, but that there is no difference in the size of prey taken. If this pattern continues into the near future, it is likely that the Huaorani will extirpate large-bodied animals in the vicinity of their village because data show that this is a clear tendency in neotropical hunting studies (Redford and Robinson 1987). The presence and abundance data clearly show signs of this happening in the hunted area near the community in contrast to the hunted area farther away. Most startling are the differences in the presence and abundance of the large-bodied mammals *Lagothrix, Pithecia, Ateles, Saimiri,* and *Alouatta,* as well as the large-bodied birds *Mitu, Pipile, Phosia, Rhamphostos,* and *Amazona,* in the two hunted areas studied.

Yost and Kelley (1983) suggested that once the Huaorani learn fully to use the shotgun, kill rates with this technology could increase. Their prediction is supported by this study. In a relatively short time, the Huaorani have basically abandoned the use of the spear and reduced the use of the blowgun, instead using the more modern technology of firearms. The kill rate, in kilograms and number of animals per hour, has increased by 15%.

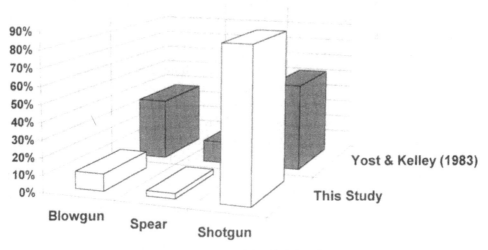

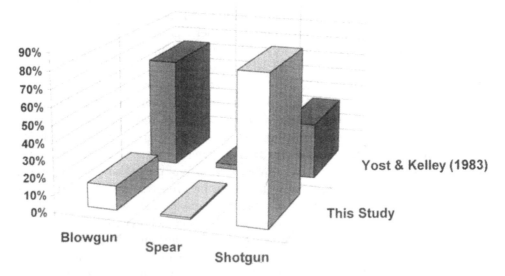

Figure 4-6. A comparison of hunting yield, biomass **(a)** and number of animals **(b)**, by weapon type for two Huaorani studies in the Ecuadorian Amazon.

It is clear that firearms are the most effective weapons used by the Huaorani. However, there were some exceptions in which other weapons were the preferred method of harvesting certain species. The blowgun was clearly the preferred weapon to hunt the icterid yellow-rumped cacique, *Cacicus cela*, whereas the tortoise, *Geochelone denticulata*, was harvested almost exclusively (97%) by hand.

Table 4-10. Comparison of Frequency of Kill by Species, According to Weapons Used, in Two Huaorani Studies

Species	Yost and Kelley (1983) Blowgun N	%	Shotgun N	%	Present Study Blowgun N	%	Shotgun N	%
Lagothrix lagothricha	457	81	105	19	80	20	311	80
Alouatta seniculus	188	76	58	24	27	34	53	66
Cebus albifrons	127	99	1	1	4	13	28	87
Ramphastos cuvieri	258	82	55	18	31	17	156	83
Penelope jacquacu	118	55	96	45	15	13	105	87

Table 4-11. Comparison of Hunting Results Between Two Huaorani Studies

Study	Hunter Hours	Animals Killed	kg Obtained	kg/ Animal	kg/Hour	Animal/ Hour
Yost and Kelley 1983	7716.3	3165	18,781	5.9	2.4	0.410
Present Study	3226.6	1546	9034	5.8	2.8	0.479

Other than these two species, the shotgun was the most effective and preferred weapon for all species hunted during the study.

SUSTAINABILITY

Sustainability implies that the resource can be used in the long term as it is in the present. To address the issue of sustainability of current hunting patterns by the Huaorani in Quehueiri-ono, several ecological and socioeconomic issues must be analyzed, including the reproductive biology of hunted species, habitat availability and proximity to conservation units, migration patterns of the Huaorani, and access to the community. When considering how to promote sustainability, development and economic alternatives also must be considered.

Reproductive Biology of Hunted Species

The recruitment potential in hunting catchment areas depends in part on the reproductive biology of the hunted species. For example, large-bodied cebid primates have low annual birth rates, long periods of infant development, long interbirth intervals, and late ages of first reproduction (Robinson and Redford 1986b), with some females reaching first reproduction at 7 years of age. As observed in this study, *Lagothrix*, *Alouatta*, and *Cebus*, all large-bodied cebid primates, were frequented killed by the Huaorani, and their populations were heavily reduced, or extirpated, in the persistently hunted area. The impact of heavy hunting on species with a low intrinsic rate of natural increase (r_{max}), such as cebid primates, tapirs, and carnivores, will be to lower these populations to levels where reproduction will be slowed, thus retarding recruitment.

Tropical wildlife management is nascent, and the population dynamics of hunted species are poorly understood. Nevertheless, Robinson and Redford (1991b) have provided a preliminary model that estimates maximum potential harvest rates for neotropical mammals. For a species with a known intrinsic rate of natural increase, and for a hunted population of known density, the model estimates maximum potential production (P_{max}) as:

$$P_{max} = (\text{density} \times \text{finite rate of population increase}) - \text{density}$$

Robinson and Redford (1991b) provide estimates of finite rates of population increase for the ten most important hunted species in this study, and these are used here. Their model suggests that harvest could take 60% of the production of very short-lived species, 40% in short-lived species, and 20% in long-lived species.

Table 4-12 compares maximum potential and actual harvest rates of the ten most important hunted species to the Huaroni in this study in the two hunted areas to determine if current harvests of these species are sustainable. The maximum potential harvest numbers and biomass are lower in the persistently hunted area because of lower densities. In the persistently hunted area, actual harvests for two of the four primate species exceed the maximum potential harvest levels (indeed, for *Lagothrix* and *Ateles* species, censuses did not record any animals present in the area). For the other two primate species, *Cebus* and *Alouatta*, actual harvests were only just under maximum potential harvests. In contrast, actual harvests of the ungulates and rodents were clearly below maximum potential levels. In the infrequently hunted area, only *Lagothrix* is clearly being overharvested, but actual harvests of *Alouatta* are approaching maximum potential levels.

These data are in accord with observations (e.g., Robinson and Redford 1991b; Bodmer et al. 1997a) that the large-bodied primates are very vulnerable to overharvesting, whereas ungulates and large-bodied rodents are less vulnerable. Species less vulnerable to overhunting are those that exhibit high levels of r_{max}, are short-lived animals, and have short generation times. The opposite is true for those species more vulnerable to overhunting, which include large-bodied primates, carnivores, and the lowland tapir.

We compared the numbers of animals and biomass hunted per km^2 for ten important hunted species for both hunted areas in an attempt to understand the hunting strategy of the Huaorani. In the persistently hunted area, 68% and 61% of the biomass and total animals taken, respectively, came from the species expected to be less vulnerable to hunting, such as the white-collared peccary, *Tayassu tajacu*. In the infrequently hunted area, just the opposite was observed, with 60% and 74% of the biomass and total animals taken, respectively, coming from the species more vulnerable to hunting, the woolly monkey, *Lagothrix lagothricha*.

Why were there such disparate hunting patterns in the two hunted areas? Is it just chance that the Huaorani hunt the more vulnerable species more frequently

Table 4-12. Comparisons of Potential Production, Potential Harvest, and Actual Harvest of 10 Important Game Species in Two Forested Areas in Quehueiri-ono

Species	Area I					Area II				
	Potential Production (No. Ind./km²)	Potential Harvest (No. Ind./km²)	Potential Weight (kg/km²)	Actual Harvest (No. Ind./km²)	Actual Weight (kg/km²)	Potential Production No. Ind./ (km²)	Potential Harvest (No. Ind./km²)	Potential Weight (kg/km²)	Actual Harvest (No. Ind./km²)	Actual Weight (kg/km²)
Tayassu tajacu	65.74	13.15	268.65	1.06	21.66	51.69	10.33	211.04	0.19	3.88
Lagothrix lagothricha	0	0	0	1.83	10.59	3.31	0.66	3.82	1.67	9.67
Mazama americana	5	2	40.46	0.15	0.3	5.51	2.2	44.5	0.15	3.03
Cebus albifrons	2.39	0.48	1.38	0.35	1.01	1.99	0.39	1.12	0.02	0.06
Alouatta seniculus	1.06	0.21	1.47	0.49	3.44	1.75	0.35	2.46	0.47	3.29
Ateles belzebuth	0	0	0	0.06	0.41	0.45	0.09	0.61	0.02	0.14
Dasyprocta fuliginosa	33.3	13.32	72.26	0.32	1.76	39.96	15.98	87.89	0.05	0.28
Agouti paca	28.5	5.7	38.25	1.3	8.72	8.15	1.63	10.94	0.2	1.34
Myoprocta acouchy	7.24	4.34	2.8	0.45	0.29	7.24	4.34	3.82	0.07	0.06
Sciurus spadiceus	74.72	44.83	30.48	0.96	0.65	74.72	44.83	30.48	0.12	0.08

in the infrequently hunted area and the less vulnerable species in the more persistently hunted area? Probably not. Large-bodied, noisy primates are diurnal, gregarious, and easier to spot and kill than the more nocturnal or crepuscular large rodents, deer, and peccaries. Many of the large-bodied primates were already gone from the persistently hunted area, and our interpretation is that the Huaorani were simply responding to what was locally available. This can be demonstrated by the fact that there were more peccaries in the infrequently hunted area, but the Huaorani hunted *Lagothrix* and *Alouatta* for the above reasons.

It appears that the Huaorani of Quehuerini-oro used a hunting strategy responsive to prey availability. They did not follow a conservation-based strategy using the reproductive biology to guide their hunts; rather, they hunted the largest and most easily acquired species first. These were the large-bodied primates, which were therefore the prime taxa to become locally extinct. After the primates had gone, or had been reduced to low numbers, the Huaorani switched to those species that produced more meat per unit hunting effort, which by chance were those less vulnerable to local extinction, such as the peccaries, deer, and large-bodied rodents.

Unfortunately, there is less knowledge of the characteristics of tropical birds, such as the cracids, toucans, and tinamous, thus precluding the same analysis. Again, some of the hunted species had disappeared from the persistently hunted area. In some cases, the densities of the hunted species were higher in the persistently hunted area than the infrequently hunted area. Perhaps the densities of these species were affected by increased food availability and/or the absence of predators.

Habitat Availability and Proximity to Conservation Units

The amount of available habitat is critical to the analysis of sustainability because all hunted species have individual space needs. In a utopian view, if large areas of habitat were available surrounding a small human community, recruitment would theoretically continue from outside the catchment area. Unfortunately, there are few such areas on Earth.

The Huaorani of Quehueiri-ono are locked into their 184 km^2 territory and do not have protected conservation units abutting their land. They are surrounded by other ethnic group territories, such as the Shuar, Quichua, and Mestizo colonists, as well as other Huaorani communities, who are using similar hunting methods and harvesting substantial amounts of animal protein from the forest.

Migration Patterns of the Huaorani

The patterns of Huaorani to migrate from one community to another, or to form a new community, are not novel to their culture (Yost and Kelley 1983). At the time of the present study, the community was in the initial stage of splitting into two. In part, the decision to split was based on the fact that wildlife was not as plentiful in the immediate vicinity of the community as in the past. The possibility of some family members migrating to a new site within their territory lends credence to there being a reduction in the wildlife populations near the Quehueiri-ono community. However, the Huaorani are not able to move outside their territory into those occupied by the indigenous Shuar and Quichua, nor onto lands occupied by more recent colonists. The Huaorani of Quehueiri-ono could move into other Huaorani territories within the Ethnic Reserve.

Access to the Community

Due to the distance from the more developed areas in the Ecuadorian Amazon and the difficulty of reaching them, Quehueiri-ono remains an isolated indigenous community. At present, the most significant sources of animal protein come from hunting and fishing. Large amounts of wild meat (10,174 kg) and fish (980 kg) were taken during the 11-month study. The Huaorani will continue to hunt at this intensity until species decline or are extirpated, unless alternative sources of protein become available. This will probably not occur on any large scale until access to the village is improved. The only viable option would be to develop a road from the Via Auca to the community. Unfortunately, road development has its own inherent problems, such as colonization, slash and burn agriculture, wild meat trade, and natural resource mining from external sources. The development of a road from the Via Auca to the Huaorani community could increase the hunting pressure in Quehuerini-ono.

The dependence on wild meat is strong because this community is far removed from supply centers such as small towns, and practically all animal protein must be drawn from the forest and waterways. The Huaorani have no options at present because development activities have been few and far between. Without domestic protein alternatives, the Huaorani will continue to exploit the forest for game animals.

Economic and Development Alternatives

The Huaorani of Quehueiri-ono will continue to hunt and fish until their socioeconomic situation changes. It is doubtful that they will revert to using traditional weapons if petroleum-related employment ceases to be an option for the village men. They will find another source of income that will allow them to purchase ammunition and replacement firearms. There is no reason to believe that they will change their current hunting patterns.

Development could help to alleviate the need for Huaorani men to hunt. Agroforestry and small-scale domestic animal production projects could provide a much-needed alternative for protein acquisition. These schemes typically are low cost and relatively sustainable. They provide a relatively fast response to a real need and can be managed, to a large extent, by community members. Without protein alternatives, there are few other options available to isolated traditional communities to manage their forest resources sustainably.

Once viable socioeconomic alternatives become available for the Huaorani of

Quehueiri-ono, a wildlife management and conservation education program for the community, such as described by Bodmer (1994 and this volume) for the Peruvian Amazon, might be considered to encourage the Huaorani to shift to such schemes. However, without some options in the near future, the sustainable harvesting of the large, upland vertebrate wildlife in Quehueiri-ono, as well as the survival of the Huaorani in this community, are at best questionable.

ACKNOWLEDGMENTS

This study was conducted under the auspices of The Sustainable Uses for Biological Resources (SUBIR) Project, financed through Cooperative Agreement No. 518-0069-A-00-1113-00, between CARE Ecuador and USAID Ecuador, and through a John D. and Catherine T. MacArthur Foundation grant to EcoCiencia. We thank these institutions for their support. Logistical support was provided by CARE Ecuador and especially through the Capitania del Puerto Francisco de Orellana and through the Brigada de Selva No. 19. We thank General Rene Yandun, General Ruben Barahona, Captain Mario Menendez, Captain Rogelio Viteri, and Captain Hernan Pone for their support in the study area. The Wildlife Conservation Society provided technical support for the development of this study. We thank D. Ruiz, W. Townsend, P. Feinsinger, and M. Crump for their invaluable assistance. B. Hayum, E. Rappe, J. G. Robinson, E. L. Bennett, C. Peres, and an anonymous reviewer made constructive comments on early versions of this manuscript. We extend our very special thanks to all of the inhabitants of Quehueiri-ono.

5

Sustainability of Aché Hunting in the Mbaracayu Reserve, Paraguay

KIM HILL AND JONATHAN PADWE

An understanding of native hunting and its impact on faunal density is critical for biodiversity conservation in the neotropics because more forested habitat is designated as indigenous reserve than all other types of conservation units combined (Rylands 1991; IUCN 1992; DaSilva and Sites 1995; Peres and Terborgh 1995). Despite a good deal of information on the overall structure of native game harvest in the neotropics (Beckerman and Sussenbach 1983; Vickers 1984; Redford and Robinson 1987) little is known about the way that harvest impacts animal populations through time and space. Some studies have suggested that hunted areas are depleted of game relative to nonhunted areas (see Hill et al. 1997 for review and discussion), but often these studies are based on census methods that are problematic. Specifically, many hunting impact studies to date (a) fail to distinguish changes in encounter rates due to evasive prey behavior from changes in animal density; (b) fail to control for the effects of variables other than hunting that also might be associated with differences in animal density; (c) do not provide independent measures of human hunting activity in areas assumed to be hunted; (d) are based on data collected exclusively on established trails that animals either avoid or use extensively depending on the amount of human activity in the study area; (e) are based on repeated transects in the same location, but inappropriately assume statistical independence of data and extrapolate to unsampled areas. We recently developed a procedure to measure the impact of human hunting on animal density that eliminates many of these problems (Hill et al. 1997).

Nevertheless, other problems remain. Most importantly, many human hunting studies do not examine the relationship between hunting and animal densities over an adequate spatial scale to draw relevant conservation conclusions about the impact of hunting on the true *unit stock* of a prey population. The unit stock is the population unit affected by harvesting. It consists of a collection of animals that interbreed, experience a common set of mortality and fertility parameters (from common causes), and show high intraunit migration relative to interunit migration. Most unit stocks of neotropical game inhabit vastly larger areas than are examined in most hunting studies; thus the scale of many hunting studies is inappropriate for use in wildlife management. Humans, like all central place foragers, are likely to deplete prey near their residential base. But how that depletion is manifested through space and time, and whether large unhunted prey populations are able to replenish depleted areas and maintain population viability, is a key issue that has not been adequately addressed.

Source-sink (SS) models are designed to examine population dynamics over a large spatial scale. SS models consider situations in which large populations are divided into subpopulations that experience negative and positive growth through time (Pulliam 1988; Pulliam and Danielson 1991). Sink areas experience annual population loss through mortality and emigration in excess of population gains through fertility. In source areas this relationship is reversed. Thus, in order for a sink population to be stable, annual immigration from a source must balance the net annual population loss. In theory, SS models can be used to determine if predator–prey relationships are stable in geographic regions larger than the observed prey catchment basin. However, because density-dependent effects on fertility, mortality, and dispersal are not well studied, SS models currently serve as more of a theoretical construct than a tool for empirical studies of harvest sustainability by predators. This concept is rarely acknowledged explicitly in prior neotropical hunting studies, but is implicit in some that examine hunted areas and protected areas together. In this chapter, we show that most hunting by our study population is performed in a 6-km radius from a permanent village site (the potential sink). We assume that the source population boundaries for the prey hunted in that zone are located at a point between the study population and other neighboring hunting populations that exploit the same prey. This assumption allows us to estimate harvest sustainability in a defined area that is biologically relevant.

The implications of game source and sink areas for determining harvest sustainability have not yet been fully appreciated by conservationists. For example, Robinson and Redford (1994b) reviewed five indices typically used as measures of game harvest sustainability:

1. Comparison of densities between harvested and unharvested locations
2. Change in animal density through time at a single location
3. Comparison of hunting yields in more and less harvested areas
4. Comparison of hunting yields through time
5. Comparison of age structure of prey populations in harvested and unharvested areas

Only the second and fourth of these points represent valid measures of sustainability when sink and source populations of prey exist. Spatial heterogeneity in animal density or age structure is expected but does not directly address the question of sustainability. Likewise, Bodmer (1994) has correctly noted that a comparison of harvest rate to reproductive rate can be used to estimate whether harvests are likely to be sustainable. However, this is only true if the prey population in a potential source area is included in the production calculation and if the prey response to density-dependent effects on life history characteristics is known (because fertility and mortality are often density dependent).

Biological resource management often has relied on equilibrium population models to determine sustainable harvest rates for commercially and recreationally exploited species. Analyses of sustainability in this chapter are based on assuming a *surplus production model* of harvestable biomass where the change in biomass through time is the difference between net production and harvest, with net production defined by the logistic growth equation. The SS modeling approach complements this model by specifying the true unit stock. These models in combination with measured demographic parameters at different population densities can be used to define a harvest rate that will not deplete the prey population through time, as well as the maximum sustainable harvest rate, and a variety of other equilibrium harvest parameters under specified harvest effort (Walters 1986). In practice, however, the use of equilibrium management methods has been problematic. Not infrequently, unforeseen circumstances (i.e., parameters and relationships not in the model) cause temporary prey depletion that can eliminate the unit stock from a geographical area or lead to dangerously low populations that cannot withstand harvest at the theoretically calculated (and sometimes legally mandated) rates. The inability of population models to capture fully all aspects of population dynamics has led to an alternative viewpoint on management. This viewpoint, termed *adaptive management*, emphasizes the need for continual monitoring and appropriate adjustments in harvest regulation to accompany any steady-state population management policy (e.g., Walters 1986; Christensen et al. 1996).

In this chapter, we examine harvest sustainability by the Aché native population in the Mbaracayu Reserve based on equilibrium assumptions. The results can be used as a guideline for guessing whether current harvest rates will be sustainable over several years, but should be complemented with monitoring and adaptive management to ensure that steady-state harvest models do not result in prey population decline due to factors unforeseen in the models. The current analysis suggests that Aché hunting does not threaten the population of any important game species in Mbaracayu. This is true despite the fact that localized depletion near the Aché settlement can be shown for some species (Hill et al. 1997). A 5-year monitoring project begun in 1994 will allow us to assess whether projections from equilibrium SS modeling are useful and whether they fully capture the dynamics between Aché predators and their prey in the Mbaracayu Reserve.

BACKGROUND

The Mbaracayu Reserve

The 60,000-hectare (ha) Mbaracayu Nature Reserve is the largest tract of undisturbed forest in eastern Paraguay. The Reserve is located at approximately 55° west and 24° south in an area that drains west to the Paraguay river. It is an area characterized by gently rolling hills composed of soft sedimentary rock. Soils are sandy, heavily leached, and of low fertility. The hills are covered with subtropical broad-leaf semi-deciduous forest, whereas some low flat valleys are filled with tall broad-blade grasses. Most of the area is at an elevation of 150 to 300 m. Rainfall totals about 1800 mm/yr on average and is characterized by extreme unpredictability in monthly pattern from year to year but with a statistical dry season from May to August (Sanchez 1973). Temperature fluctuations mark seasonality, with average daily low-high temperatures of 14 to 25°C in July and 22 to 34°C in January. Temperature extremes are approximately 41°C and –3°C, with several days of hard frost each year, which often kills the leaves on many exposed trees and shrubs.

The Reserve was purchased by The Nature Conservancy (TNC), in conjunction with the Fundación Moises Bertoni (FMB), a Paraguayan conservation NGO. In 1991, the Reserve was granted legal status as a national forest reserve by the Paraguayan congress. It was included on TNC's 1991 list of "last great places on earth" and is the home of a variety of rare and endangered animal and plant species. A crude map of vegetation zones was commissioned by the FMB in 1992 (figure 5-1).

The Mbaracayu Reserve contains about 90% of the Paraguayan animal and plant species classified as rare and endangered (FMB 1992) and was chosen as the top priority conservation site in eastern Paraguay using vegetation analysis (Keel et al. 1993). The Reserve includes a remarkable diversity of forests, rivers, mountains, caves, grasslands, and wetlands. Indeed, the most striking feature of the Reserve is its pronounced microregional diversity. Within the 60,000-ha Reserve, there are areas of mature *terra firme* tropical forest, *cerrado* (ranging from *campo sucio* to *cerradao*), grassland, palm-dominated swamps, bamboo (*Guadua*) forests, riparian flood forests, and a low drier forest type referred to as *kaati* by native Guarani speakers.

Although there are many distinguishable plant communities that form separate habitats in Mbaracayu (Hill et al. 1997), only the high forest has been systematically described. This forest type covers approximately 50% of the land surface surveyed by our research team (Hill et al. 1997). In 1987, a team of four botanists and two Aché assistants led by A. Gentry recorded all plants greater than 2.5 cm diameter at breast height (dbh) on transects that added up to a total area of 1000 m^2 in mature high forest of the north east corner of the Mbaracayu Reserve (Keel 1987). The basal area of tree species >10 cm dbh is 39 m^2/ha in this sample. The five species with highest species importance values (SIV-sum of relative frequency, relative density, and relative dominance) (Keel et al. 1993) are *Sorocea bonplandii*, *Campomanesia xanthocarpa*, *Chrysophyllum gonocarpum*, *Myrciaria baporeti*, and *Balfourodendron riedelianum*.

Common emergents in the high forest include *Tabebuia heptaphylla*, *Astronium* spp., *Aspidosperma polyneuron*, *Albizzia hasleri*, *Anadenanthera* spp., *Enterolobium contortisiliquum*, *Peltophorum dubium*, *Cedrela fissilis*, *Patagonula americana*, and *Balfourodendron riedelianum*. The maturity of the forest is indicated by the high density (103 specimens in 1000 m^2) of lianas >2.5 cm dbh, including many examples of lianas >10 cm dbh. Despite the presence of many valuable timber species

Figure 5-1. Major vegetation zones in the Mbaracayu Reserve as mapped by the Fundación Moises Bertoni (FMB 1992). The location of the Aché research team (Arroyo Bandera), as well as three small Paraguayan communities (Maria Auxiliaodra, Tendal, and Guyra Kehja) are shown. Permanent forest guard posts are also labeled.

throughout most of the Reserve, some portions of the Reserve were selectively logged for lapacho (*Tabebuia heptaphylla*) and peroba (*Aspidosperma polyneuron*) in the 1970s and 1980s.

The most important palms in Mbaracayu are *Syagrus (Arrecastrum) romanzolffiana* and *Acromia totai (sclerocarpa)*, which produce edible fiber, growing shoots, fruits, and nuts. We have counted *Syagrus* palms at a density of approximately 0.8/ha in dry forest and about 20/ha in poorly drained forest using ground transects and aerial surveys. Another extremely common species that might have important effects on the high forest animal community is *Citrus aurantium*, the common orange. This is a species of domestic origin that has spread throughout the forests of eastern Paraguay (Gade 1976) in the 450 years since it was introduced to the region by the Jesuits. It is an important food item of *Agouti*, *Cebus*, and *Tayassu* as well as numerous birds.

The Mbaracayu Reserve is one of the most important endemic bird centers in South America. Over 400 species of birds have been recorded in the Reserve in the past 2 years (Madroño and Esquivel 1995 and personal communication). The mammalian fauna of eastern Paraguay has been the subject of many inventory studies, but no studies to determine absolute or relative densities of mammals. Myers et al. (1995) have recently summarized available data and report 124 mammal species verified in eastern Paraguay. In the Mbaracayu Reserve to date, we have listed 99 species of mammals identified by visiting inventory teams or ourselves. Because these data come from several sources, and some identifications are based on single sightings, it should be considered provisional. Bats make up 23 species on this provisional list, with nonvolant mammals comprising 76 species. A provisional list of Mbaracayu mammals excluding marsupials, bats, and small rodents (< 500 g body weight) is shown in table 5-1.

The Aché

The law that created the Mbaracayu Reserve as a legal entity in 1991 states in part:

> In recognition of the prior use of the forest by the Aché indigenous community, these groups will be permitted to continue subsistence hunting and gathering in the area of the Nature Reserve, as long as they employ traditional methods, and according to that which is allowed in the administrative plan of the reserve. The participation of the local Aché community in the protection and administration of the reserve will be encouraged, and they (the Aché) will be offered permanent employment that comes about as a result of the development of scientific studies, recreation, and tourism, in the reserve and the protected areas around it. (Ley 112/91, Article 13)

For this reason, studies of Aché resource use patterns are critical to conservation planning for Mbaracayu.

Archeological data suggest that foraging peoples might have inhabited the forested areas of eastern Paraguay for at least 10,000 years. The Aché were described by early Jesuit explorers, but historical evidence shows that the Aché experienced no peaceful contact with outsiders (either Indian or peasant colonists)

Table 5-1. Provisional List of Mammals Sighted or Trapped in the Mbaracayu Reserve, 1970–1996

Scientific Name	Common Name
Edentates	**(8 Species)**
Tamandua tetradactyla	Collared anteater
Mymecophaga tridactyla	Giant anteater
Dasypus novemcinctus	Nine-banded armadillo
Dasypus septemcinctus	Seven-banded armadillo
Euphractus sexcinctus	Yellow armadillo
Cabbassous tatouay	Naked-tailed armadillo
Priodontes maximus	Giant armadillo
Tolypeutes spp.	Three-banded armadillo
Primates	**(2 species)**
Cebus apella	Brown capuchin monkey
Alouatta caraya	Black howler monkey
Lagomorphs	**(1 species)**
Sylvilagus brasiliensis	Rabbit
Large rodents	**(7 species)**
Sphiggurus spinosas	Porcupine
Coendou paragayensis	Paraguay porcupine
Dasyprocta azarae	Agouti
Agouti paca	Paca
Cavia aperea	Guinea pig
Hydrochaeris hydrochaeris	Capybara
Myocastor coypus	Nutria
Perissodactyls	**(1 species)**
Tapirus terrestris	Tapir
Carnivores	**(17 species)**
Cerdocyon thous	Crab-eating fox
Dusicyon gymnocerus	South American fox
Chrysocyon brachyurus	Maned wolf
Speothos venaticus	Bush dog
Galictis cuja	Grison
Eira barbara	Tayra
Conepatus chinga	Hog-nosed skunk
Procyon cancrivorous	Crab-eating raccoon
Nasua nasua	Coati
Lutra longicaudis	Southern river otter
Pteronura brasiliensis	Giant otter
Panthera onca	Jaguar
Felis pardalis	Ocelot
Felis wiedii	Margay
Felis tigrina	Oncilla
Felis yagouaroundi	Jaguarundi
Felis concolor	Puma
Artiodactyls	**(5 species)**
Tayassu pecari	White-lipped peccary
Tayassu tajacu	Collared peccary
Mazama americana	Red brocket deer
Mazama gouazoubira	Grey brocket deer
Mazama rufina	Dwarf red brocket deer

This list does not include marsupials, bats, or rodents with body weights of less than 500 g.

between the Spanish conquest in the 1500s and recent incursions into their territory in the 1970s (Hill and Hurtado 1996). Since the 1970s, the Aché have been "settled" on government/mission reservations but continue frequent forest treks that may last weeks or even months. The population currently consists of about 700 people living in five settlements. Previous anthropological studies of the Aché have focused on a wide variety of economic topics (see Hill and Hurtado 1989 for review) and Aché life history patterns (Hill and Hurtado 1996).

During the first half of the twentieth century, the Northern Aché inhabited a 20,000-km^2 area of eastern Paraguay centered in the upper Jejui watershed, which is now the Mbaracayu Reserve. At a population density of only 0.03 person/km^2 the Aché probably had little impact on animal densities prior to their reduction on reservation settlements. During the 1970s, Aché bands were tracked down, brought out to passable roads, and transported on missionary or military trucks to a few reservations, where they were expected to remain (Hill and Hurtado 1996, Chapter 2). Reservations consisted of only a few thousand hectares, and despite the fact that outsiders began teaching the Aché to farm, most individuals reentered the forest frequently on long foraging treks because reservation resources were inadequate to feed the population. The last of the uncontacted Aché bands took refuge in the northern section of the Mbaracayu Reserve until 1978, when they too were contacted and transported to a nearby reservation.

At this writing, two Aché reservations totaling about 500 individuals are within walking distance of the Mbaracayu Reserve, and the Aché have begun to recognize the long-term impact of their economic activities on the natural resources of the shrinking forested areas near their reservations. The Aché economy is traditionally centered around hunting mammalian game with bow and arrow, extracting wild honey, and exploiting palm starch and insect larvae. Numerous fruits are also exploited seasonally, but they constitute only a small fraction of the energy in the yearly diet (Hill et al. 1984). The current reservation economy also includes swidden agriculture (mainly manioc and maize) and some domestic animals (chickens and pigs), which have been introduced in the past 15 years. Younger Aché also have become increasingly involved in seasonal wage labor at nearby farms.

Hunting in the Mbaracayu Reserve

Both Aché and non-Aché hunters hunt in some sections of the Mbaracayu Reserve. The Mbaracayu administration has placed two guards at each of five guard stations around the periphery of the Reserve to control the entry of non-Aché hunters, and the road that bisects the Reserve is blocked by a locked gate at both ends. A three-person team of roving guards monitors illegal activity throughout the Reserve. Although park guards have been effective at eliminating large-scale disturbance (illegal timber extraction and establishment of swiddens), within the Reserve they have not been able to eliminate poaching by non-Aché hunters. Because we have no reliable data on poaching, we will examine only Aché hunting practices in this chapter. However, we do assume that regions hunted by

Paraguayans are probably sinks for some prey species and take this into account when calculating the source area for Aché prey.

The non-Aché hunters include Paraguayan and Brazilian frontiersmen—peasants as well as Guarani horticultural Indians. Guarani Indians have lived in the Mboi Jagua reservation about 7 km from the western border of the Reserve for about 30 years (and in nearby traditional villages before that) and rarely hunt in the Reserve. When they do, they usually carry only a machete and target armadillos (*Dasypus*) or tegu lizards (*Tupinambis*). Most Paraguayan and Brazilian peasants are recent arrivals to the Mbaracayu area. The Maria Auxiliadora settlement along the northern section of the western border of the Reserve has only existed for about 3 years and consists of about 80 families who are close enough to hunt in the Reserve. Recent survey data (P. Garfer, personal communication) suggest that only a small fraction of this peasant population hunts, and that number has diminished over time. However, a new settlement, Guyra Kejha, has just been established near the southeastern corner of the Reserve, and preliminary data from 1996 suggest extensive peasant hunting in that area. Peasants usually engage in night tree stand hunting with shotguns. They target tapir, deer, paca, and agouti under fruiting trees, at mud licks, or in areas that have been baited with corn. Occasionally, peasants engage in diurnal hunting in the Reserve using dogs, and they primarily target both species of peccary, armadillos, and the three species of large cracid birds found in the Reserve.

The Aché have been hunting in the Mbaracayu area for at least a century, and they began intensively hunting the northwestern portion of the Reserve when the Arroyo Bandera reservation was established near that area around 1980. Aché from the more distant Chupa Pou reservation occasionally hunt in the southwestern portion of the Reserve. The Aché hunt with bows and arrows or by hand during day hunts that originate from their reservation. They also go on extended treks inside the Reserve that generally last 3 to 5 days but can sometimes last much longer.

METHODS

Aché Game Harvest Patterns

Harvest rates for individual game species were determined from two data sets covering a span of 15 years. From 1980 to 1985, one of us (K.H.) recorded the species and weights of all game killed each day by Aché hunters on treks in the forest during trips in which we accompanied the Aché (see Hawkes et al. 1982 for details). This sample, taken from three field sessions, was not randomly distributed through the year, but does cover most months of the yearly cycle. It consists of 185 observation days in the forest, which comprised 2087 man hunting days. From August 1994 to the present, another of us (F.J.) has recorded all animals killed by members of the Arroyo Bandera reservation and (since August 1995) all man days spent hunting in the forest regardless of success. Because the observer

(F.J.) lives full time at the Arroyo Bandera reservation, and all houses are within view, with sharing of game widespread, it is unlikely that a significant number of harvested animals were unrecorded. The data from August 1994 through December 1996 cover 869 observation days. The community consisted of an average of 90 individuals, 28 of whom were potential hunters (men at least 17 years old) and 83 of whom were consumers (> 3 years old). To convert number of animals harvested of each species into biomass, we have used mean weights from our own observations or those reported by Redford and Eisenberg (1992).

Animal Densities

Game densities were estimated using a stratified sample of diurnal line transects through the Mbaracayu Reserve. Details of the field method have been reported elsewhere (Hill et al. 1997). Line transect methods are the most widely used technique to census large neotropical mammals and birds (Cant 1977; Glanz 1982; Emmons 1984; Robinson and Redford 1986b; Bodmer et al. 1988a; Peres 1990; Glanz 1991; Silva and Strahl 1991; Bodmer et al. 1994). Properly collected transect data used in conjunction with widely available analytical software can provide robust estimates of average density in an appropriately sampled area if all targets on the line are detected (Buckland et al. 1993; Laake et al. 1993).

In our study, line transect starting points were drawn without replacement from the set of all locations along the dirt road running east–west and bisecting the Reserve. About 10% of transect sample days commenced off the road, after camping overnight in the forest. Absolute locations during the study were determined using a Trimble Pathfinder Pro Global Positioning System (GPS) receiver, whereas relative distance along a transect was measured using a string box.

Transects were initially walked by a team of five native assistants and a data recorder. After the first year, native assistants recorded all data without outside help. Observers walked in parallel along a central transect line. The four native assistants spaced themselves at approximately 25 and 50 m on either side and perpendicular to the center line, and one walked directly on the center line about 5 m ahead of the data recorder. Each assistant carried a VHF radio to communicate with the data recorder. Each transect began at the specified GPS location and proceeded toward a specified compass bearing throughout the day. No transects were performed on trails, but instead the team walked through whatever vegetation was encountered.

After receiving a radio signal from the data recorder, each native observer began walking along the center transect line or parallel to it. Team members walked at a rate of about 1 km/hr. Native assistants were allowed to veer a few meters each side of the transect to verify that burrows they encountered were indeed occupied. When the data recorder reached 200 m (as measured on the string box), he instructed the field assistants to stop and report encounter data for that 200-m unit. Observers, in sequence, reported over the radio all encounters during the previous 200 m with all species (including humans) and the encounter type (defined below). The data recorder took an averaged GPS reading every 600 to 800 m along the transect in order to estimate the absolute location of the transect within the study area for later analyses. Location between GPS readings was estimated by interpolation using string box measurements. The locations of transects from June 1994 to June 1996 are shown in figure 5-2.

Native Aché research assistants were trained for 1 week prior to beginning data collection and went through periodic retraining and distance verification sessions. All had extensive experience hunting in the area and were born inside the study area. They practiced radio use, learned basic concepts of mapping, compass, and GPS use, and practiced distance measurement and verification using a range-finder, metal tape, and string box. Assistants were familiar with the western numbering system but could not read or write. Their ability to estimate distance in the forest was good. Verification of reported distances to specified points in the forest shows a regression coefficient between actual and estimated distance of 0.84 (n = 113 trials between 3 and 80 m actual distance, r = 0.94). Thus, Aché distance estimates were on average about 84% of the true distance and did not show much deviation from the actual distance. However, Aché assistants generally reported distance to the nearest animal of a social group rather than to its center. This means that absolute density estimates of animals that live in dispersed groups will be too high if the data are taken at face value (but this error can be corrected if the mean diameter of social groups is known).

Aché men insisted on working in teams to avoid forest dangers. Therefore, the method that we use generates a great deal of data per field day. However, walking simultaneous parallel transects through uncut vegetation with radio communication every 200 m is undoubtedly noisier than the standard method of a single observer walking on a cleared trail and noting encounters silently. We believe that this is not a serious problem, but empirical verification would be useful, by doing single observer transects on cut trails in the same area. Native observers are quite skilled and might detect many animals that would not be detected by observers who have not lived their whole lives in the forest. Aché observers insisted that most target species do not flee until they have direct confirmation of a predator visually or by smell. Such a response should be common in many prey species where the costs of unnecessary flight are high. Thus, we assume that noises of walking through vegetation did not spook most target species until they were near enough that they could be detected (by an Aché) when fleeing the area. The Aché concurred with this conclusion based on reading animal tracks, which suggested that few animals fled completely undetected just before observers arrived. Aché assistants also report that quiet radio conversation does not spook most target species. In any case, radio conversation was not audible to us at distances greater than 15 m, and communications took place only once every 15 minutes on average. Thus, we provisionally conclude that although there might be problems with the noise made by our transect team, these are outweighed by the multiple disadvantages of typical line transect studies that use observers who have not grown up in the survey area walking repeatedly on a few trails that are not randomly placed in the study area (see Hill et al. 1997 for discussion).

Encounters were recorded for all mammal, bird, and reptile species that are

Figure 5-2. The location of diurnal transects performed in the Mbaracayu Reserve by the Aché research team. Dotted lines show transects from May 1994 to May 1995, solid numbered lines show transects from June 1995 to June 1996. A total of 123 different transects were performed inside the Reserve and three on the Arroyo Bandera and Chupa Pou (not shown) reservations. Most transects began from the road (*dashed line*) bisecting the Reserve east—west. The northwest corner was sampled frequently in order to monitor the extent of poaching by peasants in that zone.

larger than 0.5 kg mean body weight. Two types of encounters were recorded: (a) animal seen, heard, or found in burrow; and (b) fresh signs of the animal or fresh feces encountered. All encounters of the first type are lumped together for analyses into a category that we refer to as *direct encounters*. When multiple observers detected the same animal, only the observation of the closest observer was used for analyses. We generally required observers to confirm the presence of an animal in an occupied burrow by flushing it, or by introducing a long vine and getting the animal to move inside. In a few cases, Aché assistants insisted that there was an animal inside (by certain signs, smell, etc.), even when they could not get direct confirmation, and we recorded those burrows as occupied.

Encounters of the second type are aggregated for analyses into a category that we term *indirect encounters*. These signs included tracks, feeding disturbance, territorial markers, beds and nests, urine, scent, body excretions, etc. that were judged to be less than 24 hours old. Indirect encounter patterns have been reported elsewhere (Hill et al. 1997) and are not analyzed here.

To estimate effective strip width of our line transects, and the densities of target animal species, we analyzed our data using the DISTANCE statistical package (Laake et al. 1993). This package uses a variety of transforms to estimate the shape of the detection probability function with perpendicular distance from the transect line. Detection probability is assumed to be 100% for target species located on the transect line. The integral of the best fit detection function is the estimated strip width for a species. In order to obtain a good fit for the detection function, our encounter distance data were truncated so that 5% of the observations at greatest detection distance were eliminated from the sample. Additional truncation was introduced by our field methods in which parallel transects were performed simultaneously.

Because many prey species are detected at distances greater than half the mean distance between our transect observers (12.5 m), multiple observers often detected the same animal. For these species, each observer's transect width was truncated at the mid-point between observers, and only the encounter with the closest observer was recorded. Because *D. novemcinctus* and *A. paca* were encountered in their burrows and almost never encountered at a distance of greater than 12.5 m from the transect line, the effective strip width for those species was estimated directly from the encounter frequency distribution as described above. For all other species, the effective strip width was estimated from a detection frequency distribution that was generally truncated at 12.5 m for observers not on the outside of the transect formation.

The DISTANCE program produces confidence intervals by assessing strip width and encounter variance between transects. Because our field team walked in parallel and synchronously, we aggregated the data from all five men on each day into a single transect. This led to a sample of 126 independent transects covering 2054.85 person-km. Some transects were removed from the sample for certain species when encounters with those species were reported without a perpendicular distance estimate (this happened early in the study for some species).

For social groups, the encounter rate with groups must be multiplied by the mean group size to estimate accurately the density of individuals. No data exist on group size for social animals in the Mbaracayu Reserve, and our attempts to census groups met with limited success because prey flee or adopt evasive tactics when humans are detected. Nevertheless, Aché forest knowledge and conversations with researchers at a nearby site in Yguazu Park, Argentina, and Brazil provided us with the following rough estimates of mean group size: *Cebus*, $n = 18$; *Nasua*, $n = 8$; *Tayassu pecari*, $n = 80$; *T. tajacu*, $n = 1.5$. Unfortunately, we have no variance measure around these estimates, so confidence intervals can only be estimated for group encounter rates and then multiplied by group size to produce a rough estimate of the confidence interval for density of individual animals. Encounters with other species reported in this chapter are assumed to be with individual animals, so confidence intervals and densities can be estimated directly.

Size of Hypothesized Sinks and Sources

The hunting zone that we hypothesize as the Aché sink is used by both Aché and Paraguayan hunters. The catchment area of hunting for the Arroyo Bandera reservation is considered to be a semicircular area 6 km in radius from the point at which the Aché enter the Reserve (figure 5-3). There is only one trail from the Arroyo Bandera village to the Reserve, with the border of the Reserve about 2.5 km from the village. Because women and children rarely walk more than 2 km per day off trails in the forest, and men do not participate in multi-day hunts without women and children, very few Aché hunts ever take place more than about 6 km inside the Reserve. We estimate this core hunting area to be about 56.5 km². Data support our impression of the location of the core Aché hunting zone. The encounter rate with signs of Aché hunters on random transects is 17 times higher in the 6-km radius catchment area than in the areas more distant. This catchment area also overlaps with the Maria Auxiliadora peasant hunting zone; encounter rates with signs of Paraguayan hunters were ten times higher in this zone than elsewhere in the Reserve.

The eastern hunting zone that we also hypothesize as a sink is harder to delineate and has been hunted only by Paraguayan poachers since 1995. As far as we can tell from hunter signs, it is a strip about 10 km long and 3 km wide along the far southeastern boundary of the Reserve. If the area is extensively hunted, this might be a sink of some 30 km² that competes for immigrants with the Aché hunting catchment area. Because the area is remote from our base of operations, not enough data are available to characterize hunting intensity in this area relative to other areas of the Reserve.

The potential source area for immigration into hunting zone sinks is the entire Mbaracayu Reserve, minus the western and eastern hunting zones (figure 5-3). This area shows a density of hunter signs more than an order of magnitude higher than in the area hypothesized to be the Aché sink (Hill et al. 1997). The total area is approximately 513.5 km². However, some of this area must serve as the source for the eastern hunting zone if poachers continue to operate in that area. We thus eliminate the entire southeastern quarter of the Reserve as a potential source for

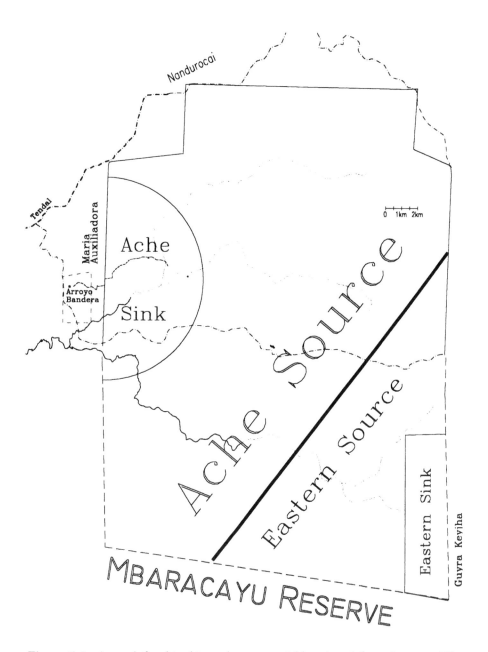

Figure 5-3. Areas defined in this study as potential hunting sinks and sources. The Aché sink begins where the Aché village trail (*dashed line*) enters the Reserve. The frequently hunted zone is a half circle of radius 6 km from that point. Most hunting by peasants from Maria Auxiliadora takes place in the same sink. The Guyra Kejha peasant population hunts in the eastern sink, which is about 10 km long and 3 km wide. The two potential source areas for the hunting sinks are the southeastern quarter of the Reserve and the remainder of the Reserve. The eastern source also extends to a large forested block outside the southern boundary of the Reserve.

prey species that immigrate into the Aché sink, leaving about 393.5 km^2 as the potential source area for species depleted in the Aché sink. The eastern source area includes 120 km^2 inside the Reserve and another 100 to 200 km^2 of forest outside the southern boundary of the Reserve. These estimates are admittedly crude but will have to suffice until better data are available.

RESULTS

Aché Game Harvest Patterns

The list of all vertebrates killed during sample periods between 1980 and 1995 is shown in table 5-2 in descending order of biomass. Nine species of vertebrates contribute about 95% of the meat in the Aché diet by weight (table 5-3). Less than 2% of the biomass in table 5-2 was taken using a shotgun; the remainder was killed with bow and arrow or by hand. Similarly, little of the game was taken with the use of dogs. The list of prey species in table 5-2 can be compared with the list of all mammals in Mbaracayu (table 5-1) to get an idea of Aché prey selectivity for mammals. A chronological trend toward an increasing importance of small hand-hunted game (*D. novemcinctus* and *A. paca*) is evident from table 5-3 and is related to the decreasing use of bow and arrow by young hunters. The increasing importance of *T. terrestris* is due to the introduction of metal-tipped arrows in the early 1990s.

Importance of Aché Hunting

The economic importance of Aché hunting is clear from table 5-2. From 1994 through 1996, Arroyo Bandera Aché obtained a mean of 12.5 kg of live game per day, which would yield about 10 kg of edible meat. This is 122 g of meat per consumer day. At local market prices it would require about $25 per day to purchase this amount of beef, well above the combined mean daily income for the entire Arroyo Bandera population. Thus, wild game could not currently be replaced by domestic meat, even if all available income were spent only on meat purchases.

The cultural importance of Aché hunting is more difficult to specify. Although the Aché have generally been willing to substitute meat from domestic animals for wild game, hunting remains important for three reasons. First, Aché children must be given a name from an animal that their mother prepares for consumption during pregnancy. Aché leaders have emphasized the importance of wild animal names in several recent tribal meetings, and there is general consensus that even rare and endangered animals must be hunted occasionally in order that important Aché names not be lost. This belief is strongly tied to an Aché mythology that sees all animals as they originally derived from Aché ancestors, and a belief that the essence of an Aché person is formed by consuming these animals. Aché names are an important link to past ancestors.

Table 5-2. Vertebrates Harvested by Aché Hunters (in Rank Order of Biomass) During Sample Periods from 1980 to 1996

Scientific Name	Common Name	Number Killed	Mean Weight[a]	Total kg Killed	All Animals (%)	Total Biomass (%)
Dasypus novemcinctus	Nine-banded armadillo	1500	3.8	5750.4	42.8	35.2
Agouti paca	Paca	390	6.7	2630.1	11.1	16.1
Cebus apella	Brown capuchin monkey	889	2.3	2032.8	25.4	12.5
Tapirus terrestris	Tapir	9	177.0	1593.0	0.3	9.8
Tayassu pecari	White-lipped peccary	55	24.9	1370.8	1.6	8.4
Nasua nasua	Coati	261	3.5	902.6	7.5	5.5
Mazama spp.	Brocket deer	27	25.8	696.6	0.8	4.3
Tayassu tajacu	Collared peccary	27	16.3	440.6	0.8	2.7
Tupinambis teguixin	Tegu lizard	77	2.3	178.8	2.2	1.1
Cabassous tatouay	Naked-tailed armadillo	24	5.4	129.6	0.7	0.8
Dasyprocta azarae	Agouti	26	2.7	70.2	0.7	0.4
Tamandua tetradactyla	Collared anteater	11	5.0	55.0	0.3	0.3
Euphractus sexcinctus	Yellow armadillo	10	5.0	50.0	0.3	0.3
Hydrochaeris hydrochaeris	Capybara	1	45.0	45.0	0.0	0.3
Eunectes murinus	Anaconda	1	40.0	40.0	0.0	0.2
Penelope superciliaris	Rusty-margined guan	44	0.8	35.8	1.3	0.2
Caiman latirostris	Caiman	7	5.0	35.0	0.2	0.2
Bothrops spp.	Bushmaster snake	33	1.0	33.0	0.9	0.2
Myrmecophaga tridactyla	Giant anteater	1	30.5	30.5	0.0	0.2
Unidentified birds	Birds	43	0.7	30.1	1.2	0.2
Alouatta caraya	Black howler monkey	5	5.8	29.0	0.1	0.2
Speothos venaticus	Bush dog	4	5.5	22.0	0.1	0.1
Coragyps atratus	Black vulture	11	2.0	22.0	0.3	0.1
Boa constrictor	Boa constrictor	1	15.0	15.0	0.0	0.1
Panthera onca	Jaguar	1	15.0	15.0	0.0	0.1
Eira barbara	Tayra	3	4.0	12.0	0.1	0.1
Sarcoramphus papa	King vulture	3	3.0	9.0	0.1	0.1
Tinamus solitarius	Solitary tinamou	8	1.0	8.0	0.2	0.0
Odontophorus capueira	Wood-quail	8	1.0	8.0	0.2	0.0
Crypturellus obsoletus	Tinamou	5	1.0	5.0	0.1	0.0
Ramphastos discolorus	Red-breasted toucan	3	1.0	3.0	0.1	0.0
Felis tigrina	Oncilla	1	2.2	2.2	0.0	0.0
Cathartes aura	Turkey vulture	1	2.0	2.0	0.0	0.0
Crax fasciolata	Bare-faced currasow	1	2.0	2.0	0.0	0.0
Sylvilagus brasiliensis	Rabbit	2	0.9	1.8	0.1	0.0
Dasypus septemcinctus	Seven-banded armadillo	1	1.6	1.6	0.0	0.0
Didelphis spp.	Opossum	1	1.5	1.5	0.0	0.0
Geochelone spp.	Tortoise	1	1.0	1.0	0.0	0.0
Ara chloroptera	Scarlet macaw	1	1.0	1.0	0.0	0.0
Pipile yakutinga	Black-fronted piping guan	1	1.0	1.0	0.0	0.0
Caluromys lanatus	Western wooly opossum	3	0.3	1.0	0.1	0.0
Ramphastos toko	Toco toucan	1	0.7	0.7	0.0	0.0
Unidentified Muridae	Mice	1	0.3	0.3	0.0	0.0
Sum		**3503**		**16,314.0**	**100.0**	**100.0**

[a]Mean weight (in kg) is taken from a sample of Aché prey, or Eisenberg and Redford (1994) in order to standardize weights for years when only a small number of individuals was recorded.

Table 5-3. Vertebrates That Contributed More Than 1% of the Biomass Harvested by Aché Hunters from 1980 to 1996

Important Species	Common Name	Proportion of All Vertebrate Biomass Harvested[a]				
		1980	1981–1985	1994–1995	1995–1996	All Years
Dasypus novemcintus	Nine-banded armadillo	0.135	0.239	0.431	0.439	0.352
Agouti paca	Paca	0.108	0.130	0.182	0.183	0.161
Cebus apella	Capuchin monkey	0.207	0.206	0.095	0.064	0.125
Tapirus terrestris	Tapir	0.000	0.000	0.123	0.181	0.098
Tayassu pecari	White-lipped peccary	0.228	0.156	0.039	0.020	0.084
Nasua nasua	Coati	0.151	0.065	0.043	0.021	0.055
Mazama spp.	Brocket deer	0.106	0.081	0.018	0.016	0.043
Tayassu tajacu	Collared peccary	0.022	0.056	0.023	0.013	0.027
Tupinambis teguixin	Tegu lizard	0.000	0.012	0.019	0.006	0.011
Sum		**0.959**	**0.944**	**0.973**	**0.944**	**0.956**

[a]Sample sizes are 1980, 2191 kg; 1981–1985, 3513 kg; 1994–1995, 4301 kg; 1995–1996, 4096 kg.

Second, the Aché food-sharing system depends heavily on redistribution of wild game (Kaplan et al. 1984; Kaplan and Hill 1985). Domestic animals do not carry the same sharing obligations as wild game, and Aché leaders recognize that food sharing is central to the Aché way of life. It is a practice that distinguishes them from their ethnic neighbors. Aché society is economically very communal, and a certain egalitarian ethic that is associated with equal shares of wild game permeates all social interaction. Aché often characterize their own people as "those who really share," and there is a strong sentiment to retain that character. There is a good deal of concern that the loss of food sharing would lead to economic stratification, something that most Aché wish to avoid.

Finally, the Aché recognize the need for hunting as a social and psychological outlet. Several leaders recently commented while justifying the maintenance of a tribal forest reserve that a major purpose of continued hunting in the future will be to "allow depressed and stressed individuals to escape the pressures and worries of the modern world" and relax in a life-style that is comfortable and enjoyable to them (figure 5-4), to seek an environment where they can feel equal to or better than the outsiders who constantly impose their superiority on the Aché. After several days in the forest, most Aché adults return with newly invigorated self-esteem, more relaxed and ready to face the difficulties of adjusting to a new world. Thus, Aché hunting sometimes serves the same purpose as weekend outings and vacations do for Americans. This might ultimately be critical to the long-term psychological health of the population.

Animal Densities

Densities in hunted and unhunted zones of the eight important hunted mammal species in order of annual biomass harvest are shown in table 5-4. (The density of T. teguixin is not estimated because this species is detectable only during a few months of each year.) The density estimates of T. pecari are somewhat

Figure 5-4. Evening meal with nine-banded armadillo (*Dasypus novemcinctus*). *Photo courtesy of Kent H. Redford.*

unreliable because this species was encountered only three times in our sample, thus we are unable to obtain a good estimate of estimated strip width for it. The significance of the association between encounter rates and distance from the Aché settlement has been analyzed, and the results have been presented elsewhere (Hill et al. 1997). These data showed that encounter rates with *D. novemcinctus, M. americana,* and *T. terrestris* increase significantly with distance from the Aché village after other variables are controlled. Encounter rates with *C. apella* might

Table 5-4. Density of Major Game Animals in the Mbaracayu Reserve and Comparison with Other Neotropical Sites

Species	Total Encounters	Mean Group Size	Effective Strip Width (m)	Total Transect Distance (km)	Encounter Ratio Indirect/Direct[a]
D. novemcinctus	205	1	3.4	1496.100	19.7
C. apella	102	18	18.2	2054.850	1.4
A. paca	40	1	3.3	1874.850	27.8
T. pecari	3	80	200.0	2054.850	12.0
N. nasua	15	8	13.3	2054.850	4.2
M. americana	91	1	13.0	2030.850	24.2
T. terrestris	33	1	16.1	2030.850	43.8
T. tajacu	14	1.5	10.0	2054.850	71.5

	Mbaracayu			Other Neotropical Sites		
Species	Density All Zones[b]	Density Unhunted Zones[b]	Density Hunted Zones[b]	Mean Density (No./km²)[c]	Minimum Density (No./km²)[d]	Maximum Density (No./km²)[d]
D. novemcinctus	19.9 (15.7–25.2)	23.35	12.03	21.90	12.00	25.00
C. apella	24.5 (15.3–39.3)	34.56	11.16	12.40		
A. paca	3.2 (2.2–4.8)	5.87	3.88	27.50	27.50	50.00
T. pecari	0.3 (0.1–0.8)	1.70	1.00	4.90	1.40	8.13
N. nasua	2.2 (1.2–4.2)	2.30	3.12	15.10	4.00	22.00
M. americana	1.7 (1.3–2.3)	1.72	1.20	10.50	1.00	1.30
T. terrestris	0.5 (0.3–0.9)	0.60	0.54	1.60	0.40	1.60
T. tajacu	0.5 (0.2–1.2)	0.69	0.36	11.90	3.35	14.50

[a]The ratio of all encounters with fresh signs of a species divided by all direct encounters with the species.
[b]Density measures are given in animals per km²; 95% confidence intervals are shown in parentheses.
[c]Mean neotropical density from various sites (from Robinson and Redford 1989, table 2.)
[d]Range of best estimates of neotropical densities (from Townsend 1995, table 7-1).

also increase with distance from the Aché village, but the trend was not statistically significant. Other prey species show no significant trends in encounter rate with distance from the Aché village.

The comparison with Mbaracayu density data from other studies suggests that *M. americana*, *T. terrestris*, *D. novemcinctus*, and perhaps *T. pecari* all show densities within the range of other reliably measured neotropical sites. *A. paca*, *N. nasua*, and *T. tajacu* are rare in Mbaracayu relative to other sites, whereas *C. apella* is abundant. Although some of these conclusions might be affected by our group size estimates, the low density of *A. paca* must be a real feature of Mbaracayu. The sample size of encounters is high enough to eliminate sample error as a cause of the low estimate, and because most *A. paca* were censused in their burrow, many of the potential problems of censusing mobile and wary prey do not apply. Aché informants state that *A. paca* is usually found only within about 100 m of surface water in Mbaracayu. Our data lend support to that view (Hill et al. 1997). Thus, much of the Mbaracayu Reserve might be unsuitable habitat for *A. paca*, leading to low average densities. The low density of *T. tajacu* is not so easily explained. The estimate we obtained is about sevenfold lower than the minimum reliable estimate from the neotropics reported by Townsend (1995). We suspect that this is partially due to the fact that *T. tajacu* are exceptionally wary in Mbaracayu and not accurately censused by diurnal transects. This can be studied further by analyzing indirect encounters with the species. The number of fresh indirect encounters (tracks, feces, etc.) with *T. tajacu* was more than 70 times the number of direct encounters. Other ungulates of similar body size such as *M. americana* or *T. pecari* showed a much lower ratio of indirect to direct encounters (table 5-4), suggesting that although *T. tajacu* is present, the species is difficult to observe using our transect methods.

Harvest Rate and Sustainability

Data on harvest rates, prey densities, and size of the Aché catchment basin and potential source areas for depleted prey species allow us to assess whether the catchment basin really is a sink, and whether the harvest rate is likely to be sustainable given the prey population in the source area. Data on harvest rates per consumer, or per km^2, also allow us to compare the specifics of the Aché economy with other areas of the neotropics. In table 5-5, we present data on the 1994 to 1996 harvest rate of each of the eight important game species, per consumer and per km^2. The mean annual number of animals harvested by the Arroyo Bandera-Aché is calculated from the total 1994 to 1996 hunting data. These data show that the Aché are typical in their dependence on *C. apella*, *A. paca*, *N. nasua*, and *T. terrestris*. Annual harvest rates per consumer are close to the mean of other neotropical studies for those prey species. The Aché harvest rate per consumer of *D. novemcinctus* is the highest ever reported thus far. Per consumer harvest rates of *Tayassu* species and *M. americana* are considerably lower than the neotropical averages, and in the case of *T. pecari*, the lowest rates in the literature review of neotropical hunters. Thus, the Aché are exceptionally dependent on armadillos, whereas peccaries are considerably less important than in most other neotropical sites. The lack of importance of white-lipped peccaries, however, is a recent phenomenon (table 5-3) and might be due to multi-annual fluctuations in peccary location and abundance. From the prevalence of traditional Aché names, we can determine that collared peccaries have never been important in the Aché economy in this century.

The annual harvest per km^2 in the Aché sink area is also presented in table 5-5. We assume that 95% of all animals are killed in the Aché sink (56.5 km^2). The total biomass from these eight species harvested was 75 kg/km^2. Because these species represent 95% of the total game biomass, we can estimate that the Aché harvest about 75 kg of vertebrates per km^2 in their primary hunting zone. Comparison with other sources again suggests that the Aché harvest a high number of *D. novemcinctus* per km^2, and a low number of *Tayassu* species per km^2. It is interesting that the Aché harvest of *A. paca* is higher than that in the other reported neotropical sites despite the fact that we measured *A. paca* at a density about eightfold lower than the neotropical average. Dietary dependence on *A. paca* (number killed per consumer year) is also higher than average among the Aché. Our observations from having hunted with a half dozen other neotropical native groups is that the Aché are exceptionally skilled at killing *A. paca* and use a technique (Hill and Hawkes 1983) to extract animals from their burrow that we have never observed elsewhere.

Table 5-6 shows the estimated number of prey for each species in the Aché

Table 5-5. Annual Harvest of Game by Aché Hunters in the Mbaracayu Reserve and Comparison to Other Sites

| Species | Annual Harvest | Annual Harvest per Consumer | | | Annual Number Killed per km^2 | | Biomass Harvest per km^2 | |
		Aché	Indigenous Mean[a]	Neotropical Range[a]	Aché[b]	Neotropical Range[c]	Aché[b]	Piro[d]
D. novemcinctus	507	6.18	0.79	0.030–4.12	8.52	4.49	32.65	
C. apella	183	2.23	2.51	0.003–8.63	3.07		7.03	
A. paca	125	1.52	0.92	0.009–4.39	2.10	1.12	14.16	
T. pecari	7	0.09	0.92	0.166–4.70	0.12	0.30–1.48	2.93	
N. nasua	49	0.60	0.59[e]	0.005–2.75	0.82	2.70	2.85	
M. americana	5	0.06	0.18	0.016–1.49	0.08	0.13–0.29	2.17	6.10
T. terrestris	4	0.05	0.05	0.009–0.12	0.07	0.08–0.09	11.89	14.10
T. tajacu	5	0.06	0.65	0.013–3.77	0.08	0.27–2.19	1.37	20.20

[a]From Redford and Robinson (1987), table 4.6.
[b]In the Aché sink area only, assuming 95% of all kills are made in that area.
[c]From Townsend (1995), table 5-2, and Bodmer (1994), p. 126.
[d]From Alvard (1993), table 6-2.
[e]Nasua taken from both native and colonist data.

hunting zone, the potential source area for the Aché hunting zone, and the whole Mbaracayu Reserve. We also calculated the annual harvest as a proportion of the estimated prey population in the Aché hunting zone, as a percentage of the prey population in the Aché hunting zone and the source area of the Reserve, and as a percentage of the total prey population estimated for the Mbaracayu Reserve. The size of prey populations in the Aché hunting zone (sink) and source areas, as well as the whole Mbaracayu Reserve, are calculated by multiplying those areas (defined above) by the animal densities in the hunted and unhunted zones (table 5-4). The data support the hypothesis that the Aché hunting zone is a sink for at least some species. *D. novemcinctus, C. apella, A. paca, N. nasua,* and *T. terrestris* are all harvested at a rate of more than 25% of the Aché hunting zone population per year, a rate that is unlikely to be sustainable. The annual harvest rate for other

Table 5-6. Annual Harvest as a Percentage of Animals Estimated to Be in Hunted Areas, Source Areas, and the Mbaracayu Reserve

| Species | Total Animals | | | Proportion Harvested 1994–1996 | | | Estimated Proportion Sustainable Harvest[b] |
	Aché Sink	Aché Source	Mbaracayu Reserve	Aché Sink	Aché S-S[a]	Mbaracayu Reserve	
D. novemcinctus	680	9187	13,030	0.745	0.051	0.039	0.396
C. apella	631	13,598	18,711	0.290	0.013	0.010	0.024
A. paca	219	2310	3350	0.570	0.049	0.037	0.079
T. pecari	57	669	959	0.124	0.010	0.007	0.280
N. nasua	176	906	1453	0.278	0.045	0.034	
M. americana	68	677	987	0.074	0.007	0.005	0.107
T. terrestris	31	236	355	0.131	0.015	0.011	0.031
T. tajacu	21	272	387	0.244	0.017	0.013	0.340

[a]Aché source and sink areas combined.
[b]Calculated from Robinson and Redford (1991), tables 27.1 and 27.2, as described in text.

species is 7% to 13% of the estimated population in the Aché hunting zone. However, when the annual harvest is divided by both the Aché sink population and the Aché source population for each prey species, no species shows a harvest rate above about 5%. Calculation of the Aché harvest rate for each species as a proportion of the total number of prey estimated in the whole Reserve provides the same general picture, namely that all species are harvested at very low annual percentages (< 4%) of the standing population.

Whether or not Aché harvest rates are sustainable is a complicated question, but we can begin to answer it by making a few simple assumptions. First, we assume that the source area really does provide immigrants to the sink areas in the Reserve. Because we can estimate the proportion of all prey in both the source and sink that is harvested annually, we need only determine whether a harvest of that proportion of the population will be sustainable over the long run. The best available estimates of harvest sustainability for neotropical mammals come from Robinson and Redford (1991b). Those investigators used estimated maximum reproduction rates with mean densities and life span information to determine a likely maximum sustainable harvest rate for a number of mammal species. Table 5-6 shows the percentage of the standing prey population that Robinson and Redford estimate can be harvested annually for each species without depleting the population through time. The sustainable harvest percentage is calculated by dividing the number of individuals that can be harvested per km^2 (Robinson and Redford 1991b, Table 27.2) by the mean density estimate of individuals per unit area (Robinson and Redford 1991b, Table 27.1) times 60% nonharvested density. The Aché data show that no species is harvested at a rate that exceeds the calculated maximum sustainable harvest percentage.

The sustainability of Aché harvest patterns through time also can be assessed with longitudinal harvest data. Figure 5-5 shows the mean return rate per hunter day from the three species, with the largest number of animals harvested over a 16-month period. No long-term decrease in harvest or return rate per hunter day is detectable; thus, it is not likely that game densities are decreasing markedly in the areas hunted by the Aché.

DISCUSSION

The data on Mbaracayu game densities and Aché annual harvest of major prey species suggest that the 450-km^2 sink and source areas for Aché prey are probably sufficient for a sustainable harvest at current rates. The only other study we know of that has calculated the minimum necessary combined sink and source for sustainability of observed neotropical hunting patterns is that of Townsend (1995). Townsend estimates that a Sirionó population of 192 consumers requires a minimum harvest area of between 221 and 1259 km^2 of forest land, depending on the species considered. However, one species (*Blastoceros dichotomus*) harvested by that community would require between 3245 and 5409 km^2 of savanna for a sustainable harvest. Without the savanna species, the Sirionó would still obtain more meat per consumer day than the Aché (and most other neotropical commu-

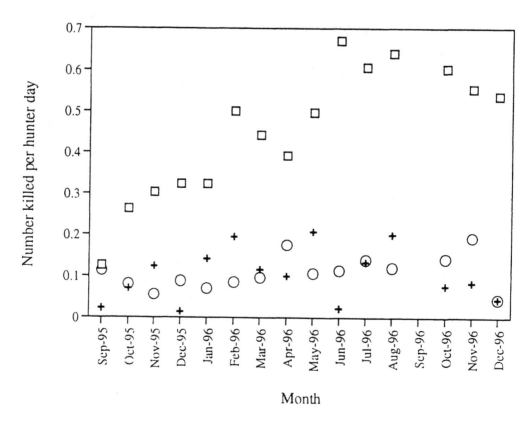

Figure 5-5. Harvest rate for Aché hunters (animals killed per hunter day) of *D. novemcinctus* (*squares*), *C. apella* (*crosses*), and *A. paca* (*circles*) from September 1995 to December 1996. The data show no evidence of longitudinal depletion for these species during the study period.

nities). Thus, taking the mid-point of Townsend's estimate, we calculate that the Sirionó need about 3.8 km² of forest per consumer to achieve a sustainable harvest at the current meat consumption rate. This is similar to the 5.5 km² per consumer that the Aché currently use in Mbaracayu. Because both estimates are close, we might speculate that *terra firme* neotropical native populations can achieve sustainable harvest if the combined sink-source areas available to them total at least 5 km² per consumer. This would imply a maximum sustainable human population density in *terra firme* forests of about 0.2 person/km² given typical meat consumption rates. With more information on the range of biomass densities in the neotropics, and human meat consumption levels, it will be possible to test this hypothesis and specify a range of likely maximum sustainable human densities that will not deplete neotropical vertebrate fauna.

Two important conservation questions arise from these analyses. First, can the results be extended to other economically less important vertebrate species? And second, do measurements of game density differences in hunted and unhunted areas provide information about sustainability of harvest rates? We believe that the answer to the first question can be partially surmised through an understanding of human hunting patterns. In general, most prey species are more difficult to locate than they are to kill. Search time for game represents the majority of all hunting time, whereas pursuit time is considerably less important (Hawkes et al. 1982; Hill and Hawkes 1983; Hill et al. 1985; Hill et al. 1987). This means that each species that is hunted is taken in approximate relation to its abundance. Species that are not profitable are not hunted at all (Hawkes et al. 1982; Hill and Hawkes 1983; Hill et al. 1985; Hill et al. 1987). Some species are more vulnerable than others, but the relationship between density and annual harvest can be clearly shown with current Aché data. Figure 5-6 shows a regression of log number of animals killed per year by log density for each of the major prey species in the Reserve. Density is associated with 80% of the variation in the annual harvest amount. In this sample, *A. paca* was exceptionally vulnerable to human predation, whereas *M. americana* was greatly underharvested relative to its estimated density. This agrees with Aché informant statements that hunters actively seek out *A. paca* habitat through the day to the exclusion of other areas, and that deer are extremely wary and thus difficult to kill with bow and arrow. If the relationship between prey density and annual harvest holds for other mammalian game species, it suggests that most species found at proportionally lower densities will simply be harvested at proportionally lower rates.

The data thus suggest that if major game species are not overharvested, perhaps other animals will also be harvested sustainably. We are concerned, however, by the implications of harvest variance due to encounter luck for rare species. If few individual animals exist in a circumscribed area, then even though a species may on average be harvested at a sustainable rate, a single year of high harvest could be enough to start the population on a decline to extirpation. For example, from transect data we estimate that there are about 30 jaguars in the Mbaracayu Reserve. Only three have been killed by the entire Aché population in the past 19 years. However, if these were all killed in 1 year and if they were reproductive aged females, this loss might eliminate the Mbaracayu population. Such a perspective suggests that a rigid harvest cap should be placed on all species estimated to contain fewer than 100 individuals in a circumscribed area. We have discussed this with Aché leaders, and they are not opposed to limits being placed on rarely encountered game animals, as long as a few can be killed occasionally for naming purposes.

Finally, we can examine the relationship between density in hunted versus unhunted areas and the sustainability of the harvest. Concerned individuals might reach unwarranted conclusions about the unsustainability of human hunting because game depletion can be shown in hunted areas. Studies that simply compare densities in hunted and unhunted areas are not instructive, however, because all central place predators are expected to deplete prey items nearby their home base. Aché reservation lands within an hour's walk of the village center have been censused by our research team three times. Those lands are extremely depleted of game, with only four mammals (three *D. novemcinctus* and one *M.*

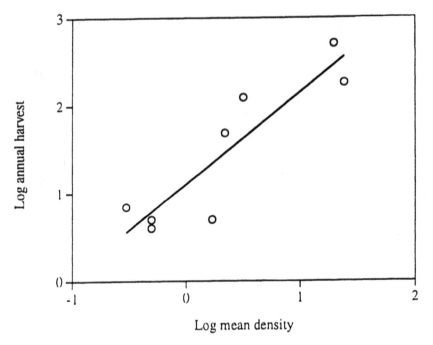

Figure 5-6. The relationship between the density of the eight major prey species in Mbaracayu and the number harvested annually by Aché hunters (r = 0.90).

americana) encountered in 51,000 m of transect. However, of the eight major game species, only *C. apella*, *T. terrestris*, and *T. pecary* showed no signs at all on reservation land. Fresh tracks, feces, or feeding signs were encountered for *D. novemcinctus*, *A. paca*, *N. nasua*, *M. americana*, and *T. tajacu*. The Aché hunt on reservation land every day of the year, and all of the eight major prey items are consistently harvested at levels many times above sustainability. Yet even after more than 15 years of hunting, five of the eight species are present within a short distance of an Aché village. Clearly, there is a source prey population that continually resupplies this area. Thus, any calculations of hunting sustainability must include that source.

Do large differences in prey density between hunted and unhunted areas indicate that the prey harvest is unsustainable? Previous analyses on a smaller subset of our transect data showed that *D. novemcinctus*, *M. americana*, *T. terrestris*, and possibly *C. apella* are characterized by increasing encounter rates with distance from the Aché, even when habitat type and other variables are controlled (Hill et al. 1997). The current analyses show a large density difference between hunted and unhunted zones for *D. novemcinctus*, *A. paca*, and *C. apella*. Comparison of actual harvest to the estimated sustainable harvest, however, suggests that none of these species is in danger of being overharvested. This is true despite the fact that densities of the two most important game species are twice as high in unhunted areas as they are in the hunted zone near the Aché village. These observations suggest that density in hunted and unhunted areas might tell us more about dispersion rates and distances of each species than they do about sustainability of the harvest.

Despite the ability to measure animal density differences between hunted and unhunted areas and estimate whether harvest rates are sustainable, we suggest that all conservation units should undergo longitudinal censusing to ensure that typical harvest rates are indeed sustainable. Because annual fluctuations in animal density are expected and not yet well understood (Glanz 1982), comparison of only two density estimates through time cannot be considered a reliable indicator of harvest sustainability. But continual monitoring will allow the discovery of a consistent population decline, which can then be counteracted by limitations on harvest rate.

Finally, data such as those presented here should induce conservation biologists to reflect on the status of humans as predators in conservation units. In the neotropics, native people are a natural predator in many conservation areas. Their hunting activity will not necessarily extirpate large or rare species if the ratio of consumers to prey is low. The simple fact that the species present today have coexisted with humans for at least 12,000 years and that humans have always been an active component of the current neotropical milieu provides common sense evidence of this. As far as can be discerned, there were no uninhabited areas of the neotropics at European contact (Denevan 1992), and there have been no large uninhabited areas in the time since European arrival (Steward 1944–49).

Conservationists also should be aware that the current set of neotropical species has never existed as a "natural community" without human activity. The earliest human populations spread through the neotropics prior to the terminal Pleistocene extinctions (Meggers 1982; Roosevelt 1994), which led to modern neotropical faunal communities. Humans have acted as predators, competitors, and seed dispersers in the neotropics for at least 12,000 years. They have disturbed and restructured forests through swidden agriculture for at least 5,000 years (Balée 1994). Humans are a natural top predator of the neotropics, their presence and activities, as practiced traditionally, are just as natural as those of jaguars, anacondas, or harpy eagles. The removal of top predators from any area is an ecological gamble that may threaten biodiversity and result in a community structure that is anything but natural (Glanz 1990; Janson and Emmons 1990; Terborgh 1990).

ACKNOWLEDGMENTS

This research was funded by two grants from the Nature Conservancy, and logistical support was provided by the Fundación Moises Bertoni. Miguel Morales, Alberto Yanosky, and Raul Gauto were instrumental in providing institutional support. Equipment was donated by Ex-Officio, Coleman, Therm-a-rest, Slumberjack, Great Pacific Iron Works, Ray O Vac, and Jansport Inc. Statistical analyses using DISTANCE software were performed by Garnet McMillan.

6

Impact and Sustainability of Indigenous Hunting in the Ituri Forest, Congo-Zaire: A Comparison of Unhunted and Hunted Duiker Populations

JOHN A. HART

Small antelope, principally duikers and the chevrotain, are among the most important game species for hunters in African forests. They are also a preferred meat sold in African markets (Colyn et al. 1987b; Aisbey and Child 1991; Juste et al. 1995; Njiforti 1996). Estimates of the volumes of bush meat, of which antelope often make up the largest proportion, that reach the urban markets of the region can be colossal (Steel 1994; Fa et al. 1995). The total value of this trade is thought to reach millions of dollars annually (Juste and Fa 1994). Despite its economic and nutritional significance, few attempts have been made to assess the sustainability of hunting in African forests (but see Auzel and Wilkie, this volume; Fimbel et al., this volume; FitzGibbon et al., this volume; Noss, this volume).

SUSTAINABLE HUNTING

To be sustainable, hunter harvest must not leave exploited populations vulnerable to extinction and must not disrupt ecosystem functioning (Robinson and Redford 1991b). Economic and biological components of hunting are intimately linked, and together they determine the trajectory of the resource use, often with devastating outcomes for exploited populations (Caughley and Sinclair 1994) and major impacts on exploiter economies (Lavigne et al. 1996).

From the biological perspective of the exploited species, sustainable harvests depend on a reproductive surplus that can be diverted to human use (Hilbourn et al. 1995). More operationally, Caughley and Sinclair (1994) imagined sustainable yields in terms of the proportion of a population that can be taken year after year yet with the exploited population size still remaining constant. As these investigators showed, sustainable yields can be achieved from a range of population sizes, even if these are below unhunted levels. Although the concept of a range of sustainable harvest levels has replaced the notion of a single maximum sustained yield (see Caughley and Sinclair 1994; Hilbourn et al. 1995), there remains the question of the overall ecological and economic impact of harvesting. Even if the population density of an exploited species remains stable, its reduced population might lead to cascading impacts on other species in the community. In addition, reduced (but stable) yields might not be economically sustainable.

Information needed to measure the sustainability of hunting includes the extent and variability of hunting, the population densities and production of the exploited species, and the response of the hunters and their game to exploitation (Robinson and Redford 1994b). Data to evaluate the sustainability of harvests directly rarely have been available, particularly for wildlife species in tropical forests. Robinson and Redford (1994b) reviewed the indices that have been used as indicators of sustainability, including comparisons of population density and age structure of harvested and unharvested populations, and the dynamics of hunting yield. Two models used to evaluate hunter harvest (Robinson and Redford 1991b; Bodmer 1995a) have provided theoretical upper limits to sustainable use. Both of these models, and the indices, depend on biological information and estimates of population parameters that are not available for most species of African forest antelope (Payne 1992). In particular, information on unhunted populations and the offtake of natural predators has never been used to assess sustainability.

This chapter takes a comparative approach to evaluate the impact of human hunting in the Ituri Forest in northeastern Congo-Zaire. I compare unhunted populations of six species of duikers, *Cephalophus* spp. and the water chevrotain *Hyemoschus aquaticus*, with populations of the same species in the same habitats that are also hunted by humans. The chapter will:

1. Assess the age structure, density, mortality, and dispersal of six species of forest duikers and the chevrotain in two areas protected from all hunting, and describe how unhunted populations are affected by variation in available forage and natural predation in the two major forest types of the region.

2. Measure offtake, age structure, and abundance of hunted antelope populations at four sites where human exploitation varied from intermittent and infrequent hunting to nearly continuous intensive hunting of antelope for commercial wild meat trade.

3. Evaluate how differences in the dispersal, natural mortality, and reproductive biology of the various antelope species are linked to differences in their vulnerability to human hunting pressure.

4. Present an assessment of the sustainability of harvest rates by investigating

population trends in exploited species in relation to hunting practices and economic context of hunter communities.

THE ITURI FOREST

The Ituri Forest is located in northeastern Congo-Zaire. The forest covers about 60,000 km² of the watershed of the Ituri River, from savanna uplands and foothills of the Albertine Rift in the east of the basin, to the gently undulating forested peneplains in the west at the confluence of the Ituri and Nepoko rivers.

The hunting studies reported here were conducted in the southern Ituri Forest in the districts of Lubero (North Kivu) between Mbunia and Biasiko, hereafter termed the Southern Ituri Study Area, and in the central Ituri Forest in the vicinity of Epulu, Mamabas District (Haut Zaire), hereafter termed the Central Ituri Study Area. Studies of unhunted populations of small ungulates were also conducted in the Epulu area (figure 6-1). Elevations of the study areas varied between 700 and 850 m, reaching about 1200 m only on the highest hills in the Mbunia area in the south. Mean annual rainfall at three recording stations in the Central Ituri Study Area averaged from 1700 to 1800 mm between 1986 and 1995

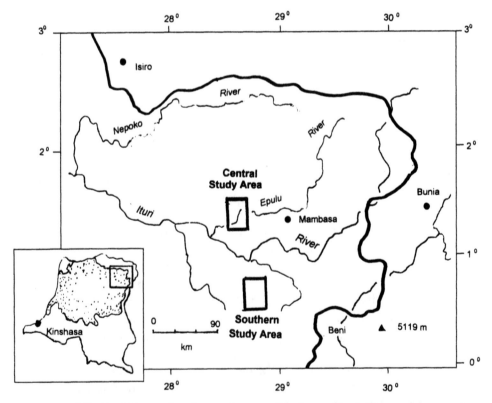

Figure 6-1. The Ituri Forest showing study areas. The heavy line indicates the approximate limit of lowland closed forest. The stippling on the inset locates the lowland moist forest zone of the Democratic Republic of Congo (Congo-Zaire).

(Hart and Carrick 1996). The Central Ituri Study Area lies within the Okapi Wildlife Reserve, established in 1992.

Forests in the Ituri Basin are classified as closed-canopy, shade-tolerant, evergreen to semideciduous moist tropical forest (White 1986). Mature forests cover between 85% and 90% of the region included in this study (Wilkie 1990). There are two main forest types in this region. The first is a distinctive monodominant forest in which a single species, *Gilbertiondendron dewevrei* (Caesapliniaceae), locally called *mbau*, comprises between 60% and 80% of the canopy trees and is well represented from the subcanopy to the understory. Stands of mbau range from a few hectares to tens and even hundreds of square kilometers in area, particularly south of the Ituri River. The second major forest type, termed *mixed forest*, is characterized by a more equitable distribution of canopy trees, and an absence of mbau, although other Caesapliniaceae can be common (T. Hart 1985; Makana et al. 1998). Mixed forest intergrades with older secondary forests in some locations, and areas of swamp forest occur in both forest types on seasonally waterlogged soils. These latter formations generally cover limited areas, less than 25 hectares (ha) in most cases. With the exception of recent fallows in locations currently occupied by shifting cultivators, stands of young secondary growth are restricted to small areas, including treefalls and blowdowns, and are caused by natural disturbance. There has been no history of logging in the region covered by this study.

UNGULATES THAT ARE HUNTED

Fourteen species of ungulates occur in the Ituri Forest. The six species of duikers and the water chevrotain that are the subject of this study are the major species hunted in the region (table 6-1). These ungulates can be classed as small (one species, the blue duiker, *Cephalophus monticola*), medium (five species, including the black-fronted duiker *C. nigrifrons*, white-bellied duiker *C. leucogaster*, Weyn's duiker *C. weynsi*, and bay duiker *C. dorsalis*, and the water chevrotain *Hyemoschus aquaticus*), or large (one species, the yellow-backed duiker, *Cephalophus sylvicultor*). The analyses that follow will use this size classification, with the middle-sized

Table 6-1. Duikers and Chevrotain of the Ituri Forest

Species	Abbreviation	Common Name	Mean Adult Weight (kg)[a]
Cephalophus monticola	MONT	Blue duiker	4.7
C. nigrifrons	NIGR	Black-fronted duiker[b]	13.9
C. leucogaster	LEUC	White-bellied duiker[b]	16.7
C. weynsi	WEYN	Weyn's duiker[b]	17.7
C. dorsalis	DORS	Bay duiker[b]	22.0
C. sylvicultor	SYLV	Yellow-backed duiker	68.0
Hyemoschus aquaticus	HYEM	Chevrotain	11.2

[a]Weights (from Hart 1986) are means of two sexes; females average 8% to 15% larger than males.
[b]These species are collectively referred to as red duikers.

duikers and the chevrotain referred to collectively as "red duikers." All forest duikers and the chevrotain are frugivorous (Dubost 1984; Hart 1986). Fruit production in the Ituri forest varies with forest type and season. Mixed forest has a more equitable seasonal distribution of duiker foods, as well as a higher abundance of trees and lianas producing fruits and seeds eaten by duikers, than does the monodominant mbau forest (Hart 1986; Hart and Petrides 1987; Makana et al. 1998). In monodominant forest, fruit availability is low except during the periods of mast seed fall of mbau. This occurs only for periods of 2 to 3 months, and at intervals of 1.5 to 2 years (Hart et al. 1989).

Small to medium-sized ungulates are the most important prey of forest felids in the Ituri Forest, with duikers and the chevrotain comprising nearly one third of prey items in diets of leopard (*Panthera pardus*) and golden cat (*Profelis aurata*) (Hart and Hall 1996).

The basic biology, including gestation, birth interval, age to first reproduction, and natural mortality, is not well documented for any of the duiker species, and it remains totally unknown for some (Ralls 1973; Van Ketelhodt 1977; Kranz 1986). The behavior and feeding ecology of duikers and the chevrotain have been described by Dubost (1980, 1984) and Feer (1989a).

THE SOCIAL AND ECONOMIC CONTEXT OF HUNTING

Hunting in the Ituri Forest, as in other African forests, is a subsistence activity that is becoming market oriented (Hart 1978; Wilkie et al. 1992; Juste et al. 1995; Noss 1995; Muchaal and Ngandjui 1995; Fimbel et al. this volume). Human occupation of the Ituri Forest dates back to at least 12,000 years BP (Mercader in press). These first occupants were hunter-gatherers, and hunting remains a principal activity of contemporary populations.

The cultures of the Ituri region have been shaped by economic and demographic fluctuations for most of this century, up to the present. This has strongly influenced both hunting and settlement patterns in the forest (Roesler 1996). Currently, the Ituri Forest is occupied by people distributed in small, nomadic or semi-nomadic settlements who subsist through shifting cultivation, hunting, fishing, and gathering. From the 1920s until independence in 1960, cash flowed into the forest communities through employment in the coffee plantations, mines, and public works sector of the colonial economy. Since independence, the economic infrastructure of the region has collapsed precipitously. Thus, in the past two decades, many forest communities have turned for cash income to gold panning, limited market crop agriculture, occasional labor employment and the sale of forest products, particularly wild meat.

The principal hunters in the central and southern Ituri Forest are the Mbuti people. The main hunting method of the Mbuti is a drive hunt with nets that target small ungulates (figure 6-2). Net hunts involve from 10 to 40 individuals (both men and women) and from 5 to 20 nets. A single drive area covers from 3 to

Figure 6-2. Blue duiker (*Cephalophus monticola*) captured in a net by Mbuti hunters. *Photo courtesy of John G. Robinson.*

17 ha. Duikers and the chevrotain comprise 78% to 92% of animals captured on net hunts (Hart 1978, 1979, 1986).

The Mbuti live in communities or bands that move between small, temporary camps at sites traditionally associated with each band (Turnbull 1965; Hart 1979). The Mbuti bands who participated in this study followed a nomadic cycle. This included at least one village-based camp on the outskirts of an agricultural settlement and a number of forest camps that were located, at the farthest, about 25 km from the nearest village and were used on a temporary basis. At village camps, Mbuti occupied themselves in a wide range of activities, including day labor as field hands as well as some hunting. In the forest camps, the primary activity of the Mbuti was net hunting, with additional seasonal exploitation of other forest food supplies, notably honey (Hart 1979). Daily net hunts usually ranged out from less than 1 km to 3.5 km from the forest camp, with hunting trips from base camps in villages ranging out as far as 5 km in a few cases.

Snares also were used in some areas, particularly in the vicinity of gardens and adjacent forest around some agricultural settlements. Snares targeted a wider range of wildlife than did nets, including porcupines, large rodents, and small carnivores, as well as duikers. Noss (1995) reported that duikers and the chevrotain comprised 72% of captures by snare hunters who used methods similar to those used in the Ituri Forest. Most snare hunters in both study areas trapped animals for their subsistence needs. This is not the case in other areas (e.g., Wilkie et al. 1992), where snare hunting is a major source of commercial meat. Shotguns had not yet penetrated the southern or central Ituri Forest by the period of this study.

Two related developments have had major impacts on the hunting practices of the Mbuti: population growth, principally by immigration from surrounding areas, and the growing importance of cash incomes. Population densities reported in the latest published national census (1974) for the administrative units covering the Ituri Forest region ranged from less than one person per km² in some of the forested units, to over 15 people per km² in the districts spanning the forest boundary (Mbuyi 1978). Immigration from the densely settled forest edge into the unsettled forest has occurred for generations, supported before 1960 by the colonial administration, but has accelerated markedly within the past 30 years (Hart 1979; Peterson 1991). A census in 1994 recorded nearly 30,000 people (about two/km²) within or adjacent to the Okapi Wildlife Reserve, in one of the largest and least settled sectors of the forest. Over 10% of these people had arrived in the area in the preceding 2-year period (Tshombe 1994). In regions near the sources of immigration, including portions of the southern study area, shifting cultivators have converted the mature forest favored by hunters into forest fallow and farmed bush; this has reduced available hunting areas (Hart 1979).

In more remote areas, deforestation is less significant. However, growing human populations and the increasing integration of hunters into the cash economy have favored a growing demand for bushmeat (Hart 1978; Wilkie et al. in press). In the southern Ituri study area, wild meat is exported from hunting camps to markets on the periphery of the area (Hart 1979). In the Central Ituri, some locally caught wild meat is bought by truck drivers and passengers traveling along the single road that crosses the forest from east to west. It is then either personally consumed or resold at their destination. In addition, many of the local restaurants at Epulu catering to the traffic offer wild meat. Most of the wild meat that is sold for cash originated from net hunters, even if it is then brought into the market by villagers.

STUDY AREAS

Areas with No Human Hunting

Studies of unhunted duiker populations were conducted between 1991 and 1996 in two study areas, Edoro and Lenda. They were established in 1986 and 1990, respectively (figures 6-3 and 6-4). Both study areas were created, and their protection, including no hunting, was negotiated with local communities, including Mbuti hunters. Prior to their creation, some hunting had occurred in parts of the study areas. Net hunters continued to operate in the forests adjacent to Lenda, although no hunts came closer than 2 km to the study areas during the study. Edoro is farther from settlements, with light hunting pressure to the south, east, and west, and unhunted forest to the north. Both study areas are associated with the Centre de Formation et de Recherche en Conservation Forestière (CEFRECOF) of the Congolese National Parks Institute, and they are regularly patrolled.

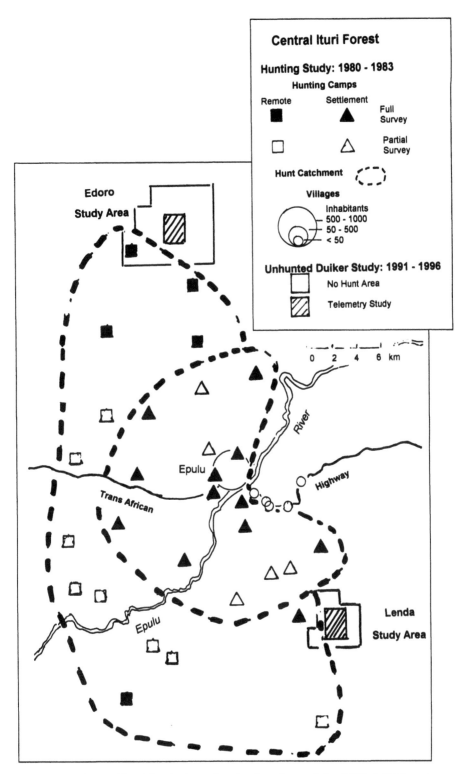

Figure 6-3. Central Ituri Study Area showing locations of the 1980–1983 hunting study and the 1991–1996 study of unhunted small ungulates. Cessation of hunting was initiated on the Edoro Study Area in 1986 and on the Lenda Study Area in 1990.

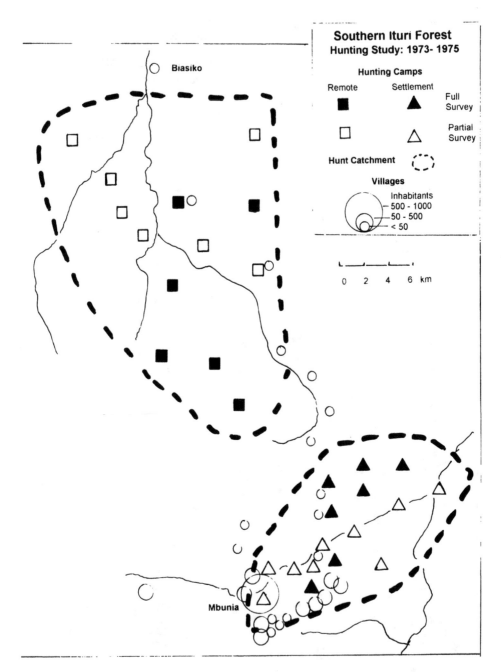

Figure 6-4. Southern Ituri Study Area showing locations of the 1973–1975 hunting study.

For the small ungulate research, a 5-km² area was selected in each study area. A path grid of 125 m was established to aid in locating radio-marked animals to conduct censuses.

Forest types, dominant canopy species, and information on the diversity and seasonal availability of fruit in each study area are summarized in table 6-2. Edoro comprises a mosaic of mixed mature forest and old secondary forest (62%), with

Table 6-2. Study Areas with No Hunting in the Okapi Wildlife Reserve, Ituri Forest

Site	Year Established	Area (km²)	Forest Types	Dominant Canopy Tree	Fruit Diversity	Period of Low Fruit Availability
Edoro Study Area	1986	50	Mature Monodominant (4%) Mature Mixed (50%) Swamp (26%) Old Secondary (12%) Young Succession (8%)	*Cynometra alexandri* (23%)	High	< 1–2 month
Lenda Study Area	1990	12	Mature Monodominant (63%) Swamp Forest (17%) Mature Mixed (13%) Old Secondary (5%) Young Succession (3%)	*Gilbertiodendron dewevrei* (65%)	Low	3–5 month in non-mast years

Source: J. Hart (1986), T. Hart (1985), Makana et al (in press).

swamp forest and young succession covering most of the remainder. Almost two-thirds of Lenda is covered in upland monodominant mbau forest, with mixed mature and secondary forest comprising 18% of the area. Swamp forest comprised 26% of the duiker study area at Edoro and 17% at Lenda. *Cynometra alexandri* is the most abundant tree at Edoro (23% of a sample of 1154 canopy trees), whereas mbau represents 65% of the canopy at Lenda.

Overall, the diversity of available fruits was higher and more seasonally equitable in the mixed forests at Edoro than in the mbau forests of Lenda. At Lenda, fruits and seeds eaten by duikers had a low availability except during brief periods of mast seed fall of mbau. Two mast seed years occurred during the 5-year study, with mbau seed available for less than 4 months during each. Pulses of mast fruit availability in an otherwise unproductive forest have been observed in other areas of mbau dominance (Blake and Fay in press).

Areas with Human Hunting

The results of the hunting studies were analyzed separately for the southern and central study areas, and between zones remote from major permanent settlement versus zones closer to settlements. Within each zone, hunting patterns and duiker dynamics were analyzed for a portion of the area, termed the hunt catchment, which is defined operationally below. Taken together, the four survey zones spanned a socioeconomic gradient from unoccupied forest to settlement frontier, as well as a gradient of hunting pressure from infrequent use to intensive hunting for a commercial meat trade. A summary of the demographic, ecological, and economic contexts of hunting in the study areas is given in table 6-3. These study

Table 6-3. Ituri Forest Research Sites for Studies of Mbuti Hunting

Study Area	Zones	Code	Study Period	Inhabitants per km²	Habitats (%)[a]				Mbuti Hunting	
					Momo-dominant forest	Mixed forest	Garden or fallow	Percent small ungulates[b]	Commercial meat trade[c]	Snare use
Southern	Settlement	S/Settled	1973–1975	5.1	15	63	22	62	Little, external market	High
	Remote	S/Remote		0.1	34	65	1	78	Major, external market	Low
Central	Settlement	C/Settled	1981–1983	4.4	15	75	10	79	Moderate, local market	Low
	Remote	C/Remote		0.2	25	75	0	92	Moderate, local market	None

[a]Estimated from ground surveys, low-elevation air photos, and satellite imagery (Central Study Area only).

[b]Total recorded catch on all hunts, including hunts not using nets.

[c]Extent of commercial trade: *Little*: Commercial exchanges involve < 20% of meat caught. Most meat traded for foods from local villagers, few cash sales. *Moderate*: Commercial exchanges involve 20% to 40% of meat caught. *Major*: Commercial exchanges involve > 40% of meat caught. Type of commercial trade: *External market*: Meat sold for cash or food to professional traders for consumption outside of the local band or band/village associations. *Local market*: Meat sold for cash to consumers or traders based in local settlements; some meat may be sold secondarily to external markets.

areas, the Mbuti hunters, and associated villagers have been described in more detail by J. Hart (1978, 1979, 1984, 1986).

Southern Ituri Study Area. Field studies were conducted in the southern study area between 1973 and 1975 and concentrated on two Mbuti hunting territories, among four in the area (figure 6-4). One hunting territory in the Southern Settlement Zone (S/Settled) was in an area of active agricultural expansion created by the arrival of immigrant settlers from eastern Kivu. Twenty agricultural settlements, containing approximately 1400 people, occurred within the hunting territory of the S/Settled band or within 5 km of its boundary. The second band, identified as the Southern Remote Zone (S/Remote) band occupied forest that contained only two village settlements totaling less than 20 households, and these were at least a day's march from the immigrant settlement frontier.

The S/Settled band comprised about 48 Mbuti adults who occupied a total of 16 camps during a 19-month study period. The S/Remote band included on average 46 adults and used 13 camps over the 12-month study period. During the study, both bands used their territories to the exclusion of neighboring Mbuti communities but could not stop hunters from the agricultural villages from using their area. In the S/Settled zone, men from many of the local settlements set snares. In the S/Remote zone several village men regularly joined Mbuti net hunts, and a small number of village hunters set snares.

About 22% of the S/Settled zone was garden or recently abandoned fallow in the vicinity of village settlements. The Mbuti rarely used these areas for net hunts because of difficulty of penetrating the dense vegetation, and general absence of duikers in recently deforested areas. Fifteen percent of the S/Settled zone was monodominant mbau forest, the remainder being mixed mature and older secondary forest, intermixed with an undetermined percentage of swamp forest. In the S/Remote zone, only 1% of the area was estimated as garden or recently farmed bush. Mbau forest comprised almost one-third of the S/Remote zone, including large stands in the north. The remainder was mixed mature and older secondary forest.

Information on the Mbuti hunting practices and their economic significance in the Southern Ituri Study Area has been summarized by Hart (1978, 1979). In the S/Settled zone, drive hunts with nets accounted for 60% of hunts and 73% of all wild meat by weight on 159 recorded hunts. Small ungulates accounted for 62% of meat caught. Small game hunting with dogs accounted for about one third of the hunts, but only 17% of wild meat. In the S/Remote zone, net drives comprised 83% of hunts and produced 88% of all meat, with small ungulates comprising 78% of the catch.

Mbuti hunters in the S/Settled zone traded 18% of their wild meat for cash or food sold by market traders. Much of the remainder was traded in a loose system of exchanges with nearby shifting cultivators for locally cultivated starch foods. Mbuti in the S/Remote zone were almost entirely dependent on commercial exchanges for their sources of starch foods. Exchanges of meat for other food with commercial traders followed 93% of hunts and accounted for 42% of all meat captured.

Central Ituri Study Area. Hunting studies in the Epulu area of the Central Ituri Forest were conducted over 27 months between 1981 and 1983. They were based in the town of Epulu, on the Trans-African Highway (see figure 6-3). During the study period, four Mbuti bands totaling approximately 325 adults used this area, as well as adjacent forest, extending out to 25 km from the settlement during the study period. Unlike the southern study area, in which each band had a discrete hunting territory, there was extensive overlap in the use of hunting camps by Mbuti bands in the Epulu area. The hunted forest around Epulu was divided into two zones based on distance from the settlement. The area from Epulu to 10 km into the forest is identified as the central settled (C/Settled) zone, whereas the area from 10 to 20 km out from the settlement is termed the central remote (C/Remote) zone. These two zones accounted for 85% of the Mbuti hunting camps known to have been used during the study. The 15% of the camps that were excluded from this study also were used by Mbuti bands that were not focal communities for this study.

Village settlements in the Central Ituri Study Area were located along the road. They included Epulu (population 850), a small commercial center and headquarters for the Okapi Capture Station of the Congolese National Parks Institute, and five smaller settlements totaling about 30 households (135 to 175 people). Although Epulu has a history as an immigrant community, with the majority of residents, including the Mbuti, tracing their roots to communities outside the area, the region itself was not an active settlement frontier during the study. Active gardens and farmed bush fallow associated with the villages covered 20 to 25 km^2, or 10% of the C/Settled zone. Agricultural clearings were only within 2 km of the road in most cases, 3 km maximum. The remaining area was covered by unbroken forest.

The Central Ituri Study Area was remote from outside markets, but some meat was sold to passing truck drivers, and there was a market for wild meat in the town of Epulu. Most of the meat that the Mbuti exchanged for cash or cultivated foods was consumed locally in Epulu and surrounding hamlets. Unlike the Southern Ituri Forest, where commercial traders entered the Mbuti forest camps to buy meat, fewer non-Mbuti entered the remote forest in the Epulu area and fewer villagers set snares. This restricted use of snares was enforced by the National Park's staff to protect the okapi, *Okapia johnstoni*, in the area.

METHODS

Unhunted Study Areas

Duiker Censuses. A team of four field assistants and 45 trained Mbuti net hunters were used to census duikers by net drives in the unhunted Edoro and Lenda Study Areas. Drive counts have been used previously to census ungulates; results have correlated with other census methods when experienced assistants are used and drives are well controlled (Downing et al. 1965). Methods used here were similar to those reported by Koster and Hart (1988) and Williams et al. (1996), but differed from those reported by Noss (1995 this volume).

The path grid in the study area was used to establish the perimeters of the drive areas, and the drives took place in the forest blocks inside. Drive locations for each census period were determined by random placement of the first drive. Drive efficiency was standardized within each study area by using the same individual drivers and duration for each count. Because of the open nature of the monodominant mbau forest at Lenda, drive areas, drive time, and the number of drivers used were higher there than at Edoro. After each drive, all captured animals were recorded, and drivers and net handlers were interviewed for reports of animals seen or flushed and escaped. Observations of the behavior of duikers flushed out on drive counts (Bowland 1990) and the movement of radio-collared animals tracked during net drives in this study indicate that, with the exception of nesting young animals, the likelihood that a duiker enclosed within the drive area would escape detection was low if drive areas were thoroughly searched. Animal densities were calculated as the total number of animals detected divided by drive area.

Data on Captured Animals. The species, sex, and weight of all captured animals were recorded, and the presence of any scars or injuries was noted. Eruption and wear of mandibular molars and premolars were classified. Reproductive condition was determined, based on the relative size and condition of testes, state of the udder, and presence of a fetus determined by palpation. All animals were tagged with a colored and numbered plastic ear tag. A sample of tagged animals was equipped with radio collars before being released.

The relative age of captured duikers was determined by a combination of mandibular tooth eruption and wear (following the method of Wilson et al. 1984) and reproductive status. Animals with no more than one erupted molar and immature sexual organs were classified as juveniles. Animals with immature reproductive organs and at least two erupted molars were recorded as sub-adults. Adults comprised all animals that were reproductively mature, based on condition of external reproductive organs (testicles or udder) and/or the presence of a fetus.

Radio-Collared Animals. An effort was made to place collars on individual animals of all small ungulate species and to maintain a sample of radio-collared animals that represented the sex and age structure of the small ungulate populations in each study area.

Radio-collared animals were tracked with the aid of trained local assistants. We attempted to locate each collared animal at least once every 3 days throughout the entire 5-year study. Any radio signals indicating that a collared animal might be dead, or had lost its collar, were followed up with direct observations of the animal or a search to locate the collar. An attempt was made to locate any animals that had left the study area by wide searches through the surrounding path grid and unmapped forest. Animals located off the research grid were followed on a regular basis if the availability of a path system permitted. Collared animals were defined as dead only if the carcass was located or if the collar was removed but showed unambiguous signs of a predator attack. With the exception of one blue duiker at Lenda, the cause of death of all collared animals that died in the study areas was determined.

Monthly radio locations were tabulated and used to define preliminary home range maps for each collared animal. An animal was defined as a long distance disperser if it moved more than 1 km from the approximate center of its previously determined home range, and if this was subsequently followed by a shift in home range location. Long distance moves with a return to previously mapped home range areas were not considered dispersal. The distance of 1 km was chosen to define long distance dispersal because animals that moved this distance would be unlikely to be detected by a net drive census in the initial home range. For all collared animals that left the path grid and could not be subsequently located, dispersal distances were classified as more than 1 km, although maximum dispersal distance could not be estimated.

For each collared animal, the number of days it was known to be alive from the collaring until death or signal loss due to emigration or failure of the collar was termed the total live animal days. Species-specific mortality rates and dispersal rates, expressed on an annual per km^2 basis, are calculated for each study area by summing total live animal days for each species as follows:

$$\text{Rate} = [\text{Number of animals dispersing or dying} / (\text{live animal days} / 365)] \times \text{mean density}.$$

Areas with Human Hunting

Data on Hunters. During each of the hunting studies, I lived in a village in the study area and made survey trips, accompanied by a field assistant, to hunting camps where we accompanied net hunts for periods ranging from 4 to 20 days. Information gathered on survey trips included camp location, individuals present, data on hunting activity and animal captures, and trade and consumption of meat. Camps were selected (a) to provide a sample that reflected seasonal and geographic patterns of hunting in each survey zone and (b) to sample sites that were used repeatedly during the study, to provide sequential estimates of offtake and animal densities. For hunting camps within the study area that I did not visit myself, I employed assistants from each of the participating bands, and in one case a local villager. They documented the locations and duration of use of the camps and provided regular reports on camp movements and hunting activities in their community. Further details on these methods have been described by Hart (1979, 1978, 1986).

Surveys of snare hunters were conducted by informal interviews with men resident in village communities. Respondents provided information on the frequency, type, and approximate numbers of snares used, access to hunting camps, seasonality of snare use, animals caught, and habitats exploited (e.g., garden or forest). Snare sets were visited with several informants to corroborate interview information. All snares observed in the forest during net hunt follows were also recorded.

Net Hunts. Data gathered by accompanying net hunts were used to determine harvest rates and to estimate densities of forest antelope. For each net drive, the following data were recorded: date, time, drive location (estimated by the paced distance and direction in relation to the hunting camp), forest type, distance traversed, drive time and drive area, and the number of drivers, dogs, and nets. Drive area was estimated by pacing the perimeter of the drive set (sometimes with the help of an assistant) and estimating the area based on an assessment of net perimeter shape, the distance of unenclosed forest at the ends of the set, and the position of drivers at the start of the drive.

In the Central African Republic, Noss (in press) also used Pygmy net hunts to census duikers. He found that drives generally covered a small area and were used to surround and capture animals that had been flushed or otherwise seen beforehand. This was rarely the case in the net drives reported in this study, which covered much larger areas, rarely enclosed a specific microhabitat, and were set without specific knowledge of what animals were present in the area. Drives in which hunters encircled detected animals (usually nocturnal bay duikers, yellow-backed duikers, or the chevrotain resting in heavy cover) were excluded from the census results given here.

After each drive, all captured animals were inspected and the number and identity of animals detected but which had escaped were recorded from interviews with the hunters. Note was made of any animals having snare scars and damaged or lost limbs from snare escapes. For any dead animals, dental and reproductive data were recorded as described for the tagged and released animals above. In addition, all dead females were examined via autopsy, and their pregnancy status was confirmed and fetuses measured.

Hunt Catchments. Hunt catchments are defined as the area of forest used by hunters during the study, and for which estimates of hunting pressure and total offtake (including snares) could be calculated. Catchment areas (km²) were estimated by measuring the area of the minimum convex polygon (Harris et al. 1990) that enclosed mapped locations of all hunting camps, with their surrounding 3.5 km radius net hunting zone. The estimated area within the polygon under active cultivation or recent fallow was excluded.

Hunt catchments, as defined here, cannot be equated with the total area available to an Mbuti hunting band. Rather, catchments represent a sample of the forest area used by the hunters and provide a basis for measuring the impact of hunters on their resource base. In the Southern Study Area, hunt catchments corresponded with the hunting territories that the two study bands used during the months they were surveyed. In the Central Study Area, two catchments were defined on the basis of distance from the major settlement of Epulu (less than 10 km, and from 10 to 20 km). The areas included in these catchments included 85% of the hunting camps used by the bands that were the focus of study. The camps were located in forest adjacent to the survey zones and were also used by neighboring Mbuti who did not participate in the study. They were excluded from the hunt catchments because total harvests could not be estimated.

Net Hunt Coverage. Net hunt coverage is defined as the area included within net drives during a year and expressed as a percentage of the total catchment area. Coverage was calculated from the sample of accompanied hunts and is given as:

(area / drive) × (drives / hunt) × (hunts / day of camp occupation).

The total number of days that each camp was occupied annually was calculated from data recorded in the field diaries by myself and the assistants. Mean values of net hunt coverage were used, with upper and lower estimates obtained using standard deviations of the measured parameters. The calculated net hunt coverage assumes no overlap in the forest area of individual drives, and is thus a maximum estimate.

Hunter Harvest. Harvests of duikers and chevrotain by net hunters were measured directly as the number of animals killed per drive area. To calculate snare harvests, first the number of snare-injured animals caught in net hunts is assessed (live animals carrying a broken snare wire or having lost a hoof or distal portion of a limb). Hence, the number of specimens of each duiker species that escape snares annually in a hunt catchment (E) is calculated as follows:

$$\text{Percentage of total annual net catch that had snare scars} \times$$
$$\text{mean density } (D) \times \text{catchment area } (A).$$

Snare kills can then be calculated using estimates of the percentage of animals that escape snares, this based on data from snare hunters in the Central African Republic, who used the same methods and materials as hunters in the Ituri (Noss 1995). Noss reported snare escape rates of 7.3% for blue duikers and 41.4% for red duikers. The number of snare kills (C) can be calculated (method of Noss) as follows:

$$\text{Percentage of snared animals that escape} =$$
$$\text{number of escapes } (E) \ / \ [\text{number of escapes } (E) + \text{snare kills } (C)]$$

These estimates assume that rates of capture and escape are constant, that surviving and uninjured animals have an equal probability of capture in net drives, and that postescape mortality is not linked to capture injuries. The first assumption is supported by the observation that snares are placed during all months of the year. We have no data to evaluate the second assumption. However, all snare-scarred animals that were caught by net hunters appeared otherwise healthy, had normal body weights, and generally were not reported by hunters to have been easy to capture or handle. The final assumption is difficult to evaluate. Hunters looked for dead animals in the vicinity of any snare where there had been an escape, especially if the animal left with the snare wire and/or portion of the anchor stake attached. Badly damaged animals therefore could be expected to be found dead nearby, whereas animals that managed to escape quickly, even if losing part of a hoof or leg, were more likely to survive. Some postcapture mortality does nevertheless occur, as confirmed by the scavenging by hunters of occasional carcasses of dead animals that had escaped snares but that had apparently died shortly thereafter. The calculation thus provides a minimum estimate of total snare offtake.

Comparisons Between Unhunted and Hunted Sites

Comparisons of hunted and unhunted populations of small ungulates focus on blue duikers, red duikers, and the chevrotain. These represented over 90% of ungulates captured by hunters. Potential unhunted duiker densities in hunted areas were calculated by multiplying mean densities of unhunted populations at Edoro and Lenda by the relative proportion of mixed and mbau forest types in the hunted area. Densities are expressed as mean values ± 95% confidence intervals (CI) and differences evaluated by standard parametric statistics when possible. Nonparametric tests were used when assumptions for parametric tests were not met (Sokal and Rolf 1981).

DUIKER POPULATIONS IN UNHUNTED AREAS

Census Results

Four capture sessions at Edoro and six at Lenda were sufficiently well controlled to provide census coverage of the study areas. Net drive areas averaged 10.8 ha at Edoro and 13.4 ha at Lenda. Between 12 and 15 drives, covering 22% to 37% of the study area, were conducted in each capture session.

Six species of duikers and the chevrotain were recorded at Edoro, and the same six species of duiker were recorded at Lenda (table 6-4). The density of all ungulate species combined was over twice as high in the mixed and old secondary forest at Edoro (44.8 animals/km^2) than in the mbau forests at Lenda (21.4 animals/km^2). Figure 6-5 shows the variability in estimates within and between

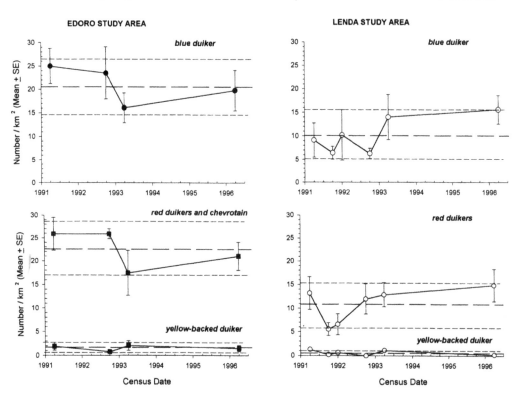

Figure 6-5. Census results of small ungulates in unhunted study areas. Mean densities (±SE) are shown for each census period, with a grand mean of all censuses (large dashes) and 95% confidence interval (short dashes) shown for the entire study period. Red duikers include four middle-sized duikers. No chevrotain specimens were censused at Lenda.

Table 6-4. Densities of Duikers and Chevrotain Censused by Net Drives in Areas of No Hunting in the Ituri Forest[a]

Study Area		Census Periods	Total Drives	Total Drive Area (ha)	Total Detected (Number per km²)								
						MONT	NIGR	LEUC	WEYN	DORS	SYLV	HYEM	Total
Edoro Study Area	Mean	4	47	508.9		20.6	2.0	5.3	11.2	2.7	1.6	1.4	44.8
	SD					3.73	1.51	0.73	2.11	1.20	0.49	0.90	6.03
Lenda Study Area	Mean	6	70	934.7		10.2	1.3	6.9	0.6	1.9	0.5	no data	21.4
	SD					3.62	0.41	2.71	0.58	1.41	0.53		6.68
ANOVA[b]	F					15.56	0.73	1.03	107.50	0.72	8.30	no test	25.35
	Probability					< 0.005	N.S.	N.S.	< 0.001	N.S.	< 0.05		< 0.001

[a]For species abbreviations, refer to table 6-1

[b]One-way ANOVA were used to test significance (Sokal and Rolf 1981). Results of one-tailed probability tests are shown, with null hypothesis being no difference between mean values at Edoro and Lenda.

census periods. Blue duiker densities at Edoro varied from 16 to 25/km^2 between sample periods, with a 5-year mean of 20.6/km^2 (95% CI 14.7–26.5/km^2). Blue duiker densities at Lenda varied from 6.3/km^2 to 15.6/km^2, with a 5-year mean of 10.2/km^2 (95% CI 5.1–15.3/km^2). Variability in estimated densities of red duikers and the chevrotain was comparable, ranging from from 17.5 to 25.9/km^2 at Edoro (5-year mean 22.6/km^2, 95% CI 17.0–28.2/km^2), and from 5.6 to 14.9/km^2 (5-year mean 10.7/km^2, 95% CI 5.9–15.5/km^2) at Lenda.

At Edoro, mean densities of both blue and red duikers were significantly lower during the first trimester (or 3 months) of 1993 than during the preceding two census periods (Kruskal-Wallace test, $p < 0.05$). This decline in observed density spanned a severe dry season, which might have affected survival, and/or recruitment. It was followed by at least partial recovery of populations, although the total number of duikers in 1996 was 17% lower than in 1991. Population densities at Lenda during the 1993 dry season did not exhibit the same decline in numbers as recorded at Edoro. Indeed, at Lenda, mean densities of blue duikers increased significantly over this same period (Kruskal-Wallace test, $p < 0.05$), whereas red duiker densities remained unchanged. Red duiker densities at Lenda were significantly lower during the 1992 dry season than either before or after (Kruskal-Wallace test, $p < 0.01$). Overall duiker densities increased by an average of 28% between the 1991 and 1996 censuses. Differences in estimated densities of yellow-backed duiker between census periods were not statistically significant ($p > 0.1$) in either study area.

Although mean densities of individual species differed significantly in mono-dominant mbau (Lenda) and mixed (Edoro) forest, the relative abundance of the three duiker size classes was comparable in both forest types. The blue duiker, the smallest species, was the most abundant, representing 46% of animals at Edoro and 48% at Lenda. Red duikers accounted for 50% of animals at both sites. However, Weyn's duiker, the most abundant of the medium-sized species at Edoro (mean 11.2 animals/km^2), was nearly absent at Lenda (mean 0.6 animals/km^2). Densities of the yellow-backed duiker were on average three times greater at Edoro (mean 1.6 animals/km^2) than at Lenda (mean 0.5 animals/km^2). Nevertheless, this species represented less than 5% of the total number of small ungulates in both forest types. The absence of the chevrotain at Lenda can be explained by the absence in the study area of larger streams; these are favored by this species as a refuge against predators. Its absence was not due to its avoidance of mbau forest, because the chevrotain was caught at other sites where large streams occurred in mbau forests.

Escape rates during the net drives varied between species but were not significantly different between mixed and mbau forest. On average, just over half of all animals detected during drives were actually captured in nets. Average escape rates were lowest for the blue duiker (35%), intermediate for the red duikers, and chevrotain (54%), and highest for the yellow-backed duiker (91%).

Mortality and Dispersal

A total of 333 animals (199 at Edoro and 134 at Lenda) were caught and tagged during eight capture sessions over the 5-year study. Of these, 120 animals were equipped with radio collars: 69 animals of seven species at Edoro, and 51 animals of four species at Lenda. Species that were collared at Edoro but not at Lenda included those that were flushed out and not caught, except for the chevrotain, which was not recorded in the Lenda Study Area. Forty animals were collared as juveniles or subadults, and many of these were followed into adulthood. Between 12 and 45 radio-collared animals (11% to 23% of total estimated populations in the study areas) were followed concurrently at any given time during the study. A total of 30,951 live animal days as recorded for the collared animals at Edoro and 26,084 live animal days for the collared duikers at Lenda.

At Edoro, between three and nine collared duikers died per year, with an average annual mortality rate of of 11.2 duikers/km^2. This is more than three times the annual death rate at Lenda, where one to five collared animals per year died. The annual mortality rate at Lenda averaged 3.4/km^2 (table 6-5). Although mortality rates for Lenda are not available for species that were not captured and collared, this is unlikely to lead to a large underestimate of total mortality within the mbau forest duiker community because the species that were not collared (Weyn's and yellow-backed duikers and the chevrotain) represented less than 15% of the total small ungulate assemblage in Lenda.

Predation, predominantly by leopards, was responsible for 20 of the 23 deaths of collared ungulates at Edoro and for nine of 12 deaths at Lenda over the 5-year study. At Edoro, the highest mortality rate was observed among the four species of red duikers and the chevrotain, with annual per capita mortality rates varying from 0.14 to 0.78 (mean 0.38). This compares with a per capita annual mortality rate of 0.11 for the blue duiker in the same forest. At Lenda, the duiker mortality rate was lower overall than at Edoro. However, the annual per capita mortality rate for red duikers at Lenda (mean for three collared species = 0.21) was higher than for the blue duiker (0.12), with the exception of the black-fronted duiker. The reduced mortality rate of black-fronted duikers might be linked to their preference for swamp habitats, which protect them from predators. Black-fronted duikers also had the lowest per capita mortality rate among red duikers at Edoro. Data for yellow-backed duikers are not available for Lenda. However, the per capita mortality rate for this species at Edoro was comparable with that for the red duikers. Trends in the mortality rates of collared animals are supported by data derived from the scat contents of leopards and golden cats in mixed forest in the Epulu area. Hart et al. (1996) found that the chevrotain, two of the middle-sized red duikers, and the yellow-backed duiker were preferred leopard prey, whereas blue duikers were killed in disproportionately low numbers relative to their abundance in the community of small ungulates.

Dispersal rates of duikers by forest type complement those of mortality. For all but the bay duiker, dispersal rates were higher in the mbau forest at Lenda than in the mixed forest at Edoro (table 6-5). Dispersal distances also tended to be greater in the mbau forest. At Lenda, 68% of dispersing animals were long-distance dispersers, and several animals moved more than once. At Edoro, only 25% of dispersals were greater than 1 km, and no collared animal shifted its home range more than once.

Table 6.5 Mortality and Dispersal in Unhunted Populations of Small Ungulates in the Ituri Forest

Species	Number Collared	Total Live Animal Days[b]	Mortality Number dead	Mortality Annual per capita rate[c]	Mortality No./ km²/ yr[d]	Dispersal > 1 km²[a] Number dispersed	Dispersal > 1 km²[a] Annual per capita rate[e]	Dispersal > 1 km²[a] No./ km²/ yr[d]
Edoro Study Area								
C. monticola	21	9984	3	0.110	2.3	0	0.000	0.0
C. nigrifrons	5	2566	1	0.142	0.3	1	0.142	0.3
C. leucogaster	8	2865	3	0.382	2.0	0	0.000	0.0
C. weynsi	23	11297	10	0.323	3.6	1	0.032	0.4
C. dorsalis	7	1868	4	0.782	2.1	1	0.196	0.5
C. sylvicultor	2	1348	1	0.271	0.4	1	0.271	0.4
H. aquaticus	3	1023	1	0.357	0.5	0	no data	no data
Total	69	30951	23		11.2	4		1.6
Lenda Study Area								
C. monticola	17	9261	3	0.117	1.1	2	0.079	0.8
C. nigrifrons	7	5068	1	0.001	0.1	6	0.432	0.6
C. leucogaster	18	8659	5	0.212	1.5	6	0.253	1.7
C. weynsi	none	no data	no data	no data	no data	no data	no data	no data
C. dorsalis	6	3096	3	0.354	0.7	1	0.118	0.3
C. sylvicultor	none	no data	no data	no data	no data	no data	no data	no data
H. aquaticus	none	no data	no data	no data	no data	no data	no data	no data
Total	48	26084	12		3.4	15		3.4

[a]First dispersal event only.
[b]Live animal days = Total number of days collared animals known to be alive 1991–1995.
[c]Annual per capita mortality rate = number animals that died / (total live animal days / 365).
[d]for animals that die or disperse, Number / km² / year = Annual rate × mean density. Density values shown in table 6-4.
[e]Annual per capita dispersal rate = number animals dispersing / (total live animal days / 365).

Age Compositions and Sex Ratios

For captured duikers, sex ratios and age at first capture are presented in table 6-6. Data for the four middle-sized red duikers and the chevrotain are grouped to facilitate comparisons with the blue duiker. There were insufficient captures of yellow backed duikers to allow analysis of population structure for either study area.

Blue duiker sex ratios were comparable at Edoro and Lenda, but the Lenda population was on average older, with adults comprising 83% of captured animals compared with only 67% at Edoro. Representation of dispersal-aged subadults was significantly lower at Lenda (7.2%) than at Edoro (26.7%) (G-test, $p < 0.05$).

In both study areas, red duiker populations included higher proportions of younger animals than did blue duiker populations. Many of these animals were juveniles, with the dispersal-aged subadults significantly less represented at Lenda (10.2%) than at Edoro (22.5%) ($p < 0.05$).

Sex ratios of blue duikers were almost identical at both Lenda and Edoro, and slightly skewed toward males. In contrast, sex ratios of red duikers were significantly different, being strongly skewed toward females at Edoro (0.82 males/ female) and toward males at Lenda (1.68 males/female) ($p < 0.001$).

Table 6-6. Age Composition and Sex Ratios of Unhunted Small Ungulate Populations in the Ituri Forest

| | Blue Duiker | | | | | Red Duikers & Chevrotain | | | | |
| | | | Percent | | | | | Percent | | |
Area	Total animals	Sex ratio (M/F)	Juvenile	Subadult	Adult	Total animals	Sex ratio (M/F)	Juvenile	Subadult	Adult
Edoro	115	1.28	6.7	26.7	66.7	82	0.82	16.9	22.5	60.6
Lenda	72	1.22	10.1	7.2	82.6	62	1.68	33.9	10.2	55.9

Taken as a whole, the differences in age composition of duikers on the two study areas are evidence for higher rates of dispersal in the mbau forests at Lenda than in the mixed forest of Edoro. This applies to both blue and red duikers. The differences in sex ratios at the two sites suggest that dispersal is skewed toward females, especially for red duikers.

Standing Biomass and Offtake

The mean total standing biomass of duikers and the chevrotain at Edoro was 435 kg/km², more than twice that at Lenda, 202 kg/km². Red duikers and the chevrotain as a group represented the major portion of the small ungulate standing biomass at both sites (70% and 72% at Edoro and Lenda, respectively). At Edoro, natural mortality resulted in an average annual offtake of 25% of all animals or 31% of the ungulate biomass (136 kg/km²/year). At Lenda, average natural mortality removed 16% of animals, or 36 kg/km²/year. This is 18% of the standing biomass annually (table 6-7). Most of the total natural mortality was of red duikers and chevrotain (84% at Edoro and 87% at Lenda), mainly due to predation by large cats. Animals dispersing more than 1 km accounted for only 4% of total small ungulate standing biomass at Edoro, and 19% of standing biomass at Lenda. At both sites, red duikers accounted for most of the long-distance dispersal.

DUIKER POPULATIONS AND HUNTING IN THE ITURI FOREST

Catchment Areas and Net Hunt Coverage

Estimates of catchment areas in the Southern Study Area are based on a total of 29 hunting camp locations recorded over a 20-month period. Estimates of catchment areas in the Central Study Area are based on 31 hunting camp locations over a 24-month period (table 6-8).

Estimates of net hunt coverage in the southern study area are based on data from (a) a total of 193 hunts from seven hunting camps in the S/Settled catchment, including two sites that had had at least three separate occupations (defined as

Table 6-7. Mortality and Dispersal of Small Ungulates as a Percentage of Mean Biomass Standing Crop in Unhunted Study Areas in the Ituri Forest

| | | Edoro | | | | | Lenda | | | | |
| | | | Mortality | | Dispersal (> 1 km) | | | Mortality | | Dispersal (> 1 km) | |
Taxon	Mean weight (kg)[a]	Standing crop (kg/km²/yr)	(kg/km²/yr)	Percent	(kg/km²/yr)	Percent	Standing crop (kg/km²/yr)	(kg/km²/yr)	Percent	(kg/km²/yr)	Percent
Blue duiker	4.0	82.4	9.2	11.2	0.0	0.0	41.2	4.8	11.7	3.2	7.9
Red duikers & chevrotain	13.5	305.1	114.8	37.6	17.6	5.8	145.8	31.1	21.3	35.0	24.0
Yellow-backed duiker	30	48.0	12.0	25.0	no data	no data	15.0	no data	no data	no data	no data
Total		435.5	136.0	31.2	17.6	4.0	202.0	35.9	17.8	38.2	18.9

[a]Mean body weights are calculated based on age composition of populations in the study areas.

Table 6-8. Net Hunt Survey Data, Catchment Area and Net Hunt Coverage in the Ituri Forest Hunting Studies

Study Area	Zone	Catchment			Hunt Surveys					Net Hunt Coverage (Mean ± SD)		Annual Coverage (Mean ± SD)	
		Survey Dates	Total camps used	Area (km²)	Camps surveyed	Days monitored	Days hunting occurred	No of drives	Drive area (ha)	Drives/ hunt	Hunts/ month	km²/ yr	Percent catchment
Southern	Settled	7/73–2/75	16	188	7	330	193	656	7.8[a] (5.5–10.0)	3.6[a] (2.4–4.8)	17.6[a] (14.4–20.8)	59.3 (57.2–64.3)	31.5 (30.4–34.2)
	Remote	4/74–12/74	13	475	6	90	77	300	10[b] (9.0–12.0)	3.9[b] (3.4–5.0)	25.6[b] (24.3–26.9)	119.8 (112.8–136.0)	25.2 (23.7–28.6)
Central	Settled	6/81–2/83	16	207	11	80	51	284	6.1[a] (2.7–9.5)	5.5[c] (4.5–6.6)	17.5[a] (12.0–22.0)	70.5 (68.7–84.3)	34.1 (33.2–40.7)
	Remote	7/81–5/83	15	364	5	26	23	115	8.2[a] (4.8–11.5)	5.1[c] (4.4–5.7)	14.7[c] (10.0–18.3)	73.8 (71.3–85.8)	20.2 (19.6–23.6)

[a]Values with the same letters designate means which are not statistically different ($p < .05$) using Tukey-Kramer unplanned comparisons of means (Sokal & Rolf 1981).

multiple use) during the study; and (b) 77 hunts from six sites in the S/Remote catchment. None of the C/Remote hunting camps had multiple use during the study period. In the central study area, net hunt data were gathered on (a) 51 hunts at 11 camps in the C/Settled catchment, including two with multiple occupations; and (b) on 23 hunts at five locations in the C/Remote catchment, including one camp location with multiple occupations during the period.

Mean drive areas varied from 6.1 ha to 10 ha in the different survey zones, with largest mean drive areas in the S/Remote zone. Net hunt frequency (hunts per month) was also highest in this zone, leading to a mean annual net coverage of almost 120 km². Highest percentages of catchment area included in annual net hunt coverage were recorded for the S/Settled (32%) and C/Settled (34%) zones.

Census Results

The same six species of duikers and the chevrotain caught in the unhunted study areas also were caught in all four hunting zones surveyed. Estimated mean small ungulate densities ranged from 9.2/km² to 27.2/km² (60 to 210 kg/km²) across the four survey zones. Densities in the S/Settled zone were significantly lower than in the other three areas, and densities in the S/Remote zone were significantly lower than in the Central Study Area.

The blue duiker was the most abundant species in all areas, with densities in the S/Settled zone being significantly lower than in other areas. Densities of red duikers and the chevrotain were also lower in both southern survey zones than in the Central Study Area. Yellow-backed duiker densities were low in all survey areas, although with significantly lower densities in the S/Settled area than the Central Study Area (table 6-9).

Changes in duiker densities over time at five camp locations with multiple occupations are shown in figure 6-6. Also included are potential unhunted population densities estimated for each site, based on percentage forest composition of the area. None of the duiker populations surveyed declined consistently over the sample periods. However, densities of hunted animals were significantly lower (t test, $p < 0.05$) than hypothetical unhunted densities for at least one period at all sites. Blue duiker densities were significantly lower than unhunted densities during all survey periods at the S/Settled sites of Telebongi and Masange (figure 6-6A).

The response of red duikers and the chevrotain to hunting was even more dramatic (figure 6-6B). Densities in hunted areas were lower than hypothetical unhunted densities at all times and sites in both S/Settled and C/Settled zones. The differences were significant in all cases except one hunting period at Kare camp in the C/Remote zone. In the S/Settled zone, average densities were only one tenth of those that would be expected if the area had remained unhunted.

Hunter Harvest

Net Hunts. The mean total annual harvests from net hunts ranged from 465 to 1446 animals per year across the four catchment areas (table 6-10). Harvest

Table 6-9. Densities and Biomass of Hunted Small Ungulate Populations in the Ituri Forest

Area	Catchment		Density (No./km²)[a]				Mean Biomass (kg/km²)			
			Blue duiker	Red duikers & chevrotain	Yellow-backed duiker	Total	Blue duiker	Red duikers & chevrotain	Yellow-backed duiker	Total
Southern	Settled	mean	6.9[a]	2.2[a]	0.1[a]	9.2[a]	27.6	29.7	3.0	60.3
		95% C.I.	(3.2–9.6)	(1.3–2.1)	(0.0–0.3)	(7.5–10.9)				
	Remote	mean	11.7[b]	3.9[a]	0.3[b]	15.9[b]	46.8	52.7	9.0	108.5
		95% C.I.	(7.5–15.9)	(0.8–7.0)	(0.0–0.8)	(11.3–20.5)				
Central	Settled	mean	14.8[b]	8.2[b]	0.7[b]	23.7[c]	59.2	110.7	21	190.9
		95% C.I.	(10.0–19.6)	(6.7–9.7)	(0.2–1.2)	(18.8–28.6)				
	Remote	mean	17.8[b]	8.7[b]	0.7[b]	27.2[c]	71.2	117.5	21	209.7
		95% C.I.	(11.5–24.1)	(5.8–11.6)	(0.3–1.1)	(19.9–34.5)				

[a]Mean values shown with 95% confidence intervals in parentheses. Values with the same letters designate means which are not statistically different ($p < .05$) using Tukey-Kramer unplanned comparisons of means (Sokal & Rolf 1981).

Table 6-10. Harvests of Small Ungulates by Mbuti Net Hunters in the Ituri Forest[a]

Area	Hunting Zone	Harvest per Unit Net Hunt Coverage (number/km²/yr)				Total Harvest in Catchment			
		Blue duiker	Red duikers and chevrotain	Yellow-backed duiker	Total animals	Blue duiker	Red duikers and chevrotain	Yellow-backed duiker	Total animals
Southern	Settled	6.0[a] (5.6–6.1)	1.8 (1.7–1.8)	0.03 (0.03–0.04)	7.8 (7.3–8.0)	356 (320–392)	107 (97–116)	2 (1–3)	465 (418–511)
	Remote	8.8 (8.0–9.0)	2.8 (2.6–2.9)	0.03 (0.03–0.04)	11.6 (10.6–11.9)	1054 (904–1224)	335 (293–394)	4 (3–5)	1393 (1200–1623)
Central	Settled	9.9 (7.0–10.4)	4.3 (3.0–4.5)	0.2 (0.1–0.3)	14.4 (10.1–15.2)	698 (481–877)	303 (206–379)	14 (7–25)	1015 (694–1281)
	Remote	13.2 (9.3–14.1)	6.3 (4.5–6.8)	0.1 (0.06–0.2)	19.6 (13.8–21.1)	974 (663–1210)	465 (321–583)	7 (4–17)	1446 (988–1802)

Values shown are means and standard deviations based on estimates of coverage provided in table 6-8.

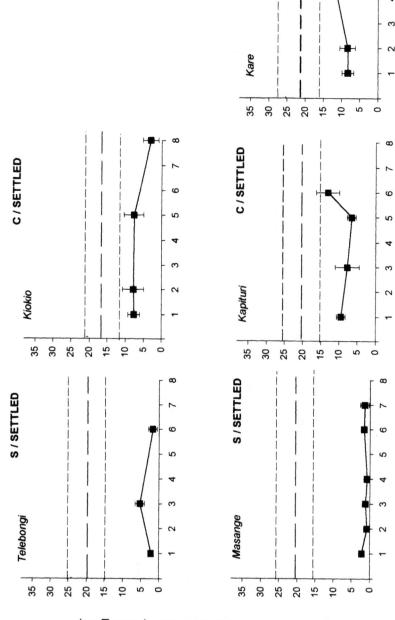

(A)

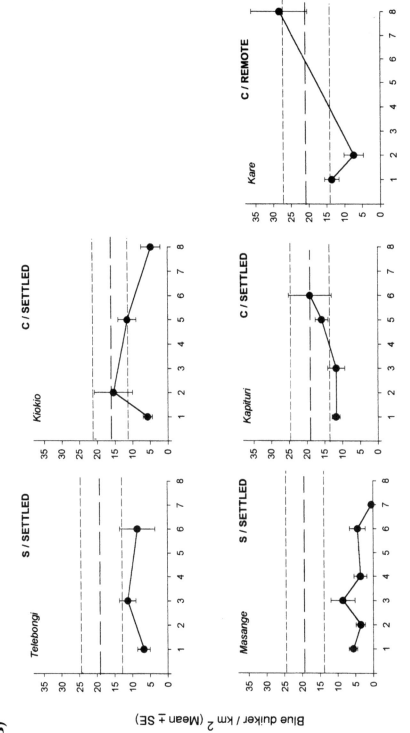

Figure 6-6. Sequential censuses of **(A)** blue duikers and **(B)** red duikers and chevrotain in hunt catchments (capital letters) at hunting camp locations (italics) which had at least three sequential periods of use by hunters during the study period. Relative dates of hunting activity and censuses are given on a trimestral (3-month) basis, with the first trimester being the period when the camp was first used during the study period. Mean (±SE) estimates of density are shown for each census period. Potential mean unhunted densities (long dashes) and 95% confidence intervals (short dashes) are shown for each location, extrapolated from density estimates at Edoro and Lenda multiplied by the proportional representation of mixed and monodominant forests in the hunting area around each camp.

rates of blue duikers were lowest in the S/Settled zone, intermediate in the S/Remote zone and C/Settled zone, and highest in the C/Remote zone. Harvest rates of red duikers and the chevrotain were highly variable, with very low yields in both catchment areas in the Southern Study Area, and substantially higher harvest rates in the Central Study Area, particularly in the remote zone. Harvest of yellow-backed duikers was very low overall, although higher in the central study area than in the south.

Snares. In the S/Settled zone, a total of 25 duikers caught during net drives during the study had scars resulting from escaping snares. These included four blue duikers, representing 0.9% of the annual blue duiker catch, and 21 red duikers, representing 13.8% of the annual catch of these four species. Five duikers (one blue duiker and four bay duikers) caught on net drives in the S/Remote zone also had snare scars. In the C/Settled zone, two bay duikers with net scars were caught. No snare-scarred duikers were caught in the C/Remote zone.

Calculations of total snare harvest (table 6-11) use the snare escape rates of 7.3% for blue duikers and 41.4% for red duikers that Noss (1995) observed in the Central African Republic. Calculated harvests are supported by the results of hunter surveys. The highest estimated annual snare harvest, 152 blue duikers and 81 red duikers, was recorded in the S/Settled catchment. This was also the area with the highest number of snare hunters. Approximately 65 village men from 11 settlements in or adjacent to the S/Settled catchment reported that they regularly used snares during the study period. In villages in or adjacent to the S/Remote catchment, only 14 people reported setting snares. The estimated annual snare harvest totaled 35 blue duikers and 18 red duikers for this area. In the Central Study Area, an estimated 15% of village men in the village of Epulu and surrounding settlements (25 to 30 animals) were known or suspected to use snares, although these were illegal. All who agreed to be interviewed on the subject reported using snares around gardens or within a few kilometers of the settlement, and not in the more distant forest. This was corroborated by the low record of snare encounters in survey net drives in the C/Settled catchment and by their absence in the C/Remote catchment. A minimum of 13 red duikers as harvested annually in the C/Settled catchment. If the proportional snare harvest of blue and red duikers was comparable in the southern and central study areas, then at least 25 blue duikers were also harvested annually by snares from the C/Settled catchment. Estimates of snare harvest for the C/Remote catchment cannot be made because snare-scarred animals were never encountered there. Snare offtake was likely to have been small in this area.

Total Harvest. Estimated total annual harvests of small ungulates from survey catchments varied from 3.0 to 5.2 per km^2 (19.3 to 37.3 kg/km^2). The highest total harvest was in the C/Settled catchment, but the total harvest was also high in the S/Settled and C/Remote catchments. Snare hunting contributed almost one third of total offtake in the S/Settled catchment, where snare hunting was most intensive. In the C/Settled catchment, snare hunting contributed only 4% of the total harvest (table 6-12).

Table 6-11. Estimates of Harvest of Small Ungulates by Snares in the Ituri Forest

			Blue Duiker						Red Duikers						
				Animals with snare scars caught on net hunts		Annual catchment totals			Animals with snare scars caught on net hunts		Annual catchment totals	
Area	Catchment	km² (A)	Density (D)	Total Observed	Percent catch (N_E)	Snare escapes (E)	Snare harvest (C)	Density (D)	Total Observed	Percent catch (N_E)	Snare escapes (E)	Snare harvest (C)
Southern	Settled	188	6.9	4	0.9	12	152	2.2	21	13.8	57	81
	Remote	475	11.7	1	0.05	3	35	3.9	4	0.7	13	18
Central	Settled	207	14.8	0	no data	no data	no data	8.2	2	0.2	9	13
	Remote	364	17.8	0	no data	no data	no data	8.7	0	no data	no data	no data

Table 6-12. Average Total Small Ungulate Harvests by Hunters in the Ituri Forest

Hunt	Catchment	*Numbers/km²/year*				*Biomass (kg/km²/year)*			
		Blue duiker	*Red duikers & chevrotain*	*Yellow-backed duiker*	*Total*	*Blue duiker*	*Red duikers & chevrotain*	*Yellow-backed duiker*	*Total*
Net hunts	S/Settled	1.9	0.6	0.01	2.5	7.6	8.1	0.3	16.0
	S/Remote	2.2	0.7	0.01	2.9	8.8	9.5	0.3	18.6
	C/Settled	3.4	1.5	0.07	5.0	13.6	20.3	2.1	36.0
	C/Remote	2.7	1.3	0.02	4.0	10.8	17.6	0.06	28.5
Snare	S/Settled	0.8	0.4	no data	1.2	3.2	5.8	no data	9.0
	S/Remote	0.07	0.03	no data	0.1	0.3	0.4	no data	0.7
	C/Settled	0.1	0.06	no data	0.2	0.5	0.8	no data	1.3
	C/Remote	0.0	0.0	0.0	0.0	0.0	0.0	0.0	0.0
Total	S/Settled	2.7	1.0	0.01	3.7	10.8	13.9	0.3	25.0
	S/Remote	2.3	0.7	0.01	3.0	9.1	9.9	0.3	19.3
	C/Settled	3.5	1.6	0.07	5.2	14.1	21.1	2.1	37.3
	C/Remote	2.7	1.3	0.02	4.0	10.8	17.6	0.06	28.5

Age Structure of Hunted Duikers

The age composition and sex ratios of the duiker harvests are presented in table 6-13. Sex ratios of blue duikers were close to unity in the samples of harvested animals from all study areas. Age distributions of the harvest were also similar between sites, with adults representing between 72% and 76% of all animals caught.

Sex ratios of red duikers were skewed toward females in all harvested samples except for that from the C/Remote zone, where males dominated the catch. The proportion of adults in the harvested sample was lower for red duikers than for blue duikers, and also more variable. Only 39% and 42% of red duikers caught were adults in the S/Settled zone and C/Settled zone, respectively. A higher proportion of adults was recorded in the remote catchments in both study areas.

DYNAMICS OF UNHUNTED DUIKER POPULATIONS

Duiker populations at the two unhunted sites had very different dynamics from each other. Numbers fluctuated over time and were not synchronized, and differences in rates of dispersal and predation were marked, even for the same species. At Edoro, the total density of small ungulates was on average twice that of Lenda. Predation rates were higher and dispersal rates lower. These differences might be attributed in part to differences in food availability and productivity in the two forest types. Fruit availability was higher and seasonally more equitable in the mixed forests at Edoro than in the mbau forests at Lenda. Higher duiker densities would be expected in the more productive mixed forests than in the mbau forests, thus favoring higher predator activity and rates of predation. Higher levels of

Table 6-13. Age Composition and Sex Ratios of Harvested Small Ungulate Populations

Area	Zone	Blue Duiker					Red Duikers & Chevrotain				
		Total animals	Sex ratio (M/F)	Percent			Total animals	Sex ratio (M/F)	Percent		
				Juvenile	Subadult	Adult			Juvenile	Subadult	Adult
Southern	Settled	311	1.00	8.6	19.2	72.2	91	0.70	37.5	23.9	38.6
	Remote	317	1.02	10.5	15.2	74.3	85	0.79	28.6	14.3	57.1
Central	Settled	125	1.08	8.0	19.2	72.8	52	0.62	42.3	15.4	42.3
	Remote	142	0.89	4.2	19.7	76.1	66	1.20	22.7	27.3	50.0

predation might increase the probability that dispersing animals would not travel far to locate available home ranges. Differences in ranging behavior of individual species and forest understory are also likely to affect predation and dispersal. The open understory of the mbau forest might make stalking by predators more difficult, thus reducing predation rates in this forest type; human hunters had problems operating in this forest type for this reason.

Not all duiker species exhibited the same shifts in density between forest types, suggesting that factors other than food availability alone are structuring the community. Nevertheless, the differences in dynamics between blue duikers and red duikers were consistent. Per capita rates of predation and long distance dispersal were lower for blue duikers than for red duikers in both forest types. Red duikers exhibited greater differences in susceptibility to predation and likelihood to disperse between the two forest types than did blue duikers. The reasons for this are not clear, but it would suggest that red duikers as a group are more susceptible to variations in hunting pressure and habitat fragmentation than are blue duikers.

Immigration into the study areas could not be observed directly, but high rates of long-distance dispersal from Lenda, especially for red duikers, suggest that mbau forests might be population sources for adjacent areas, at least in some seasons. Differences in sex ratios of red duikers between Lenda and Edoro suggest that long-distance movements in the mbau forests are skewed toward females. Whether these represent dispersing animals seeking more productive breeding territories remains to be determined. The low rates of long-distance dispersal at Edoro imply that this was not a population source for neighboring areas, at least during the study period. The decline in populations of both blue and red duikers at Edoro between 1992 and 1993, and their slow recovery toward 1991 levels 3 years later (more complete for blue duikers than red duikers), would suggest that Edoro might have been a population sink.

If the dynamics of duiker populations at Lenda and Edoro are at all typical, then the prey base of net hunters is unlikely to be in equilibrium in any one location, and prey abundance at any one site might be unpredictable. This is probably the basis for the mobility of hunters. Decreasing forest area, and use of hunting camps on ever shorter cycles, are likely to destabilize the Mbuti net hunting system. If duiker populations are also maintained by source-sink dynamics, then connectivity in the landscape that permits dispersal and immigration would also be important in maintaining stability of hunter offtake. The effects of forest fragmentation due to clearing thus might not only affect duiker populations by reducing available forest areas, but could lead to even greater reductions in populations in the remaining forest patches if dispersal into and out of forest fragments were also reduced (McCullough 1996a). Compared with red duikers, blue duikers appear to maintain populations with lower levels of long-distance dispersal. Thus, population densities of blue duikers at any given location are less likely than those of red duikers to be affected by hunting pressure or habitat loss in neighboring areas.

COMPARISON OF HUNTED AND UNHUNTED POPULATIONS

Hunting Intensity

Hunting intensity using nets can be measured as a function of the area of forest actively searched by hunters over the course of a year and the rate at which animals are harvested (figure 6-7). The former is estimated as the annual net hunt coverage quantified as a percentage of hunt catchment, and the latter is estimated by the proportion of the catchment standing crop removed. Among the four catchments studied, hunting offtake was lowest in the C/Remote catchment. Annual net coverage accounted for about 20% of the catchment area, whereas hunter offtake averaged about 15% of the standing biomass of duikers. Snare hunters did not operate in this zone, and hunters averaged just 14 days of hunting per month during nomadic visits. Hunting intensity was also low in the S/Remote catchment, with net drives covering about one quarter of the catchment annually and about 18% of the standing biomass of duikers removed by net hunters, with an additional harvest by snare hunters of about 1% of the biomass. The Mbuti band operating in

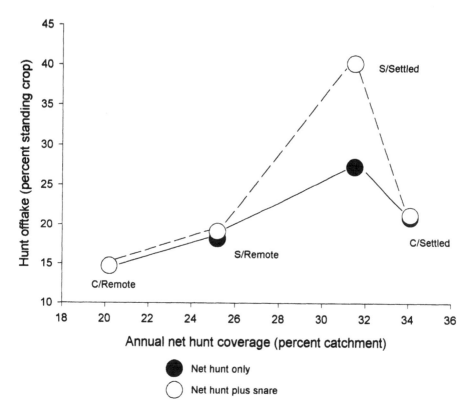

Figure 6-7. Net hunt coverage (measured as a percentage of hunt catchment area) and small ungulate offtake (measured as a percentage of standing crop) harvested by net drives (*black circles*) and by net drives plus snares (*stippled circles*) in four hunt catchments in the Ituri Forest.

the S/Remote catchment hunted more frequently (mean = 25.6 days/ month) than did net hunters in the C/Remote zone, but the intensity of hunting in the area was reduced by their use of a very large catchment (475 km²). Snare offtake in this area comprised less than 1% of standing biomass. Hunting was most intense in the settlement zones in both the Southern and Central Study Areas. Over 30% of the estimated catchment areas were covered during net drives in these areas. Total offtake was very high in the S/Settled catchment, with an estimated 27% of standing biomass harvested annually by net hunters and an additional 13% taken by snares. Twenty-two percent of standing biomass was harvested by nets in the C/Settled catchment, and 1% was taken by snares.

Population Density

With the exception of blue duiker in C/Remote catchment, mean densities of hunted duikers were significantly lower (t test, $p < 0.05$) than populations expected if unhunted (figure 6-8). Lowest densities were recorded in the most intensively hunted settlement catchments in both the southern and central study areas. Red duikers (figure 6-9) were affected more severely by hunting than were blue duikers, with population densities in the S/Settled catchment only 10% of unhunted levels, and C/Settled populations 22% of unhunted expectations. White-bellied duikers, in particular, became rare in heavily hunted areas, representing just 3% of the total number of duikers censused in the S/Settled catchment, and 7.5% and 8% of totals in the C/Settled and S/Remote catchments, respectively. By comparison, white-bellied duikers comprised 12% of the duiker community at Edoro, and 32% at Lenda. Among red duikers, the bay duiker appeared better able to withstand hunting.

In contrast, blue duiker populations were only reduced to 34% of unhunted levels in the intensively exploited S/Settled catchment. In the C/Remote catchment, densities of hunted and expected unhunted blue duikers were not significantly different. Hence, as hunting pressure intensified, blue duikers represented an increasingly greater percentage of the total ungulate community. Blue duikers comprised about 50% of total duiker and chevrotain community in the two unhunted study areas, 66% of total animals in the little exploited C/Remote catchment and 79% of animals in the heavily exploited S/Settled catchment.

Data on the yellow-backed duiker are few, but it appears that changes in densities of hunted populations were comparable with those of red duikers. Hunted populations in the central forest, where snaring pressure was low, averaged about 50% those of the unhunted areas. In the more intensively hunted S/Settled catchment, where snaring also occurred, observed population densities were only 6% of the potential unhunted levels.

Although differences in duiker densities and composition of hunted duiker communities were correlated with differences in hunting pressure, the extremely low densities recorded in the S/Settled zone might have been affected by additional factors, particularly habitat fragmentation. An estimated 22% of the hunting territory of the S/Settled Mbuti band was classified as garden or recent

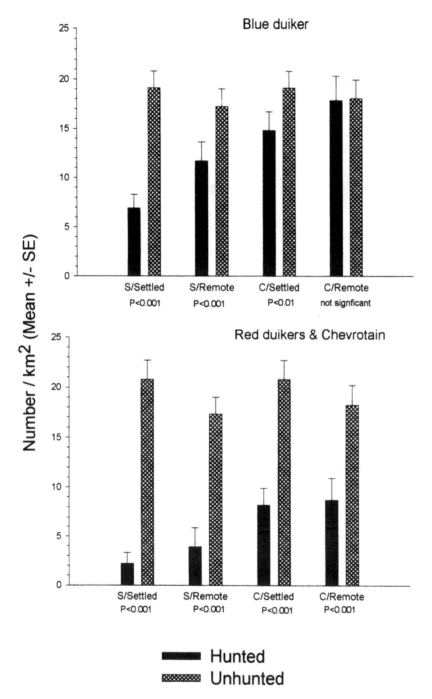

Figure 6-8. Mean densities (± SE) of small ungulates in hunt catchment areas in the Ituri Forest. Unhunted population estimates extrapolated from density estimates at Edoro and Lenda multiplied by the proportional representation of mixed and mbau forests in the hunt catchment. Student's *t* test probabilities are shown. Null hypothesis is no difference between hunted and unhunted densities.

Figure 6-9. Mbuti hunter with Weyn's duiker (*Cephalophus weynsi*) captured on a net drive in the Ituri forest. *Photo courtesy of John Hart.*

fallow, distributed around 16 villages scattered through the area. Species showing the largest decline in numbers were yellow-backed and white-bellied duikers. These had the highest dispersal rates and might have been adversely affected by the mosaic of garden and farmed bush.

Hunter Versus Predator Offtake

In the unhunted study areas, leopards and (to a lesser extent) golden cats accounted for 87% and 75% of small ungulate mortality at Edoro and Lenda, respectively. Leopards are widely present through the Ituri Forest, and although estimates of leopard densities are not available, leopard signs were observed in all four catchments during the studies. Thus, it is likely that both humans and the cats harvested duikers from the areas where the hunters operated.

Annual hunter offtake (both nets and snares) of red duikers averaged from about 12% to 20% of standing biomass in the moderately hunted study areas, but over 45% in the S/Settled catchment. Leopard predation on red duikers in the un-hunted areas was also highly variable, representing about 18% of standing biomass in the mbau forest and about 33% in the mixed forest at Edoro.

Annual hunter offtake of blue duikers ranged from 14% to 22% of biomass in the moderately to lightly hunted catchments and 39% of biomass in the inten-sively hunted S/Settled catchment. This compares with an approximately 9% offtake of standing biomass removed by leopards and golden cats in the unhunted study areas (figure 6-10).

It is unlikely that predation rates by natural predators remain unaffected by human hunters exploiting the same prey base. In India, prey shifts by tigers have been observed where populations of large prey have been depleted by human hunters, and tigers increase their relative intake of smaller prey (Karanth and Sunquist 1995). In the Ituri Forest, the leopard has a very eclectic diet. However, large and medium duikers are favored prey (Hart et al. 1996). Hence, competition between large cats and humans for duikers is probable and will affect the wider sustainability of hunting.

Age Composition

Differences in age composition of exploited and unhunted populations of red duikers were commensurate with high per capita probabilities of mortality associ-ated with elevated levels of harvest. Heavily exploited populations such as those in the S/Settled, S/Remote, and C/Settled catchments had more young animals and fewer adults than did those in the less exploited C/Remote catchment, and espe-cially in the unhunted study areas (figure 6-11). For potentially long-lived species such as red duikers, reductions in numbers of reproductively active animals could render the population vulnerable to a major population decline.

Blue duiker populations did not exhibit the same age shifts in response to harvest as did the red duikers. There were no consistent differences in the propor-tion of prereproductive age classes, which ranged from 24% to 28% of the popu-lation under a wide range of harvest levels. The proportion in the unhunted Edoro population (33% prereproductive) was similar. Only the unhunted Lenda popula-tion deviated from this trend, with a high proportion of adults. This appeared to be linked to high rates of subadult emigration from the mbau forest.

Blue duikers were more capable than red duikers of maintaining population numbers under exploitation. This might be linked to more rapid growth rates and, in particular, to enhanced rates of maturation of subadult animals. Females in harvested populations of duikers responded to increased levels of harvest and decreased population density by reaching sexual maturation at a younger age. Less than 5% of two molar females were sexually mature and classed as adults in the unhunted Edoro and Lenda samples, whereas over a third of this age class had achieved reproductive status in the intensively hunted populations in the S/Remote catchment (figure 6-11). The more heavily exploited S/Settled catch-

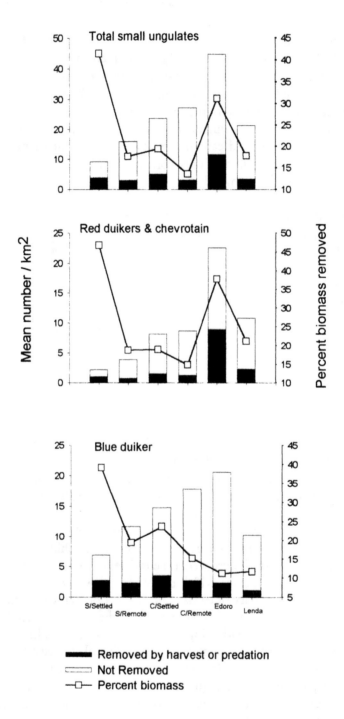

Figure 6-10. Annual hunter harvests (snares plus nets) in four hunt catchments and predator offtake in unhunted study areas (Edoro and Lenda). Harvests and offtake are represented by stacked histograms as total animals (mean number/km²), and offtakes are represented by a line and point plot as percentage of total biomass. Removals in hunt catchments are for human hunters only and do not include possible removals by natural predators from these areas.

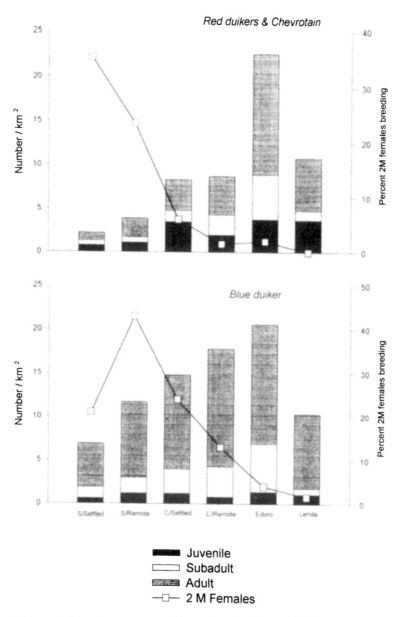

Figure 6-11. Age structure of hunted and unhunted duiker populations (stacked histograms) and proportion of young, two-molar females in the sample that had achieved sexual maturity (line and point plot).

ment did not show as dramatic an increase in the age of first reproduction, perhaps indicating constraints on maturation rates at very low population levels.

Red duikers also exhibited accelerated rates of early maturation in females under low population densities, but accelerated maturation does not appear to be as well developed in these species as in the blue duiker. Thus, red duikers would be less able than blue duikers to compensate for declining densities with increased reproduction as populations become depleted.

IS HUNTING SUSTAINABLE?

Sustainability of hunting can be evaluated directly by observing trends in exploited populations over time and by examining offtake rates.

Population Trends

Population densities of duikers at sites hunted repeatedly over the study period did not show a consistent decline at any of the five sites surveyed (see figure 6-6). Hunting had not led to the total extirpation of duikers in any of the survey zones, nonetheless, some populations were severely depleted. Densities of red duikers at two sites in the C/Settled zone were less than 50% of potential unhunted levels, whereas those in the S/Settled zone were even more depleted. Densities at Telebongi were on average just 15% those of extrapolated unhunted levels. At Masange, in the same zone, densities of red duikers averaged less than one-tenth of potential unhunted levels over a 21-month period, during which time the area was hunted during six different periods. Although biologically extant, these heavily depleted populations of duikers might be ecologically extinct in heavily hunted areas, compromising the prey base of large cats and possibly also seed dispersal.

Blue duiker populations were significantly more resistant than red duikers to the same levels of hunting pressure. In the C/Settled zone, blue duiker densities averaged over 60% and 73% expected unhunted levels at two sites with repeated use by hunters. In the S/Settled zone, densities averaged nearly 30% unhunted levels at Telebongi, whereas at Masange, density averaged about 25% expected unhunted levels.

At two camps, Kapituri in the C/Settled zone and Kare in the C/Remote zone, population densities of both red and blue duikers were higher during the final period that hunters were present than at any other time during the study. The response of populations at Kare in particular show that if surrounding populations are sufficiently high, and periods between hunting sufficiently long, hunted populations can recover to unhunted levels. The Kare camp was in an area of only lightly hunted forest and in a zone used only by net hunters and not snare hunters.

Offtake Rates

Little information is available on the sustainability of offtake rates for tropical forest ungulates. For closely managed populations of ungulates in temperate areas, usually in the absence of natural predators, offtakes of 15% to 30% of the standing population are reliably sustainable (e.g., Eltringham 1984; Beasom and Roberson 1985). In the Ituri, if leopards and golden cats are harvesting 18% and 9% of the standing biomass of red duikers and blue duikers, respectively, then hunted harvests of 45% (red duikers) and 39% (blue duikers) of the standing biomass in heavily hunted areas clearly are of concern. Even in moderately hunted areas, where red duiker offtake is 12% to 20% of standing biomass and blue duiker offtake is 14% to 22%, the sustainability is questionable.

SOCIOECONOMIC DYNAMICS OF HUNTING

Variation in hunting practices between the four study zones reveals that exploitation of wildlife is very dynamic, with marked differences in the economic context of hunting even between neighboring communities of the same ethnic group. This section discusses several factors that affect hunting in the southern and central Ituri and their implications for the sustainability of harvests in this region.

Commercialization of Hunting

Mbuti bands in the S/Remote zone had the largest involvement with the commercial meat trade during the study. Compared with the other Mbuti communities, this band devoted the most effort to net hunting. Their mean drive areas, hunt frequency, and net hunt coverage all were greater than for other bands that were studied (Hart 1978). These Mbuti, along with those of the S/Settled zone, also used dogs to increase the efficiency of their drives. Dogs were used less frequently in the Epulu area.

Access to a large hunting territory was important in permitting the S/Remote band to attain the high harvests required by the commercial meat trade. The ability to hunt in remote areas was made possible by commercial traders who paid porters to transport food into the deep forest, thus allowing hunters to operate far from agricultural settlements.

Although expansion into new forest areas might permit hunters the high yields demanded by the market trade, it is clearly not a sustainable long-term strategy because the area of forest accessible to Mbuti communities is ultimately limited. Indeed, conflicts between bands over claims of trespass and rights to hunt in some areas of the forest had already surfaced in southern study area during the study.

The Mbuti hunters harvested at high rates so they could purchase the foods they needed from the traders (Hart 1978). This might favor sustainability if, as small ungulate populations decline, the Mbuti were to reduce hunting in favor of other activities. However, the prices of meat and trade foods were actually elastic. The amount of food that the meat could purchase remained stable as meat supply decreased. Thus, there was an incentive to continue hunting even though stocks were reduced. Commercial hunting will thus deplete duiker populations rapidly over large areas of forest.

Nets Versus Snares

Yields from net hunting are ultimately affected by the constraints of available time and labor. As ungulate stocks decline, these constraints can be overcome by using more efficient hunting technologies (e.g., snares in the S/Settled zone). Noss

(1995) found that snares produced a more efficient harvest of wildlife (return per unit effort) than nets. For net hunts, animals not present within any given drive area will not be caught by that drive. Snares represent a potential continuous source of mortality to animals over weeks or even months, even though the time inputs of hunters are small. In time, an animal in whose home range a snare occurs runs an increasing probability of encountering it. Thus, snares are an efficient method of hunting even when wildlife densities are low and can lead to even greater depletions of populations than can nets.

The indiscriminate use of snares also represents a danger to species that are rare and vulnerable, but whose harvest rates do not merit the use of the technology. The white-bellied duiker is a species that becomes increasingly rare under snaring. White-bellied duikers could be seriously overhunted if snares continue to produce acceptable yields of more common game species.

Snares caught a higher percentage of the total harvest in the S/Settled zone than in any other survey zone: an estimated 34% of the total harvest. Mbuti of the S/Settled zone also hunted using spears and dogs to catch brush-tailed porcupines, *Atherurus africanus*, mongooses, and Gambian rats (*Cricetomys emini*). Nearly one third of recorded Mbuti hunts in the S/Settled catchment were for these small animals, which constituted 35% of the total catch of this band. Almost none was sold to commercial buyers, and a large portion was used by the Mbuti for their own subsistence.

The declining potential of net hunting might have increased the dependence on small animals. Other factors also might have favored this change. The mosaic of secondary vegetation surrounding many of the settlements of the S/Settled zone might have produced an abundance of species such as rodents and mongooses, which are better captured by dogs and spears than by net drives. In addition, as small ungulates decreased in availability, the Mbuti spent increasing time near villages giving them a greater range of subsistence options, including planting of gardens.

Immigration and Sedentary Life-styles

The S/Settled zone was in an active settlement frontier, and during the 26-month study, immigrant farmers arrived and two new communities were established. Densities of some wildlife species increased in gardens and recent secondary vegetation (FitzGibbon et al. 1996). However, this is not the case for duikers and the chevrotain, whose diets are dominated by fruits and seeds of mature and old secondary forest trees and lianas. In addition, creation of a patchwork of clearings and forest islands creates barriers to dispersal, and limits species such as the white-bellied duiker, which requires large contiguous tracts of forest.

Human population growth and forest conversion in the S/Settled zone has thus reduced suitable habitats for duikers. Hunters without access to new areas of forest will either be forced to intensify hunting in the remaining areas, turn to more efficient means of hunting, or ultimately shift their economic orientation away from hunting. At that point, populations of some species would probably be very low and possibly even locally extinct.

MONITORING AND MANAGEMENT
OF TRADITIONAL HUNTING

Results of the studies reported here can be used to provide guidelines for the management of small ungulate hunting in the Ituri Forest, as well as a perspective on sustainable hunting in tropical forests worldwide.

The major conclusion of this study is that for forest-dependent species such as the duikers, protection of large forest areas from settlement and fragmentation is the most important prerequisite before any hunting can be considered potentially sustainable. The sustainability of hunting for duikers in the Ituri Forest has several significant constraints. Most importantly, hunting is unlikely to be sustainable unless hunters can operate in large catchments, with low intensities of hunting and long periods before hunted areas are reused. Hunting methods should have clear limits in time inputs so that hunting pressure is sensitive to declining yields. In the Ituri Forest, net hunting can meet these conditions, and communal net drives are a potentially sustainable hunting method.

However, several further caveats must be stated. First the advent of commercial wild meat trade in remote areas that would otherwise remain only lightly exploited, if at all, is likely to lead to major declines in hunted species. Second, management of hunting must protect the most vulnerable species. In the Ituri Forest, red duikers are more vulnerable to overharvesting than are blue duikers under the same hunting regimen. Among red duikers, species such as the white-bellied duiker are more vulnerable than others. Controls on hunting must prevent unsustainable offtake of these species, even though other more resilient species might be exhibiting only minimal impacts. Further knowledge of the biology and ecology of the most vulnerable species is essential.

Other significant conditions affecting the sustainability of hunting, but for which data are limited, include the impact of reduced small ungulate densities on ecosystem processes, in particular the dynamics of predators that share a prey base with hunters, and the effects on the forest of loss of seed dispersers. Studies such as that by Fragoso (1997) would provide data needed to formulate concrete proposals for the management of hunting.

Joshi and Gadgil (1991) and McCullough (1996b) have argued that traditional hunting by indigenous peoples might be managed by controls on the spatial extent of hunting through the creation of a network of refuges in a matrix of hunted forest. The results of this study lend qualified support to this suggestion. Dispersal appears to be potentially important in maintaining small ungulate populations; however, it is not clear how large source areas should be in relation to sinks, or their spatial relationships. Differences in blue and red duiker dispersal indicate that a system of refuges adequate to maintain hunted populations of one species might fail to maintain those of another. Finally, the dynamics of unhunted populations indicate that a source population at one time could be a sink at some other time so fixed refuges might not be effective, especially if they are small in area.

Several factors, however, favor spatial controls on hunting, and in particular the establishment of fully protected zones. The results of these studies revealed that a significant portion of the standing biomass of ungulates is exploited by large predators, especially leopards. Even low levels of hunter offtake have a potentially significant impact on the cats' prey base. Maintenance of forest predators, particularly large cats, might depend on relatively unhunted populations of ungulates. Fully protected areas also might offer the only effective means to conserve species such as white-bellied duikers, which are especially vulnerable to even low levels of hunting.

Finally, the potential of refuges to maintain duiker populations in adjacent exploited areas is favored by the dynamics of the mbau forest. Although this relatively unproductive forest supports lower densities of duikers than do mixed forests, the high dispersal rates from mbau forest indicate that it might be acting as a source for adjacent areas. The mbau forest is not favored by Mbuti hunters and feline predators, in part because its open understory makes hunting difficult. Mbau forests, however, have a limited distribution and are unsuitable habitats for some species, particularly the Weyn's duiker, which is abundant in mixed forest but is almost absent from large areas of mbau forest.

Realistic management of hunting in African forests will probably use a variety of approaches. Interventions must reflect the biology of the species concerned and be culturally acceptable in the communities in which they are applied. They also must be politically and administratively feasible. In Central Africa, there is little precedent to guide choices at present, and management must advance in an experimental manner, often with compromises. This will necessitate monitoring and development of rapid, easily analyzed methods that can be understood and applied by local hunters themselves to track exploited populations through time. Agreements with hunters on when, where, and how populations will be protected might be one way to start, with monitoring provided by serial censuses, supported by control studies in unhunted areas. Recent work by Marks (1994, in press) among traditional hunters in Zambia has shown that wildlife managers working with local hunters can accurately monitor wildlife populations. It remains to be seen if these same collaborators can effectively manage traditional hunting as well.

ACKNOWLEDGMENTS

The field work reported in this chapter was supported by the T. J. Watson Foundation (1973–1975), U.S. Man and the Biosphere Program (1981–1983), and the Wildlife Conservation Society (1986–present). The Institut Congolais pour la Conservation de la Nature provided authorization for field work in the Epulu area and has aided in protection of the Lenda and Edoro Study Areas. The impetus for the analyses reported here resides as well in the encouragement of my Congolese collaborators to provide guidance concerning the "problem of subsistence hunting" in the Okapi Wildlife Reserve.

I would particularly like to acknowledge John Robinson and Elizabeth Bennett for providing the encouragement and venue to bring this manuscript out of its long gestation and in particular to have provided critical editorial and analytical guidance at various points in its development. I would also like to thank Terese Hart, who participated in all but the earliest studies reported here and who read and criticized earlier drafts. David Wilkie and Richard Bodmer also read the paper critically and provided many suggestions that improved the clarity of analysis. Finally, the field work involved in these studies would not have been possible without the major contributions of many colleagues and collaborators, both Congolese and expatriate. To all of these individuals, I convey my generous thanks.

APPENDIX 6-1: Experimental Net Drives—Lenda and Edoro Study Areas

Studies of duiker populations on the two study areas were initiated in 1991 and continued until 1996. At both sites net drives were organized with a team of 45 to 50 Mbuti hunters who were hired to capture duikers and assist in the tagging and radio collaring of animals. Management of the Mbuti hunting team and assistance in data collection on the hunts were provided by trained Mbuti and villager field assistants from the local communities who were full-time employees of the project.

A total of eight standardized hunts, consisting of 12 to 14 drives each, was organized on both study areas over the course of the five-year study. Drives were conducted by enclosing predetermined sections of the path grid with 18 to 20 hunting nets with 15 to 20 minutes allotted to each drive conducted by 22 to 30 drivers. The location of the drives was determined to systematically cover the study areas during each hunting period. Hunters were trained to report all observed escapes as well as captures. For all small ungulates captured, sex, age, tooth eruption, and reproductive condition were noted by myself or a trained research assistant. All captured animals received an ear tag. A subset of animals, including both sexes and all age classes, received radio collars as well. Only three animals of a total of 310 captured over the course of the study died during capture and handling.

7

Threatened Mammals, Subsistence Harvesting, and High Human Population Densities: A Recipe for Disaster?

CLARE D. FITZGIBBON, HEZRON MOGAKA,
AND JOHN H. FANSHAWE

It is increasingly recognized that the conservation of threatened ecosystems depends on the support of local people. Unless these people benefit from the habitat and the wildlife it supports, they have no long-term incentive to protect it (Bell 1987; Ehrlich and Wilson 1991; Bodmer et al. 1994). In theory, wild-meat harvesting is one way in which local communities could benefit from forest conservation, off-setting costs such as reduced timber harvests (Balmford et al. 1992). The problems of first determining and then enforcing sustainable harvesting levels for a range of prey species are substantial, however, and have rarely been tackled in tropical forest habitats. Although a wildlife harvesting program might be feasible where human pressure on forests is relatively low, the problems become particularly severe in areas where human population density is high and competition for forest resources substantial. Under these circumstances, a more realistic solution might be to ban hunting and promote other forms of use, such as ecotourism, which can bring revenue and employment to local communities. In this study, we investigate the impact of wildlife harvesting on the mammal community of Arabuko-Sokoke Forest, a reserve facing intense human pressure on the Kenya coast, and determine whether a sustainable harvesting program is a realistic management option.

Coastal forest was formerly extensive along the Eastern African coast, but has been largely removed to make way for agriculture and tourist developments and to provide wood for fuel, building, and carving (Hamilton 1981; Howell 1981). The largest remaining area in East Africa, Arabuko-Sokoke Forest (372 km²), is an important site for biodiversity conservation (Burgess et al. 1996), notably for six threatened bird species (Collar et al. 1994). It provides habitat for a number of rare and threatened mammal species, including Ader's duiker (*Cephalophus adersi*), the golden-rumped elephant-shrew (*Rhynchocyon chrysopygus*) (Nicoll and Rathbun 1990; FitzGibbon 1994), the Sokoke bushy-tailed mongoose (*Bdeogale crassicauda omnivora*), and the lesser pouched rat (*Beamys hindei*) (FitzGibbon et al. 1995a). As a result, conservation of the forest's mammal populations is a high priority.

Over the past century, the whole coastal belt has experienced a rapid increase in population density (currently estimated at 3.8% per year), vastly increasing the demand for natural resources and for settlement land. Arabuko-Sokoke is currently a forest reserve, surrounded by villages and farmland on all sides. It is jointly managed by the Forest Department and Kenya Wildlife Service. Some limited use of timber and forest products is allowed, and although harvesting of wild animals is illegal in Kenyan forest reserves, local communities of Mijikenda and Sanya are known to harvest a wide range of mammal and bird species (Stiles 1981; Mogaka 1992). Wildlife harvesting is primarily for home consumption, although a small amount is sold within the local community (Mogaka 1992).

METHODS

The study was conducted in Arabuko-Sokoke Forest, Kenya (figure 7-1), between January and April 1992. The forest covers 372 km² and consists of three main habitat types: (a) woodland dominated by *Brachystegia spiciformis* (approximate area 67 km²), (b) *Cynometra-Brachylaena* woodland (approximate area 242 km²), and (3) *Afzelia* forest (approximate area 63 km²) (Kelsey and Langton 1984). Annual rainfall is dominated by a long rainy season in April, May, and June. Mean rainfall ranges from below 600 to more than 1000 mm per annum.

The Sanya people are considered to be the indigenous inhabitants of the forest and were traditionally hunter-gatherers, harvesting fruits, medicinal plants, and honey, as well as wildlife. They lived in the forest until its gazettement in the 1950s, when they settled around the forest boundary taking up crop cultivation. The Mijikenda people arrived more recently in the early 1900s and are now subsistence farmers, primarily cultivating maize, cassava, and coconuts.

The study comprised two separate research activities. First, wildlife harvesting rates were determined by interviews with households living around the forest; second, prey densities and the impact of harvesting on prey were determined by a series of population surveys in different parts of the forest. Data on prey harvesting rates were collected. They were derived from a survey of 51 households living adjacent to the forest edge and 24 households living approximately 2 km from the forest boundary, between March and May 1991 (see FitzGibbon et al. 1995b for further details). Questions on hunting activity were included as part of a general socioeconomic interview conducted by Hezron Mogaka. In addition, 16 regular hunters kept records of how often they went hunting and trapping, the

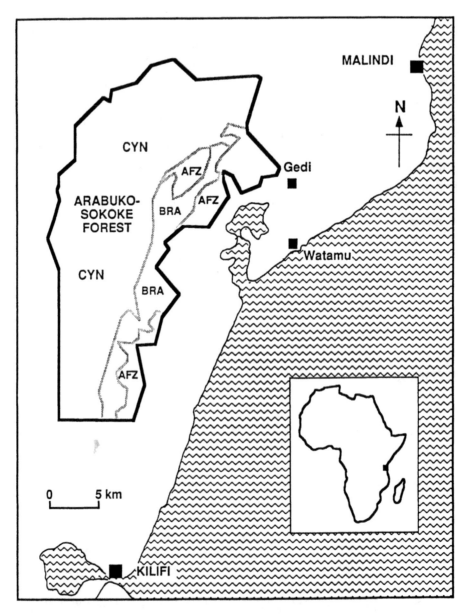

Figure 7-1. Map of Arabuko-Sokoke Forest and surrounding area showing the main vegetation zones (BRA, *Brachystegia* woodland; CYN, *Cynometra-Brachylaena* woodland; AFZ, *Afzelia* forest).

number of traps set, trapping success rates, and the prey species caught. Recent surveys have shown that 2660 households live within 2 km of the forest boundary, 334 households of which are adjacent to the forest.

Although we were able to determine how many animals were removed by regular hunters/trappers each year, we were not able to determine the number of animals harvested by the irregular trappers, which therefore had to be estimated. Because regular trappers reported making an average of 2.4 trips per week and irregular trappers about one trip every 2 months, each irregular trapper was estimated to harvest 1/19 of a regular trapper's yearly catch. Consequently, the 1151 trappers who reported hunting irregularly were considered equivalent to 61 regular trappers. Only households living adjacent to the forest were included in the estimate of primate offtake rates because Syke's monkeys and baboons are primarily trapped by these households to reduce crop raiding (personal observation). Although hunting and trapping were illegal and interviewees therefore may have understated their involvement, estimates of traps set are similar to the number of traps actually found in the forest during surveys (see FitzGibbon et al. 1995b), suggesting that estimates of offtakes are at least of the correct order of magnitude.

To determine prey densities, a total of 28 1-km transect lines were cut in the forest, distributed between the three main habitat types. Eight were in the *Brachystegia* woodland, eight in the *Afzelia* forest, and twelve in the *Cynometra* forest. In each habitat, half of the transects were on the forest edge, where trapping and hunting intensity was predicted to be high, whereas the other half were positioned near the center of the forest. The survey focused on a few key groups, namely primates (*Papio* spp. and *Cercopithecus* spp.), duikers (*Cephalophus* spp.), bushpigs (*Potamochoerus porcus*), elephant-shrews (*Rhynchocyon* and *Petrodomus* spp.) and squirrels (*Heliosciurus* and *Funisciurus* spp.), for two reasons: they constitute the bulk of the prey taken, and their densities can be estimated either directly from sightings or indirectly from nests, dung piles, and feeding sites.

The occurrence of animal signs was recorded for each transect. An observer walked along the transect line recording the number of duiker dung piles and golden-rumped elephant-shrew nests within 3 m on either side (FitzGibbon and Rathbun 1994) and the number of four-toed elephant-shrew (*Petrodomus tetradactylus*) paths that crossed. All transects were checked twice, once by a tracker who marked all the nests, dung piles, and paths seen, and again by one of the authors (C.D.F.), ensuring that virtually all visible nests and dung piles were found. In addition, after each 10 m of transect, the observer recorded whether there were any signs of recent bush pig activity (holes, feeding sites, dung, etc.) within 5 m on either side of the previous 10 m of transect. Three 200-m sections (0–200 m, 400–600 m, and 800–1000 m) were searched per transect, and the results from these subtransects were averaged to provide an abundance index score for each mammal group (the average number of dung piles or nests recorded per 100 m of transect for duikers and golden-rumped elephant-shrews; the number of trails crossing per 100 m of transect for four-toed elephant-shrews; and the number of 10-m sections in which activity was recorded per 100 m of transect for bushpigs). There are four duiker species recorded in Arabuko-Sokoke Forest (Ader's duiker, blue duiker [*Cephalophus monticola*], common duiker [*Sylvicapra grimmia*], and red duiker [*Cephalophus natalensis*]), as well as suni [*Neotragus moschatus*], but the range of fecal pellet sizes meant that it was not possible to distinguish the different species on this basis.

An alternative method was used to determine the relative abundance of primates and squirrels. An observer returned to each transect early in the morning (always before 1000 hr, usually before 0800 hr) several days after the cutting, and walking along the path recorded the presence of any primates (yellow baboons [*Papio cynocephalus*] or Syke's monkeys [*Cercopithecus mitis*]) or squirrels (red-legged sun-squirrel [*Heliosciurius rufobrachium*] or red-bellied coast squirrel [*Funisciurus palliatus*]), their approximate distance from the path, and the number of animals in the group. Great care was taken to determine the size of all primate groups encountered close to the transect line, but this was not possible for groups further away, and the number of groups encountered within 50 m of the transect was used as the abundance measure. We used 50 m as the cutoff distance because we could be certain of detecting all groups within this distance regardless of which habitat type was being surveyed. Each transect was walked on three separate mornings approximately a week apart, and the data from the three visits were averaged.

For some species (baboons, Syke's monkeys, both elephant-shrews), it was possible to estimate population densities, either directly or by using conversion factors derived from data collected from intensive study sites, whereas for others only relative abundance could be determined (see FitzGibbon et al. 1995b for further details).

Following the method of Bodmer (1995a), the preferences of hunters were calculated using Ivlev's index of selectivity, which compares availability (A) with use (U) by (U – A)/(U + A) (Ivlev 1961). Availability was determined from the density estimates (number of animals counted/km^2), whereas use was estimated from the harvest rates determined from interviews (number of animals harvested/km^2).

Immediately after the 28 transects were cut, the number of traps within 15 m of each transect line was recorded. Two local people familiar with the traps used in the forest walked at approximately 5 and 10 m parallel to the line and recorded any traps seen, as well as the type and state of the trap, particularly whether it appeared to have been set within the past 24 hours.

In order to provide approximate estimates of maximum sustainable harvest rates for the species harvested in Arabuko-Sokoke, we used the method described by Robinson and Redford (1991b). For each species, the model requires information on the population density at carrying capacity and the maximum rate of population increase, and it assumes that the density that produces the maximum sustainable yield is at 0.6 K (see FitzGibbon et al. 1995b for further details). The maximum rate of population increase (r_{max}) was estimated using Cole's (1954) equation:

$$1 = e^{-r_{max}} + be^{-r_{max}(a)} - be^{-r_{max}(w+1)}$$

where a is the age of first reproduction, w is the age of last reproduction, and b is the annual birth rate of female offspring (the reproductive parameters used in calculating r_{max} have been described in FitzGibbon et al. (1995b).

SIGNIFICANCE OF WILD MEAT HARVESTS FOR LOCAL COMMUNITIES

Local communities harvest a wide range of mammal species from the Arabuko-Sokoke Forest, ranging in size from small mice, rats, and elephant-shrews to primates such as yellow baboons and large ungulates (e.g., the bushbuck [*Tragelaphus scriptus*]) (table 7-1), although these are now rare in the forest. Both active hunting and trapping take place. All species can be caught in traps, but baboons, Syke's monkeys, and ungulates, such as duikers and bushbuck, are also hunted directly.

Of the households interviewed, 62.7% of those living adjacent to the forest performed some sort of hunting and/or trapping in the forest, whereas 33.3% of those living 2 km from the forest did so. Approximately 60 people hunt regularly in the forest, each hunter taking one to four hunting trips per week and setting 10 to 25 traps per day, and at least 500 people hunt irregularly, depending on the supply of crops from their farms and the availability of work. Regular hunters are estimated to harvest 1077 kg of meat per hunter per year by hunting and trapping.

Wildlife is the main source of animal protein for those people living close to the forest (Mogaka 1992), and it is the poorer households who have little alternative source of food that do most of the harvesting. The mean annual income of those households that performed some hunting and/or trapping was 9000 Kenya shillings (KShs) ($225); among those that did not, it was 24,000 KShs ($600; N = 40 households) (Mogaka 1992). Wild animals are also harvested to reduce crop raiding. Mice, rats, Syke's monkeys, baboons, and bushpigs cause extensive damage to crops, many households losing between half and three-fourths of their planting. Thus, it is not surprising that trapping is now particularly focused on these species, and estimates of offtakes might include some animals that are not entirely forest dependent but spend part of their time in the agricultural land surrounding the forest.

Most of the meat is for home consumption rather than for sale (the actual percentage has not yet been quantified), but regular hunters do sell some or all of their catch locally. The value of wild meat in 1992 was $0.25/kg, so regular hunters could earn approximately $275/year by selling the meat they obtain, a high figure considering that the average annual per capita income of people living around the forest at that time was only $38 (Mogaka 1992). The low level of commercial harvesting probably results from the fact that wild meat harvesting is

Table 7-1. Number of Animals Caught by Regular Hunters and Trappers from Arabuko-Sokoke Forest Each Year

	Mean No. Trapped by a Regular Trapper	Mean No. Hunted by a Regular Hunter
Four-toed elephant-shrews	47	0
Golden-rumped elephant-shrews	26	0
Bushpigs	6	0
Aardvarks	5	5
Mongooses	5	0
Syke's monkeys[a]	8	11
Baboons[a]	3	8
Duikers	1	1
Squirrels	4	0

The occasional bushback and buffalo are taken, as well as several species of rodents, including the giant pouched rat *Cricetomys gambianus*.

[a]Baboons and Syke's monkeys are primarily killed by those living adjacent to the forest.

illegal in Kenya, and therefore bushmeat markets as found in west and central African countries do not exist. In addition, many of the people inhabiting local towns, such as Malindi, are Muslim Swahilis who consume little wild meat.

HARVESTING METHODS AND CHOICE OF PREY

Hunters use bows and poisoned arrows, spears, and catapults, and they often have dogs. Guns are rarely available. A wide range of drop traps and other types of traps are used, but snares predominate. Traps are usually specific to a particular species, as a result of both their design and appropriate placement. For example, small snares are placed along four-toed elephant-shrew trails, and drop traps are placed outside the burrows of giant pouched rats. Traps have to be checked regularly (at least once per day) to ensure that carcasses are not stolen by other trappers or eaten by ants, baboons, or other wildlife. As a result, traps are primarily placed within 1 km of the forest edge (figure 7-2). In general, active hunting was more successful overall, yielding an average of 5.5 kg meat per hour spent compared with 2.9 kg per hour from trapping, but trapping was more reliable. Although 34% of hunting trips were unsuccessful, it was rare for all the traps set by a trapper to be unsuccessful. On average, about a third of traps yielded catches each day.

What influences the selection of prey by hunters in Arabuko-Sokoke? In terms of numbers, small elephant-shrews are the most common prey species, whereas bushpigs and baboons comprise the greatest biomass (table 7-2). Four-toed elephant-shrews are also the prey selected most often (figure 7-3) in terms of Ivlev's index of selectivity, whereas the primates are selected least often. Differential harvests reflect in part the ease with which different species can be captured. For

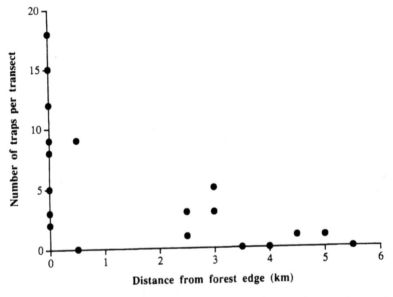

Figure 7-2. The effect of distance from the forest edge on the total number of traps recorded per transect ($r = -0.715$).

Table 7-2. Total Number and Biomass of Animals Harvested by Hunters and Trappers from Arabuko-Sokoke Forest Each Year

	Approximate Weight/ Animal (kg)	*Total Number Harvested from Forest*	*Total Biomass Harvested from Forest (kg)*
Four-toed elephant-shrews	0.2	5687	1137
Golden-rumped elephant-shrews	0.5	3146	1730
Syke's monkeys[a]	5.0	1202	6010
Aardvarks[b]	55.0	905	49,775
Bushpigs	80.0	726	58,080
Baboons[a]	16.0	683	10,928
Mongooses[c]	1.5	605	908
Squirrels	0.3	484	145
Duikers	6.0	181	1086
Total		13,619	129,799

[a]Syke's monkeys and baboons are primarily killed by those living adjacent to the forest.

[b]*Orycteropus afer.*

[c]*Herpestes, Ichneumia,* and *Bdeogale* spp.

example, four-toed elephant-shrews are easy to catch because they make clear trails through the leaf litter, along which traps can be set. In contrast, golden-rumped elephant-shrews make no such trails and are therefore much more difficult to catch. The differences also reflect preferences for particular wild meat; the meat of ungulates, for example, is preferred to that of primates.

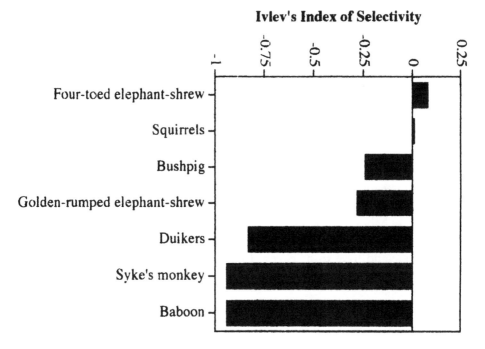

Figure 7-3. Hunters' preferences for different prey species as measured by Ivlev's index of selectivity. Increasing values between 0 and 1 indicate increasing positive selection, whereas negative values indicate increasing avoidance of that species.

Overall, the number of individuals harvested per species correlates positively with abundance estimates ($r = 0.894$, $n = 7$), but not with body size ($r = -0.324$, $n = 9$) or reproductive rates (r_{max}) ($r = 0.562$, $n = 7$). The number of species for which relevant reproductive and abundance data are available is, however, limited. People selected more smaller species and those with higher reproductive rates (correlation between Ivlev's selection index and reproductive rate, $r = 0.948$, $n = 7$, $p < 0.05$, and negatively with body size, $r = -0.831$, $n = 9$, $p < 0.05$). There was no significant relationship between prey density and the selection index ($r = 0.443$, $n = 7$, $p = $ NS).

In general, human hunters prefer large-bodied prey, and although some studies show that hunters do select larger species (e.g., Alvard 1993b; Bodmer 1995a; Colell et al. 1994), no such selection was found here. However, most studies where selection for larger bodied prey have been reported are of commercial hunting systems where much of the meat is for sale at market and/or has to be transported long distances. Although hunters and trappers in Arabuko-Sokoke did say that they preferred to eat the meat of duikers and large ungulates, it is likely that the difficulty of catching these animals made the effort generally not worthwhile. In contrast, because the small four-toed elephant-shrews were so easy to trap (success rates often exceed 30% per trap night for traps set along newly cleared trails), by focusing on this species and setting perhaps eight to ten traps, a meal of some sort was guaranteed, even though it did not provide large quantities of high-quality meat. Large numbers of small rodents are also harvested for the same reason. Hunting provides a larger proportion of large prey than trapping, and although hunting was more profitable in terms of harvested biomass/hr of effort, it is logistically more complicated. Most people hunt in groups and usually with a small pack of dogs, and hunting normally takes half a day or so. The probability of being caught by forest rangers is relatively high. In contrast, traps can be set in a matter of minutes, trapping is relatively inconspicuous and can be done alone, and the chances of being caught are lower. Thus, the ease with which smaller prey can be caught and the inverse relationship between r_{max} and body size (Fenchel 1974; Western 1979; Hennemann 1983) probably explains the selection by hunters and trappers of small species with high reproductive rates and the relationship between harvest levels and prey abundance.

The influence of prey abundance on the prey selection differs, depending on whether they are hunting or trapping (most people do a bit of both). Hunters find it worthwhile to shoot at any suitable animal they encounter, whether a relatively abundant baboon or an uncommon bushbuck. However, trappers rarely find it worthwhile to set traps for uncommon species because traps are generally species specific and have to be checked regularly. Thus, although active hunters continue to harvest rare species, trappers tend to focus on the more abundant species.

Other factors that influence harvest rates are the extent to which a species is considered a crop pest, attraction to baits, and group size. Primates and bushpigs are major crop pests and therefore targeted by farmers living adjacent to the forest, where they are attracted to traps using baits of maize and nuts. Primates are also highly susceptible to active hunting because they live in groups; therefore, more than one animal can be killed in one hunting foray.

IMPACT OF HARVESTING ON PREY POPULATIONS

By determining the total number of people hunting and trapping in the forest, and the composition of their yearly catch, it is possible to estimate that over 5000 four-toed elephant-shrews, 3000 golden-rumped elephant-shrews, 1000 Syke's monkeys, and 600 baboons are currently harvested each year (see table 7-2) (FitzGibbon et al. 1995b).

If harvest rates and population densities are averaged for the whole forest, then a comparison of current harvest with approximate estimates of maximum sustainable harvests suggests that the primates are overharvested, whereas current harvests of elephant-shrews, squirrels, and duikers are probably sustainable (FitzGibbon et al. 1995b) (table 7-3). Unfortunately, it was not possible to estimate densities or maximum sustainable harvests for either bushpigs or aardvarks in Arabuko-Sokoke. No significant differences in bushpig abundance were detected between low and high hunting pressure transects (figure 7-4), but the very high numbers of bushpigs removed are cause for concern. Using the density (3.1 individuals/km^2) and production (1.9 individuals/km^2) estimates for bushpigs from Fa et al. (1995) suggests that current harvests from Arabuko-Sokoke are twice the maximum sustainable harvests. However, no data are available to determine whether densities in Arabuko-Sokoke are comparable with those reported by Fa et al. (1995) in Equatorial Guinea. No data are available on the density of aardvarks in forests, and no estimation of maximum sustainable harvests is therefore possible, but it seems impossible that such reportedly high harvests of this species can be sustainable. Overharvesting has already reduced the densities of bushbuck and buffalo (*Syncerus caffer*) to such low levels that hunting and/or trapping them is rarely worthwhile, although they are still hunted if encountered.

What causes these differences in susceptibility to overharvesting? One very important factor is the relative vulnerability of the species to trapping or active hunting. Trapping can substantially reduce population densities, for example, reducing the abundance of four-toed elephant-shrews and squirrels by 41% and 66%, respectively (FitzGibbon et al. 1995b) (figure 7-4). However, trapping is concentrated around the forest edge. Traps have to be checked at least once a day;

Table 7-3. Current Population Densities (1992), Total Annual Harvest Rates (from Both Hunting and Trapping), and Maximum Sustainable Offtake Levels (All in Number of Individuals/km^2) for Those Mammal Groups for Which Density Estimates Could Be Calculated

	Current Density	*Current Harvest*	*Maximum Sustainable Harvest*
Four-toed elephant-shrews	391	15	274
Golden-rumped elephant-shrews	59	8	20
Syke's monkeys	58	3	1
Baboons	16	2	<1
Duikers	63	1	3
Squirrels	11	1	6
Bushpigs	3	2	1

See FitzGibbon et al. (1995b) for calculation of maximum sustainable harvest rates. Data on density and production levels for bushpigs taken from Fa et al. (1995).

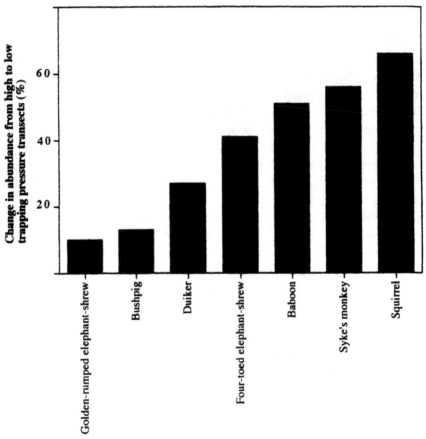

Figure 7-4. Percentage increase in mammal abundance indices from transects of high to low trapping pressure.

consequently, it is not worthwhile to set traps more than a couple of kilometers from the forest edge, with the result that 60% of the forest is relatively unaffected. For those species that are only vulnerable to trapping, therefore, the center of the forest acts as a refuge. In contrast, active hunting occurs throughout the forest, and species vulnerable to hunting, particularly the large ungulates and the primates, therefore have no refuge.

An additional factor that has a profound effect on susceptibility to overharvesting is reproductive rate. Thus, although four-toed elephant-shrews are preferentially selected and large numbers are taken, their densities are relatively unaffected on account of their ability to reproduce rapidly. In contrast, even though relatively low numbers of primates are taken, the effects on the population are substantial because of their low reproductive rates.

CHANGES IN HUNTING PATTERNS AND PRACTICES

What has caused the changes in the extent and form of wildlife harvesting that have resulted in the overexploitation of some prey species? Clearly the increase in human population density has been a major contributing factor. In addition, a change in the way that the forest is harvested is important. The Sanya people, who are considered to be the indigenous inhabitants of the forest, concentrated on the larger mammals, for example elephants and buffalo, and never hunted primates. The Mijikenda people, who arrived more recently in the early 1900s, were less discriminating, hunting a wide range of species, including both baboons and Syke's monkeys. Although the Sanya people tended to be more mobile, moving on when wildlife became scarce, the fact that the local people are now resident means that harvesting is more concentrated near areas of settlement. There is also some suggestion that the forest was divided into hunting territories, each group of households hunting a particular area. This system has broken down so that it is no longer worthwhile for hunters to protect wildlife stocks in their area. In addition, the fact that hunting and trapping are illegal might increase pressure on wildlife populations; people feel that they might be prevented from harvesting in the future, and therefore might as well get as much from the forest while they can now while management is poorly resourced and controls are comparatively lax.

ARE HUNTING AND MAMMAL CONSERVATION COMPATIBLE IN ARABUKO-SOKOKE?

If hunters are concerned with conserving mammal populations, they should focus on those species that are best able to withstand harvesting, i.e., those that have high reproductive rates (Alvard 1993b; Bodmer 1995a), and those that are abundant. The positive correlation between prey selection and r_{max}, and between harvests and prey abundance, suggest that this is exactly what hunters are doing. However, it is unlikely that hunters are foregoing opportunities to kill larger animals, which have lower reproductive rates, for the sake of conservation. Now that the system of hunting territories has generally broken down, there is little incentive to conserve prey resources. Although hunters would prefer to harvest larger animals such as bushbuck and bushpigs, which provide much more meat, the difficulties of doing so make the smaller prey a more feasible option for most people. The positive correlation between prey selection and r_{max}, and between harvests and prey abundance, result from the fact that more abundant species tend to have higher reproductive rates; it is easier to trap more abundant species, which are therefore harvested at greater rates. In addition, the species that has the highest r_{max} of those included in the study, the four-toed elephant-shrew, occurs at the highest density and is the easiest to catch, on account of its habit of making trails.

Clearly, trapping of small mammals such as elephant-shrews is sustainable at present and provides an important supply of protein for many people living around the forest, particularly when other food is in short supply. There is little evidence that duikers are affected at current harvesting levels, although it is possible that different species are affected to different extents. Harvesting of primates needs to be controlled, however, if populations are to be maintained. The low densities of bushbuck and buffalo suggest that hunting them needs to be stopped altogether. In addition, further data are required on population sizes of bushpigs and aardvarks, and the harvest levels of these two species need to be confirmed. Arabuko-Sokoke Forest is a critical site for mammal conservation, and although current harvesting is not apparently affecting populations of the golden-rumped elephant-shrew, its impact on the Ader's duiker and bushy-tailed mongoose populations is not known. Sightings of these species are, however, now extremely rare.

In addition, although none of the species negatively affected by harvesting are threatened worldwide, loss of any one of them would reduce the conservation value of this forest, and primates certainly contribute to attracting tourists. It is believed that ecotourism will be an important source of income for forest management in the future: a nature trail and visitors' center have already been established. If a balance between traditional use and the value of the forest for other activities, like ecotourism, is to be struck, it is clear that some limitation to the wild meat harvest is required.

To some extent, current hunting patterns suggest that wildlife harvesting in Arabuko-Sokoke might lend itself to wildlife management. At present, people are primarily selecting those species that have the highest productivity and are least affected by harvesting. The type of harvesting that needs to be reduced is that which targets species such as primates, which are selected less often. As Bodmer (1995a) points out, the success of management programs will depend on their acceptance by hunters, and it is easier to persuade people to reduce harvests of least preferred species. In Arabuko-Sokoke, primates are primarily harvested near to the forest edge because they raid and cause considerable damage to crops, not because they are a highly desirable source of meat. Thus, if more effective ways of deterring crop raiding could be found, primate trapping would be less of an issue. However, if new data suggested that it was necessary, the control of bushpig and aardvark harvests might be less popular because they currently provide such a large proportion of the prey biomass.

The most sensible management option for Arabuko-Sokoke in the short term seems to be to stop all active hunting, while allowing trapping to continue close to the forest edge. This would still enable households adjacent to the forest to control crop raiding and obtain supplies of wild meat, while leaving much of the forest undisturbed. In addition, all trapping of bushbuck and buffalo could be stopped to allow populations to recover. Because most traps are species specific, it should be possible for forest guards to destroy traps set for this prey group. Steps to facilitate the efficient introduction of controls might include (a) dialogues with hunters to explain the risk to key species and the need selectively to remove traps set for them and (b) training for forest patrols.

The probable consequence of these changes will be increased pressure on the smaller species, primarily the elephant-shrews, particularly if it becomes clear that trapping is now condoned. Current densities of the four-toed elephant-shrew are high, however, and populations can withstand increased offtake levels. Although golden-rumped elephant-shrews are relatively rare, and Arabuko-Sokoke supports the main population of this species (FitzGibbon 1994), four-toed elephant-shrews are widely distributed and not in danger of extinction. Although golden-rumped elephant-shrews are caught in traps designed for four-toed elephant-shrews, it is unlikely to occur at levels sufficient to threaten the golden-rumped elephant-shrew population as a whole.

As long as it is not worthwhile for trappers to travel long distances to set traps, people will continue to trap mainly around the forest edge. Thus, their activities are, at least to some extent, self-regulating. As densities of prey close to the forest edge decline, trapping becomes less profitable because of the costs of transportation. The need to check traps regularly makes trapping far from people's homes uneconomical. The central untrapped portion of the forest will therefore act as a refuge and should be sufficient to maintain prey densities at the forest edge. In the long term, a more sophisticated harvesting regimen might need to be established, but with the restricted resources currently available to forest managers, it would probably be too difficult to implement and enforce. In the short term, the specific arrangement suggested here is probably the one most likely to succeed. An effective monitoring program is clearly required to determine the effects of trapping at the edge of the forest on different species according to ranging patterns and densities, and to detect changes in harvesting levels and prey population densities.

In conclusion, the results of this study have shown that an understanding of the biology of prey species and the factors that influence a hunter's prey choice are central to the management of subsistence hunting. The lack of commercial hunting in Arabuko-Sokoke, for example, makes the harvesting of smaller prey, which reproduce relatively rapidly and can support high offtake rates, more acceptable than in other wildlife harvesting areas, where larger carcasses are in greater demand. Only if we learn enough about the organization and sustainability of each harvesting system can we hope to manage subsistence hunting effectively. Sympathetic and flexible management is most likely to gain the acceptance of those people who use the forest and who will have to make sacrifices in order to ensure sustainability, and therefore be successful in the long term.

ACKNOWLEDGMENTS

We thank the Kenya Government for permission to undertake this study as part of the Kenya Indigenous Forest Conservation Project (KIFCON) and the joint National Museums of Kenya—International Council for Bird Preservation (ICBP) Arabuko-Sokoke Forest Project. We also thank Anthony Githitho, District Forest Officer, Kilifi, for allowing us to work in Arabuko-Sokoke Forest, and David Changawa, Francis Charo, and David Ngala for extensive help in the field. The study was funded by the Overseas Development Administration, Frankfurt Zoological Society, and BirdLife International (formerly ICBP).

8

Hunted Animals in Bioko Island, West Africa: Sustainability and Future

JOHN E. FA

Although there have been attempts to develop indices and models to evaluate sustainability of hunting in tropical forests (Robinson and Redford 1994b), practical solutions of how to achieve this are lacking. Developing these solutions will require measuring the impact of hunting as well as the needs of the consumer population, measurements complicated by other factors such as accelerating deforestation. Wildlife preservation, although partly related to the management of wildlife populations, ultimately has more to do with the management of hunters and consumers or, in other words, with the problems of supply and demand. This is the main theme in this chapter.

Wild animals are exploited for food (bushmeat) in many parts of the world (Prescott-Allen and Prescott-Allen 1982). Throughout the tropics, a large number of species are of major nutritional and cultural importance in the diet of forest-dwelling and urban human populations. Particularly in West Africa, and most notably in the humid forest zones, consumption of wild meat is especially intense because of the scarcity of domestic livestock as well as it being a preferred food. Here, bushmeat represents a substantial proportion of the primary protein intake—more than 60% of the dietary protein in some ethnic groups (Chardonnet et al. 1995). The significance of bushmeat in the diet of countless people in the region has doubtless aggravated pressures on local faunas. Modern hunting techniques and rising human populations can now rapidly lead to overexploitation, to the increasing rarity and even extinction of some species.

Generally, subsistence hunting for consumption at the village level might not pose a severe risk to prey populations where human numbers are low and enough habitat is available. However, wildlife harvests can be maintained only if hunters know how the different species respond to hunting pressures and which species are vulnerable to overexploitation (Bodmer et al. 1994). On the other hand, commercialization of wild meat, at levels now seen in several western and central African countries, can rapidly develop into uncontrolled exploitation by professional hunters (hunters whose primary income is derived from the sale of wild meat) responding to the ever-increasing demands of the fast-growing towns and cities. The contribution that this type of resource extraction makes to the gross national product (GNP) of some countries puts the scale of this problem into perspective. In the Ivory Coast, for example, the wild meat trade is worth an estimated US$117 million annually. In Liberia, out of an annual total wild meat harvest worth US$42 million, half is taken by commercial hunters. Similar figures are available for Nigeria and Ghana (Feer 1993). Such estimates rarely appear in the economic statistics of these countries because bushmeat is not considered a product of economic value.

Although the traditional approach for conserving wildlife has been to create reserves that provide reservoir territories for some species and a surplus that could be hunted (Shaw 1991), it is also necessary to establish workable rules for the acquisition and distribution of the benefits from wildlife at the local and national levels (Hofer et. al 1996). In any case, the creation and management of protected areas has all too often provoked hostile attitudes among local communities, to the extent perhaps that it is now only realistic to perceive wildlife as having to pay its way to survive (Eltringham 1994). In relation to bushmeat species in West Africa, Martin (1983) claims that conservation areas have longer term viability if linked to multipurpose uses including game cropping. Whichever the view held, it is only through reducing conflicts between local communities and wildlife managers that the pace will be set for sustained benefits to be obtained from wildlife in the long term (Robinson and Redford 1991c).

Bioko Island (formerly Macias Nguema Biyogo and previously Fernando Poo), has a unique and important collection of fauna and flora (Juste and Fa 1994). Bushmeat is a vital source of protein and cash for the people of Bioko. This chapter reviews the extent and impact of hunting on the wildlife populations in the island. Administratively, Bioko forms part of the insular sector of the Republic of Equatorial Guinea; the country's capital, Malabo, is located here. Equatorial Guinea is an independent republic in West Africa, consisting of a mainland section (Rio Muni) on the western coast and the coastal islets of Corisco, Elobey Grande, and Elobey Chico, as well as the islands of Bioko and Annobón (Pagalu) in the Gulf of Guinea. Bioko was discovered in 1471 by the Portuguese, and in 1778 Portugal ceded the island to Spain. Later (1827–1844), Great Britain maintained a naval station here and also administered the island. In 1844 the Spanish settled in Rio Muni, and by 1904 Fernando Po and Rio Muni were organized into the Western African Territories, later known as Spanish Guinea. On October 12, 1968, the territory became the independent republic of Equatorial Guinea, with Francisco Macias Nguema as President. Nguema appointed himself President for life in 1972. Extreme dictatorial and repressive policies led to the flight of an estimated 100,000 refugees to neighboring countries. In 1979, Nguema was overthrown in a military coup, tried for treason, and executed.

Against this historical background, there have been few studies on the biodiversity of the island until a major effort, initiated by Spanish biologists, started in 1980 (Castroviejo et al. 1986, 1994a). This chapter summarizes the available data on the use of wildlife on the island collected in collaboration with the Spanish program (Juste et al. 1995; Fa et al. 1995) and further explores whether the present levels of unregulated hunting are likely to be sustainable in the future. It also reviews the practicalities of controlling wildlife offtake at sustainable levels, and the development of alternative sources of protein for local consumption.

NATURAL AND ANTHROPOGENIC BACKGROUNDS

Habitats and Climate

Bioko is a continental shelf island, rectangular in shape (69 × 32 km), 2017 km^2 in area, and 32 km from the Cameroon coast. In geological terms, the island was formed recently, probably in the Lower Tertiary period, and shaped during the Pleistocene period. This is reflected in its dramatic landscape, which is characterized by deep valleys and dominated by two main volcanic massifs, the Caldera de Luba in the southwest and Pico Basilé in the north. Elevations range from sea level to a maximum altitude of 3011 m at the summit of Pico Basilé; the Southern Highlands rise to 2261 m at the Caldera de Luba in the southwest and to 2009 m at the Pico Biao in the southeast.

Bioko's vegetation is structured along altitudinal rings (figure 8-1) and includes formations dominated by Guineo-Congolian rain forest species with Afromontane elements appearing at higher altitudes (FED/DHV 1989; Juste 1989). At low elevations (0–800 m) the main vegetation type is tropical rain forest, although most was cleared and transformed into cocoa agrosystems or other food plantations in the early 1920s (Fa 1992a; Perez del Val 1996). This is still the most widespread habitat on the island, and the majority of the cocoa plantations that have been abandoned since independence (1968) are now covered with secondary forest. At present, about 40% of the island is composed of abandoned or semi-abandoned plantations (Pl/Sf), other crop lands (Oc), or vacant cattle pastures (Pst). Original lowland rain forest (Lrf) is mostly found along the southern part of the island, where the vegetation cover is considered monsoon forest (Mof) because of the prevailing high seasonal rainfall. Such forests have tall emergents (50 m) and a high tree canopy (30 m) with a sparse understorey. *Ficus* spp. are the most dominant trees, together with *Chrysophullum africanum*, *Chlorophora excelsa*, *Richinodendron africanum*, and *Crotogyne manniana*.

At higher altitudes (800–1800 m), montane forest (Mf) appears as a result of higher precipitation (3000–4000 mm), lower average temperatures (15–23°C), and high relative humidity. Within this altitude, at least 5000 hectares (ha) of relatively flat land in Moka in the south are abandoned cattle pastures from colonial times. Montane forest is superficially similar to the lowland forest, but its flora is less rich and its trees smaller. In Bioko, tree ferns (*Alsophila camerooniana* and *A. manianna*) dominate this forest type.

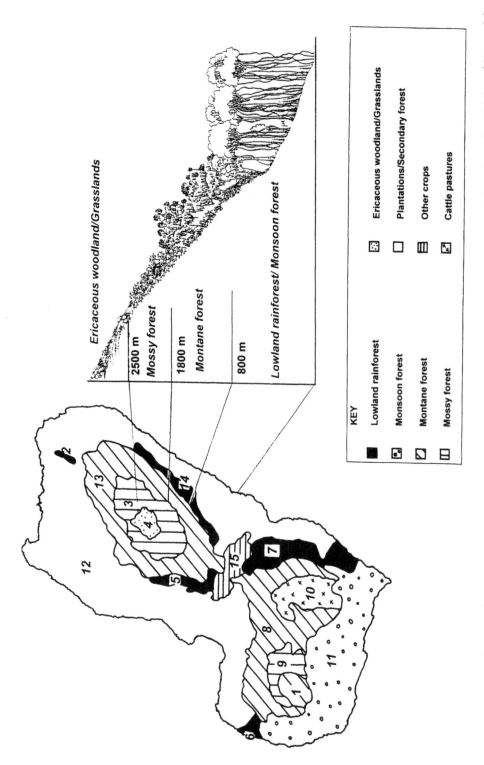

Figure 8-1. Distribution of the main vegetation type blocks in Bioko island. Vegetation map modified from Juste B. (1991) and Perez del Val (1997). Vegetation profile adapted from Castroviejo et al. (1986).

At altitudes of 1800 to 2500 m, trees of the Araliaceae family (*Polyscias fulva*, *Schleffera mannii*, and *S. barteri*) dominate the vegetation, known as mossy forest (Msf; 133.9 km²). Temperatures here vary between 8°C and 30°C. This vegetation type covers only 4% of the island, mostly in the Pico Basilé, with another 7% of the total coverage in the island found in the Pico Biao and Caldera de Luba in the south. Mossy forest gives way to ericaceous woodland (Ew) between 2500 and 2700 m, where *Aguaria salicifolia*, *Blaeria mannii*, and *Philippia mannii* are dominant; this formation is found in only 900 ha. Here, *Hypericum revolutum* and *Crassocephalum mannii* are also common. Toward the summits of the main peaks the vegetation changes again and consists of grasslands (Gr), with temperate plant genera that include *Helichrysum*, *Geranium*, *Clematis*, *Senecio*, and *Solanum*.

The climate of Bioko island is typically tropical equatorial, strongly influenced by the north–south movements of the intertropical convergence zone. Consequently, there is a distinct rainy season (March–July). Weather conditions are tied to the topography of the island so that the southern part of the island may receive over 10,000 mm of rain annually, whereas the north averages just 2000 mm (Nosti 1947). Temperatures near sea level vary from 17°C to 34°C.

Human Population

The indigenous ethnic group is the Bantu-speaking Bubis, but since the 1930s the Fang (Spanish Pamues) have migrated from the mainland. The human population has increased from 28,000 in 1932 to 62,000 (1990 census), 50,000 of whom reside in the capital city and in the four largest towns. Most of the population lives in the northern half of the island (figure 8-2) in villages that are about 2 to 3 km from the coast. Few villages are situated in the midlands and highlands; about half of Bioko has no permanent settlements. The human population density is about 93 inhabitants/km² in the north, declining to eight to ten inhabitants/km² in the southwest and southeast. Villages vary in size from 44 to 363 inhabitants in a sample of 13 villages studied by Mas et al. (1995), but most average around 100 inhabitants. The population growth rate was estimated by Juste and Cantero (1991) as 2.7% for the year 2000.

A road connects most of the inhabited areas of the island and is passable year round. Most villages are on average about 6 km apart. Only Ureka, on the southernmost part of the island, is not accessible by road and is about 35 km from the nearest village. Other than villages, there are the *ipatiosî*, the workers' residences in the cocoa plantations. The main occupation of the rural population is cocoa production supplemented by subsistence farming and hunting. The staple foods are bananas, yams, and millet. The monetary system is based on the franc system (500 Cefa francs [CFAs] equalled US$1 in 1995).

The prevailing socioeconomic conditions in Equatorial Guinea are undoubtedly among the lowest of any West African country. Estimated life expectancy is only 50 (Juste and Cantero, 1991), largely due to nutritional deficits (FAO/PNUD 1984) and a high incidence of disease (see Mas et al. 1995 for the prevalence of onchocerciasis). Per capita GNP earnings for the country were around US$800 per year (1995 estimates).

Faunal Communities

Bioko's fauna, though rich, is relatively impoverished when compared with equivalent continental ecosystems (Fa 1992a). This is probably a result of the relaxation trends that ensued when Bioko became isolated from the mainland as a result of rising sea levels since the Pleistocene period (12,000 BP). Patterns of species richness and endemism reflect this biogeographic history (Jones 1994). Among vertebrates, two bird species are single-island endemics and 43 are endemic subspecies (Perez del Val et al. 1994). Eighteen mammal subspecies are endemic to Bioko, including four of the nine primates: Preuss's monkey (*Cercopithecus [l'hoesti] preussi insularis*), red-eared monkey (*Cercopithecus [cephus] erythrotis erythrotis*), drill (*Mandrillus leucophaeus poensis*) (figure 8-3), black colobus (*Colobus satanas satanas*), and Pennant's red colobus (*Piliocolobus pennanti*) (Butynski and Koster 1994). One reptile and one amphibian species are also known to be endemic to the island (Jones 1994).

The most obvious difference between Bioko and the Mount Cameroon area on the mainland is the absence of very large bodied species such as the forest elephant (*Loxodonta africana cyclotis*), forest buffalo (*Syncerus caffer nanus*), and chimpanzee (*Pan troglodytes*). The mammalian and bird faunas of Bioko are typical of the Cameroon faunal region, with additional montane elements. Bioko's mammalian community is comparatively poorer than that of the Mount Cameroon habitat island (Gartlan 1989; Gadsby and Jenkins 1992) where artiodactyls, carnivores, and primates are better represented. Mount Cameroon has 13 primate species compared with nine in Bioko, eight carnivores in contrast to three, and five ungulates, of which only two now are found in Bioko. From evidence provided by Nosti (1947), Basilio (1962), Eisentraut (1973), and more recently by Castroviejo (1995), who discovered the use of a piece of buffalo skin in traditional ceremonies in Moka, it is now certain that the forest buffalo once lived in Bioko.

The most evident feature of Bioko's fauna and flora is the characteristic elevational gradient in species abundance and richness. Typically, there is a progressive, gradual decline in total species number from the lowland to the montane areas. This is either because the range of many rain forest species does not extend into the mountainous areas or because, for historical reasons, some species have not reached these areas. It is likely that at one time undisturbed animal populations (especially duikers and primates) would have occurred in greater numbers than on the continent because of the absence of such large predators as the leopard (*Panthera pardus*) and golden cat (*Felis aurata*) (see Prins and Reistma 1989; White 1994). Man's colonization of the island in the Neolithic period changed all this radically, especially with the mass migrations from the continent and the extensive modification of the lowland areas in recent times (Castroviejo et al. 1994b). Since the arrival of humans on Bioko, the forest buffalo has disappeared, and it is likely

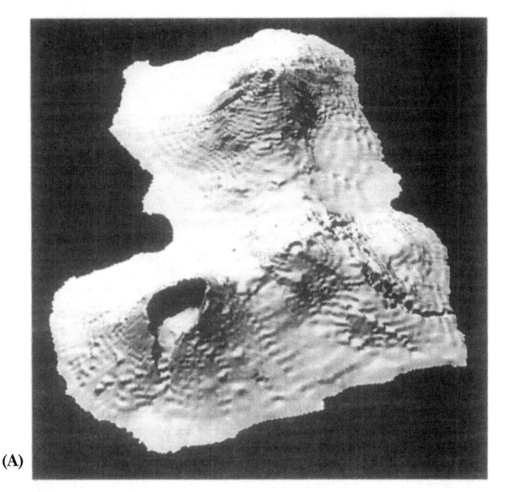

(A)

Figure 8-2. (A) Terrain and **(B)** location of the main human population settlements in Bioko. Only locality names mentioned in the text are included.

that the red forest hog (*Potomachoerus porcus*) and the giant forest hog (*Hylochoerus meinertzhageni*), if present at one time (see Basilio 1962), also might have been hunted to extinction. Butynski et al. (1995a) dispute the historical presence of these two species.

Estimates of animal densities in the different habitat types in Bioko are rare, but some transect surveys of primates have been undertaken by various investigators (table 8-1). The data generated are the frequencies with which groups of anthropoid primates were seen per km of transect walked. These frequency-of-sighting measures provide a simple comparison between sites, although they are affected by differences in detectability of groups arising from variations in vegetation structure and behavior of the different species (Oates 1996a). Nevertheless, these encounter frequencies can be used as an index of the status of primate species and, by extension, the status of other species throughout the island. The working assumption is that primates are adequate indicators for other species (at least those that are hunted), and that variations in species densities are determined by habitat-specific carrying capacities and local hunting pressure.

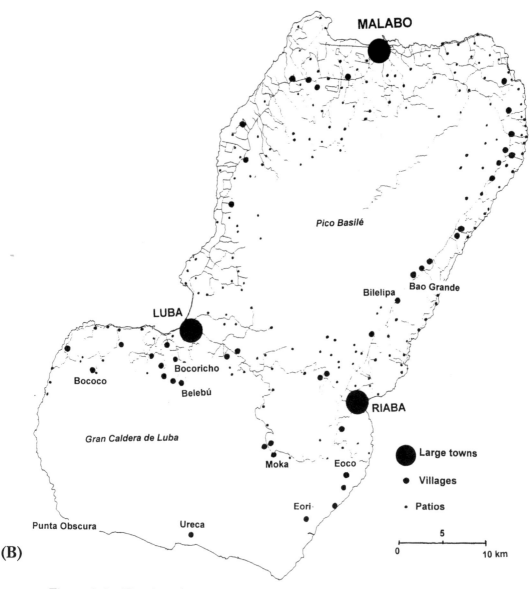

MALABO

Pico Basilé

Bilelipa Bao Grande

LUBA

Bocoricho

Bococo

Belebú

Gran Caldera de Luba

RIABA

Moka Eoco

Eori

Punta Obscura Ureca

● Large towns

● Villages

· Patios

5

0   10 km

(B)

Figure 8-2. (Continued)

Data from the six primate surveys point to the Gran Caldera de Luba as having by far the highest faunal densities. This is explicable in terms of the inaccessibility of the area; there is only one route to the site, which takes 2 days to cover on foot, and it is thus rarely visited by hunters. Animal populations here have probably remained relatively undisturbed for long periods (although see Hearn and Berghaier 1996 for a review of expeditions to the area). The Caldera's primate populations (1.2–3.3 encounters/km), include all seven species found in Bioko and compare most favorably with the highest densities recorded in Africa (Oates 1996a). Outside the Caldera, densities decrease to 0.96–2.1 encounters/km for

Figure 8-3. Drill (*Mandrillus leucophaeus*) in forest on Bioko Island. *Photo courtesy of Liza Gadsby.*

Table 8-1. Sighting Rates of Anthropoid Primates (Groups/km) in Bioko Habitats

Locality	Total km Censused	Cer	Cni	Cpo	Cpr	Mle	Ppe	Csa	Overall	Source
		Primate Sighting Rate[a]								
Pico Basilé	86.0	ND	ND	ND	ND	ND	ND	ND	0.22	Butynski & Koster 1994
Caldera de Luba	24.0	ND	ND	ND	ND	ND	ND	ND	2.18	Butynski & Koster 1994
	92.0	0.8	—	0.30	0.01	0.10	0.40	0.40	1.96	Schaaf et al. 1990
	ND	S	—	S	—	S	1.20	0.50	3.25	Gonzalez-Kirchner 1991
	26.3	0.42	—	0.19	—	0.04	0.49	0.27	1.18	Hearn & Berghaier 1996
Rio Epola	41.5	1.01	0.12	1.01	—	—	0.48	0.10	2.12	Schaaf et al. 1992
Ariha	100.19		0.03	0.05	0.15	—	0.01	—	1.10	Maté & Colell 1995
S.W. Bioko	51.0	ND	ND	ND	ND	ND	ND	ND	0.96	Butynski & Koster 1994
All Bioko	373.0	0.13	0.01	0.04	0.03	0.008	0.03	0.02	0.29	Butynski & Koster 1994

[a]Cer, *Cercopithecus erythrotis*; Cni, *Cercopithecus nictitans*; Cpo, *Cercopithecus pogonias*; Cpr, *Cercopithecus preussi*; Mle, *Mandrillus leucophaeus*; Ppe, *Piliocolobus pennanti*; Csa, *Colobus satanas*; ND, no data given; S, seen but not counted. Dashes indicate that the species was not seen or was absent.

sites in the Southern Highlands (lightly hunted) and dramatically to 0.22 for the Pico Basilé areas (table 8-1). Primate densities are thus significantly higher in the Southern Highlands (35.4–89.8% higher) than in the Pico Basilé or other forest areas. The crowned monkey and red colobus currently are not found in Pico Basilé (Butynski and Koster 1994).

The patterns of distribution and abundance of duiker populations reflect the survey data for primates. Butynski et al. (1995b) noted that the blue duiker (*Cephalophus monticola*) is widespread throughout Bioko and is most common in the Caldera (0.1 encounters/km), but less abundant in Pico Basilé and in the Southern Highlands. Ogilby's duiker (*Cephalophus ogilbyi*) is also more common in the Caldera (0.2 encounters/km), with populations heavily reduced in other areas. Both species are found from lowland forest to high elevations (Butynski et al. 1995b).

Table 8-2 is an attempt to estimate population sizes of the most commonly hunted mammals in the island. This approach is still rudimentary and cannot substitute for more detailed long-term censuses. Nonetheless, to provide a first approximation of potentially sustainable harvests, this exercise has an important heuristic value. The starting point was the average population densities reported by Fa and Purvis (1997), drawn from a number of tropical forest sites in Africa. These data do not distinguish habitat types and assume that animal densities in montane and mossy forests, for example, are equivalent to those in lowland rain forests. Densities for each area (Pico Basilé, Caldera de Luba, and Southern Highlands) were then estimated by multiplying these average densities by the observed decrease in abundance, as shown in table 8-1. Thus, densities in the undisturbed Caldera de Luba were taken as being equivalent to the average densities reported by Fa and Purvis (1997), whereas Southern Highlands values were multiplied by 0.65, those in Pico Basilé by 0.10, and those in the Plantations/Secondary Forest regions by 0.08. From these data, the total biomass of mammals estimated for pristine habitats (i.e., the Caldera) is probably in the region of 143 kg/km^2. Most of this is composed of primates (70.3%), largely red colobus. This contrasts with mammal biomass figures for mainland forests, which are close to 1000 kg/km^2 (Prins and Reitsma 1989; White 1994) due to the inclusion of large herbivores such as pigs and elephants.

Wildlife population estimates for the two main areas (Pico Basilé and Southern Highlands) illustrate the importance of the Caldera and surrounding areas. Practically all species of primates on the island have their main population strongholds in the Southern Highlands. Within these areas, the monsoon forest in the Southern Highlands and the montane and mossy forests in the Pico hold the largest populations. However, these estimates show that minimum viable populations for some species might be seriously compromised. For example, populations of drills, Preuss's monkey, and the crowned monkey, all between 2000 and 3000, might not be sufficient to be viable in the long term. Added to this, the likely extinction of the remaining small populations of these species in Pico Basilé might be imminent because fewer than 500 drills and Preuss's monkeys are thought to remain. The drill and Preuss's monkey are two of the most endangered primates

Table 8-2. Estimated Population Sizes for the Main Hunted Mammal Prey Species in Bioko Island According to Habitat Types and Distribution Areas

| | | Southern Highlands | | | | | | | Pico Basilé | | | | Other Areas | |
| | | Area No. and Habitat Type | | | | | | | | | | | | |
Species	Observed	1 Mf	7 Lrf	11 Mof	8 Mf	9 Msf	6 Lrf	2 Lrf	14 Lrf	5 Lrf	13 Mf	3 Msf	12 Pl/Sf	Totals
Cercopithecus erythrotis	24.7	494	1091	4717	2847	585	—	—	78	54	568	246	—	10,680
Cercopithecus nictitans	22.7	319	1003	4335	—	—	—	—	71	50	—	—	—	5778
Cercopithecus pogonias	14.4	288	636	2750	—	—	—	—	—	—	—	—	—	3386
Cercopithecus preussi	9.9	186	—	—	1141	234	—	—	—	—	227	99	—	1887
Mandrillus leucophaeus	6.7	128	296	1279	772	—	—	—	21	15	154	67	—	2732
Piliocolobus pennanti	156.3	87	6906	29,848	18,015	3699	—	—	492	—	—	—	—	59,047
Colobus satanas	20.4	2021	901	3896	2351	483	—	—	64	45	469	203	—	10,433
Atherurus africanus	55.0	264	2430	10,503	6339	1302	48	124	173	121	1264	548	3605	26,721
Cricetomys emini	134.0	711	5921	25,590	15,445	3171	117	303	422	294	3079	1335	8784	65,172
Cephalophus monticola	22.6	1733	999	4316	2605	535	20	51	71	50	519	225	1481	12,605
Cephalophus ogilbyi	13.0	292	574	2483	1498	308	—	29	41	29	299	129	852	6534
Totals		6235	20,757	89,717	51,013	10,317	185	507	1433	659	6579	2852	14,722	204,975

Observed average densities are derived from Fa and Purvis (1997). Area numbers refer to those in figure 8-1.

in Africa (Oates 1996b), whereas the red-eared monkey, the black colobus, and Pennant's red colobus are also considered at risk (Lee et al. 1988).

HUNTING

Hunting Activity

Local people harvest a wide range of vertebrate species (table 8-3). Twenty-one mammals (out of a resident 65 species), two birds (out of 187 species), and two reptiles (out of 55 species) are regularly hunted in Bioko. In the forest areas, giant snails (Castroviejo 1995) are collected, as are sea turtles along the southern coast

Table 8-3. Species Hunted in Bioko Island

Species	Common Name	Body Mass (g)	Diet	Pl/ Sf	Lrf/ Mof	Mf	Msf	Ew/ Gr
Mammals								
Eidolon helvum	Straw-colored fruit bat	535	FH	X	X	X	X	
Rousettus aegyptiacus	Egyptian fruit bat	140	FH	X	X	X	X	X
Cephalophus monticola	Blue duiker	5000	FH	X	X	X	X	
Cephalophus ogilbyi	Ogilby's duiker	19,000	FH	X	X	X	X	
Dendrohyrax dorsalis	Tree hyrax	3000	HB	X	X	X	X	
Phataginus tricuspis	Tree pangolin	2500	MY	X	X	X	X	
Cercopithecus erythrotis	Red-eared monkey	4273	FO		X	X	X	
Cercopithecus nictitans	Putty-nosed monkey	8700	FO		X	X		
Cercopithecus pogonias	Crowned monkey	1346	FO		X			
Cercopithecus preussi	Preuss's monkey	9650	FO			X	X	
Piliocolobus pennanti	Pennant's red colobus	7989	HB		X	X		
Colobus satanas	Black colobus	12,000	FG		X	X	X	
Mandrillus leucophaeus	Drill	14,925	FO		X	X		
Galago alleni	Allen's squirrel galago	423	FO		X	X	X	
Euoticus elegantulus	Elegant needle-clawed	450	FO		X	X	X	
Atherurus africanus	Brush-tailed porcupine galago	4000	FO		X	X	X	
Cricetomys emini	Giant pouched rat	2000	FO		X	X	X	
Anomalurus beecrofti	Beecroft's anomalure	650	FG		X	X	X	
Anomalurus derbianus	Lord Derby's anomalure	775	FG		X	X	X	
Funisciurus leucogenys	Red-cheeked rope squirrel	250	FG		X	X	X	
Heliosciurus rufobrachium	Red-legged sun squirrel	325	FG		X	X	X	
Paraxerus poensis	Green squirrel	120	FG		X	X	X	
Protoxerus stangeri	African giant squirrel	770	FG		X	X		
Birds								
Corythaeola cristata	Great blue turaco	1000	FO			X	X	
Ceratogymna atrata	Black-wattled hornbill	2000	FO			X		
Reptiles								
Python sebae	African python	2500	CA		X	X		
Varanus niloticus	Nile monitor	1000	CA			X		

CA, carnivore; FG, frugivore-granivore; FH, frugivore-herbivore; FO, frugivore-omnivore; HB, herbivore-browser; MY, myrmecophage.

Taxonomic names follow Kingdon (1997). Data from Colell et al. (1994), Fa et al. (1995), Juste and Perez del Val (1995), Fa and Purvis (1997), and Kingdon (1997).

(Castroviejo et al. 1994a). Hunters use guns and dogs as well as a wide range of traps in virtually all of the island's habitats. The cable snare is the predominant trap (Colell et al. 1994). As in other African countries (Noss 1995), cable snaring has now replaced most other traditional snaring and trapping techniques.

According to Castroviejo (1995), most families in rural and even urban areas in Bioko hunt either professionally or for sport. Because there are no community-based or government-imposed closed seasons, hunting takes place throughout the year, although not always at the same intensity. The rainy season sees a decrease in hunting activity because of the heavy rains and swollen rivers. In villages where there are farms, some hunters might turn to clearing, burning, and planting during this time of the year. Outside the wet season, hunting intensity is relatively constant (Juste et al. 1995). During any season, time dedicated to hunting depends to a large extent on the individual hunter's preference. Most hunters shoot and trap away from villages (Colell et al. 1994), but some may venture into more remote areas.

Castroviejo (1995) argues that most of the hunting in Bioko is undertaken by part-timers, from government officials to members of low-income families, who live in Malabo and along the villages between the capital and Banabé, and also in Luba and Riaba. These hunters operate on weekends or days off and hunt to take meat to the family table or to sell in order to recoup ammunition, gun hire, and transport costs. Most hunt in the Pico Basilé area and surroundings. Because wildlife is relatively scarce around the Pico region, in contrast to sport hunters, most commercial hunters operate in the south of the island. These professional hunters, mostly immigrant Fangs from Rio Muni, set up permanent camps from which hunting parties start and end. In 1991, Juste (1992) reported a minimum of four hunting camps within the Southern Highlands protected area (Eori, Belebú, Eoco, and Punta Obscura). These camps were located about 3 to 4 hours' walk from the nearest village. Some professional hunters are known to live in villages (Moka, Beleb, Batete, Bao Grande, Bilelipa) from which they then enter the forest to hunt. Hunting excursions are usually night events (Castroviejo 1995). According to Colell et al. (1994), hunters indicated that they had to cover increasingly greater distances to find wildlife. This is shown by greater yields at greater distances from villages (see figure 8-4). This pattern arises, as seen in other studies in Africa, because animal populations closest to human habitation are the first to be depleted (Lahm 1993).

Prior to independence, the number of shotguns on the island was certainly higher than it is now, owing to their primary use to control squirrel predation on cocoa pods. However, the guns also were used to hunt other species, namely antelopes and monkeys; in Ureka alone (300 inhabitants at the time), 25 guns were known to be present. With dictatorial rule, all guns were prohibited in 1974. The actual number of hunters and guns active throughout the island presently is difficult to estimate, but it is certainly less than the pre-1974 period. Shotguns are still not easily available because of costs and political control. Certain restrictions are still in force with regard to the purchase or importation of guns and ammunition. Nevertheless, in Malabo alone, Juste (1992) and Castroviejo (1995) estimated that

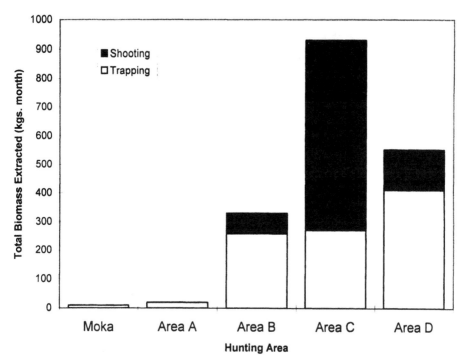

Figure 8-4. Variation in biomass of animals hunted with increasing distance away from the Moka villages in southeastern Bioko. Data from Colell et al. (1994). Area *A* = 8 km from the villages; area *B* = 10 km; area *C* = 12 km; area *D* = 14 km.

over 200 shotguns were in use at any one time. Data collected in the Moka villages (southeast Bioko) indicate that most men from 18 to 60 years old hunted regularly with guns and snares (Colell et al. 1994); all hunters do not own guns, but some are able to hire them from their owners. The picture that emerges, extrapolating from the age structures of villages (Mas et al. 1995) as well as from the distribution of the human population (see figure 8-2), is that most of the island is now subject to very intense hunting pressure. Nonetheless, hunting intensity will vary according to differences in accessibility of the surrounding terrain, traditions, and socioeconomic status of the inhabitants.

Hunting Success

Snares are not species specific and catch virtually all of the terrestrial medium-sized prey species. Shotgun hunting, in contrast, is directed at the more arboreal monkeys and the two colobus, drill (usually with the aid of dogs) and Ogilbyi's duiker. Large numbers of snares are set out in extensive areas of forest. In an area of about 100 km² around the Moka villages, a total of 4200 snares was set throughout the year (Colell et al. 1994). Similar trap densities are likely to be found in southwest Bioko, manned by hunters from Belebé, Bocoricho, and Bococo, but trapping is probably less intense in the north of the island. The number of traps set tends to increase as one moves away from inhabited areas, presumably matching the availability of wildlife (figure 8-5), up to the distance where the time to check traps regularly is too great, at which point the number of traps declines. Hunting with guns is more important in those areas furthest away from the villages (Colell et al. 1994). The density of snares in Bioko seems much higher than that reported for Congo-Zaire (Almquist 1992), Cameroon (Infield 1988), Gabon (Lahm 1993), and Congo (Wilson and Wilson 1991) where individual hunters might maintain 15 to 500 snares each. In these countries, professional hunters are the only ones who keep the high numbers of snares typical in Bioko.

The only available data on hunting success in Bioko come from hunter surveys in the southeastern part of the island (Colell et al. 1994). Hunters hunt with both shotguns and snares. In this 4-week study, a total of 245 animals (19 species, 428.5 kg of meat) was taken in 69 monitored hunting trips. Hunters harvested around 2.9 animals per trip (or 25.7 kg of wild meat per trip) with shotguns. Slightly more animals were caught in traps (3.6 animals per trip), although these comprised significantly less meat per trip (14.1 kg of meat per trip). Capture success varied according to a species' propensity to be shot or snared (figure 8-5). For instance, 148 rodents (mostly *C. emini*) were caught in snares; a rate of 0.04 animals per trap. This contrasts with the rates calculated for ungulates, 0.02 animals per trap (mostly *C. monticola*) and for primates (0.003 animals per trap). These data do not include the number of animals that escape with injuries, or those that decompose or are eaten by scavengers. In the only study on the subject, in Bayanga, Central African Republic (CAR), Noss (1995) reported up to 40% wastage for some species.

Weekly yields from snare trapping and shotgun hunting average 17.9 kg per hunter (Colell et al. 1994), but Castroviejo (1995) believes it can be up to 70 kg, although this figure is unconfirmed. Assuming that all marketable meat is sold, annual yields may average 240,000 to 540,000 CFAs per hunter (Castroviejo 1995). Infield (1988) calculated around 350,000 CFAs per Cameroon hunter per year and 360,000 for hunters in the CAR (Noss 1995). Average weekly yields from hunting exceed the official minimum wage.

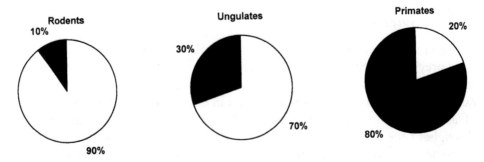

Figure 8-5. Proportion of the main mammal species groups trapped and shot, from data in Colell et al. (1994). Dark sections of the pie charts denote the proportion of animals hunted with shotguns.

BUSHMEAT TRADE CIRCUIT

The bushmeat trade circuit in Bioko does not differ from those in other studied African countries. Producers (the hunters) take their quarry to intermediaries (taxi drivers), who supply the end retailer in the markets (figure 8-6). Some meat actually flows to villagers (where it is consumed), and a proportion of the amount taken by the taxi drivers is sold to passengers before it ever gets to town (figure 8-7). Castroviejo (1995) points to the trade circuit being run by Fangs at both the production and commercialization ends, and therefore essentially controlling the bulk of the trade to the Malabo market. The business triangle, hunter–transporter–retailer, is prearranged and in some cases may involve relatives working together. Women, around 20 of them according to Castroviejo (1995), are the principal processors and distributors of bushmeat in the market. They are the ones who purchase the meat from the hunters (via the taxi drivers) and sell to the public. Unlike the larger continental areas (e.g., Congo-Zaire, Colyn et al. 1987a, 1987b; CAR, Kalivesse 1991; Ghana, Addo et al. 1994), in Bioko commercialization of bushmeat is less complicated because there are few middle persons involved. This is a consequence of the shorter distances (maximum 80 km) between the source area and market. An important corollary to this is that more fresh meat and less smoked meat appear in markets daily, although this is likely to change as prey becomes scarcer and hunting trips longer.

Bushmeat in Bioko is not cheap (table 8-4). Market prices for bushmeat recorded in 1991 and more recently in 1996 point to prices ranging from $1/kg to around $24/kg. Although there are no standard prices or cuts for large animals, primates are more expensive than rodents and ungulates, with the exception of Ogilbyi's duiker. From prices recorded in villages and markets at two different times, the most expensive are the larger primates, the drill, red colobus, and black colobus, in that order. Prices differed by as much as 30% between village and market. An average increase of around 19% in prices was detected between market prices for 1991 and for 1996. An increase of more than 50% was detected for the drill and for the crowned monkey.

BUSHMEAT CONSUMPTION RATES AND PATTERNS

In a study undertaken by Juste et al. (1995), 12,974 carcasses, the equivalent of 111,879.63 kg of dressed meat, were counted in a year in the Malabo market. Ungulates (two species) accounted for 36.7%, followed by primates (seven species, 26.2%) and rodents (two species, 21.2%). The most numerous were the blue duiker and the giant pouched rat (*Cricetomys emini*). Three species, the blue duiker, giant pouched rat, and brush-tailed porcupine, made up almost 70% of all bushmeat on sale, both in terms of numbers as well as biomass. Abundance (defined as average daily number of carcasses of species appearing in the market) and availability (defined as the percentage number of days during which the species was seen in the market) of bushmeat species varied with season. Because of the typical heavy rains during the wet season, fewer carcasses (10 carcasses/day) were

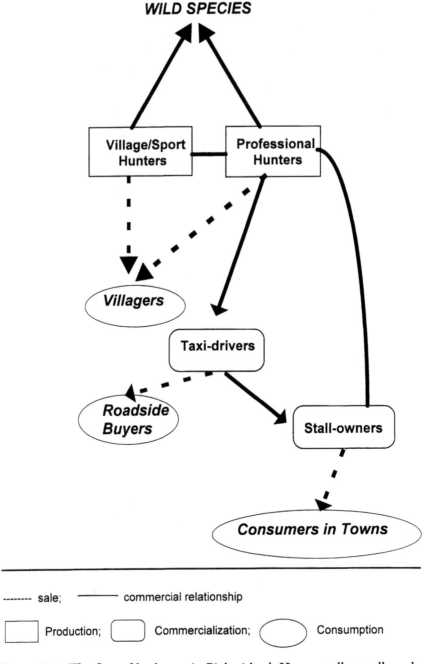

Figure 8-6. The flow of bushmeat in Bioko island. Hunters sell to stall vendors, who receive the quarry from taxi drivers, who get paid per trip.

observed in market between April and July with a minimum in May. In contrast, more wild meat was seen (around 30 carcasses/day) after September. The greatest availability of any species (40–50%) occurred in the September to December period, but a trough of 30% to 35% appeared from March to July. The relationship between abundance and availability of wildlife species in the market is poly-

Figure 8-7. Blue duiker (*Cephalophus monticola*) and red-eared monkey (*Cercopithecus erythrotis*) being sold on the roadside in Bioko. *Photo courtesy of John Fa.*

nomial. The three more abundant species in the market are species that not only are found in large numbers for sale, but which buyers also can find regularly in the market (figure 8-8).

There are essentially two classes of animals that enter the commercial market: (a) the large-bodied antelopes and monkeys, which are normally hunted with shotguns and provide a good rate of economic return (kg/time spent hunting or kg/cost of hunting, including shotgun hire, shells, and transport costs), and (b) the smaller animals that are normally snared, and therefore easy to catch, but not as valuable. Animals of the first class are more likely to enter the market, whereas those in the smaller class are more likely to be consumed by the hunter and his household. This pattern has been confirmed by data from Congo-Zaire (Colyn et al. 1987a, 1987b) and Gabon (Lahm 1993). The only information on this subject for Bioko shows that most antelopes (92%) that were taken to southeastern Bioko

Table 8-4. Prices[a] of Some Bushmeat Species Based on Animals Obtained from Villages and Malabo Markets

Scientific Name	Village[b] Average Price[c]	Price per kg of Carcass	Malabo Market[d] Average Price	Price per kg of Carcass	Malabo Market[d] Average Price	Price per kg of Carcass
Artiodactyla						
Cephalophus monticola	1500	0.30	2750	0.55	—	—
Cephalophus ogilbyi	5000	0.26	7000	0.37	—	—
Rodentia						
Atherurus africanus	2000	0.50	3500	0.88	—	—
Cricetomys emini	500	0.25	1150	0.58	—	—
Primates						
Mandrillus leucophaeus	4000	0.27	6000	0.40	12,000	0.80
Piliocolobus pennanti	2750	0.34	3250	0.41	6000	0.75
Colobus satanas	1250	0.10	3750	0.31	6000	0.50
Cercopithecus erythrotis	1250	0.29	2750	0.64	3500	0.82
Cercopithecus nictitans	2000	0.23	2750	0.32	3500	0.40
Cercopithecus pogonias	1250	0.93	2750	2.04	3500	2.60
Cercopithecus preussi	1250	0.13	2750	0.28	3500	0.36
Pholidota						
Phataginus tricuspis	2000	—	4500	—	—	—
Birds						
Corythaeola cristata	1000	—	—	—	—	—

[a]In Cefa francs: 500 FCFA = *ca* US$1.
[b]M. Colell, J. Juste, personal communications.
[c]Average price for a whole carcass of the species.
[d]Hearn and Berghaier 1995.

villages were sent to market, but less than 10% of the smaller game (e.g., *Cricetomys*, *Atherurus*, *Phataginus*) was consumed in the village (Colell et al. 1994). However, the general trend is for the proportion of animals entering villages (a measure of hunting impact on each species) (Colell et al. 1994) to reflect the proportion appearing in the market (Juste et al. 1995) (figure 8-9). Three species fall outside this main trend: the two colobus monkeys, which are more abundant in the market, and crowned monkeys, which appear less than expected in the market.

In comparisons across bushmeat markets in western and central Africa, the species composition of animals for sale varies (table 8-5). Most markets rely on ungulates, primates, and rodents, and to a lesser extent on carnivores (average around 3%). Although in the lightly hunted rural sites (e.g., Bayanga, and two sites in Gabon, sites 2 and 3) over half of all recorded carcasses were duikers and antelopes, in sites that have been heavily exploited, the percentage contribution of ungulates decreases to less than 20%. Moreover, there is a significant inverse relationship between the number of ungulates and rodents in markets (figure 8-10). This picture emerges because in sites with high human population densities, particularly those near large urban areas (Kisangani in Congo-Zaire and cities in

A) Luba   B) Riaba

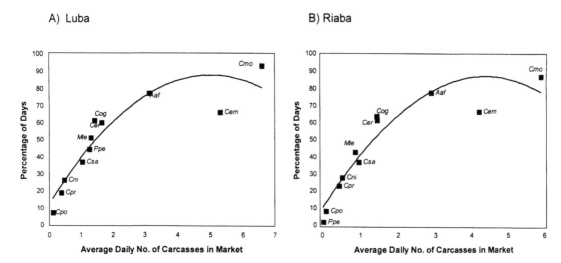

Figure 8-8. Prey species appearing in the Malabo market: Luba section **(A)** and Riaba section **(B)**. The Luba and Riaba market sections refer to the western and eastern regions of the island, respectively, from which the wild meat has been hunted. Daily availability and average number of carcasses appearing per day are plotted to show the relationship between abundance and regularity of prey in the market. Data from Juste et al. (1995). Species codes: Cer, *Cercopithecus cephus erythrotis*; Cni, *Cercopithecus nictitans*; Cpo, *Cercopithecus pogonias*; Cpr, *Cercopithecus preussi*; Ppe, *Piliocolobus pennanti*; Csa, *Colobus satanas*; Mle, *Mandrillus leucophaeus*; Cem, *Cricetomys emini*; Aaf, *Atherurus africanus*; Cmo, *Cephalophus monticola*; Cog, *Cephalophus ogilbyi*.

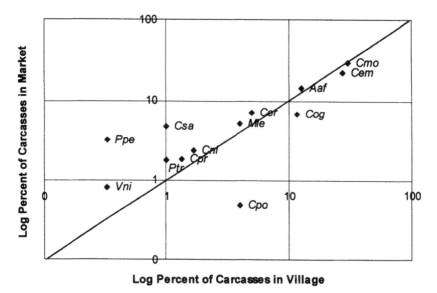

Figure 8-9. Relationship between the percentage number of carcasses of species hunted in the Moka villages and that appearing in the Malabo markets. Data from Colell et al. (1994) and Juste et al. (1995). The expected line is included. Species codes: Cer, *Cercopithecus cephus erythrotis*; Cni, *Cercopithecus nictitans*; Cpo, *Cercopithecus pogonias*; Cpr, *Cercopithecus preussi*; Ppe, *Piliocolobus pennanti*; Csa, *Colobus satanas*; Mle, *Mandrillus leucophaeus*; Cem, *Cricetomys emini*; Aaf, *Atherurus africanus*; Cmo, *Cephalophus monticola*; Cog, *Cephalophus ogilbyi*; Ptr, *Phataginus tricuspis*; Vni, *Varanus niloticus*.

Table 8-5. Mammal Group Composition of Selected Market Sites in West and Central Africa

		Percent Composition of Main Mammal Groups for Sale						
Site	Market Type	All ungulates	Duikers	C. monticola	Primates	Rodents	Carnivores	Source
Bayanga	Villages	78	71	32	10	0	0	Noss 1995
Bioko	Malabo market	33	33	29	25	37	0	Juste et al. 1995
Congo 1	Roadside	47	47	6	8	35	3	Wilson 1990
Congo 2	Urban markets	39	38	3	3	20	1	Wilson 1990
Gabon 1	Libreville market	40	34	1	20	27	2	Steel 1994
Gabon 2	Urban markets	51	43	15	15	12	3	Steel 1994
Gabon 3	Village market	81	70	15	12	6	0	Lahm 1993
Nigeria 1	Urban markets	32	26	0	9	48	4	Martin 1983
Nigeria 2	Urban markets	18	11	0	16	61	1	Anadu et al. 1988
Rio Muni	Bata market	41	22	15	22	30	2	Juste et al. 1995
Zaire 1	Kisangani market	13	12	8	21	44	15	Colyn et al. 1987
Zaire 2	Urban markets	44	0	0	16	36	0	Colyn et al. 1987

Nigeria), an increase in rodent numbers in markets indicates that forest ungulates and primates have already been exterminated. This is corroborated by data from Falconer (1991) for Ghana and by Wilkie (1989) for Congo-Zaire, where species hitherto unimportant in the bushmeat trade (e.g., the grasscutter [*Thrionomys swinderianus*] and the giant pouched rat) gain prominence when ungulates and primates become rare. Given the choice, hunters will hunt large ungulates first, then primates, and lastly rodents and small ungulates in line with increasing effort and decreasing economic returns.

Bioko appears similar to the more exploited area markets because of the emphasis on rodent meat (37% of all carcasses counted). This is probably, at least at present, not so much due to overexploitation of the ungulates and primates, but a reflection of the poorer ungulate biomass typical of the island. This might be forcing Bioko hunters to dedicate more time to snaring rodents, although economic limitations in the hiring of guns and purchase of ammunition also might be a factor. The overall result, however, is that pressure on the blue duikers is likely to be much greater compared with the continental sites; this species accounts for nearly 90% of the ungulate meat in Bioko markets and almost a quarter of the total bushmeat biomass (figure 8-11). In mainland African markets, blue duikers contribute up to half of the ungulate meat.

THE DEMAND FOR PROTEIN IN BIOKO

Despite the seemingly large number of animals on sale in the Malabo markets, production might not be enough to supply the protein needs of the population. Recent human nutrition studies in Bioko confirm very low protein intakes and a high incidence of malnutrition (Frantz et al. 1984). The deficit in protein for the entire country is judged to be enormous. A rural development study carried out in

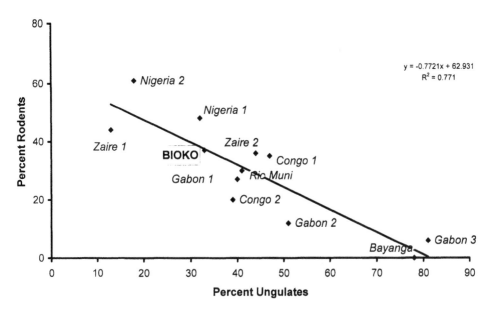

Figure 8-10. Relationship between percentage number of ungulates and rodents appearing in bushmeat markets in central and west Africa. Data sources as in table 8-5.

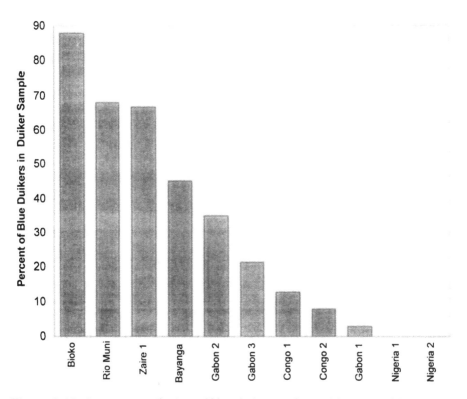

Figure 8-11. Percent contribution of blue duikers to the total biomass of duiker meat appearing in west African and central African markets. Data sources as in table 8-5.

1984 (FAO/PNUD 1984) calculated that total protein (animal and plant) intake in the country was around 30 g per inhabitant per day, about half as much as the recommended level. An estimated 0.03 tons of animal protein per inhabitant per year would be required to meet the minimum daily recommended allowance (minimum U.S. RDA is 0.28 kg of meat per day for an average man; see Wilkie et al. in press). These figures are also confirmed by the bushmeat biomass available per inhabitant in the Malabo region (assuming that this is the major protein source). If the 111,879.63 kg of dressed meat recorded in the Malabo markets by Juste et al. (1995) are divided by the 30,000 inhabitants in the region (assuming a carcass yield of around 60%; Feer 1993), there would be a mere 6 g of meat per inhabitant per day available. The reality, however, is more likely to be that only certain sectors of the population can buy bushmeat. If most meat (certainly the more expensive primates and ungulates) is only accessible to the higher income social strata, as observed in Nigeria by Martin (1983), this bias underlies an even greater problem of lack of animal protein for the entire population.

THE IMPACT OF HUNTING: IS IT SUSTAINABLE?

If hunting is to continue in Bioko, harvest estimates for the different species within each habitat block are necessary. Production values calculated using Robinson and Redford's (1991b, 1991c) model for the 11 most important hunted mammal species show that the maximum expected sustainable harvestable numbers and biomass are around 4729 and 23,000 kg, respectively. Most of this biomass would be made up by the two most productive species: *P. pennanti* and *C. emini*, but *C. ogilbyi* can produce another 15% of the biomass (table 8-6). By habitat, the most productive are the monsoon forest and montane forest blocks in the Southern Highlands. Calculated yields according to the main areas in the island are 19,866 kg of wild meat from the Southern Highlands, 1540 kg from Pico Basilé, and 1277 kg from the plantation belt, a total of 22,672 kg. Within the natural habitats, most of the available bushmeat would come from the seven primate species and from the more productive rodents and blue duiker. Of significance is the amount that can be produced within plantations and secondary forest. Although these make up a mere 6% of the estimated production, these areas might be the only ones in which hunting could be allowed.

Overharvesting of wildlife species will have a negative impact on populations. The extent of this impact on populations in Bioko was first estimated by Fa et al. (1995) by comparing figures generated by the Robinson and Redford model with market data provided by Juste et al. (1995). Average densities and life history data were taken from the literature (Fa and Purvis 1997). Fa et al. (1995) calculated extraction rates for all mammal prey species in Bioko at around 6.49 animals/km^2 or 55.94 kg/km^2. By species, rates varied between 1.59 animals/km^2 for the Ogilbyi's duiker to 0.10 animals/km^2 for the pangolin. Within Equatorial Guinea, the total volume of wild meat harvested in Bioko is significantly higher than that taken in a larger catchment area in Rio Muni—a difference of over 60% more carcasses in Bioko (Fa et al. 1995; Juste et al. 1995). However, harvest rates for Bioko are substantially lower than in continental sites for which data are available. For example, extraction rates in Korup, Cameroon (Infield 1988), were 271,000 kg of wild meat per year (217 kg/km²) in 1987 to 1988, whereas figures for subsistence hunters in the Arabuko-Sokoke forest, East Africa (FitzGibbon et al. 1996) were around 130,000 kg of game meat per year (a harvest rate of 350 kg/km²). These differences in harvesting levels between Bioko and the continental sites are difficult to interpret because it is unknown whether the rates for the island site reflect overharvesting or whether the continental sites are always likely to be more productive because of the higher biomass of ungulates and primates present.

The calculated percentage deviation of actual from sustainable harvest levels (actual harvests divided by the maximum sustainable harvests calculated from the Robinson and Redford model) averaged 4.98 times greater than the sustainable harvest in Bioko as estimated by Fa et al. in their 1995 study. This contrasted dramatically from figures obtained for the continental area of Rio Muni. By species, offtakes ranged from 28 times greater than the sustainable harvest for the crowned monkey, to 0.96 times less for highly prolific species like the giant pouched rat. Exploitation was regarded as unsustainable for four out of the seven hunted primates and for Ogilbyi's duiker (30.7% of all recorded species). A recalculation of potential harvests, according to the more accurate habitat block data given in this study, shows that only hunting rates for the red colobus and the pouched rat are in the sustainable range. Deviations from sustainable harvests are largest for the drill and crowned monkey (figure 8-12). Among the primates, only hunting of the red colobus, usually the most abundant and most folivorous colobus species (Oates 1996a), might be sustainable.

The deviation from sustainable harvest figures reported herein are most unsettling because offtake estimates used were for the period 1990 to 1991, and there is some evidence that hunting might have increased since then. Moreover, harvest figures were minimum estimates because only market meat was considered. If village consumption were to be taken into account, the overall effect of hunting could be even more dramatic (Colell et al. 1994). This is not an unrealistic scenario given recent reports that the Riaba market site in Malabo (which is supplied by meat from the Pico), active in 1990 to 1991 (Juste et al. 1995), has seemingly disappeared, perhaps in response to a decreasing number of hunted animals from Pico Basilé (Hearn and Berghaier 1996).

PROVIDING ALTERNATIVES: BETWEEN THE DEVIL AND THE DEEP BLUE SEA

Steps toward conserving Bioko's biodiversity can only succeed if people's needs are considered. Not to address this issue would be to misunderstand completely the island's central problem, i.e., how to provide meat to the 50,000 people in urban centers and the 12,000 in rural areas. Assuming a consumption rate of 0.28 kg of meat per person per day, derived from studies of rural Liberians (Anstey

Table 8-6. Annual Potential Harvests (No./km²), Potential Harvestable Numbers (No.), and Harvestable Biomass of Dressed Meat (kg) in Each Area for the Main Hunted Mammal Species in Bioko Island

| | Habitat and Area No. | | | | | | | | | | | | |
| | Southern Highlands | | | | | | | Pico Basilé | | | | Other | |
Species	Mf 1	Lrf 7	Mof 11	Mf 8	Msf 9	Lrf 6	Lrf 2	Mf 13	Msf 3	Lrf 14	Lrf 5	Pl/Sf 12	Totals
Ceropithecus erythrotis													
No./km²	0.50	0.33	0.33	0.33	0.33	—	—	0.05	0.05	0.05	0.05	—	2.01
Harvestable no./km²	10	22	96	58	12	—	—	12	5	2	1	—	218
Harvestable biomass area	43	95	411	248	51	—	—	49	21	7	5	—	930.89
Cercopithecus nictitans													
No./km²	0.30	0.19	0.19	—	—	—	—	0.03	0.03	0.03	0.03	—	0.81
Harvestable no./km²	6	13	57	—	—	—	—	7	3	1	1	—	88
Harvestable biomass area	52.14	115.18	497.83	—	—	—	—	59.90	25.96	8.20	5.72	—	764.92
Cercopithecus spogonia													
No./km²	0	0	0	—	—	—	—	0	0	—	—	—	0.04
Harvestable no./km²	0	1	3	—	—	—	—	0	0	—	—	—	5
Harvestable biomass area	0.47	1.03	4.44	—	—	—	—	0.53	0.23	—	—	—	6.70
Cercopithecus preussi													
No./km²	0.13	—	—	0.08	0.08	—	—	0.01	0.01	—	—	—	0.33
Harvestable no./km²	3	—	—	15	3	—	—	3	1	—	—	—	25
Harvestable biomass area	25	—	—	145	30	—	—	29	13	—	—	—	241.95
Mandrillus leucophaeus													
No./km²	0.16	0.06	0.06	0.06	—	—	—	0.01	0.01	0.01	0.01	—	0.37
Harvestable no./km²	3	4	17	10	—	—	—	2	1	0	0	—	38
Harvestable biomass area	48.00	58.32	252.07	152.14	—	—	—	30.33	13.15	4.15	2.89	—	561.05
Piliocolobus pennantii													
No./km²	3	1	1	1	1	—	—	0	0	0	0	—	8.78
Harvestable no./km²	56	91	394	238	49	—	—	47	21	6	0	—	902
Harvestable biomass area	449.55	728.30	3147.83	1899.87	390.13	—	—	378.74	164.17	51.86	0.00	—	7210.45
Colobus satanas													
No./km²	0.54	0.17	0.17	0.17	0.17	—	—	0.03	0.03	0.03	0.03	—	1.35
Harvestable no./km²	11	12	51	31	6	—	—	6	3	1	1	—	122
Harvestable biomass area	129.00	143.00	617.00	372.00	76.00	—	—	74.00	32.00	10.00	7.00	—	1461.71

Table 8-6. (continued)

| | Southern Highlands | | | | | | Pico Basilé | | | | | Other | |
Species	Mf 1	Lrf 7	Mof 11	Mf 8	Msf 9	Lrf 6	Lrf 2	Mf 13	Msf 3	Lrf 14	Lrf 5	Pl/Sf 12	Totals
Atherurus africanus													
No./km²	5.41	0.47	0.47	0.47	0.47	0.07	0.07	0.07	0.07	0.07	0.07	0.06	7.79
Harvestable no./km²	108	32	139	84	17	1	0	17	7	2	2	48	456
Harvestable biomass area	432.96	128.31	554.57	334.71	68.73	2.54	1.04	66.72	28.92	9.14	6.37	190.35	1824.36
Cricetomys emini													
No./km²	32	2	2	2	2	0	0	0	0	0	0	0.28	44.08
Harvestable no./km²	650	156	676	408	84	3	1	81	35	11	8	232	2345
Harvestable biomass area	1299.26	312.61	1351.13	815.47	167.45	6.18	2.53	162.57	70.46	22.26	15.52	463.77	4689.22
Cephalophus monticola													
No./km²	3.42	0.39	0.39	0.39	0.39	0.06	0.06	0.06	0.06	0.06	0.06	0.05	5.37
Harvestable no./km²	68	26	114	69	14	1	0	14	6	2	1	39	354
Harvestable biomass area	341.71	131.81	569.69	343.84	70.61	2.61	1.07	68.54	29.71	9.39	6.54	195.55	1771.06
Cephalophus ogilbyi													
No./km²	0.81	0.22	0.22	0.22	0.22	—	0.04	0.04	0.04	0.04	0.04	0.03	1.90
Harvestable no./km²	16	15	66	40	8	—	0	8	3	1	1	22	180
Harvestable no./km²	308.26	288.11	1245.26	751.57	154.33	—	2.33	149.83	64.94	20.52	14.30	427.43	3426.88
Totals													
No./km²	47	5	5	5	5	0	1	1	1	1	1	0.00	72.84
Harvestable no./km²	932	373	1613	952	193	—	2	197	85	27	14	341	4729
Harvestable no./km²	3129.88	2001.57	8651.08	5063.54	1008.53	—	6.96	1069.85	463.73	—	—	1277.11	22,672.26

Robinson and Redford (1991) developed a simple model to provide estimates of potential harvest rates for different neotropical forest mammals. This required the use of measures of population density and the intrinsic rate of natural increase from which maximum potential harvest for the different species could be estimated. A figure emerges for the maximum sustainable harvest expected if production is at a maximum and harvesting is to be sustainable. The harvest model also uses life span as a good index of the extent to which harvest takes animals that would have died anyway (species were divided into three categories: long-lived [ù10 years]; short-lived [5–10 years] and very short-lived [<5 years]). Robinson and Redford (1991) assumed that harvest could take 60% of the production in very short-lived species, 40% in short-lived, and 20% in long-lived ones. Life history data on the Bioko mammals used herein can be found in Fa et al. (1995). Estimates were derived using Robinson and Redford's 1991 model (see Fa et al. 1995).

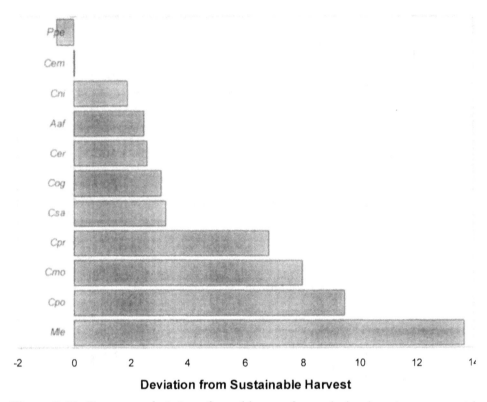

Figure 8-12. Percentage deviation of actual harvest from calculated maximum potential sustainable harvest for mammal species hunted in Bioko.

1991) and close to the U.S. RDA, more than 4 million kg of meat would have to be produced in the island every year. Sustainable levels of hunting could produce only 23,000 kg of meat. Nevertheless, part of the shortfall could be derived from alternative sources such as rearing game species and/or livestock. The latter is rarely suggested by biologists and conservationists, where the overwhelming consensus is that extensive livestock ranching in the tropics encourages destruction of forests. To them, the idea of introducing this practice as a solution to defaunation is anathema. Uncontrolled, extensive ranching can no doubt accelerate forest clearing for more grazing areas, and Myers' (1981) "hamburger connection" thesis has generated a psychosis about rearing beef in tropical lands. It is often assumed, perhaps incorrectly, that existing nondomesticated animal species "will feed the human race" (Feron 1995). Yet, although the rearing of "mini-livestock" (animal species smaller than sheep, goats, pigs, rabbits, or poultry) (Hardouin 1995) could fill another feeding niche in rural and peri-urban situations, it is doubtful, at least at present, that these meats could be produced at levels that would divert pressure from wild sources. The sustainable production of forest rodents, such as the giant pouched rat (Asibey 1974; Ajayi 1971, 1974) or the grasscutter (Asibey 1974; Jori et al. 1995), held as the panacea to the African food problem since the early 1960s (see Asibey 1974), has still not taken off. No substantial advances have been made, in quantitative terms, either in the domestication or in the exploitation of these animals in any country in West Africa. The farming of wild meat species for the market is expensive and probably not ecologically sound; but, most ominously, it will not satisfy the demand for meat now. Feer's (1993) and Jori's (1996) analyses of the feasibility of producing wild meat rests on comparisons of potential meat returns and not on actual returns. Thus, even though wild species may have a potentially higher productivity than domestic stock, per individual animal, their size means that a West African zebu or a tropical pig can render between 253 and 4757 times more meat than can potential bushmeat species such as cane rats, blue duikers, and bay duikers (*Cephalophus dorsalis*). Even if the expertise and technology for rearing such animals in large numbers had been resolved (knowledge of reproductive physiology, diet, and disease resistance) to produce the equivalent to the amount of wild meat extracted in Bioko in 1991 (100,000 kg of dressed meat per year), as a minimum around 6000 bay duikers (17.0 kg weight at slaughter; Feer 1993), 22,000 blue duikers (4.2 kg weight at slaughter; Feer 1993), or 24,000 cane rats would have to be ready for slaughter every year. This is unrealistic both on economic grounds and because of biological constraints. Critics of wild game farming share this skepticism and point to the limitations in reproductive output and behavior of wild species (Emmons 1992).

Given that the mass production of bushmeat is not likely to occur to satisfy the dearth of protein for Bioko, solutions could be found in the sensible rearing of more traditional livestock. This could be implemented in tandem with controlled hunting and management of forest resources, especially of the more prolific species such as rodents and blue duikers. Rearing domestic stock as a bushmeat substitute as opposed to farming wild animals could then be a more pragmatic answer to the prevailing emergency situation. All this has to be seen in the context of traditions and social structures, mitigated by the actual acquisitive powers of the island's human population. The importance of bushmeat as a mere commodity available to only those who can afford it needs to be understood. Otherwise, the new protein source might just replace bushmeat in certain social sectors, only to push its consumption to another lower income level.

In the past, Bioko has produced substantial quantities of domestic meat. Prior to independence, it was self-sufficient in beef and dairy products. Livestock raising almost disappeared during the 1970s due to widespread emigration and land devastation. Most livestock was raised in the Moka highlands, of which around 5000 ha of disused cattle pasture can still be found. Cattle ranching in Moka followed directly from the Decree of 11 July 1904, which set out property rights in Equatorial Guinea and authorized the concession of land for exploitation. In 1905, the Compañia Transatlántica began fencing fields, and soon about 60 cattle from Rio de Oro were being farmed. By 1907, a freehold concession of 150 ha was granted in Moka to the Compañia, even though the pastures had already been established (Moreno 1947). This settlement's objective was to provide the company's ships with meat and supply the European market. The cattle were driven to the beach at Riaba, and from there they were transported by boat to Malabo. The first available data for herd counts in Moka are for 1927 and 1928, when a total of 1001 (Valdes 1928) and 1202 head (Anonymous 1928), respectively, were counted. Business prospered, and in the 1946 census there were already three main cattle breeding ranches owned by Europeans in the Moka valley, covering a total of 4403 ha (Anonymous 1955). The largest ranch, belonging to the Ganaderia Agricola Ecuatorial, covered 2200 ha. The Compañia Colonial de Africa had a similar extension of 2183 ha and Marcelino Puente 20 ha. There was no rational cattle exploitation until the 1930s because there were too many different breeds and probably little control of stabling, diet, or cross-breeding (Valdes 1928). Between 1949 and 1955, large imports of "dwarf" cattle from neighboring Nigeria and Cameroon were made (Anonymous 1955). This animal is immune to tripanosomiasis and could be kept in low-lying areas. However, official bulletins (Anonymous 1954) indicated that cattle ranching was not to be allowed in low-lying areas to prevent the spread of sleeping sickness. The possibility of setting up a cattle ranching farm near Malabo at higher altitude (e.g., Pico Basilé) (Anonymous 1955) was also investigated at the time.

Data of livestock slaughtered in Malabo for the colonial period show that around 100,000 kg of meat was produced annually (e.g., 829 cattle; 124,000 kg in 1950). These figures do not distinguish between locally produced meat and imports, but meat production was nevertheless considerable. Since independence, there have been isolated attempts at meat production, but only after 1979. In 1986 the African Development Bank advanced a loan of US$12.6 million to revive cattle ranching in Moka. A further study of exploitation levels and distribution networks is urgently required.

CONCLUSIONS

The protection of areas must be seen in line with providing local people with alternatives (Eilers 1985). Embracing the task of protecting the environment for people and wildlife can be dictated by sectoral approaches because there is a clear interdependence between ecology and economics. Little success would be achieved in a place such as Bioko by outlawing hunting or blocking access to protected areas. Nonetheless, for wildlife to survive at all, it is necessary to set out measures to safeguard it in prescribed areas. A first step is to put into effect the already decreed protected areas of Pico Basilé and Bioko Sur (Southern Highlands). This is vital to ensure the conservation of the endemic fauna through hunting controls because these areas are the only places left where it can survive. Protective boundaries set down in the existing legislation (Castroviejo et al. 1986, 1994a; Fa 1992a, 1992b) must be demarcated, kept, and safeguarded. Some areas within the parks could be exploited on a rotational basis, with fallow zones serving as reproductive refuges. This possibility needs to be explored further because remaining land must be large enough to sustain viable populations. Faced with the momentous task of meeting the basic needs of the human population, species conservation can only proceed by promoting and enlisting native cooperation. The way forward may be to advocate a "Biosphere Reserve" concept which would include multiple uses of core areas, buffer zones, restoration areas, and stable cultural areas. To assume that, as conservationists, our role should be limited to direct "preservation of nature" is sectoral and has no long-term prospects. Agencies that administer protected areas cannot pursue policies of wilderness protection, education, and research, and exclude an approach that would involve the sustainable use of resources by native peoples. Some might consider allowing controlled exploitation too risky a proposition because it requires so much organization. Yet to exclude people from their source of livelihood without providing alternatives is irresponsible and, as such, doomed to failure in the long term. Romantic notions of village-based rearing of untried "farm exotics" are similarly unacceptable. Rather, we must capitalize on the wisdom and expertise of native people, harness their support and cooperation, and make every effort to balance their requirements with the need to protect the environment. In places like Bioko, the population is much too large for small-scale solutions. Equally, solutions to economic problems cannot be sought in isolation from environmental issues. A case in point is the proposal by the World Bank Rehabilitation Project to introduce large numbers of shotguns to control squirrels that cause the loss of up to 45% of the cocoa crop because rodenticides are ineffective and too expensive. This one-sided solution to the cocoa problem would no doubt cause increased exploitation of all wildlife. What is required are programs of a high degree of competence, responsibility, and commitment from scientists, multilateral and unilateral aid organizations and governments alike.

A major concern for the future is to create alternatives for the professional hunters in Bioko to divert or provide substitutes for the significant financial incentives that commercial hunting affords them. Hunting is a major source of revenue for many families who use these earnings to purchase commodities (Wilkie et al. 1992). It would be simplistic to suppose that merely by increasing the supply of beef or other meats to Bioko, protein would become available to the general populace and the threat to wildlife would subsequently disappear rapidly. Availability is by no means accessibility. Sustainable beef production could provide an ecological safeguard to wildlife exploitation only if it is widely available. To move away from bushmeat as the prime source of meat is not easy when cultural resistance to beef and dairy products is taken into account. Prices would have to decrease, or the purchasing powers of the entire population would have to increase. This might just be political naiveté at present. However, community involvement in the whole process of linking nutrition to conservation might be a way forward, as suggested by Addo et al. (1994) for Ghana.

Although sustainable use of wildlife has long been a concern in biology (Caughley 1977; Hilbourn et al. 1995), we still face major difficulties implementing actions to make conservation compatible with local economic development. This inability to combine development with the exploitation of renewable resources has recently prompted some authors to question the validity of such a linkage (see Oates 1995). Others, like Noss (1997a), point to the need for major socioeconomic changes in communities before conservation can be integrated with their basic needs. Whatever the option, the present study has shown that there is still much to be learned before we completely throw out the sustainability baby with the bathwater.

ACKNOWLEDGMENTS

I am most grateful to Dr. Javier Juste for his encouragement and friendship in the assembly of this review. Dr. Dietrich Schaaf provided feedback on a number of topics included herein. Liza Gadsby and Dr. Jaime Perez del Val supplied me with photographs, and the latter contributed historical information on meat production and cattle ranching in Bioko. Stuart Lenton provided invaluable help with maps.

9

Differential Vulnerability of Large Birds and Mammals to Hunting in North Sulawesi, Indonesia, and the Outlook for the Future

TIMOTHY G. O'BRIEN AND MARGARET F. KINNAIRD

Sulawesi is biologically the most distinctive island of the Indonesian archipelago. Lying east of Borneo in the center of the biogeographic region of Wallacea, Sulawesi is characterized by an unusual blend of Asian and Australasian fauna. Sulawesi's complex geological history and degree of isolation from surrounding islands have resulted in the highest levels of endemism in the archipelago (Whitten et al. 1987a). Sulawesi harbors 380 of Indonesia's 1500 bird species; approximately 96 species (25%) and 14 genera (10%) are endemic to the island (Holmes and Phillipps 1996). The mammalian fauna is equally remarkable. Of 127 mammal species on Sulawesi, 79 (62%) are endemic to the island; this figure increases to 98% if bats are excluded. The uniqueness of the island's vertebrate fauna firmly places Sulawesi among the most important centers of global biodiversity.

Wildlife populations on Sulawesi, particularly those in the north, occur at remarkably high densities relative to other large islands of the region, such as Sumatra, Borneo, and New Guinea (MacKinnon and MacKinnon 1981). For example, the endemic Sulawesi red-knobbed hornbill (*Aceros cassidix*) is found at a density of 51 animals/km^2 in the Tangkoko-DuaSudara Nature Reserve of North Sulawesi, the highest density reported for any forest hornbill in the world (Kinnaird et al. 1996). Sulawesi crested black macaques (*Macaca nigra*), despite dramatic population declines during the 1980s and 1990s, still are found at densities as high as 67 animals/km^2 in North Sulawesi (Rosenbaum et al. 1998), estimates that are more than two times higher than those for long-tailed macaques (*Macaca fascicularis*) and 10 times higher than estimates for closely related pig-tailed macaques (*Macaca nemestrina*) in lowland Sumatran rain forests (Whitten et al. 1987b). Possible reasons for these high densities include reduced predation pressure and a high proportion of tree species bearing fleshy fruits. The large predatory cats and birds of prey characteristic of islands west of Wallacea are not found on Sulawesi, and with the exception of humans and pythons (*Python reticulatus*), Sulawesi's larger vertebrates live in a predator-free environment. Unlike the islands of Sumatra and Borneo, Sulawesi's canopy is not dominated by tree species of the family Dipterocarpaceae, which are characterized by dry, hard fruits (Meijer and Wood 1964) and high levels of secondary compounds (Waterman et al. 1988). In North Sulawesi, large numbers of trees produce an aseasonal abundance of fleshy fruits favored by frugivorous birds and mammals, particularly fig trees (*Ficus* spp.), and this might explain the unusually high wildlife densities (Kinnaird and O'Brien 1995a, 1995b; Kinnaird et al. 1996).

Historically, Sulawesi's wildlife populations have been affected by two human activities: habitat conversion and hunting. Humans have been hunting, felling trees, and modifying forests since their arrival some 30,000 years ago. The agricultural activities of early farmers were small in scale and probably took place in localized plots near their dwellings (Bellwood 1980, 1985). It was not until iron tools were introduced approximately 1500 years ago that humans acquired an efficient means for large-scale forest clearance. The process of forest conversion accelerated greatly in the early 1970s, when government incentives promoted commercial logging, transmigration, and estate crop projects. These activities have degraded or converted forest lands, resulting in the loss of more than 67% of productive wet lowland forest habitat. Today, Sulawesi is a landscape of natural forest islands in a sea of cultivation and development (Whitten et al. 1987a). The fragmentation and opening of Sulawesi's vast forests has had two effects: it has increased the amount of forest edge adjacent to human settlements, and it has made remote forest interiors more accessible by reducing the distance from the edge to the center of small fragments, and by building roads into forested areas.

The evolution of hunting in Sulawesi followed a trajectory similar to agricultural activities. Cave deposits indicate that early hunters preferred ground-living species such as pigs (*Sus celebensis*), babirusa (*Babyrousa babirussa*), anoa (*Bubalus depressicornis*), and macaques (*Macaca* spp.), but there is no evidence that hunting by these early humans caused extinctions on Sulawesi (Whitten et al. 1987a). Over time, an increasing human population put greater pressure on the resource base, particularly on prey that were easily captured using snares, ground traps, or spears, or were run down with dogs. Logging and forest fragmentation also facilitated hunting due to logging roads and areas becoming more accessible due to their smaller size (Bennett et al. this volume).

Despite the rapid economic development of the past two decades, the people of Sulawesi continue to use wild animals as a food source, and hunting for subsistence and commercial markets is still practiced today. Although present-day hunting techniques are similar to those used by early humans (with the exception of the air gun), population growth and the development of commercial markets for wild meat have resulted in serious wildlife population declines in North Sulawesi (Clayton and Milner-Gulland this volume; Hedges 1996; O'Brien and Kinnaird 1996). For example, Lee (1995, this volume) showed that hunting for subsistence and commercial markets has had devastating effects on populations of Sulawesi crested black macaques in North Sulawesi, and Clayton and Milner-Gulland (this volume) predict that continued commercial hunting for babirusa will result in their extirpation from all but the most remote areas.

Can Sulawesi's wildlife community persist under today's hunting pressures? Are the present rates of harvest sustainable? In this chapter we examine how differential vulnerability of exploited species affects their potential for sustained harvest by evaluating the impact of hunting on a community of large fruit-eating birds and mammals in the Tangkoko-DuaSudara Nature Reserve, North Sulawesi. We then discuss the potential for sustainable harvests given the current situation, and briefly summarize the outlook for the future of Sulawesi's unique wildlife.

TANGKOKO-DUASUDARA NATURE RESERVE: A CASE STUDY

The Tangkoko-DuaSudara Nature Reserve (hereafter referred to as Tangkoko) is an 8867 hectare nature reserve lying at the northernmost tip of Sulawesi (figure 9-1). It was established as a forest reserve by the Dutch colonial government in 1919 and was retained as a nature reserve after Indonesia gained independence, making it one of the oldest protected areas in Indonesia. Tangkoko encompasses a wide range of habitats within a relatively small area, although the main habitat is broadly classified as lowland rain forest (IUCN 1991). There are three volcanoes in the reserve: a parasitic tuft cone (450 m) that resulted from an eruption in 1839, Mt. Tangkoko (1100 m), and the twin peaks of DuaSudara (1351 m). The reserve has six villages on its borders, inhabited by primarily Christian people who cultivate coconut and other crops, fish the coastal waters, and hunt within the reserve (O'Brien and Kinnaird 1996; Lee this volume). Although North Sulawesi is one of the wealthiest provinces in Indonesia, wealth is unevenly distributed. Such unequal distribution of wealth is partly responsible for maintaining two forms of hunting: market hunting to supply wealthy urban dwellers, and subsistence hunting to provide protein in the poorer rural areas. Hunting pressure remains high due to adherence to traditional diets and increasing human population.

The wildlife community of the Tangkoko area was already well known and exploited at the time the famous English naturalist Alfred Russell Wallace visited North Sulawesi in 1861 (Wallace 1869). Wallace traveled to Tangkoko to collect specimens of babirusa, anoa, and maleo birds (*Macrocephalon maleo*), species known to be abundant at the site at that time. Tangkoko's volcanic black beach was famous as one of the largest nesting beaches for maleo birds, an endemic genus of megapode that incubates its eggs in volcanic soils and beaches. Even then, the nesting colony was heavily exploited by local inhabitants, and by 1915 the beach

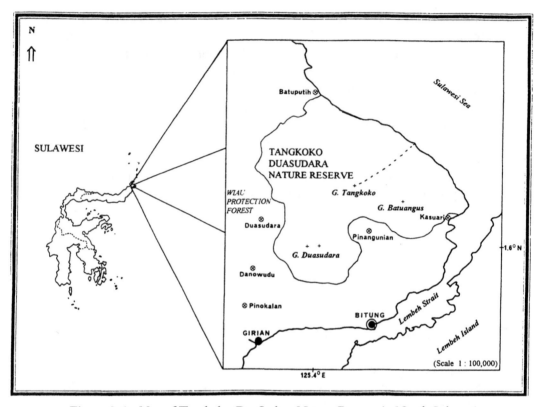

Figure 9-1. Map of Tangkoko-DuaSudara Nature Reserve in North Sulawesi.

colony was completely wiped out (MacKinnon and MacKinnon 1981), leaving only a few smaller inland colonies.

As well as maleo eggs, hunters in Tangkoko appear to prefer Sulawesi pigs, crested black macaques, and bear cuscus (*Ailurops ursinus*) (figure 9-2). These species are large (>5 kg) and more common than the anoa. People claim that wild pig meat tastes better than domestic pig meat, and crested black macaque is traditionally served at important meetings, social gatherings, and Christmas. Forest rats and the nocturnal dwarf cuscus (*Strigocuscus celebensis*) also are commonly eaten. Hunting methods tend to be nonselective; most hunters use snares and traps, although we have observed hunters carrying airguns and occasionally rifles. Bear cuscus are harvested by hand from trees by either climbing the tree and cutting the cuscus' prehensile tail so it falls, or cutting the tree down. In many areas, dogs are used during hunting, but this method is less common in Tangkoko. Ground-dwelling forest birds, such as red junglefowl (*Gallus gallus*) and Tabon scrubfowl (*Megapodius cuminggi*) are caught in snares using drift fences to guide the birds to the snare site. Canopy-dwelling red-knobbed hornbills (2.5 kg) are occasionally taken using guns.

MacKinnon and MacKinnon (1981) provided the first density estimates for a range of Sulawesi wildlife. In 1978 and 1979 they conducted monthly censuses of most mammals and conspicuous birds in Tangkoko. They reported extraordinary

Figure 9-2. Sulawesi bear cuscus (*Ailurops ursinus*). *Photo courtesy of Margaret Kinnaird.*

densities of crested black macaques (300 birds/km^2), an anoa population of 50, and healthy populations of Sulawesi pig, bear cuscus, Tabon scrubfowl, red junglefowl, and Sulawesi red-knobbed hornbill (figure 9-3). The MacKinnons reported that the maleo bird was in trouble due to egg harvesting and that the babirusa was probably extinct in the reserve, and during their 18 month study, they documented an unexplained 50% decline in a population of Sulawesi tarictic hornbills (*Penelopides exarhatus exarhatus*).

Methods

From April 1993 to March 1994, we repeated the MacKinnons' surveys in Tangkoko (O'Brien and Kinnaird 1996). We conducted monthly censuses on a subset of the original trails and used the same technicians employed by the Mac-Kinnons, thereby minimizing biases common to such longitudinal comparisons. We gathered data on nine species known or believed to be hunted by local people, using standard line transect methods (Burnham et al. 1980). Each month, two observers walked four transects, once in the morning and once in the afternoon. Transect lengths ranged from 4.65 to 5.9 km, and the total length of transects walked was 42 km/month. We estimated population densities using DISTANCE (Laake et al. 1993) for common species and fixed-width transect estimates for uncommon species (encountered less than 20 times/month). Monthly estimates were averaged for overall density estimates. We developed density estimates for the northern half of the reserve for comparison with the MacKinnon data, and then divided the data into separate estimates for the northeast (heavily hunted) and northwest (lightly hunted) sections. The northwestern part of the reserve is

Figure 9-3. Sulawesi red-knobbed hornbill (*Aceros cassidix*). *Photo courtesy of Margaret Kinnaird.*

better protected from hunters due to tourism activity (Kinnaird and O'Brien 1996) and proximity to reserve headquarters. The southern and eastern parts of the reserve are poorly patrolled and, according to local informants, heavily hunted. Vegetation in the northeast and northwest is broadly similar.

POPULATION CHANGES

There were marked changes in population density over 15 years for most of the nine species (table 9-1). With the exception of the Sulawesi pig, mammals experienced more dramatic population declines than birds: black macaques, bear cuscus, and anoa all suffered population declines of more than 75%. In contrast, only two of five bird populations declined, whereas three increased in size. Sulawesi taritic hornbill and Tabon scrubfowl populations doubled in size over the past 15 years,

Table 9-1. Intrinsic Rate of Increase (r_{max}), Change in Population Density over 15 Years (%), Density Estimates for Northeast (Heavily Hunted Area), Northwest (Lightly Hunted Area), and Overall Population Density (km²) for Nine Species of Birds and Mammals in the Tangkoko-Duasudara Nature Reserve

			Density		
Common Name	r_{max}	*% Change*	*Northeast*	*Northwest*	*Overall*
Crested black macaque	0.145	−75	46.0	72.2	58.0
Anoa	0.312	−90	0.04	0.32	0.05
Sulawesi pig	0.620	+ 5	5.56	6.8	12.0
Bear cuscus	0.297	−95	0.44	0.48	2.0
Maleo	0.341	−90	0.08	0.24	0.2
Tabon scrubfowl	0.400	+ 100	2.68	1.0	6.2
Red junglefowl	1.098	−50	2.32	1.34	1.5
Red-knobbed hornbill	0.186	+ 15	53.6	51.8	51.0
Sulawesi tarictic hornbill	0.333	+ 100	7.0	5.45	2.8

whereas red junglefowl and maleo experienced population declines of more than 50% during the same period.

Species exhibiting the greatest declines over the 15-year period were those species with densities lower in the heavily hunted area than in the lightly hunted area (table 9-1). Overall, densities of mammals were uniformly higher in the lightly hunted zone than in the heavily hunted zone, with the density of Sulawesi pigs showing the least difference between the two areas. Among birds, only the maleo had higher densities in the lightly hunted area. For hornbills, the differences between areas were slight, but for the ground-dwelling Tabon scrubfowl and red junglefowl, the densities were substantially higher in the heavily hunted area.

There are three possible explanations for the patterns of population change observed between the late 1970s and early 1990s: (1) habitat loss around the reserve; (2) habitat loss or deterioration within the reserve; and (3) hunting (O'Brien and Kinnaird 1996). None of these explanations are mutually exclusive. If habitat loss around the reserve was the primary force driving population changes, then populations should increase initially due to the influx of "refugees" from the surrounding areas, followed by a relaxation and eventual return to normal population densities. Current densities would be similar to or higher than those reported by the MacKinnons. Alternatively, if habitat degradation within the reserve was the primary factor affecting density change, we would expect species that are capable of exploiting early successional habitats, such as black macaque and Sulawesi pig, to increase. Forest specialists, on the other hand, would suffer population declines as forest habitat deteriorated. Although we could not rule out the effects of habitat deterioration, our data indicated that habitat changes alone were insufficient to explain the changes observed. Thus, we suggest that the patterns of population changes are better explained by differences in vulnerability to hunting.

DIFFERENCES IN VULNERABILITY

For a community of Amazonian mammals, Bodmer (1995b) and Bodmer et al. (1997a) demonstrated that r_{max} is a good estimator of a species' vulnerability to hunting. Caughley (1977) defined r_{max} as the exponential rate at which a population with a stable age distribution will grow when resources are not limiting. Values of r_{max} depend on the demographic characteristics of the population, such as age at first reproduction, age-specific survival, and fecundity. r_{max} is a useful indicator of a species' innate ability to respond to stochastic events or exploitation that result in loss of a portion of the population. Bodmer (1995b) found that species with low r_{max} values showed considerable differences in population densities between heavily hunted and lightly hunted areas, whereas species with high values showed little difference in their densities.

We developed estimates of the intrinsic rate of increase (r_{max}) for the nine study species using Cole's equation (Cole 1954) (table 9-1). The parameters used for calculating Cole's equation were derived from field data, information on captive populations of the same or closely related species, and the literature. Our data (figure 9-4) do not show patterns similar to those presented by Bodmer (1995b). Among birds, wide disparities arise that cannot be explained by r_{max}. Although there is a weak relationship for mammals, only the Sulawesi pigs and black macaques conform to expected trends. Pigs have a relatively high r_{max} and comparable densities between areas, whereas black macaques have low r_{max} and densities are almost 40% less in the heavily hunted area compared with the lightly hunted area. Black macaques, and possibly bear cuscus, anoa, and maleo, are sensitive to

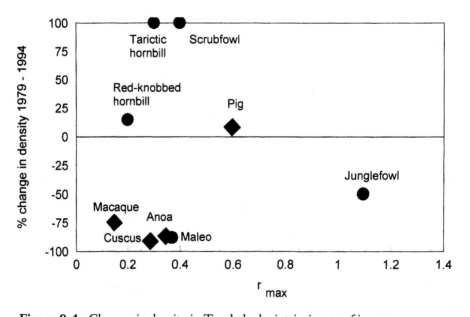

Figure 9-4. Changes in density in Tangkoko by intrinsic rate of increase.

hunting pressure because of low r_{max} values, but statistical trends for either the bird or mammal community might be obscured by the small sample size in this study.

Behavioral differences between species render some more susceptible to specific forms of hunting than others. Ground-dwelling birds and mammals are especially vulnerable to hunters using ground traps and snares, and to being chased down by dogs. Although traps can be constructed to exclude nontarget species, snares and dogs are nonselective and indiscriminately take species and all age and sex classes. Clayton and Milner-Gulland (this volume) report that snares set for Sulawesi pig commonly capture babirusa, and we have observed black macaques in snares set for pigs and deer.

Communal nesting by birds at restricted locations allows efficient exploitation of eggs, nests, and breeders. Egg harvesting, in particular, is a serious threat to many populations of megapodes in Indonesia (Jones et al. 1995; Argaloo and Dekker 1996). The communal nesting of maleos has permitted efficient exploitation of eggs and resulted in drastic declines in the species throughout Sulawesi. Exploitation of maleo nesting grounds is more severe than reported for other communal nesting megapodes because the breakdown of traditional harvesting systems has led to scramble competition among egg collectors (Argaloo and Dekker 1996; P. Jepson personal communication). In contrast, Tabon scrubfowl eggs are harvested when possible, but nests are dispersed throughout the forest and cannot be exploited as efficiently as the maleo.

Canopy dwellers are perhaps the hardest group to hunt efficiently, and selective hunting techniques are necessary. Hunters use air guns, shotguns, rifles, specialized nets, traps set on branches in the canopy, and glue traps to capture birds or to raid nests to harvest chicks. With the exception of guns, these techniques require climbing high into the canopy and entail significant risks to the hunter. Species such as hornbills, pigeons, and fruit doves are less vulnerable to hunting because rifles and shotguns are illegal in Sulawesi and are not readily available. Cost considerations might also restrict the use of rifles and shotguns; cartridges are expensive (as much as US$1.00 per shotgun shell) and not widely available except to the police and the military. A mid-canopy resident that is effectively harvested is the bear cuscus; slow movements and tendency to sit quietly rather than flee allow hunters to capture them easily.

ESTIMATES OF HUNTING PRESSURE AND SUSTAINABLE HARVEST RATES

Sustainable harvest models are based on the premise that if the production of new individuals exceeds natural mortality, then there exists a proportion of the population that may be harvested without causing a population decline. If estimates of population size and rate of population growth are available, we can calculate how much of the annual production may be harvested safely. We follow this approach to compare average annual harvest rates with the expected sustainable harvest rates of large birds and mammals in Sulawesi. We used models by Caughley (1977, pp. 172–179) to estimate harvest rates. Caughley's models are based on two relationships:

$$b = 1 - e^{-H} \tag{1}$$

and

$$HN = r_{max}N(1 - N/K) \tag{2}$$

where b = average hunting mortality rate, expressed as a proportion of the total population size; H = instantaneous hunting rate; HN = sustainable harvest rate for a population of size N; r_{max} = maximum rate of increase; and K = carrying capacity. Equations (1) and (2) can be combined and rearranged to:

$$N_{t+1} = N_t (1 - b) \, e^{r_{max}(K - N_t)/K} \tag{3}$$

which describes the change in population size from time t to time $t + 1$ for a species experiencing logistic population growth in an environment with carrying capacity K that is subject to an average hunting rate b. The model assumes populations are growing logistically at r_{max}, or that the population is growing at r_{max}, discounted by the population size relative to carrying capacity. At very small population sizes, the population grows at rate r_{max}, but growth slows as N approaches K. When $N = K$, population growth is 0. The model also assumes that carrying capacity is known and constant for each species. Finally, we assume that instantaneous harvest rates (H) are inversely proportional to density or population size. As the population size approaches carrying capacity, $(K - N)/K$; therefore, H approaches 0. At K, the population replaces itself, but is not expanding, so any harvest will result in a population decline. The model follows the logic that sustainable harvest rates are maximized when population size is low relative to the carrying capacity (usually 0.5 to 0.6K) because production is maximized.

We used estimates of r_{max} and population estimates based on densities from 1979 and 1994 (table 9-1) for N_t. Carrying capacity is rarely known for a population (Caughley 1977), and most sustainable harvest models assume that a species' population size is at some proportion of K (often 0.65 to 0.9K; Robinson and Redford 1991b). We modeled species populations assumed to be at different proportions of K (e.g., 90%, 50%, and 10% of K). We then estimated the average annual hunting mortality rate, b, by iteratively evaluating a range of mortality rates within a range of carrying capacities and choosing the values of b and K that gave the best fit between expected and observed population densities for 1979 and 1994.

Data for crested black macaques and taritic hornbills were used to test the model because we had an intermediate data point for macaques from 1987 (Sugardjito et al. 1989) and because taritic hornbills were an example of an expanding population. The best fit model for crested black macaques assumed that the population was at 90% of carrying capacity in 1979 and an average hunting mortality rate of 17.8% per year (table 9-2; figure 9-5). The model overestimates

Table 9-2. Estimated Number of Animals Harvested per km² from Tangkoko in 1979 and 1994, Based on Simulation Results

			Harvest	
Common Name	N/K[a]	h[b]	1979	1994
Crested black macaque	0.9	17.8%	54.4	10.4
Anoa	0.9	31.4%	0.182	0.16
Sulawesi pig	0.5	26.0%	2.85	3.0
Bear cuscus	0.8	34.8%	13.6	0.68
Maleo	0.5	38.6%	0.76	0.076
Red junglefowl	0.9	46.0%	1.35	0.675
Tabon scrubfowl	0.5	0.0%	0.0	0.0
Red-knobbed hornbill	0.85	0.0%	0.0	0.0
Sulawesi tarictic hornbill	0.5	0.0%	0.0	0.0

[a]N/K is population size expressed as a percentage of carrying capacity.
[b]Estimates of annual harvest.

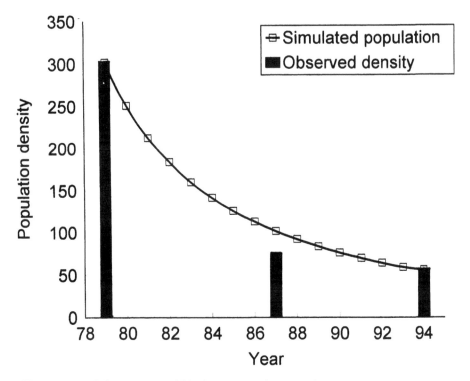

Figure 9-5. Sulawesi crested black macaque densities for 3 years (1979, 1987, and 1994) and simulated population densities under an annual harvest of 17.8% of the population.

the population size in 1987 but accurately estimates the 1994 population. Because Sugardjito et al. (1989) did not provide standard deviation estimates, it is difficult to know if the modeled and measured estimates are significantly different for 1987 or if differences result from higher harvest rates in the early 1980s. For the territorial tarictic hornbills, we assumed the population was at 50% of carrying capacity in 1979 because the MacKinnons reported a 50% decline during their study and the population doubled in density by 1994. The model predicts no hunting mortality for the species (table 9-2). Thus, the model appears to work well with unharvested as well as harvested species. When we applied the model to the other species, we found that the species that did not suffer dramatic observed population declines (hornbills and Tabon scrubfowl) suffered no hunting mortality in the model, as expected. The other species were being harvested at average annual rates of 17.8% to 46% of the population (table 9-2).

We next asked if the estimated average annual harvest rates could be considered sustainable. We calculated the sustainable harvest rates for populations at 90%, 50%, and 10% of carrying capacity over the range of r_{max} values for all nine species. The resulting three lines (figure 9-6) may be thought of as the upper and lower limits as well as the middle distribution of sustainable harvest rates, expressed as a proportion of population size or density. We then plotted the estimated annual harvest rates for the hunted species (figure 9-6; for hornbills and Tabon scrubfowl, $h = 0$ and is not shown). The results show that estimated annual harvest rates for black macaque, bear cuscus, anoa, and maleo are well above sustainable harvest rates for similar populations with low r_{max} values, irrespective of

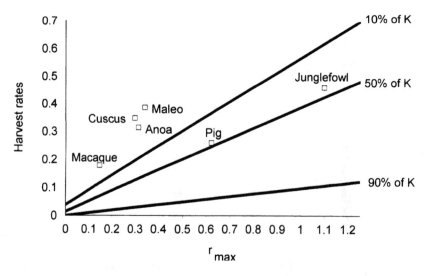

Figure 9-6. Estimates of sustainable average annual hunting rates (*lines*) for populations at three levels of carrying capacity across a range of r_{max} and the estimated average annual hunting rate for seven species of birds and mammals at Tangkoko (*open squares*). Tabon scrubfowl and the hornbills are not shown because their estimated harvest rates are zero.

K. The estimated harvest rate for anoa, for example, is more than twice as high as a sustainable harvest rate for an anoa population that was at 0.5K in the late 1970s. The estimated annual harvest rates for Sulawesi pig and red junglefowl closely match the sustainable harvest rates for comparable populations at 50% of carrying capacity. This indicates that if Sulawesi pigs and red junglefowl populations are indeed near 0.5*K*, then the harvest of these species is probably sustainable.

These results are consistent with other similar sustained yield models. Applying Bodmer's (1995b) model to the crested black macaque data, we estimate a sustainable harvest of five monkeys/km². This value is similar to our estimate of 4.0 monkeys/km² for a Tangkoko macaque population at 50% of carrying capacity (see figure 9-5). Results of Clayton and Milner-Gulland (this volume) for Sulawesi pigs indicate that a sustainable harvest can be achieved with a population at approximately 37% of carrying capacity. Both models indicate that the current harvest rate of Sulawesi pig at Tangkoko is sustainable.

OUTLOOK FOR THE FUTURE

We have demonstrated that harvest levels of macaques, maleo, anoa, and bear cuscus are well above sustainable levels. Overharvesting of these species has resulted primarily from the maintenance of traditional diets as demand increased and supply declined. Rapid conversion of lowland forests in the past 25 years (Whitten et al. 1987a) has reduced the habitat available for wildlife populations, and an increasing human population has caused additional hunting pressure in the remaining forests. In the case of the babirusa (O'Brien and Kinnaird 1996, Clayton and Milner-Gulland this volume); and perhaps anoa (Hedges 1995), heavy exploitation coupled with low reproductive rates have resulted in local extirpation in Tangkoko and extreme rarity in other areas (Clayton and Milner-Gulland this volume). Although laws were passed in 1970 and 1991 to protect the babirusa and other species such as the crested black macaque, continued lack of enforcement in protected areas has allowed hunting to proceed unimpeded. Between 1980 and 1995, for example, only one case of poaching in Tangkoko was prosecuted.

Laws protecting wildlife might actually prevent effective management of game populations in situations when the government is unwilling to recognize wildlife regulations or enforce severe penalties. Additionally, laws regarding use of nature reserves and national parks need to be reconciled with actual use of these areas. Tourism, for example, is increasingly popular in Tangkoko and could generate needed revenues to combat poaching activities, but because intensive tourism is prohibited in Indonesian nature reserves, appropriate tourist management plans cannot be implemented (Kinnaird and O'Brien 1996). Zoning of protected areas, local participation in park management, and other solutions to overexploitation will not work without an enforceable legal framework.

Unless intervention occurs, some species will probably go extinct in the near future. In the absence of effective law enforcement, only a willingness on the part of hunters and marketeers to forgo attempts at maximizing short-term yields and profits and to take only those species that can withstand the harvest pressure (i.e., Sulawesi pig) will achieve ecologically sustainable harvest levels. Conservation of wildlife will undoubtedly entail paying a short-term cost (forgone profits) to assure a long-term gain for conservation (Alvard 1995a). The need to conserve the resource base today to ensure future harvests is not intuitively obvious to a hunter currently enjoying healthy profits or putting protein on the table. The prospect of North Sulawesi's hunters refraining voluntarily from hunting endangered species is low.

Education, examples, and motivation by the local resource managers are required to generate a concern for modifying hunting behavior. Administrators responsible for making decisions regarding wildlife need to be well informed, to have good communication skills, and be able to instill motivation. Unfortunately, in Sulawesi, few natural resource managers have been trained in communication skills, and interaction with local communities is rare. The interface between the hunters, marketeers, and wildlife managers occurs primarily at the level of park guards and local villagers. If park guards are to be entrusted with changing behavior and attitudes of local hunters, it is paramount that they too be well trained and highly motivated.

Financial and human resources for wildlife management in tropical countries usually are too meager to ensure success without the cooperation of the local government, local communities, hunters, and marketeers. Currently, less than 1% of the conservation budget in Indonesia is spent on protection of parks and reserves (MacAndrews and Saunders 1997). Authorities in Indonesia must be encouraged and helped financially to invest more resources in their parks. In the near term, successful management of exploited wildlife populations in North Sulawesi ultimately rests on controlling hunters and creating incentives to encourage better stewardship (e.g., involvement of local people in tourism and management activities). Macaques, babirusa, bear cuscus, and anoa populations should not be hunted, whereas hunting of pigs, rodents, and other more resilient species may be encouraged if hunting methods are selective (Clayton and Milner-Gulland this volume; Alvard this volume). Ideally, people of North Sulawesi should abandon their dependence on wild meat completely, and the government should enforce wildlife protection laws. Realistically, if changes in attitude do not occur soon in Sulawesi, a fitting epithet for the loss of its endemic mammals and birds may be that "they tasted good."

ACKNOWLEDGMENTS

This work was supported by the Wildlife Conservation Society in collaboration with the Indonesian Directorate General for Nature Preservation and Forest Protection (PHPA) and Puslitbang Biologi of the Indonesian Institute of Science (PB/LIPI). Funding was provided by the Wildlife Conservation Society, The National Geographic Society (grant no. 4912–92), and the Wenner-Gren Foundation for Anthropological Research (grant no. 5543). We thank Effendy Sumardja (PHPA), Soetikno Wiryoatmodjo (PB/LIPI), Dedi

Darnaedi (PB/LIPI), and Romon Palete (PHPA) for their kind help and encouragement throughout our time in Sulawesi, and Kathy MacKinnon who suggested that we work in Tangkoko. Farenheid Pontonudis, Aman Dorongi, and Finche, PHPA guards, are thanked for assisting with data collection. Robert Lee and Graham Usher provided valuable friendship and critical feedback. Finally, we thank Andy Dobson for the harvest simulation and, as always, valuable suggestions on the manuscript.

10

The Impact of Traditional Subsistence Hunting and Trapping on Prey Populations: Data from Wana Horticulturalists of Upland Central Sulawesi, Indonesia

MICHAEL ALVARD

Wildlife use has increasingly become recognized as a way native and local people glean value from tropical forests (Bodmer et al. 1994). Recent studies also have shown that traditional subsistence hunters do little actively to mitigate their impact on prey populations and often drive prey species to local extinction (Kaplan and Hill 1985; Hames 1987, 1991; Olson 1989; Smith 1991; Alvard 1994, 1993b, 1995b; Kay 1994; Fitzgibbon et al. 1995b; Alvard et al. 1997). The implication of these studies is that hunters value the forest and its products only up to the point that it provides short-term returns for themselves and their families. This problem is not specific to local or indigenous people. There is strong theoretical and empirical evidence that suggests there is little incentive for individuals voluntarily to conserve open-access resources they might otherwise find immediately useful (Olson 1965; Hardin 1968; Williams 1966; Dawes 1980; Rogers 1991; Hodson et al. 1995).

Although proactive conservation by native people who harvest open-access prey is unlikely, overhunting is by no means inevitable. Human groups that lack behavior designed to conserve will not necessarily overhunt (Alvard 1993a). Studies that have shown evidence for overexploitation by subsistence hunting populations, for example, usually do so only for a limited number of the total prey species hunted. That is, some species are hunted sustainably, whereas others are not (Vickers 1988, 1991; Fitzgibbon et al. 1995b). Alvard et al. (1997), for example, showed that although Piro hunters in Amazonian Peru were overhunting many of the large primate species, they were hunting collared peccaries sustainably.

Awareness that hunters can have differential impacts on different species is important because it moves the debate beyond arguing for or against the idea that indigenous peoples are "natural conservationists" (Alcorn 1993; Redford and Stearman 1993). Such variation provides clues to factors that determine the nature of sustainable harvesting in general. It is also important because although subsistence hunters hunt a wide variety of species, often only a small number provide the bulk of their sustenance. For example, in the Piro case, four species—tapir, collared peccary, deer, and capybara—provide 86% of all mammalian game harvested (Alvard 1993a).

Conservation biologists wishing to balance local peoples' needs with conserving the animal resources they use are faced with a challenge. They need to identify prey species that can provide sufficient sustainable value for hunters yet whose harvest maintains the biodiversity of the site in question. One way to do this is to examine both the impact that hunters have on their prey, as well as the contribution that each prey species makes to their economy.

This chapter examines these issues with data on the hunting and trapping practices of the Wana, a traditional indigenous group living in the highlands of Central Sulawesi, Indonesia. Data on prey numbers in hunted and unhunted areas are also examined. The Wana, who number in the thousands, practice rice and manioc swidden horticulture, and hunt a wide variety of wildlife. This chapter examines the large game (>5 kg) harvest, with a focus on the Wana's impact on two species: the Sulawesi pig *Sus celebensis* and the Tonkean macaque *Macaca tonkeana*.

Pigs and monkeys are chosen as the focus of this chapter for a number of reasons. The pigs are the most important large game animal for the Wana, both in terms of numbers taken and contribution to the diet. Along with the pigs, the macaques are among the most common large mammals, at least in unhunted areas. As primates, they are also among the most vulnerable to overhunting by the Wana. Because these species are relatively common and easily observed, the data collected on their encounter rates are of sufficient quality to make comparison between hunted and unhunted areas. As discussed below, other prey species were rare (anoa and deer) or absent from the study area (babirusa). Lastly, these two types (suids and primates) are among the most hunted and most important prey types for subsistence people living in rain forest environments and vary considerably in their ability to withstand hunting.

The Wana hunt and trap many smaller species, including birds, rodents, reptiles and amphibians. For example, preliminary analyses indicate that at least 74 types of birds, 12 types of rodents, and 17 types of bats are regularly or occasionally harvested (Alvard 1996). Species identification continues for these prey types, and the nature of the Wana small game harvest will be detailed in a subsequent paper.

SITE LOCATION AND WILDLIFE

The study was conducted in Morowali Nature Reserve, in the province of Central Sulawesi, Indonesia (figure 10-1). Sulawesi, formerly known as Celebes, lies between the islands of Borneo and New Guinea, and has long been noted for its fascinating biogeography (Wallace 1869). Morowali Nature Reserve (*Cagar Alam Morowali*) was established in 1980 and encompasses approximately 225,000

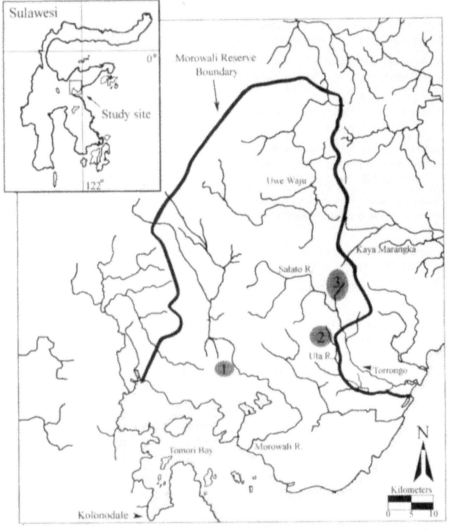

1 - Babilusa (unhunted site)
2 - Anyu Lempe (unhunted site)
3 - Posangke/Uwe Tondo (hunted site)

- Approximate Wana Range

Figure 10-1. Map of Morowali Nature Reserve and surrounding area. Study sites are indicated. Kolonodale is the terminus of the road from the provincial capital. Torrongo is the nearest government recognized village to Posangke. Uwe Waju and Kaya Marangka are others areas inhabited by Wana.

hectares of upland and coastal areas in Central Sulawesi. The status of *Nature Reserve* legally limits the types of activities that can take place within the Reserve's boundary. No one, including the Wana, has the legal right to live within the boundary of the Reserve. In addition, because the Wana do not live in government recognized villages, the Morowali Wana are currently living contrary to Indonesian law. Things may change if the status of the Reserve is changed to that of *National Park (Taman Nasional)*, as has been discussed (Schweithelm et al. 1992).

Vegetation in the Reserve includes mangroves and lowland alluvial forests along the southern coastlines, lowland rain forests, and montane rain forests at the higher elevations (WWF 1980a; Schweithelm et al. 1992). The interior is extremely rugged and mountainous. The Reserve has three peaks over 2200 m in altitude, and Mt. Tokala rises to 2593 m immediately to the east of the study site of Posangke. The environment surrounding Wana settlements is a highly anthropogenic one, consisting of gardens, areas in various stages of secondary growth, gallery forests along water courses, and vast meadows of *Imperata cylindrica*, known commonly as alang-alang grass. This grass is a species that invades and thrives in soils depleted from long-term swidden gardening in tropical regions (Eussen 1980). Primary forests are found along the higher slopes of the valleys up to the summits of the peaks.

There is a marked wet season from approximately April to August, and another from December to January. The dry months during the field season were from September to November. The year 1995 was a particularly wet year, with over 6000 mm of rain (Alvard, unpublished data). The combination of intense rain and the Wana practice of cutting swidden on steep slopes results in extensive erosion. Landslides were common in the Posangke area in 1995 and have been reported for other areas of the Reserve populated by Wana (WWF 1980a).

The Reserve contains a suite of Sulawesi fauna, which has been described as one of the most distinctive in Indonesia because of the high level of endemism (Musser 1987; Whitten et al. 1987a). Ninety-seven percent of Sulawesi's nonvolant mammal species are endemic. Large-bodied endemic species include the piglike babirusa (*Babyrousa babyrussa*), two species of dwarf buffalo or anoa (*Bubalus quariesi* and *B. depressicornis*), and the Celebes pig (*Sus celebensis*). One of the seven endemic species of Sulawesi macaques (*Macaca tonkeana*) also occurs in the Reserve. Tarsiers (*Tarsius spectrum*) are the only other primates living in the Reserve or on the island. One species of deer (*Cervus timorensis*) was introduced from Java. Both species of anoa and the babirusa are listed in CITES Appendix I.

Anecdotal references to the fauna of Morowali note that animals are rare compared with other areas of Sulawesi, such as Lore Lindu National Park located approximately 170 km to the northwest. The apparent low faunal density has been attributed, with no evidence, to the hunting practices of the Wana (WWF 1980a; Schweithelm et al. 1992). In contrast, Whitten et al. (1987a) argue that the low density of vertebrate fauna might be attributable to the poor-quality, ultrabasic soils that characterize Morowali. This question remains unresolved, although the data to follow suggest that both factors probably contribute to low densities of large mammals at Morowali.

THE STUDY GROUP

The Wana inhabit the northeast portion of the Reserve and are noticeably absent from the remainder except for a small number of isolated communities to the west. Linguistically identical to the Ta'a people living along the lowland coasts, the Wana have adapted to the interior mountainous rain forest environments. They practice swidden dry rice and manioc horticulture. Sago is also cultivated, as is millet for some groups. Hunting and trapping of wild animals is common. Domestic chickens, and rarely a domestic pig, are kept. The Wana have maintained relative isolation from much of the outside world. Culturally intact, most adult Wana speak no or very little Indonesian, have little or no interaction with the cash economy, and maintain a traditional religious belief system (Kruyt 1930; Atkinson 1989).

The Wana's isolation in the remote valleys of the uplands works to insulate them from the often politically tumultuous coast (Atkinson 1989). The lowlands have historically been an area of unsettled political animosities. It is unclear for how many years the Wana have inhabited the area, although it has probably been for a number of centuries. It is clear that before significant western contact, the Wana paid nominal tribute to both the sultanates of Ternate and of Bugnku. The Wana are ethnographically known from a study of shamanism by Atkinson in the late 1970s (Atkinson 1989) and from a general ethnography of the Posangke Wana written by the Dutch missionary Kruyt (1930).

This study focuses on a Wana population located along one segment of the upland course of the Salato river valley and its tributaries in an area called Posangke. The sample of Wana from which the harvest sample was collected consisted of 153 individuals as of March 1, 1995. The entire population of Posangke consists of approximately 370 individuals living in a number of watersheds and valley systems that all feed the Salato river. The sample was selected by traveling up the Salato river to an area that was approximately central in the Posangke area. An area was chosen that would allow a suitable number of Wana houses to be visited on a rotating basis every 2 to 3 days. The sample area covers all of one watershed (drained by the Sumi'i river into the Salato River) and the upper part of the next watershed (drained by the Uwe Kiumo River to the east).

Several thousand Wana live in and around the Reserve (Alvard 1996). The Posangke Wana are as representative of a typical Wana subsistence pattern as can be found today. There are Wana with both more and less outside contact than the Posangke Wana. The amount of contact seems to be related to the distance from the coast. Patterns of resource use also vary with geographic location. For example, I observed that some Wana in Uwe Waju, an area to the north of Posangke, grow millet. There is also anecdotal information that some Wana to the north do more trapping than the Posangke Wana, although Atkinson has reported that in the late 1970s the Posangke Wana were known for their hunting and trapping (Atkinson, personal communication).

The Wana of the Posangke region inhabit a niche ranging in altitude from about 250 m to about 600 m, although a number of the houses in the sample were located above 1000 m. The Wana do not live in centralized villages, but rather in scattered households of five to 15 people. Household mobility is high; families changed their residence location often during the course of the study. The new house locations, however, are almost always less than 1 km from the old house site (Alvard 1997). Land is not owned by individuals or kin groups; for the Posangke Wana, land is an open-access resource. After the harvest, gardens go fallow and are planted in the future by whoever decides the soil is suitable.

The Wana obtain most of their animal protein and fat from hunting and trapping and have a varied diet. Prey include the smallest mice and birds, bats, reptiles and amphibians, as well as the larger ungulates and primates. Trapping for both small and large game is widespread and accounts for most of the meat consumed on a day-to-day basis. Rodents and pigs are the most frequently trapped prey. Blowpipes and darts, sometimes poisoned with the resin of a tree (*Antiaris toxicaria*; the Wana call the poison *impse*), are used primarily for bats and birds, although primates and other small mammals are occasionally taken as well. For the largest terrestrial game such as the anoa, hunters use spears and dogs. Although river organisms such as fish, eels, and crayfish are taken in the dry season, the rivers seem to be relatively unproductive.

METHODS

The data presented in this chapter were collected from January 1995 to March 1996 at three locations within Morowali Reserve (see figure 10-1). Two sites are in uninhabited and unhunted portions of the Reserve, whereas the third is an area hunted and trapped by Wana. One unhunted area, Babilusa, is located on the western side of the Reserve, whereas the second unhunted area, Anyu Lempe, is to the east. Both of these sites are undisturbed primary forest in hills adjacent to rivers—the Morowali River for Babilusa and the Ula River for Anyu Lempe. The terrain at both sites is rough, steep, and broken by small water courses. Rattan poachers reportedly work the Anyu Lempe area on occasion, but informants report that it is too difficult to reach from the Salato Valley for significant economic activity or hunting. Babilusa is near the site of a small Wana settlement abandoned approximately 20 years ago. The people moved to the site of Kayapoli located 10.2 km down the Morowali River. The inhabitants of Kayapoli report that the site near Babilusa was inhabited for a short time and that no hunting has occurred in the area since then.

The site affected by hunting is the Uwe Tondo area of Posangke, located adjacent to the area inhabited by the Wana study group. The area was chosen because it is primary forest and relatively undisturbed from Wana garden activities. It is also on terrain that allows transect data to be collected, yet is close enough to a Wana settlement to be affected by hunting and trapping activities. Most of the accessible areas near Posangke are heavily disturbed by Wana gardening, whereas most of the undisturbed forest is located on very steep slopes above the valley floor and are unsuitable for transects.

Line transect methods were used to determine relative encounter frequencies and prey densities at the three areas (Buckland et al. 1993). Transect trails were cut, measured, and marked with surveyor's tape indicating distance every 50 m. The observers walked slowly (0.8–1.1 km/hr) along the marked and measured transect lines and recorded the following information for all animals encountered: time, species, estimate of group size, location on the transect, distance of the animal from the observer, and angle of the animal from the transect line. Observers consisted of two trained biologists and a number of local Wana assistants who aided in the detection and identification of the animals. Data were collected in teams consisting of one biologist and one Wana assistant. To minimize observer effects, the two biologists rotated among the three sites on a regular basis. Data on sample size, transect length, and location are presented in table 10-1.

Morning transects started at approximately 0600 hr. Each afternoon transect was conducted following a morning transect. Observers rested for approximately 2 to 3 hours and returned on the same trail. Although early morning and late afternoon transects are preferred for absolute density estimates (Robinette et al. 1974), this is not a critical issue for determining relative encounter rates between sites.

Logistic regression analysis was used to test for differences in the rate of encounter between the sites or, more specifically, the probability of encountering a prey type during any particular hour of transect. This method is particularly well suited for examining data where the dependent variable is dichotomous (encounter or no encounter). When the dependent variable is dichotomous, the assumptions of standard linear regression are not met, and logistic regression is the appropriate method (Cramer 1991).

Although the number of kilometers sampled was relatively large, the samples of actual prey encounters were too small to calculate densities at each of the three sites. This is presumably because prey densities are low. As a solution, data were pooled from those sites that did not have significantly different encounter rates, and densities were calculated for the pooled data.

The analytic methods used to calculate density were taken from Buckland et al. (1993). The perpendicular distance from the transect line for each observed group was calculated. A detection function was fitted to the distribution of distances; this function describes the probability of detecting a group that is distance x from the line (Buckland et al. 1993). This was accomplished using the computer program DISTANCE (Laake et al. 1993). This function is transformed to give a density estimate of groups. The primary assumption of this method is that animals directly over the transect line are always detected, and particular care was taken to

Table 10-1. Transect Sample Description

Site	Total Transect Length (km)	Total Hours Walked	Total km Walked
Babilusa (unhunted)	10.0	465.42	368.9
Anyu Lempe (unhunted)	11.1	532.70	540.3
Uwe Tondo (hunted)	7.1	421.74	465.3

ensure that this requirement was met. Because estimated group size did not correlate with distance from the transect line for either pigs ($r = -0.2607$, $p = 0.229$) or for macaques ($r = -0.1734$, $p = 0.351$), group density estimates could simply be multiplied by mean group size to estimate population density.

Concurrent with the transect data, data were collected to determine the nature of the Wana offtake from hunting and trapping. Households in the sample were visited every 2 or 3 days, and the adults present were interviewed concerning all the animals killed since the previous visit. Species and sex were noted for all animals killed. Date, hunter, location of the kill, and weapon also were identified. The mandibles of most prey were collected, and the relative ages (adult or subadult) of animals were determined by examining dentition. Animals whose molars had not fully erupted were classified as subadults.

WANA HUNTING TECHNIQUES

To capture large game, the Wana use two techniques: active hunting with spear and dogs, and trapping. Blowguns are used primarily for small game such as birds, bats, and squirrels. Hunters trapped the majority of pigs with either one of two types of trap: the *kayoro* or the *saiya*. Kayoro are common snare traps and accounted for 40% of the pig kills during the field session. The hunter bows over a sapling or length of bamboo with a length of cord tied on the end. The other end of the cord is tied into a loop, placed on the ground and attached to a trigger device that keeps the trap in place and set. When the leg of an animal disturbs the trigger, the sapling quickly straightens, tightening the loop around the leg of the prey, trapping it. The hunters usually kill the trapped animal with a spear (figure 10-2). The Wana also occasionally trap anoa and deer with *kayoro*, although none were reportedly set during the field session with these prey types in mind. All the *kayoro* observed were placed along pig trails.

Saiya are traps that disable and kill by causing the animal to impale itself on a picket of bamboo staves (figure 10-3). Forty-four percent of the harvested pigs were killed with *kayoro* traps. The trap employs a trip line that, when activated by the passage of an animal, causes a log to fall. The falling log frightens the animal, causing it to leap forward to its death transfixed on the staves. Hunters placed the majority of *saiya* along pig trails, often on steep slopes to facilitate the pig's journey toward the stakes. Both types of traps are frequently placed near the previous year's gardens. Such gardens are thick with undergrowth and still contain a substantial amount of unharvested manioc that attracts the pigs. Hunters also killed a number of pigs using dogs and spears (12%), as well as by hand or with other trap types (4%).

Hunters captured all the anoa with spear and dogs. Hunters report two tactics. The first uses the dogs to find, harass, exhaust, and finally corner the anoa, which is then dispatched with the spear. The second, and more dangerous, method is to provoke the animal into impaling itself on the spear. Two of the monkeys taken were killed by traps and one was killed by blowgun. The one deer was trapped by a *kayoro* set for pigs.

Figure 10-2. Wana hunter spearing a Sulawesi pig (*Sus celebensis*). Spears are commonly used to dispatch trapped and cornered prey. *Photo courtesy of Michael Alvard.*

Figure 10-3. Bamboo stakes of a Saiya pig trap, which is often placed along pig trails. *Photo courtesy of Michael Alvard.*

IMPACT OF THE WANA

The Harvest

Table 10-2 presents the large game harvested by the Wana. The Celebes pig and, to a lesser degree, the anoa were the focus of the Wana's large game interests. Macaques and deer were killed rarely, and a babirusa was never killed. The overall number of large game killed was not great compared with other traditional horticulturalists. The Wana killed 0.35 ungulates (pig, anoa, and deer) and 0.016 primates (macaque) per person per year. For comparison, using data published on native Amazonian groups, Redford and Robinson (1987) calculated that 1.53 ungulates on average were killed per person per year ($N = 20$ studies of different ethnic groups) and 2.3 primates were killed per person per year ($N = 21$ groups).

Pigs provided 58% of the total large game harvest by weight, whereas 40% came from anoa. Anoa were taken in smaller numbers than pigs but contributed disproportionately to the harvest because of their larger size. The total amount of large game harvested by gross weight per person per day was approximately 38 g for the Wana. This is a small amount compared, for example, to 225 g for the Piro and 233 g of large game per consumer per day harvested by the Machiguenga (Alvard 1993a). Both these groups are native hunters studied by the author in Peru (Alvard 1993a). For the Peru cases, consumers were defined as 3.5 years of age or older, and weights were for field dressed carcasses.

Table 10-3 presents the age and sex distributions of the offtake. For both anoa and pigs, almost half of the animals killed were subadults. In the case of anoa, 43% of the kills were subadults. For pigs, 47% of those harvested were subadults. Neither for anoa (chi-square = 1.5, $p = 0.213$) nor for pigs (chi-square = 0.77, $p = 0.380$) was the sex ratio of the harvest significantly different from 50:50. It is unknown whether in either case the harvest is proportional to the age/sex profile of the prey population.

Table 10-2. Large Game Kills at Posangke: January 1, 1995 to March 31, 1996 (455 days)

Common Name	Wana Name	Latin Name	Estimated Adult Body Weight (kg)[a]	Number Killed	Estimated Gross Weight (kg) Harvested
Celebes pig	*wawu*	*Sus celebensis*	35	57	1522.5
Anoa	*menzo muyoku*	*Bubalus* sp.	150	9	1050.0
Macaque	*wonti*	*Macaca tonkeana*	10	3	13.5
Deer	*lago*	*Cervus timorensis*	60	1	60.0
Babirusa	*tamarari*	*Babyrousa babyrussa*	50	0	0.0

Consumer population = 153 (31 adult male hunters). For estimate of gross weight harvested, subadults were assumed to be half of adult weight.

[a]Macaque, Bauchot and Stephan 1969; pig, Silva and Downing 1995 and estimate of author; anoa, Nowak and Paradiso 1983; deer and babirusa, Silva and Downing 1995.

Table 10-3. Age and Sex Profile of Large Game Harvest

		Age			Sex	
Prey	Adult	Subadult	Unknown	Male	Female	Unknown
Pig	28	25	4	31	22	4
Anoa	4	3	2	7	2	—
Deer	1	0	—	1	0	—
Macaque	1	2	—	0	3	—

Encounter Rates

Comparisons between encounter rates for hunted and unhunted areas in Morowali show that pigs were equally common in all three areas. Observers encountered pig groups on average 1.12 to 2.14 times per 100 hours of transect, with no statistically significant differences between sites (tables 10-4 and 10-5). There was also no significant difference in the size of pig groups encountered at the three sites (tables 10-6 and 10-7). The combined mean was 3.95 animals per group. The lack of any significant difference in encounter rates for the hunted area suggests that the Wana pig harvest is sustainable. The density estimate from the pooled sample consisting of the three sites is 1.54 pigs per km^2. This is much lower than reported at the other site for which data are available. O'Brien and Kinnaird (1996) reported 12 *S. celebensis* per km^2 at Tangkoko Nature Reserve, North Sulawesi.

In contrast, the macaques show considerable local depletion in the hunted area of Posangke (see tables 10-4 and 10-5). Macaques were encountered 10 times more often at Anyu Lempe and Babilusa than they were at the hunted area. There was no significant difference in the size of encountered macaque groups at the three sites (see tables 10-6 and 10-7). The combined mean was 6.56 animals per group.

The samples from the two unhunted sites were pooled, and the density was estimated to be 2.2 *M. tonkeana* per km^2. Again, this is much lower than other sites in Sulawesi. O'Brien and Kinnaird (1996) reported 58 *M. nigra* per km^2 at Tangkoko Nature Reserve, North Sulawesi. Bynum (1995) reported 27 per km^2 for wild *M. tonkeana/M. hecki* hybrids at the site of Kabun Kopi in Central Sulawesi near the town of Palu.

The data also show that pigs are more heavily hunted than macaques relative

Table 10-4. Encounter Rates for Pigs and Macaques at the Three Morowali Sites

Site	Prey	Number of Encounters	Encounters per 100 Hours
Babilusa (unhunted)	Macaque	11	2.36
	Pig	10	2.14
Anyu Lempe (unhunted)	Macaque	21	3.94
	Pig	6	1.12
Uwe Tondo (hunted)	Macaque	1	0.24
	Pig	8	1.89

Table 10-5. Results of Logistic Regression Analyses.

Test	Prey	Chi-square	p
Uwe Tondo vs. Anyu Lempe	Pig	0.958	0.3276
	Macaque	16.120	**0.0001**
Uwe Tondo vs. Babilusa	Pig	0.072	0.7878
	Macaque	8.910	**0.0028**
Anyu Lempe vs. Babilusa	Pig	1.657	0.1979
	Macaque	1.243	0.2647

p values in **bold** indicate statistically significant differences in encounter rates between sites.

Table 10-6. Group Size of Pigs and Macaques Encountered During Transects at the Three Morowali Sites

Site	Prey	Mean Group Size	n
Babilusa (unhunted)	Macaque	7.18	11
	Pig	3.11	9
Anyu Lempe (unhunted)	Macaque	6.30	20
	Pig	5.00	6
Uwe Tondo (hunted)	macaque	6.00	1
	Pig	4.14	7

Data on group size were not obtained for one pig encounter each at Babiliusa and Uwe Tondo, and one monkey encounter at Anyu Lempe.

Table 10-7. Results Comparing the Size of Encountered Groups

Test	Prey	t	p
Uwe Tondo vs. Anyu Lempe	Pig	–0.530	0.6065
	Macaque	—	—
Uwe Tondo vs. Babilusa	Pig	–0.767	0.4558
	Macaque	—	—
Anyu Lempe vs. Babilusa	Pig	–1.580	0.1381
	Macaque	0.674	0.5054

to their abundance (as indicated by rates of encounter). Table 10-8 standardizes the number of each species killed per the number encountered during the transects. Pigs are hunted more heavily than macaques, yet hunting has less effect on their population.

No other large animals—deer, anoa, or babirusa—were encountered on any transect, hunted or unhunted. This is probably partly a reflection of the low density of animals in general in Morowali. It also may be a reflection of the heterogeneous distribution of some species. Anoa, for example, are rare, exceptional among the Asian wild cattle in their preference for primary forest, and are not attracted to disturbed habitats as are pigs (Musser 1987; Whitten et al. 1987a).

Table 10-8. Harvest Rate Compared with Encounter Rate

Prey	Number Killed	Number of Groups Encountered per 100-hr Transect	Mean Group Size	Mean Number of Animals Encountered per 100 Hours	Number Killed per Number Encountered per 100 Hours of Transect
Pig	57	1.89	3.95	7.47	8.57
Monkey	3	0.24	6.56	1.57	1.91

The upland species (*B. quariesi*) is reported to prefer high mountain ridges (Wirawan 1981). This is supported by the observation that all the anoa killed by Wana were taken from altitudes above 1200 m and in very rough and mountainous terrain. As mentioned above, Wana informants report that no babirusa are present in the Posangke area. Encounters with babirusa were expected, however, at the two unhunted sites. Informants report their presence at both sites. A babirusa cranium was found at the Anyu Lempe site, and tracks were observed at both unhunted sites.

BIOLOGICAL AND SOCIOECONOMIC FACTORS AFFECTING HUNTING RATES

Primates and members of the infraorder Suina (pigs and peccaries) are among the most important mammalian prey types for traditional subsistence hunters in tropical rain forests (Beckerman and Sussenbach 1983; Vickers 1984; Redford and Robinson 1987). In the neotropics, primates and peccaries are the mammals most frequently harvested by natives (Vickers 1984; Redford and Robinson 1987). Although primates are often taken in the greatest numbers, ungulates and particularly peccaries often provide the most meat (Beckerman and Sussenbach 1983; Vickers 1984; Redford and Robinson 1987; Alvard 1993a). Pigs are an extremely important source of hunted meat for traditional groups in Southeast Asia as well (Caldecott 1988). It is widely known that both wild and domesticated pigs are a central focus of the traditional economies in New Guinea and Indonesian Irian Jaya (Rappaport 1968; Ellen 1975; Dryer 1985). Pigs and monkeys also represent extremes with respect to vulnerability to overhunting. Primates are notoriously susceptible to local depletion (Mittermeier and Cheney 1987), whereas pigs and peccaries display exceptional resilience to subsistence hunting (Bodmer et al. 1988b). These patterns are supported by the Wana data.

A number of factors affect the Wana's harvest rates of macaques and Sulawesi pigs. One reason that the monkeys are locally depleted at Posangke is probably related to their reproductive rate. Although population density is also an important factor, reproductive rate is positively correlated with density (Robinson and Redford 1986a; Thompson 1987). The maximum intrinsic rate of increase (r_{max}) is a measure of how fast an animal population can increase if unlimited by resources (Cole 1954). Although there are no data for the two species in question, primates reproduce more slowly for their body size than do suids. Published values for r_{max} for nonhuman primates range between 0.07 for the gorilla to 0.87 for the Senegalese galago (Ross 1988). Ross (1988) reported r_{max} values of 0.15 to 0.23 for five other species of the genus *Macaca*.

Wana hunters find two fetuses inside pregnant female *S. celebensis*. Assuming a pattern similar to other suids (Hayssen 1993), Cole's equation returns an r_{max} = 0.69 for *S. celebensis*. This is a reproductive rate at least three times that estimated for *M. tonkeana*. This result assumes age at first reproduction to be 1 year, that two offspring are produced per year, and that this species has a life span of 10 years.

Another critical factor affecting hunting rates is the size of the human consumer population. Because more people require more food, the amount of prey taken will be proportional to the number of consumers, all other factors being equal (Alvard 1995a). This, of course, assumes that prey productivity can support increasing demand. If the consumer population increases, harvests can increase to levels that are not sustainable and will eventually lead to overhunting of one or more species. This can occur with no change in hunters' behavior and with no increase in the per capita harvest rate (Alvard 1995a). One prediction from foraging theory is that as prey in the diet become depleted, species that were otherwise ignored will enter the diet (Krebs 1979). The long-term rate of population growth for the Wana has not yet been estimated, but it is probably very low, as is common for traditional populations without access to modern medicines (Salzano et al. 1967). The overall mortality rate in Posangke was very high in 1995, at 98.9 deaths per 1000 per annum (Alvard 1996). If the Wana population were to increase due to government-provided medical care, game harvests are predicted to increase accordingly.

Related to absolute consumer population size is population density and mobility. The Wana have high residential mobility; the average family changed house location 4.2 times during the course of the study. The absolute area of land used by each family was relatively small, however. The mean size of the area encompassing all of one family's gardens, traps, and house locations was only 0.61 km² (Alvard 1997). The area that encompasses all of the houses, gardens, and traps for the entire sample of 153 animals was 10.7 km². Population density within this area is 14.3 people/km². It should be noted that this is an ecological density (Haila 1988); the area used to calculate the density includes only the habitat used by the Wana of Posangke.

A portion of the habitat in the Reserve is almost inaccessible to hunters because of the severity of the terrain. This is particularly true near Posangke, where nearby peaks reach 2600 m. Hunters report that steep and eroding slopes, rocky uneven terrain, and mud make travel and pursuit of prey very difficult, and gardening unproductive. This is probably why the Wana of Posangke concentrate their subsistence activities in the river and stream bottoms and lower slopes of the mountains. This is perhaps also why the Wana rely on trapping near gardens rather that active hunting in the forests.

These high and steep areas may act as prey refugia and help to maintain some species in the Reserve. A number of researchers have pointed out the importance of such areas for creating conditions that allow predator-prey coexistence (Mech 1977; Sih 1987; Pulliam 1988). Although transects were not performed in these rugged areas, informants have reported higher densities of pigs and primates, as well as anoa.

Another factor that can affect hunting rates is access to markets (Redford 1992). The ability to sell meat can significantly increase its value to subsistence hunters who have little other access to cash (Fa et al. 1994). Market access motivates hunters to harvest more than they would otherwise for subsistence purposes. Commercial hunting also can develop and provides substantial profits for rural people (Caldecott 1988; Wilkie et al. 1992). Although the Posangke Wana do have access to a market 1 to 2 days away by foot, they do not sell any of the meat that they harvest, at least partly because the majority of their lowland neighbors are Muslims who are prohibited from consuming pigs and monkeys.

Much of the work examining the impact of humans on prey tends to focus on hunting harvests (Robinson and Redford 1991a). Habitat loss due to the agricultural activities of the hunters can play an important role as well. Landscapes in Wana settlement areas are characterized by considerable disturbance resulting from garden production. The amount of land cleared each year for gardens is small; the study sample cleared 24.35 ha in 1994 (unpublished data). This is approximately 0.16 ha per person, none of which was cleared from primary forest. The apparently long-term presence of the Wana, however, has created a very anthropogenic environment characterized by a mosaic of secondary forest, gardens, and alang-alang grasslands. Primary forests exist on the margins of settlement areas and in areas of the Reserve uninhabited by Wana.

The data presented here suggest that Wana forest disturbance may be partly responsible for both maintaining pig populations and depleting macaque numbers. The pigs might actually prefer the disturbed habitats provided by garden production. Pigs raid manioc gardens and are a regular target of Wana complaints. This is a relatively common pattern for some areas of the world (Greenburg 1992; Jorgenson 1995a; Suárez et al. 1995). Linares (1976) argued that garden hunting is a common pattern in the American tropics. Rodents, peccary, and deer were among the species attracted to gardens at the site she studied in Panama. The Wana place almost all of the pig traps at strategic locations along pig trails near garden sites (unpublished data). Monkeys also raid gardens in Indonesia (Supriatna 1991), but were not a common problem for the Posangke Wana. This is perhaps because the Wana grow little corn, a crop favored by macaques.

THE FUTURE OF SUSTAINABILITY

The transect data indicate that there are no significant differences in pig densities between the hunted and unhunted sites. The conclusion drawn from this analysis is that the Posangke Wana subsistence strategy has led to a sustainable pig harvest.

This is probably due to a combination of a relatively small consumer population, low offtakes relative to pig production, and a pattern of Wana habitat disturbance favorable to pig feeding ecology.

In contrast, the data do not point to the same pattern for macaques. Measurements of low prey densities in hunted areas do not, by themselves, prove that hunters are responsible, or that hunting is not sustainable. However, the observation that macaques are rare in hunted areas compared with unhunted areas is strong evidence that Wana subsistence activities have led to local depletion of the macaques. In addition, loss of habitat due to forest disturbance cannot be ruled out as an important factor to explain the low macaque numbers.

It is difficult to know how many macaques were killed in the past, so it is difficult to gauge what impact their loss is having on the Wana today. Informants report that although primates have been killed in the years previous, they have not been taken in numbers appreciably larger than they were the year of the study. This suggests that the depletion of the Posangke macaque population was not a recent event. If the observed primate harvest is representative of the average contribution they make to the diet, removing primates from the Wana's diet would have a negligible nutritional effect. In contrast, the pigs and anoa harvested by the Wana provide a substantial dietary benefit. Although the rice that they cultivate provides the Wana with much of the protein they consume, the fat provided by pigs and anoa is a crucial component of their diet (Speth and Spielmann 1983).

Whether these patterns will hold into the future is unknown. If the Wana maintain any aspect of their subsistence hunting ecology in the future, pigs are the most sustainable of their prey options. If, however, they follow the path of other traditional native groups, entering into contact with the outside world, the Wana's socioeconomic strategy is likely to change dramatically. For example, construction of a road scheduled to bisect the northern end of the Reserve will bring settlers, transmigrants, and perhaps a new market for meat. The road, and what it will bring to the area, represents the greatest threat to the Morowali ecosystem. In the absence of the road, subsistence hunting can continue to provide benefits to the Wana, but it must be managed if conservation goals are also to be met. One complicating factor is that since the Reserve was formed in 1980, the Wana residing inside have technically done so illegally. Because hunting is prohibited by law, informants distrust outsiders and report fear of being relocated to lowland settlements.

One solution would be to exclude the areas inhabited by the Wana from the Reserve and create buffer zones in these areas. This idea was proposed by the World Wildlife Fund in 1980 prior to the establishment of the Reserve, but was not adopted. The data presented here show that Wana areas lack the macaque populations common to the rest of the Reserve. The areas inhabited by Wana are also significantly disturbed by gardening activities; only a small proportion of original forest is present. Morowali Reserve would lose little pristine forest by creating buffer zones from areas currently inhabited by Wana. If given secure tenure to their land, the Wana would perhaps gain enough trust of government officials to cooperate in a management scheme.

Part of this plan would include convincing the Wana to accept a restricted diet breadth limited to pigs and small prey in exchange for secure land tenure. The Wana would lose little by forgoing monkey harvests, although the same cannot be said for anoa. Anoa provided 41% by weight of the large game harvested during the sample period. Persuading Wana hunters not to pursue anoa will be the most difficult challenge for Morowali's managers. If harvests restricted to pigs were limited to subsistence needs and not for sale, such a plan might prove workable provided that migrants into the area were not also given permission to hunt. Such schemes would take time, money, education, and enforcement but might be one way to provide sustainable value for Wana hunters as well limit negative impacts to Morowali Reserve.

ACKNOWLEDGMENTS

The research was funded by The Nature Conservancy, The National Geographic Society, The Wenner-Gren Foundation, The Indonesian Field Office of the Nature Conservancy, and the Faculty of Social Science at SUNY-Buffalo. I thank Dr. Jatna Supriatna at the University of Indonesia for sponsoring my research. Thanks also to Fred Rawski, Martarinza, Nurul Winarni, and Jabar Lahadgi for assistance in the field. I also thank Soeprapto Hadi Pranoto and Iksan Tengkow at the Indonesian Forestry Department (PHPA); Marty Fujita, Nengah Wirawan, Mulyanto, Duncan Neville, and Totok Hartono of The Nature Conservancy; and Dewi Soenariyadi at the Indonesian Institute of Science (LIPI). I extend special thanks to the Wana of Posangke.

PART II

Sociocultural Context Influencing Sustainability

11

A Pound of Flesh: Social Change and Modernization as Factors in Hunting Sustainability Among Neotropical Indigenous Societies

ALLYN MACLEAN STEARMAN

Current research describing wildlife exploitation by native peoples of the neotropics has tended to move away from romanticizing indigenous resource use, or what Redford (1991) has called the *ecologically noble savage* perspective, to explanations that derive from epiphenomenal factors (Alvard 1993). Hence, rather than attributing resource conservation to an environmental ethic presumably embedded in cultural traditions, researchers are coming to understand that limiting factors such as slow rates of population growth, low-density settlements, nomadism, and the use of traditional technologies more often must be credited with preventing the overexploitation of resources such as wildlife.

Nonetheless, the most basic life-styles of traditional neotropical peoples are increasingly being threatened by the outside world, a phenomenon that is now both pervasive and irreversible. Whether patterns of apparently sustainable wildlife use are epiphenomenal or ideological in origin might become hardly more than an interesting theoretical sidebar as the forces of modernization radically affect long-standing indigenous relationships to the environment. In order to understand how culture change and modernization may seriously impinge on the success of efforts to promote wildlife conservation, I will examine five factors that I believe present the greatest challenge to sustainable hunting in the neotropics.

These factors are sedentarism, population growth, market involvement, techno-logical enhancements, and incursion and/or circumscription. All may be consid-ered the result of externalities in that they are precipitated by outside forces and all interact in synergistic ways. I will draw from my 15 years of fieldwork among the Sirionó and Yuquí of the Bolivian Amazon, from my long-term experience in that region that began in 1964, and from other studies of Amazonian peoples to analyze these issues. Finally, I will discuss the prospects for sustainable hunting as indigenous societies participate more fully in the modernization process.

SEDENTARISM

Anthropologists have long known that *sedentary forager* is an oxymoron. Yet many of those working in one capacity or another with native peoples often fail to grasp this most fundamental concept: that the sedentary existence that invariably follows contact and acculturation of nomadic or semi-nomadic hunter-gatherers places them at odds with their subsistence base. In human evolutionary terms, increasing sedentarism accompanied the development of first incipient and then highly specialized intensive agriculture. Once people began producing food surpluses, constant movement to track resource availability, particularly game animals, became less critical to survival. However, for early horticulturalists, those who cultivated small gardens, hunting remained an important subsistence activity. Because domestic animals required continual care and protection, competed with croplands for penning and/or foraging space, and consumed large amounts of feed (which often placed them in competition with humans for crop resources), the farmer relied on hunting to supplement daily protein needs. This pattern has changed little in areas such as Amazonia, where indigenous horticulturalists still depend on bush meat to supply their animal protein needs.

The critical link between hunting and mobility, even among semi-sedentary horticulturalists, is evident in the patterns of modern native peoples. For example, localized game depletion rather than soil exhaustion or weed invasion was shown to be a major contributing factor in determining the movement of Siona-Secoya horticultural villages (cf. Vickers 1983); and among the Yanomamö of Venezuela and Brazil, game depletion and the stresses occasioned by meat scarcity cause intravillage tensions to rise, eventually resulting in village fissioning and relocation (Good 1987). Although many native Amazonians may be described as *farmers* because they live in villages and produce crops, they are in reality *forager-farmers*, an adaptation that depends on the ability to relocate the settlement as resources decline. Because a dependence on domestic animals as a primary source of animal protein is unknown, only in those native horticultural communities where aquatic resources are plentiful is true sedentarism possible; and as Harris and Ross (1987) noted, "access to productive aquatic resources inclines to an increase in seden-tarism, since such resources are relatively immobile; increased sedentarism will usually mean greater development of horticulture [Carneiro 1968], while a more sedentary, horticultural lifestyle will differentially diminish access to the spectrum of available game animals." However, most horticultural peoples of Amazonia living in interfluvial zones not only continue to depend on wildlife for animal protein, but make use of a wide range of forest products to assure their well-being (Irvine 1987; Johnson and Baksh 1987; Gragson 1989; Balée 1994; Townsend 1995, 1996). The accessibility of these resources in adequate amounts most often requires mobility.

The pattern that typically emerges when foraging or horticultural peoples are contacted by missionaries or government agencies interested in the integration of native or local peoples into modern state society is that of settling them into permanent communities. This strategy is particularly common when indigenous groups present some threat to the goals of nation states. Often this threat comes in the form of hostile encounters with settlers, loggers, ranchers, miners, oil companies, or others wishing to exploit the resources of an area. Hence *pacification* is frequently part of the acculturation process that also may include proselytizing; and for those engaged in these processes, experience has shown that they can be most expeditiously accomplished by setting up permanent settlements where acculturation may occur unfettered by distance or dispersion. However, when sedentarism is imposed on traditional peoples, it has far-reaching effects on their subsistence strategies, social and economic organization, and ultimately, on their survival as ethnic groups.

The Sirionó and Yuquí of Bolivia are two linguistically and culturally related groups that demonstrate many dimensions of the effects of sedentarism on nomadic foragers. Prior to contact and settlement on mission stations, they traveled in small groups of fewer than 20 people; they knew little specialization, following the pattern of most foragers where the division of labor is based primarily on sex and age; and their living structures were simple, as was their technology, which consisted of a few baskets, hammocks, and bows and arrows—whatever could be easily transported from camp to camp. Although the Sirionó were reported to have practiced some horticulture prior to European contact (Holmberg 1969) descriptions of this activity indicate that it was very sporadic and that their "fields" comprised only the small space of bare ground created by a tree fall. These very small plots were planted with manioc, sweet potatoes, and when seed was available, maize, but could hardly be considered a consistent or dependable source of food. The Yuquí did not farm at all. Both groups were kin-based societies with leadership vested in an elder male having some power but by no means able to impose his will on his peers. This leadership was fluid and consensual, with decision making commonly a group activity. More importantly, much of Sirionó and Yuquí traditional life was based on repetition of known experience so that these people worked and lived with each other comfortably, seldom having to deal with innovations of great magnitude (Stearman 1984, 1987; Townsend 1995, 1996).

With the advent of peaceful contact that came to the Sirionó in the late 1930s and the Yuquí in the mid-1960s, their relationships with their environment and their own people radically changed. In both cases, missionaries and the Bolivian military intervened to cease hostilities that existed between these native peoples and the settlers who were moving in around them. Peaceful contact was accomplished by settling both ethnic groups on mission stations and beginning the process of acculturation that, from the perspective of outsiders, would render these people into productive Bolivian citizens.

The missionaries attempted to convert the Yuquí and Sirionó to Christianity with the doctrine of salvation going hand in hand with the European ideology that productive work is synonymous with settled agriculture. Although the Sirionó farm on a more regular basis than the Yuquí, perhaps only because at the time of their contact the use of coercive practices to encourage such behavior was common, farming is not an activity highly valued by either group.

First, the life patterns of foragers are qualitatively different from those of farmers. Tropical hunters and gatherers carry out their work in an environment of forest cover or intermittent grassland, often resting during the heat of the day as do their prey. They tend to move over the landscape at a leisurely pace as they move from patch to patch.

Second, although there appear to be great inconsistencies among studies done of hunter-gatherers and their actual energy expenditures and well-being (Lee and DeVore 1968; Hill and Hurtado 1996), most Yuquí and Sirionó comment that when compared with a foraging existence, whose hardships they do not glorify, the life of the farmer is one of boredom, drudgery, and irritating deferred gratification.

Third, dietary diversity among hunter-gatherers and incipient horticulturalists also tends to be higher than that of settled agriculturalists, who do not have access to the breadth of natural resources available to people who forage for food (Fleuret and Fleuret 1980; Dewey 1981; Flowers 1983; Haaga et al. 1986; Cohen 1989). Researchers also have determined that a shift from foraging to farming may be accompanied by a decline in overall health and human nutrition (Cohen and Armelegos 1984). And although tropical hunter-gatherers might experience periods of scarcity and hunger, actual starvation is rare.

Fourth, in foraging and many horticultural societies, prestige is accrued through hunting and the provision of meat, not from gathering plants or planting crops that present no danger, require no chase, and rarely are considered a scarce resource. Thus, there is little social capital accrued from farming, and for the Yuquí and Sirionó, little reason to devote much interest or initiative to this activity. As a consequence, when the Yuquí and Sirionó set out small gardens, they do so half-heartedly, and not infrequently abandon what they consider to be scars on the landscape before a crop is ever harvested. Without a dependable source of crops to feed domestic animals, most Sirionó will keep only a few free-range chickens or a pig that also must feed itself. The Yuquí have never shown much interest in raising animals, eat those they are given before they have a chance to reproduce, and lump all farming activities together as having low prestige value regardless of whether meat may be produced. Hunting, on the other hand, brings a Yuquí or Sirionó man status among his peers and the social rewards that extend to his spouse and children (Stearman 1987; 1989a).

Central place hunting such as that currently practiced by the Sirionó and Yuquí cannot be practiced for long without depleting the game in the immediate settlement area. Typically, semi-sedentary farming peoples who must practice this type of hunting tend to follow a logical progression outward in concentric rings until game depletion reaches a distance from the village that requires too much time and energy investment in travel to make the trip worthwhile. At that juncture, the village typically is relocated. Horticultural peoples who must remain stationary during critical phases of the planting season or who are for some other reason limited in mobility frequently resolve the problem by periodic trekking (Henley 1982; Werner 1983, 1990). Trekking provides protein in relatively large amounts during certain times of the year, supplemented by a diet high in carbohydrates during the remainder. In writing about the Mekranoti-Kayapó of Brazil, Dennis Werner concluded that "the explanation for trekking in lowland South America may lie in the concept of dietary optimization" (1983), and that the Mekranoti trek primarily for protein maximization. Trekking also tends to satisfy ritual needs for prestige among males, providing a time when men can accrue honor by hunting. The availability of large amounts of game, particularly of favored species, also serves to strengthen social bonds through the sharing of meat.

As modernization ensues, however, the opportunities for mobility such as trekking decrease. For example, men and women leave their communities to engage in seasonal or periodic wage labor. Children must adhere to a school year that is determined in the capital city by government officials. Consequently, treks tend to become less frequent, are shorter in duration, cover less distance, and are governed by an external calendar that might not coincide with optimal hunting periods. Both the Sirionó and Yuquí continue the practice of trekking; but because of the exigencies of participating in a national society as outlined above, treks are becoming much less common.

Although there is evidence that a frequently hunted area around a village might serve as a population sink for wildlife in surrounding areas (Pulliam 1988), sedentarism also leads to population growth of the human predators, which tends to work against the source-sink effect.

With expanding populations, fissioning of villages may occur, and new settlements may be established. The reasons behind the fissioning of villages might be attributed to both social and environmental factors (Gross 1983; Hames 1983; Good 1987), but the end result is the same: previously un- or underexploited areas come into production. Although there are cases in which hunting near settlements apparently has continued for many years, the number and size of game animals harvested decreases over time and the suite of prey taken changes to favor those animals with high reproductive rates and tolerance of human activity (Irvine 1987; Jorgenson 1993; Stearman 1990). Thus, the effects of sedentarism on wildlife populations may range from total extirpation, either locally or over an increasingly larger area, to a shift in the composition of species of animals that are better able to withstand relatively high levels of hunting pressure.

Finally, the availability of fish as an alternative to game meat also must be considered as a factor affecting wildlife populations being exploited by sedentary peoples. Robert Carneiro was one of the first anthropologists working with Native Amazonians to recognize the significance of fishing in supporting large, sedentary native Amazonian populations, as well as in relieving pressure of semi-sedentary horticultural peoples on game animals (Carneiro 1968). Wagley also commented on the significance of fish resources in maintaining the Tapirapé of Brazil once they were contacted and became sedentary; and in the following excerpt from his book, *Welcome of Tears*, Wagley's description of this indigenous group in 1953 bears uncanny resemblance to the Yuquí of the 1980s:

> It must be remembered that in 1939–40 the Tapirapé lived far from the river. They were "foot Indians" and most of them could not swim. They ate fish only for a short period in the dry season. By 1953, to my surprise, I found Tapirapé boys and girls who lived in New Village swimming for hours in the Araguaia River. Some men were by then excellent canoemen. Most people could swim well, though the older people, of whom there were but few, could not. As late as 1965 the Tapirapé did not know how to build dugout canoes, but they owned canoes which they had purchased from their Carajá neighbors. Men fished as they had in the past, along the banks of the rivers and lakes with bow and arrow, but now they also fished with hook and line. Where hunting had once provided them with most of their protein needs, now fishing is more important. In much less than a generation, they have changed from "foot Indians" to "canoe Indians" (Wagley 1983).

During my 1983 fieldwork among the Yuquí, I, too, was surprised to discover that of the total meat take for this group of people who define themselves as hunters, 57% by weight was fish. It was clear that like the Tapirapé described by Wagley, the Yuquí's ability to continue to subsist primarily from foraging despite being sedentary for more than two decades was dependent not only on food subsidies provided by the mission but also on river resources. Fish from the Chimoré River were being taken with new technologies, hook and line and nets, provided by the mission. With the addition of river fish to their diets, the Yuquí were able to combine fishing with hunting in order to secure adequate amounts of animal protein. Unfortunately (as discussed in the section on Incursion and/or Circumscription), a serious decline in these resources as the result of exploitation by outsiders has seriously affected Yuquí subsistence.

POPULATION GROWTH

Group size is a function of available resources, and population controls are necessary to assure that people do not outstrip their food supply. Overexploitation of the resource base by traditional foraging peoples is due to epiphenomenal factors such as disease, injury, and food scarcity (Hames 1991; Alvard 1993a) and social customs which limit population size (Wagley 1951; Birdsell 1958, 1968; Sahlins 1968, 1972; Murphy and Murphy 1972; Holmes 1995). Cultural factors such as the purposeful spacing of children affect not only the birth rate, but are critical to the survival of the group since a woman cannot reasonably travel carrying more than one child at a time (Friedl 1975). Abortion continues to be quite rare in traditional societies because it is so dangerous to the life of the pregnant woman. And in the larger scheme of things, an adult woman of child-bearing age represents a much greater investment than a newborn whose chances for survival are typically less than even. Consequently, many societies practice various forms of infanticide (Holmberg 1969; Wagley 1951; Chagnon 1968; Stearman 1989b; Gregor 1987; Holmes 1995). Infanticide is an effective means of controlling population size since preference is normally given to the killing of females who ultimately determine the rate of population growth. Other forms of population control include postpartum taboos that prohibit sexual activity between a recently-delivered married couple (Wagley 1951; Gregor 1985; Kensinger 1995). Some of these taboos may last as long as 3 years, becoming a form of spacing as well as population control (Kensinger 1995). And finally, prolonged lactation has been shown to inhibit ovulation in women (Ellison 1990).

Typically, when people shift from foraging to farming, population size increases rapidly. Farmers need an ample labor supply, and children often serve this function; and with food surpluses, offspring can always be fed, albeit perhaps not well. Hence, having large families becomes a goal of most farmers, and those who succeed in having males first and then females tend to be the most successful in terms of labor requirements. Sedentarism also works to the advantage of women in the successful conception and bearing of children. With higher daily caloric intakes, women's fertility rates increase, and with cultural standards now encouraging large families, more is invested in children. Sedentarism allows for children to be closely spaced, and lowered mobility means less wear and tear on both pregnant women and their young offspring.

If native societies can survive the initial chaos caused by contact and the disease that may follow from contagion by outsiders, populations tend to grow rapidly. Anthropologist Charles Wagley, writing about the Tenetehara and Tapirapé of Brazil comments:

> In many cases, these differences in population trend following European contact may be explained in terms of the nature of contact with Europeans to which the different native groups have been subjected. Epidemics have varied in frequency and in intensity among native groups and the systematic exploitation of native peoples through slavery and other forms of enforced labor has taken a heavier toll upon native population in some areas than in others. On the other hand, introduced crops, domesticated animals, new instruments, and new techniques have sometimes raised the aboriginal subsistence level and made possible an expansion of population (Wagley 1951).

The Yuquí population increased in number from 73 in 1983 to 150 in 1996. With access to modern medicines which lower infant mortality, and the influence of missionaries who discourage infanticide, the Yuquí population is growing at the rate of 8% per year, giving this group a conservative doubling time of about 12 years. When combined with sedentarism, circumscription, and market involvement, this unprecedented population increase has for some time now contributed to the lowered success rate of Yuquí hunters. Among the Sirionó as well, population in the village has grown from 304 individuals in 1987 to 459 in 1993, largely due to natural increase. This growth rate has a doubling time of about 10 years (Stearman 1987; Townsend 1995, 1996).

The growth of Yuquí and Sirionó populations places more pressure on game resources in an absolute sense because more people must be fed. However, there are also indirect effects of population increase that have a significant impact on wildlife. As the numbers of people increase, the need for material possessions expands as well, and more people own a greater variety of goods. Social and economic distinctions begin to appear in what were formerly egalitarian societies. Envy, a driving force of consumerism, is now commonplace. The upward spiral of conspicuous consumption is not a pattern limited to industrial societies. Although farming is not esteemed by either the Sirionó or Yuquí, it is increasingly viewed as a means of producing a surplus that can be converted to consumer goods. With ever larger numbers of people, more land will need to be cleared to produce crops, and this can only have negative consequences for wildlife as habitat is reduced.

Finally, population increase must be considered from the perspective of non-native populations that are threatening the integrity of indigenous lands. This topic will be covered in the section on Circumscription and/or Incursion, but bears mentioning at this point. The populations of neotropical countries continue to grow at high rates. Although a good case can be made for resolving many of the current economic problems faced by these nations through extensive social and economic reform, such responses will only forestall the inevitable conflict over access to finite resources such as land. Large intact tropical landscapes, privately or publicly owned, not only will become more scarce and harder to obtain, but virtually impossible to protect. Native peoples will become at once victims and despoilers as they are trapped by the same economic forces that drive the nation states to which they belong. Limiting population growth—whether from the standpoint of indigenous peoples overexploiting their existing resource base because of changing life-styles or the growing masses who are expanding into native lands to feed themselves—is an issue that is not comfortably addressed by anyone. Inevitably the specter of genocide, ethnocide, imperialism, classism, and racism that have so consistently haunted the history of our own species will be raised.

MARKET INVOLVEMENT

The integrity of traditional life-styles, often characterized in terms of their harmony with nature, depends to a large extent on isolation. Contact with the outside world transforms indigenous subsistence patterns that were based on self-sufficiency into ones that become permanently linked to a consumer economy. This is a process of social change known by many euphemisms: acculturation, integration, development, modernization. When the Sirionó and Yuquí were settled onto mission stations, their needs began to shift from those required by traditional adaptations to those defined by outsiders. As noted above, many of these needs are now being met by increasingly intensive exploitation of the local environment including wildlife. Regardless of whether conservationists and indigenists might view these introductions exploitive, destructive, or maladaptive, they are nonetheless real and permanent. It is also naive to believe that most indigenous peoples do not welcome many of these changes; and it is certainly unfair to expect them to hold to a conservation ethic, whether ours or theirs, that denies them access to services and products that industrialized peoples take for granted. People such as the Yuquí, Sirionó, and Aché of Paraguay do not romanticize their past as do many westerners who would like to portray their precontact life as idyllic. With regard to the Aché, Kim Hill and Magdalena Hurtado take strong issue with those who view the life of the forager as one of leisure and abundance:

> The "original affluent society" myth suggesting that our ancestors easily met their daily needs before we became greedy and began to desire more than that which is necessary is an idea that tells us more about late twentieth century anthropological thought than it does about the lives of foraging peoples. The "original affluent society" concept has no basis in empirical reality or biology, but it is also a cruel hoax because it leads members of modern societies to avoid the empathy or guilt that they should feel when considering the plight of people living under difficult conditions (Hill and Hurtado 1996).

Consumerism among indigenous peoples is often presented, albeit not always unfairly, as an example of the worst the industrial world has to offer: plastic containers that litter the landscape, junk food and candy wrappers, boom boxes, and expensive sneakers. But although we might criticize the value of these items to the lives of indigenous peoples and even to our own, they also arrive on the same boat or plane with food, clothing, blankets, mosquito nets, tools, books, newspapers, and, perhaps most important, modern medicines. Commenting on the Panare of Venezuela and their increased sedentarism and proximity to market towns, Paul Henley noted that "The Panare give various reasons for the change that is taking place in their settlement pattern but undoubtedly the most important of these is the desire to have readier access to industrially produced goods and medicines" (1982).

The efficacy of traditional medicine is well documented, and it is common knowledge that many modern pharmaceuticals derive from substances first identified by traditional healers. The potency of modern medicine, particularly the antibiotics, is rarely equaled by most native cures. This is particularly true when indigenous peoples are confronted with introduced diseases whose prevention and treatment remain the specialized domains of western medicine. Where living and dying traditionally were in the hands of shamans and the spirit world, access to modern medicine not only changed the world view of indigenous peoples about the nature of sickness and health, but made them as dependent as are we on the economic system that provides modern health care.

This same dependency now extends to areas such as education. Native peoples understand that in order to survive as ethnic groups, they must have the knowledge and leadership skills that will enable them successfully to defend themselves in a complex political environment. For traditional communities, education is expensive. The Sirionó are provided with several teachers whose salaries are paid by the government; but the community must feed and house these individuals. In addition, books and other school supplies must be purchased for each student, a significant expense for those families with several school-aged children. Hunting and the by-products of hunting often meet these growing educational needs, not only by provisioning the teachers, but by providing cash or trade items for school books, pencils, erasers, and paper.

Contact with the outside world through trade networks also brings about new social relationships that open the forest to those who hunt for sport or commercial gain. As native peoples form trading alliances, it is not uncommon for these trading partners to gain access to forest reserves or homelands that previously had remained the domain of the native inhabitants. In my experience and those of others who have worked with indigenous peoples, social bonds with traders, although often highly exploitive and contentious, are actively nurtured. Because of the long distances and often harsh and dangerous conditions they must endure, in these societies men rather than women are typically the itinerant traders. As an outgrowth of the trading partnerships established with indigenous peoples, itinerant traders might make use of these alliances to hunt commercially or for sport on indigenous lands. These privileges may extend outward to any family member or friend of the trader. Thus, there may be periods when the forest is heavily exploited by outsiders whose access to these areas only exacerbates the problem of increasingly limited game resources. Native peoples commonly complain of these abuses, but are often unwilling to threaten those necessary social and economic ties they have established with outsiders.

To a large extent the Yuquí and Sirionó have persisted in their hunting traditions because of their history of living in a dependency relationship on mission stations that have provided various types of subsidies to sustain them (food, medicine, clothing, tools). Yet, these special circumstances are in themselves instructive and cautionary. Sirionó and Yuquí reluctance to shift from foraging to farming is reinforced not only by the strong influence of culture and ideology described above, but by the inputs of externally produced food that enable foraging to continue as a subsistence activity. Both groups receive surplus food products from the state, missionary organizations, or both in the form of international food donations. However, the arrival of these products is sporadic and undependable and requires a cash payment of transport costs, placing both groups at risk of food shortages. This situation also has created a dependency relationship that threatens not only the cultural integrity of the Sirionó and Yuquí, but undermines their future ability to provide for their own needs.

The Sirionó and Yuquí are also performing wage labor more regularly, which separates members of the group, interrupts normal interactions, makes individuals vulnerable to permanent displacement, and increases the desire for exotic consumable and nonconsumable goods. Although cash inputs that are converted to food might have some positive effect on the local wildlife populations, this effect can only be temporary given increased numbers of people to feed and habitat loss due to circumscription (Stearman 1990, 1995; Townsend 1995, 1996). In spite of food being available through donations or purchase, wildlife continues to be a major food source, particularly when external subsidies fail to appear; and as the Sirionó and Yuquí become more enmeshed in a consumer economy, wildlife is increasingly used as an economic fallback to meet these needs as well (Stearman and Redford 1992; Stearman 1995; Townsend 1995, 1996). The developing pattern of practicing unregulated hunting with a growing population in an expanding economy can only lead to unsustainable wildlife exploitation.

Participation by indigenous peoples in a cash economy is correctly interpreted as a major threat to the preservation of large tropical landscapes. However, this participation is often trivialized, in the most perverse form of ethnocentrism, by interpreting it as corrupting indigenous societies with the trinkets of western consumerism. The right of native peoples to enjoy the same benefits that living in the modern world offers the peoples of industrialized nations ethically cannot be denied. In the same vein, it goes against all that is known about the nature of human culture to expect indigenous peoples to remain unchanged, holding onto life-styles that no longer adequately meet their needs. Even prior to contact with western culture, native societies were not static, responding to change in ways that would increase their adaptive success. Much of the participation in the market economy must be interpreted as a similar adaptive response as modern indigenous peoples seek ways to improve their chances for survival in a more complex and threatening world than previously experienced.

TECHNOLOGICAL ENHANCEMENTS

Firearms, lights, and outboard motors are now the help mates of what once were traditional hunters (figure 11-1). Hunting among native peoples is not a sport but a means of survival that in today's world is defined in different terms than under aboriginal conditions. Native hunters use every innovation at their disposal to bring down their prey; for the subsistence hunter there is not, and cannot be, any concern about giving one's prey a "fighting chance." As noted earlier, the definition of subsistence itself is also changing, and this too must be considered. Subsistence now implies purchasing power, not simply production for consumption. Hunting has taken on new dimensions as a means of provisioning a family, and technology is an integral part of this.

Technological enhancement in hunting means that those who take advantage of tools to improve their game take also become tied to the market system that provides these improvements. A shotgun requires a substantial initial investment to acquire, and although it may last for years, it also demands the purchase of ammunition to remain functional. In the context of the third world economies in which most neotropical peoples live, the purchase of ammunition requires a significant capital outlay on a regular basis. It is not uncommon for a hunter to pay as much as US$1.00 per shell when bought in small quantities from an itinerant trader. Few can afford to buy in bulk at reduced prices; and the life of the shell, even in plastic casings, is short in the humid tropics, making long-term storage inadvisable. These same conditions apply to the purchase of fuel for outboard motors, flashlight batteries, commercial traps that need parts for repair, and so forth.

Figure 11-1. The old and the new. Two Yuquí hunters, one with the traditional bow and arrow, and the other with a modern firearm. *Photo courtesy of Allyn Stearman.*

Thus, hunters who formerly relied solely on nature as their source of hunting implements now require regular and frequent contact with the outside world in order to feed and provision their families. Trips to trading centers to replenish hunting equipment often involve transportation, food, and lodging costs and entail the purchase of other now important staples such as salt, sugar, kerosene, and medical supplies. Hunters find that in order to meet these demands, they must hunt for profit as well as food, and this might mean entering the bush meat market, or the legal or illegal trade in live animals and animal skins. Thus, wildlife that might have been captured rarely or not at all becomes the prey of the hunter dependent on meeting the costs of technological enhancements.

The Sirionó, not unlike many other indigenous peoples, have long-standing relationships with local traders that now span several generations (Stearman and Redford 1992). These individuals, in this case both men and women, trade shotgun shells, 22-caliber bullets, new and used weapons, traps, machetes, axes, flashlight, batteries, and even hunting dogs for the products the Sirionó acquire in the forest: animal skins (legal and illegal species), rhea feathers, honey, medicinal products, building materials such as palm thatch that are later shipped out in ox carts or trucks, fruit, fresh and salted bush meat, live birds and mammals for the pet trade, and virtually anything else that might have some current market outlet. Traders place "orders" for certain items and will extend credit to the Sirionó, often in the form of staples or ammunition, to assure that these orders are filled. In this way, the social and economic contracts between traders and indigenous people continue in virtual perpetuity.

Native peoples need little instruction to understand the value of technological innovation. Following a debate on the issue of the efficiency of the shotgun versus the bow, in 1979 anthropologist Raymond Hames published the results of an experiment he conducted to determine which was the more efficient weapon. Hames found that the shotgun had a considerable advantage over the bow. In analyzing the relationship of shotguns to blowguns and spears among the Waorani, Yost and Kelly came to the same conclusion (1983). These results do not come as a surprise to any indigenous hunter. Yuquí and Sirionó hunters very clearly articulate the advantages of hunting with a shotgun or rifle and revert to the bow only when ammunition is unavailable.

In an experiment I conducted among the Yuquí in 1988 involving shooting at a target consisting of a life-sized drawing of a collared peccary, the Yuquí, using traditional bows, could not hit the target at distances greater than 20 m. This has a number of implications for traditional hunting. The greatest accuracy was achieved by Yuquí bowmen at a distance of 10 m or less, or approximately the average open space available for a clear shot in the forest they inhabit. Because an arrow is easily deflected by leaves, the Yuquí rarely attempt a shot from a distance greater than 10 m. With a shotgun or rifle, however, this distance can be vastly exceeded. Thus, a firearm presents a number of advantages beyond just killing power (lowering wounding rates and the time and energy expended in tracking a wounded animal). Modern firearms allow hunters to expand into hunting niches that may require more than one traditional weapon to exploit successfully, or simply cannot be efficiently exploited at all with traditional weapons. The shotgun in particular offers the greatest effectiveness in terms of the diversity of prey taken, being useful on arboreal as well as terrestrial species.

The efficiency of modern firearms is further enhanced when combined with other innovations. Flashlights extend hunting beyond sunset and to those nights when a moon is not present. Flashlights also encourage stand hunting (building a platform above areas where game animals are likely to feed), which is most effective at night and hence was a technique rarely practiced by the Sirionó prior to the availability of flashlights. Sirionó hunters commonly hang a hammock in a tree above a place where animals are known to feed and quite literally lie in wait. When a sound is heard, the hunter fixes a beam of light on the prey below and shoots. Several animals may be taken in the same spot in a single night using this technique. Horses, outboard motors, bicycles, and even motorcycles or four-wheel–drive all-terrain vehicles extend the range of native hunters equipped with firearms, and can improve a hunter's chance of bringing down large, swift game. The Sirionó often hunt in open grassland, inhabited by the large marsh deer (*Blastoceros dichotomus*) and rheas (*Rhea americana*) and are very successful in harvesting these animals during the dry season, when horses, bicycles, or even occasionally a motorcycle borrowed from a trader can be employed.

For the native hunter who must feed his family and kinfolk, and who also must be able to purchase consumer items, these new technologies, often in combination with traditional methods of hunting (Yost and Kelly 1983), offer greater assurances of a successful hunt. From the perspective of wildlife conservation, unrestricted use of firearms and other technological enhancements that improve hunting efficiency by native peoples pose the very real threat of local extinctions. When extrapolated to a larger scale, the offtake of wildlife by native and other local peoples as a consequence of the combination of modern technology, population growth, and market pressures is prodigious. Redford (1992) estimated that in the Brazilian Amazon alone, 14 million mammals a year may be harvested by rural peoples; and with birds, reptiles, and animals fatally wounded added to this list, this number could reach 57 million animals per year.

INCURSION AND/OR CIRCUMSCRIPTION

The establishment and protection of native homelands remains a charged and controversial issue throughout the neotropics. The process of validating, documenting, legislating, mapping, and defending indigenous territories is daunting at best, and at worst, it is never successfully completed. Governments, private interests, multinational corporations, small farmers and ranchers, loggers, miners, and others interested in the resource rights to lands occupied by native peoples often conspire singly or in concert to prevent or postpone the recognition of these lands as indigenous homelands. Meanwhile, these lands are being surrounded or invaded by those seeking to profit from them. Circumscription of and/or encroachment on indigenous territories means habitat destruction, competition for wildlife and other natural resources, and often closer proximity to markets that promise outlets for forest products.

In the Sirionó case, cattle ranching on unimproved native pasture, which requires extensive range lands, has left intact much of the local habitat. In this instance, ranching has impeded circumscription by colonists in this region. Still, even the Sirionó are now beginning to feel the impact of increased settlement by farmers inhabiting the forested areas of their hunting range (Townsend 1995, 1996).

The impact of settler incursion on Yuquí subsistence and well-being has been substantial. The growing of coca for the illegal cocaine trade has brought highland Bolivian colonists into areas near the Yuquí settlement as the market for this product expands. The coca plant requires very little care, so most coca growers do not remain permanently in the area, choosing instead to return to their homes in the highlands between harvests to engage in other economic activities. Because harvesting is a short-term activity, coca farmers plant virtually no food crops, depending instead on products that are imported or provided by the forest and rivers. Consequently, the presence of colonists in the Yuquí territory has brought about not only the clearing of large tracts of wildlife habitat for the growing of coca, but direct competition with the Yuquí for fish and game.

Following large-scale incursion by colonists between 1983 and 1988, Yuquí protein intake dropped from 88 g per day to only 44 g. The data revealed that this was attributable to an astounding decrease in the harvest of river fish, from 1055 kg in 1983 to 59 kg in 1988. The causes of this reduction appear to have been commercial fishing, which began with the opening of a market road from the colonies to the river, and the use of dynamite by colonists who as former tin miners soon discovered that by making use of their skill with this dangerous technology, large numbers of fish could be killed quickly (Stearman 1990).

Although Yuquí game takes remained relatively stable between 1983 and 1988, with a small and not statistically significant reduction in the total harvest by weight, the suite of animals being taken in 1988 differed greatly from those taken in 1983. Because of dietary preferences, colonists tend to pursue a much narrower range of game animals (Redford and Robinson 1987), focusing on the larger ungulates such as peccaries (*Tayassu pecari* and *Tayassu tajacu*), tapir (*Tapirus terrestris*), and deer (*Mazama americana* and *Mazama gouazoubira*), and the larger rodents, agouti (*Dasyprocta variegata*) and paca (*Agouti paca*). Colonists also kill in large numbers, but do not eat, capybara (*Hydrochaeris hydrochaeris*), which is considered a crop pest by settlers but is a favored game animal of the Yuquí. The loss of large game animals and fish resulted in what appears to be overhunting by the Yuquí of smaller, less desirable species that are probably being exploited at unsustainable rates. In 1983, five species (red brocket deer, tapir, capybara, collared peccary, and white-lipped peccary) accounted for 63% by weight of all game taken. Although about equivalent by weight, the 1988 game take comprised larger numbers of smaller animals. In 1983 the proportion of small animals (< 5 kg) to medium (5–20 kg) and large (× 20 kg) comprised 62% of the total take; by 1988 they made up 88%. Animals harvested in 1983 numbered 156, representing 27 taxa. In 1988 the Yuquí captured 348 animals representing 44 taxa.

Additional data gathered in 1991 during the same time of the year, collected during a period of 20 days, indicate that this trend continues. Extrapolations from the 20-day sample to the 56 days sampled in 1988 (extrapolation factor is 2.80) show that although the difference is not statistically significant, the per diem game captures by weight per hunter continue to drop: during 1988, Yuquí hunters averaged 63.87 kg of game (live weight) per day; in 1991, they took 63.15 kg per day. However, extrapolating from the 168 animals harvested during 20 days in 1991, to attain this average hunters would have to have taken 470 animals. This represents a 35% increase from the number taken in 1988, corresponding almost exactly to the percentage increase in the number of men hunting (36%, or from 19 in 1988 to 26 in 1991) and in the amount of game taken by weight (35%, or from 1213.70 kg in 1988 to 1641.92 kg in 1991). Of the 168 individual animals captured during the 1991 sample period, about 88% weighed under 5 kg, also continuing the 1988 pattern.

Another interesting result from the 1991 data is the species composition of animals harvested during the sample period: only two species, tortoises (*Geochelone* spp.) and armadillos (*Dasypus novemcinctus*) comprised 40% of the animals harvested (39 armadillos and 29 tortoises). Most of the larger, more preferred game animals had dropped out of the sample. In addition, during the 56-day sample in 1988, only three armadillos were taken. In the 20-day sample from 1991, 39 armadillos were captured, which if extrapolated to the 56-day period would bring the total armadillo catch to 109 animals. It is hardly likely that this difference is entirely density dependent, reflecting a change in encounter rates of armadillos. On the contrary, I propose that these data are the result of changes in prey choice. Armadillos are not a high prestige animal among Yuquí hunters because they are usually hunted by hand, and hence can be taken by women and children, and they are dug out of a burrow rather than pursued in an active hunt. Consequently, when more preferred game is available, armadillos are ignored. Virtually the same conditions apply for tortoises. What appears to be occurring in the case of the Yuquí is a forced shift in prey choice due to the depletion of preferred species. And finally, the 1991 data also reveal a continuing problem with the fish supply. In a repetition of the 1988 findings, the Chimoré River produced only 53 kg of fish during the extrapolated 56-day sample period.

The implications for the future of Yuquí foraging are clear. As their population increases and settlers progressively encroach on their lands and streams, hunting and fishing will become less profitable. The Yuquí are having to take larger numbers of smaller animals in order to meet their animal protein needs, a pattern that cannot continue indefinitely. Fishing, once a dependable alternative to hunting that reduced pressure on game and supplied large amounts of animal protein, no longer continues to provide nutritional backstopping. Moving the village to an unhunted area is also unlikely if not impractical given the investment in infrastructure (clinic, school, housing, etc.) and its proximity to market centers. Trekking to the interior of their territory—remote, difficult to access, but as yet unpopulated—still remains a viable option. However, the growing demands of wage labor and education conspire to keep the Yuquí from making regular and extended treks to improve their diets. Food subsidies provided by the state and mission also are not a secure source of nutrition, and a recent health survey (Hjerpe unpublished manuscript) indicates that some of the Yuquí are now showing signs of malnutrition and increased susceptibility to illness. These are difficult problems that confront many people like the Yuquí and will not be easily resolved.

CONCLUSION

Conservationists continually grapple with the contradictions of a "use it or lose it" strategy in their struggle to conserve biodiversity in tropical regions, hoping that this approach will be more politically feasible than proposals to exclude all human activity from parks and reserves. In the Amazon as in other tropical areas of the world, indigenous territories often coincide with those relatively undisturbed large landscapes that conservationists seek to protect. In what seems a natural alliance between native cultures seeking rights to their homelands and powerful conservation organizations attempting to set aside conservation areas, indigenous peoples have positioned themselves as the most appropriate stewards of these reserves. According to current conservation strategies, "saving the rain forest" and the biodiversity within must include the people who inhabit these regions as well (Plotkin and Famolare 1992; Reed 1995, 1997). Yet no one, most of all native peoples themselves, can agree on what this means or how it should be accomplished. Most do agree, however, that solutions must continue to be sought and alternatives attempted if biological and cultural diversity are to persist in any meaningful way.

Debates among various interest groups concerning the "natural" place of people in the Amazon range from those who contend that the forest we now see is largely the result of human interaction (Bennett 1962; McNeely 1994) and therefore, by implication, humans should be viewed as a natural component of the ecosystem (Baleé 1989), to those who essentially believe that most human activity is detrimental to neotropical forests (Peres 1994). Conservationists, anthropologists, and members of numerous nongovernmental organizations committed to protecting both biodiversity and the native inhabitants of neotropical forests are seeking some workable middle ground that implies trade-offs (Redford and Mansour 1996). The impediments to reaching a compromise that preserves some acceptable (also as yet undetermined) level of biodiversity while at the same time giving native peoples the opportunity to participate in a wider society have been examined in the previous pages and are formidable but not insoluble. Currently, there are many promising and innovative projects underway being carried out by individuals and organizations who are dedicated to the premise that such solutions can be found (Redford and Mansour 1996).

Wildlife conservation strategies focusing on sustainable hunting pose serious management problems for those interested in preserving the biodiversity of neotropical forests. In terms of the effects of human activity, there is an enormous qualitative difference between the exploitation of plants versus that of animals. Plant products often can be harvested without the need to destroy the source; this is not the case with animals. Plants also stay put and ownership can be established, mitigating at least some of the effects of the "tragedy of the commons" problem that plagues wildlife management (Robinson and Bennett this volume). Therefore, it is not surprising that many of those who tout "sustainable resource use" as an easy solution to current development problems are looking at systems such as agroforestry and extractive reserves that hold the immediate promise of leaving the forest standing. Yet the well-being of neotropical forests ultimately will depend on the health of their animal populations that perform important ecological functions in maintaining these forests (Emmons 1989; Redford 1992). Although achieving truly sustainable resource use is problematic in any context, developing systems for sustainable hunting in the neotropics will be particularly challenging. Although wild plant foods can be easily substituted one for another or by domestic crops, game remains the primary source of animal protein for most of rural Amazonia. In those promising extractive reserves, for example, rubber tappers in Brazil are meeting their daily protein needs by hunting, an expensive subsidy from nature that permits these individuals to engage in extractive activities and make a profit.

As conservation biologists, ecological anthropologists, development workers, and others seek innovative ways of making neotropical hunting sustainable, they will need to consider carefully how the external forces discussed in this chapter will impinge on the outcome. Each situation will bring complex challenges and problems that cannot be addressed with simplistic choices or some predetermined bureaucratic template for action. None of this will come without significant economic and social investments, and the peoples of the world who are most able to respond must be convinced that it is their responsibility to do so. Ecological decisions also cannot be made without taking into account the people who will act on them, or in some cases, choose not to act. In the very long term, I believe that hunting will be replaced by other patterns of subsistence as communities continue to evolve in a rapidly changing world. The question remains as to whether this place will be reached as the result of timely and sound decision making or simply because there is nothing left to hunt. If we are to follow the former course, we must continue to promote the development of humane and feasible conservation strategies that recognize the value of humans and wildlife and assure the persistence of both.

12

Wildlife Conservation and Game Harvest by Maya Hunters in Quintana Roo, Mexico

JEFFREY P. JORGENSON

Maya hunters have harvested wildlife in Mexico, Belize, and Guatemala for several thousand years. For most of this time, wildlife conservation has not been a major concern because the harvest rate was accommodated by the natural production of the main hunted species. Now, however, biologists and anthropologists are realizing that harvests cannot continue indefinitely at their current rates, and the Maya themselves are aware that current harvests in many cases exceed production rates, but they are at a loss as to how to alter their hunting patterns to conserve the main game species. Given these circumstances, it is imperative to examine the sustainability of subsistence hunting in the Maya Forest and to develop strategies that both conserve the wildlife species and meet the cultural and nutritional needs of the people who depend on these resources.

This chapter addresses the following questions from the point of view of Maya subsistence hunters in the State of Quintana Roo, Mexico:

- What is the importance of wildlife species to rural Maya communities?
- What is the impact of hunting on these species?
- Are present hunting rates sustainable?
- What are the biological, social, and economic conditions that affect hunting rates?
- What approaches might make hunting more sustainable?

STUDY AREA AND SETTING

The study took place at Ejido X-Hazil y Anexos, Quintana Roo, Mexico, between 1989 and 1990. The study area covered 552.95 km² (for additional information, see Jorgenson 1993, 1995a). Harvest data were obtained from hunters at the Maya Indian village of X-Hazil Sur (19°23'30"N, 88°05'00"W; village population = 1040), one of three villages on the *ejido* (total population = 1680). *Ejidos* are properties where landless subsistence farmers have the legal right to use and profit from the natural resources of the area (Gordillo 1988). Although these villagers are highly acculturated, they also maintain many of their traditional beliefs and activities. Only a few non-Maya live on the *ejido*.

The mean annual temperature is about 26°C. The area typically has one dry season (December to May) and one wet season (June to November); total rainfall during 1990 was about 1300 mm (Jorgenson 1993). About 88.52% of the *ejido* was categorized as late secondary forest, 6.07% as plots and gardens, 5.18% as early secondary forest, and 0.23% as other (Jorgenson 1993). The study area has been occupied since about 1915 by Maya Indians whose main subsistence activity is shifting cultivation, primarily corn (*Zea mays*).

The harvest of wildlife species must be considered within the context of other subsistence and economic activities practiced by the Maya at X-Hazil Sur. The main subsistence activity is swidden agriculture. Forested areas are cleared and gardens planted that are composed of corn, beans (*Phaseolus* spp.), squash (*Cucurbita* spp.), and numerous other crops (Redfield and Villa Rojas 1962; Webber 1980; Dachary and Arnaiz Burne 1983; Villa Rojas 1987). Most trees survive the clearing process and readily resprout (Rewald 1989), and this facilitates the recovery of the forest and its use by wildlife after the garden is placed into fallow.

Gardens typically are about 2 hectares (ha) in size, are located near the village, and are tended by adult men. Women rarely assist their husbands in tending the garden, but young boys frequently help their fathers beginning between the ages of 10 and 15. Gardens usually produce a single harvest of corn in December to January, but other crops may continue to produce for 1 to 2 years (e.g., tubers known locally as camote [*Ipomoea batatas*]; and macal [*Dioscorea* spp.]). Although tending a garden is not a daily, full-time task at X-Hazil Sur, the work requires careful planning and must be conducted in a sequence that closely conforms to the weather (Noguez-Galvez 1991).

Maya women also are involved in agricultural production (Elmendorf 1976; Redfield and Villa Rojas 1962; Webber 1980; Villa Rojas 1987). Women at X-Hazil Sur generally care for small numbers of pigs, turkeys, or chickens. Some of the meat is consumed within the household, and the rest is sold locally. Normally this money belongs to the woman, but it is usually spent on daily household expenses. Maya women also tend small gardens near the house. Some of the herbs, fruits, and vegetables are consumed by the household, and the rest are sold locally, often door-to-door by small children. The sale of meat and produce provides only a limited income to women because these products are commonly produced by most area residents.

X-Hazil Sur men also harvest forest products for sale in local markets, and thus earn cash to purchase manufactured goods. On the *ejido*, the main economic activity for men is to extract latex from sapodilla (*Manilkara zapota*) trees growing wild in the forest (A.B. Jorgenson 1992; Barrera de Jorgenson 1993). Chicle latex is used to produce chewing gum and has been exported to manufacturers in the United States, Japan, and several European countries for more than 100 years (Otañez Toxqui and Equihua Enriquez 1981; Dachary and Arnaiz Burne 1983). Men also harvest trees for lumber (Edwards 1986; Murphy 1990), including mahogany (*Swietenia macrophylla*), which is economically valuable (Negreros 1991; Snook 1993), and *chechem* (*Metopium brownei*), which is used for railroad ties in Mexico. Timber harvesting occurs during the dry season. Workers often harvest game in the forest while processing trees.

Maya Indians in Quintana Roo have cared for sheep, horses, and cows since about 1519 (Hamblin 1984, 1985). Today, about eight groups of men herd cattle on the *ejido*. These herds are relatively small (five to ten head of cattle per herd). About once a month a cow is butchered in the village and the meat sold locally for approximately US$3.33/kg. Small numbers of goats and domestic rabbits also are raised. Domestic livestock thus supplements the meat obtained through consumption of wild meat.

Men and women at X-Hazil Sur also tend small stores (five stores in the village offer the basic necessities), weave hammocks, sew *huipiles* (embroidered white cotton dresses worn by women), and care for special garden plots where irrigated fruits and vegetables are grown to be sold outside the village as part of a government community-development program. The importance of these activities to X-Hazil Sur residents is limited in that transportation costs are high, outside buyers usually pay low prices, and residents of other villages in the area also produce the same crops.

Some X-Hazil Sur residents work outside of the *ejido* for wages. For example, five to ten men work in the nearby town of Felipe Carrillo Puerto as masons and construction laborers. Also, young men and women increasingly are obtaining training that will enable them to leave the *ejido* and work in the nearby vacation sites of Cancún and Cozumel.

There are two important considerations regarding subsistence and economic activities (Barrera de Jorgenson and Jorgenson 1995). First, these activities occur throughout the year. For example, after the end of the chicle tapping season a man might switch to logging or gardening, only to resume chicle tapping later in the year. At no time are these people inactive for more than a few days at a time. Second, X-Hazil Sur residents usually engage in several activities simultaneously in the forest. For example, a man might weed his garden during the morning, evaluate nearby areas as potential garden sites in the early afternoon, and gather firewood on his way home in the late afternoon. Rather than limit hunting, these activities increase the opportunities to harvest wildlife because Maya hunters usually carry their shotgun or rifle while in the field (figure 12-1). Given that many game species tend to concentrate their movements around gardens, they are frequently and rather easily harvested as the Maya walk or ride their bicycles along trails between the various work sites.

Figure 12-1. Maya hunter in his maize field, armed for opportunistic hunting. *Photo courtesy of John G. Robinson.*

IMPORTANCE OF WILD SPECIES TO RURAL MAYA COMMUNITIES

Maya hunters do not harvest all of the wildlife species found on the *ejido*. Instead, they concentrate on those medium and large species of mammals and birds commonly found in gardens or the adjacent forest. However, several exceptions were noted. Among mammals, tapirs (*Tapirus bairdii*) and armadillos (*Dasypus novemcinctus*) routinely are not hunted. According to hunters, the tapir is too difficult to kill and transport, whereas armadillo meat contains undesirable fat nodules (G. Yeh Poot, personal communication). Among the birds, parrots (Order Psittaciformes) and toucans (Family Ramphastidae), despite their relatively large size and great abundance, are not taken.

A total of 584 wild animals were reported to be taken by hunters at X-Hazil Sur during the 17 months from June 1989 to October 1990 (table 12-1) (Jorgenson 1993, 1995a, in press). Mammals comprised 66% (*n* = 385) and birds 34% (*n* =

Table 12-1. Reported Number of Animals Taken, Mean Weight, and Total Weight of Wildlife Taken by Maya Hunters at X-Hazil Sur, Quintana Roo, Mexico, June 1989 to October 1990

Common Name	*Scientific Name*	*Total Number Harvested*	*Mean Weight (kg)*	*Total Number Weighed*	*Total Weight (kg)[a]*
Mammals					
Pocket gopher	*Orthogeomys hispidus*	53	0.4	51	22.3
Paca	*Agouti paca*	47	5.8	47	274.8
Agouti	*Dasyprocta punctata*	35	2.8	35	96.8
Coati	*Nasua narica*	167	3.0	167	504.9
White-lipped peccary	*Tayassu pecari*	3	31.4	3	94.3
Collared peccary	*Tayassu tajacu*	40	17.2	36	618.5
Brocket deer	*Mazama americana*	16	15.6	16	250
White-tailed deer	*Odocoileus virginianus*	24	32.2	22	709
Total mammals		385		377	2570.6
Birds					
Thicket tinamou	*Crypturellus cinnamomeus*	13	0.4	13	4.9
Great curassow	*Crax rubra*	13	3.1	13	40.1
Plain chachalaca	*Ortalis vetula*	167	0.4	167	64.9
Ocellated turkey	*Agriocharis ocellata*	6	3.3	6	19.7
Total birds		199		199	129.5
Mammals and birds					
Total mammals and birds		584	4.7	576	2700.1

[a]Total weight was determined by summing the weights of the individual prey items that were weighed. The degree of precision varied between species as different scales with assorted capacities and graduations were used.

199). No reptiles, amphibians, or insects were collected by hunters for personal consumption, although honey was consumed when encountered in the forest, and small quantities of fish were taken intermittently in nearby sinkholes.

Eight mammalian and four avian taxa were hunted for subsistence (Jorgenson 1993, 1995a, in press). The coati (see table 12-1 for scientific names) was the most frequently hunted mammal, followed by the pocket gopher and the paca, whereas the plain chachalaca was the most frequently hunted bird (see table 12-1).

The wildlife taken by hunters provided residents of X-Hazil Sur with a substantial amount of meat. The total body weight of the 584 animals was 2700 kg (see table 12-1). Of this total, 95% was from mammals and 5% was from birds. Three mammalian taxa combined—white-tailed deer, collared peccary, and coati—provided 68% of the total weight. This meat was primarily consumed by the individual hunters and their immediate families (about 400 people). Although small quantities also were sold locally at about US$3–4/kg (similar in price to pork and beef), the direct economic importance of the wild meat was of minimal importance to Maya hunters.

The primary species hunted by Maya at X-Hazil Sur were mammals and birds. This contrasted with other indigenous groups in Mexico that—in addition to birds and mammals—take substantial quantities of fish, insects, reptiles, and amphibians. For example, Lacandon Maya Indians in the state of Chiapas consume a wide variety of amphibians and reptiles (Góngora-Arones 1987; March 1987), whereas Mixteca Indians in the state of Oaxaca consume large quantities of iguanas (*Ctenosaura pectinata*) and crabs (*Cardisoma crassum*) (Parra Lara 1986). Mestizos throughout Mexico also consume a wide variety of fish, reptiles, and crustaceans (Reyes Castillo 1981; Mellink et al. 1986; Santana et al. 1990). This suggests that most of the Maya hunters at X-Hazil Sur were selective in their consumption of wildlife.

IMPACT OF HUNTING ON GAME SPECIES

To evaluate thoroughly the direct impact of hunting on wildlife, it is necessary to compile data on animal densities and the age-sex structure of hunted populations. Unfortunately, these data are not available for X-Hazil Sur. Wildlife censuses were conducted in the study area, but the resulting densities were so low (fewer than one animal per km^2 for most species versus 10 to 100 reported elsewhere in the neotropics) (Jorgenson 1993), that statistical comparisons were not conducted. Data on the number of animals harvested and their age-sex structure were compiled, however, and can provide indirect evidence on the sustainability of hunting. These kinds of data are important in determining the relationship between what is hunted and what is potentially available (Redford and Robinson 1990). In the case of X-Hazil Sur, data on age, sex, and reproductive condition are important in determining the biological impact of hunting on the wildlife community.

Males and females were not taken in equal proportions. For all mammals combined, significantly more females were taken than males (table 12-2). However, for birds, more males were taken than females (see table 12-2). The ratio of males to females taken varied between species. For mammals, more females were taken than males for seven of the eight taxa (see table 12-2). The mean sex ratio for all mammals combined was 1:1.3 (males:females). For mammals, the greatest statistical disparity between males and females hunted was for the white-tailed deer (sex ratio 1:3.8). The ratio of males to females taken also was significantly different for the agouti (sex ratio 1:1.8) and pocket gopher (sex ratio 1:1.7). For birds, more males were taken than females for three taxa, but the only significant difference was for the plain chachalaca (sex ratio 1:0.6). The mean sex ratio for all birds combined was 1:0.6.

The age class of species hunted was also determined. The proportion of adults, subadults, and young taken by hunters varied between species. For all mammals combined, adults were hunted more frequently than subadults or young (55% adults, 43% subadults, and 2% young; see table 12-2). The greatest difference by species was for the pocket gopher (87% adult and 13% subadult), but was also high for the brocket deer (75% adult and 25% subadult) and white-tailed deer

Table 12-2. Sex and Age Classes of Hunted Animals at X-Hazil Sur

Common Name	Males	Females	Unknown	Chi-square	p^a	Adults	Sub-adults	Young
	Sex (No. of Animals)					*Age Class (% Animals)*		
Mammals[b]								
Pocket gopher	19	33	1	3.7692	< 0.100	86.8	13.2	0
Paca	22	25		0.1915	< 0.900	70.2	29.8	0
Agouti	12	22	1	2.9412	< 0.100	65.7	25.7	8.6
Coati	80	86	1	0.2169	< 0.900	57.5	37.7	4.8
White-lipped peccary	1	2		—[c]		66.7	33.3	0
Collared peccary	19	20	1	0.0256	< 0.900	67.5	30	2.5
Brocket deer	10	6		1	< 0.500	75	25	0
White-tailed deer	5	19		8.1667	< 0.005	66.7	25	8.3
Mammals combined	168	213	4	5.315	< 0.025	69.5	27.5	3
Birds[b]								
Thicket tinamou	7	4	2	—[c]		92.3	7.7	0
Great curassow	7	6		0.0769	< 0.900	84.6	15.4	0
Plain chachalaca	100	57	10	11.777	< 0.005	89.2	10.2	0.6
Ocellated turkey	3	3		—[c]		66.7	33.3	0
Birds combined	117	70	12	11.813	< 0.005	83.2	16.7	0.1

[a]Krushkal-Wallis chi-square approximation.
[b]Individuals with sex undetermined were not included in this analysis.
[c]Test not performed because at least one of the cells included fewer than five animals.

(67% adult, 25% subadult, and 8% young). Adult birds also were hunted more frequently than subadults or young both and for each of the four taxa (83% adults, 17% subadults, and 0 young; see table 12-2). Few subadult or young birds were hunted.

At X-Hazil Sur, despite the apparent selection for specific age-sex classes, hunters did not report any active hunting preference on the basis of age or sex. Rather, hunters reported that the harvest reflected differences in species' vulnerability during the year as birds and mammals searched for food, water, and mates. It is clear that animals of all age-sex classes are potential prey for hunters.

Only a few studies have documented the age or sex of animals taken by subsistence hunters. Age and sex are summarized, however, for three studies in lowland Peru. Pacheco (1987) showed that for both mammals and birds, slightly more females than males were taken by forestry technicians conducting a tree survey, but the study was short (13 days) and sample sizes small (14 species, $n = 46$ animals). At a second site, Alvard and Kaplan (1991) reported that adults comprised 57% of pacas hunted, 83% of brocket deer, and 74% to 76% of peccaries. At a third site, Bodmer (unpublished report) reported that males were hunted more frequently than females for the collared peccary (1:0.66, males:females, $n = 164$ animals), grey brocket deer (*Mazama gouazoubira*; 1:0.75, $n = 28$), and paca (1:0.57, $n = 174$), whereas females were hunted more frequently than males for the white-lipped peccary (1:1.10, n = 166), red brocket deer (1:1.31, $n = 60$), agouti (1:1.31, $n = 97$), and acouchi (*Myoprocta* spp.; 1:1.17, $n = 13$), whereas hunt by sex was equal for the capybara (*Hydrochaeris hydrochaeris*; 1:1.00, $n = 10$).

Reproductively active females were harvested by hunters at X-Hazil Sur. Among mammals, 100% of the white-lipped peccary females were pregnant, 27% of agouti females, 20% pacas, 17% brocket deer, 15% collared peccaries, and 2% coatis (figure 12-2). About 32% of the female white-tailed deer and 20% of the female pacas hunted were lactating. None of the harvested pocket gophers or white-tailed deer was determined to be gravid.

Among birds, one of four female thicket tinamous and three of 57 female plain chachalacas were carrying eggs (see figure 12-2). None of the great curassows or ocellated turkeys that were harvested was determined to be gravid, but the reproductive condition of many birds and mammals was difficult to determine due to damage that resulted from abdominal shots by hunters that destroyed the reproductive organs.

At X-Hazil Sur, hunters harvested female mammals regardless of their reproductive condition. Although most female mammals were not gravid or lactating, three species had relatively large proportions of females that were pregnant or caring for young (see figure 12-2): white-lipped peccary (100% gravid), paca (40% gravid or lactating), agouti (32%), and white-tailed deer (32%). These species

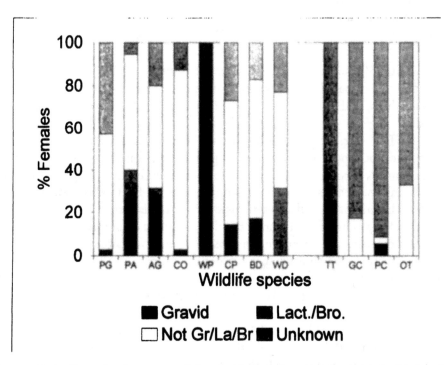

Figure 12-2. Reproductive status of female birds and mammals. Animals (% females) were categorized as gravid, lactating/brooding (Lact./Bro.), not gravid/lactating/brooding (Not Gr/La/Br), or unknown. Abbreviations correspond with common name: PG = pocket gopher, PA = paca, AG = agouti, CO = coati, WP = white-lipped peccary, BD = brocket deer, WD = white-tailed deer, TT = thicket tinamou, GC = great curassow, PC = plain chachalaca, and OT = ocellated turkey. Sample sizes shown in Table 12-2.

were among the main game species by number of animals and total body mass (see table 12-1).

The importance of the reproductive condition of the wildlife is that the total number of harvested animals might not reflect the total number of animals potentially lost by the wildlife population. For example, female white-tailed deer normally bear one or two young per year (Kleiman et al. 1997). If a pregnant or lactating female is harvested, the local deer population potentially would lose two or three animals (mother plus young) if the young are unable to survive on their own. The situation may be more critical for the coati, the most frequently hunted mammal. Although precise data are lacking, it appears that coatis at X-Hazil Sur bear up to six or eight young annually per litter. Thus, harvesting a pregnant or lactating coati means that the population could lose seven to nine animals. Killing a male deer or coati, however, would mean the loss of only a single animal because other males are available to mate with the females.

In summary, it appears that at X-Hazil Sur different factors affect harvest patterns with respect to sex, age, and reproductive condition. Among the game mammals, there is no significant sexual dimorphism by species except that male deer of both species have antlers (Leopold 1977). Among the game birds, male (black plumage) and female (reddish-brown plumage) great curassows differ by color, and male ocellated turkeys differ from females by the presence of spurs and the prominence of skin protuberances on the head (Peterson and Chalif 1973; Leopold 1977). Thus, hunters generally are unable to determine the sex of wildlife before killing it and would be unable to select for or against a certain sex except in those specific cases above.

With respect to age class, all species undergo increases in body size as they mature, as well as changes in hair or feather color and general body appearance, so most hunters can readily differentiate adults from subadults or young. In general, the data supported the idea that hunters are selecting adult animals. This can be interpreted as an attempt by hunters to obtain large animals that would provide more meat. It would not be reasonable to interpret this to mean that adults were more abundant than subadults or young in the forest.

With respect to reproductive condition, pregnant or lactating females generally cannot be readily differentiated from males during the few seconds in which the hunter must decide whether or not to shoot. These females, however, might be more vulnerable to hunters by being slightly slower than males to escape or less vigilant while feeding due to distractions from their young. Most hunters are keenly aware of the specific period of the year when females of each species give birth or are caring for young. Despite this knowledge, most hunters at X-Hazil Sur would shoot an animal even though there was a high probability of it being a pregnant or lactating female. Thus, the harvest of reproductively active females has a greater impact on the population than the harvest of males.

Little has been written about the reproductive condition of animals taken by hunters in the neotropics. In Mexico, Santana et al. (1990) reported that subsistence hunters at the Sierra de Manantlán Biosphere Reserve refrained from hunting during specific periods to avoid harvesting pregnant or nursing birds and mammals. For lowland Peru, Bodmer (1989) reported that reproductively active ungulates were harvested by hunters; 41% of collared peccary females hunted were reproductively active, 62% of white-lipped peccary females, 40% of red brocket deer, 33% of grey brocket deer, and 37% of lowland tapir (*Tapirus terrestris*). It is probable that subsistence hunters throughout the neotropics take substantial numbers of gravid, lactating, or brooding females, but until further studies are conducted, the impact on wildlife populations cannot be assessed.

SUSTAINABILITY OF HUNTING RATES

It is difficult to determine if present harvest rates by the Maya hunters at X-Hazil Sur are sustainable. Some data exist regarding (a) the extent and variation of hunting patterns and (b) the population status of game species at X-Hazil Sur (Jorgenson 1993, 1995a, in press). Little information exists, however, regarding the productivity of hunted species or the response of hunted species to hunting. According to Robinson and Redford (1994b), these data are necessary to evaluate the sustainability of harvests. In the absence of these data, indices or indirect methods may be used to assess sustainability.

Maya hunters have practiced subsistence hunting in Mexico, Belize, and Guatemala for at least 4000 years (Adams 1977, 1991). Archeological studies have characterized the diet of the ancient Maya in Mexico and Guatemala and the important role of game (Hamblin 1984; Pohl 1976). Anthropological studies have described the diet of present-day Maya in the region (Nations and Nigh 1980; Murphy 1990). These studies document the continued importance of meat from wild species in the diet of the Maya. These studies show that Maya hunters today are harvesting more or less the same species that they have been taking for more than 4000 years. This suggests that historically, on a regional basis, hunting as practiced by the Maya has been sustainable. But this does not answer the question of whether hunting at X-Hazil Sur is sustainable today.

A preliminary analysis of hunting at X-Hazil Sur suggests that wildlife harvests might be sustainable for some species, but not so for other species. Using a model that incorporated harvest rates, production, and mortality (Robinson and Redford 1991a), it was determined that Maya hunters at X-Hazil Sur were harvesting very little of the apparent game production, perhaps only 1% to 10%. Although this could suggest sustainable hunting, it was probable that the biological parameters used to calculate production and mortality—taken from various neotropical sites—were not appropriate for X-Hazil Sur, and thus overestimated potential production rates. Specific production and mortality data applied to this model would probably show that a much greater proportion of the wildlife production is being harvested than the preliminary estimates.

Anecdotal data for X-Hazil Sur suggest that certain species now are less abundant than before (Jorgenson in press). According to local hunters, at least six species were once more common but are now in danger of being extirpated. Three such species were harvested during 1989 to 1990 (Jorgenson 1993): white-lipped peccary (three animals), great curassow (13 animals), and ocellated turkey (six animals). Three other species were not harvested: spider monkey (*Ateles geoffroyi*), howler monkey (*Alouatta pigra*), and guan (*Penelope purpurascens*). Although local hunters attribute the population declines for these species to overhunting, it is possible that other factors, such as timber harvesting and chicle tapping, might be responsible for these reductions. Regardless of the cause, harvest of these species by Maya hunters is declining or has stopped altogether.

BIOLOGICAL, SOCIAL, AND ECONOMIC CONDITIONS AFFECTING HUNTING

Biological as well as social and economic conditions affect hunting patterns by the Maya. Among the principal biological conditions are species-specific behaviors and changes in the movements and aggregation of wildlife brought about by the annual cycle of seasons and the occurrence of major storms.

The rainy and dry seasons occur on a regular basis throughout the year at X-Hazil Sur. The seasons vary in timing, duration, the amount of precipitation, and daily minimum and maximum temperatures. These variations affect the wildlife in different ways, including the availability of food and water. White-tailed deer, for example, are hunted primarily during May to July, at the beginning of the rainy season (Jorgenson 1993). According to hunters, female deer enter gardens to consume plants with a high moisture content that cannot be easily found elsewhere in the forest. Pacas and agoutis are more frequently hunted during the end of the rainy season. According to hunters, pacas and agoutis range widely at this time in order to locate and consume locally abundant fruits, especially those of the sapodilla tree. Other game species are also hunted during specific times of the year in a pattern that is repeated year after year according to the passage of the seasons.

Hurricanes occur on a regular basis during the months of June through November in the Caribbean. During 1988, Hurricane Gilbert passed over the northern part of the Yucatán Peninsula, flooding low areas, destroying crops, and blowing down trees. The damage was severe and widespread (Anonymous 1988; Wilder 1988). The following year, forest fires blazed throughout the region during the dry season, burning the abundant litter that had accumulated since the hurricane. According to hunters, wildlife suffered greatly during this period due to a food shortage in the affected areas because many species of trees and shrubs were completely denuded of leaves, flowers, and fruits (J. Poot Ake personal communication). The frequency of hunting outings by the Maya after the hurricane decreased due to the heavy accumulation of litter, which made it difficult to stalk animals. Thus, hurricanes can have both beneficial and negative effects on wildlife but generally reduce hunting rates for several months following the storm.

The behavior of game species varies according to the species-specific annual cycle of reproduction, and this indirectly affects hunting rates. Male white-tailed deer, for example, congregate and fight among themselves to gain reproductive access to the females during the months of September and October. This fighting behavior, however, makes males highly susceptible to Maya hunters, who are able to approach the male deer without being readily detected. Male great curassows likewise are extremely vulnerable to hunters during the months of January and February because these birds call out repeatedly from highly visible perches in tall trees in the forest to attract mates. Other species also have biological characteristics that make them vulnerable to hunters at various times during the year.

Among the principal social and economic conditions that affect Maya hunting patterns are traditional cultural practices, as well as efforts by many Maya to participate in modern activities outside the *ejido*. These activities, like the biological conditions, both promote and reduce wildlife harvest rates.

Traditional cultural practices are especially important to the Maya. Although Maya hunters do not have a special ceremony or rites of passage through which a boy becomes a man, they do harvest wildlife that will be consumed during religious ceremonies prior to planting their gardens in May or June, and again a few months later when the corn crop begins to mature. However, during this study, the amount of wildlife harvested for these ceremonies was minimal, and meat from domestic animals often was used instead.

Socioeconomic conditions are changing for the Maya of X-Hazil Sur, and in general they favor less overall hunting. As indicated earlier, Maya men participate in several economic or culturally important activities, including gardening, chicle tapping, harvesting timber, and observance of special religious ceremonies. Men taking part in these activities usually do not have much free time during which they could hunt. As a result, the number of kills per hunter is low (about seven per hunter during 17 months) (Jorgenson 1993, 1995a).

Canned meat and domestic animals were readily available at X-Hazil Sur through local stores and small cattle ranches. The availability of these alternative protein sources greatly affected the amount of wildlife taken by hunters. At least two effects were noted. First, hunters could depend on purchased meat, making it less essential to harvest wild meat during every hunting outing. Second, the opportunity to purchase meat or eggs in the village gives the hunter the choice of not taking relatively small or poor-tasting species of wildlife. Given the ready availability of canned meat and domestic animals, and recognizing the low harvest rates, subsistence hunting at X-Hazil Sur clearly reflected the wide use of protein sources other than wild game.

Given their highly acculturated status, and their role in a market economy through wage labor and the purchase and sale of goods, one might expect that Maya hunters would alter their hunting rates on a short-term basis (seasonally or annually) according to changing economic needs. This does not happen, however, at X-Hazil Sur. Hunters, on an individual basis, either pursue wildlife consistently at regular intervals (perhaps two or three outings per week) or essentially give up this activity, perhaps harvesting one or two animals a year (Jorgenson 1995b). For example, several adult men reported that they had given up hunting temporarily in order to take a part-time job in a nearby town or to dedicate more time to their gardens. When the job or garden work was completed, perhaps after 1 or 2 years, these men were unable to resume hunting successfully because they had lost many of their outdoor skills. Many of these men, for example, could no longer recognize with confidence the den, track, or call of a particular species. Likewise, they did not know which areas animals currently were using to obtain food or water because the men were not aware of current conditions in the forest (e.g., garden locations, fruiting trees, water holes, or distribution of recent work sites). Many men also indicated that they no longer wanted to be out in the forest at night or to endure rain and insect bites in order to attain a kill. Although these skills could easily be relearned over 3 to 6 months of frequent hunting outings, most former hunters appeared unwilling to dedicate the time or effort necessary to do so. Thus, although short-term hunting rates probably do not vary greatly, changing economic conditions at X-Hazil Sur are gradually and, most likely, permanently reducing hunting rates by the Maya on a long-term basis.

ENHANCING THE SUSTAINABILITY OF HUNTING

Hunting can be regulated in various ways to make it more sustainable. At the national or federal level, government officials in Mexico have instituted country-wide regulations regarding hunting seasons, bag limits, and the areas where wildlife can be legally harvested (Barrera de Jorgenson and Jorgenson 1995; Jorgenson 1995a). In some cases, these regulations are based on sound biological information. In many other cases, however, such information is lacking, and the regulations therefore represent little more than a best guess by officials about how to manage the wildlife species subject to harvest.

The successful management of hunting is further complicated by the need to have an enforcement capability. Game wardens or game enforcement officers in several parts of Mexico check hunters to ensure that regulations are being followed, with violators subject to severe legal penalties. In most parts of Mexico, including Quintana Roo, however, wardens are few in number and generally are unable to ensure compliance with hunting regulations. Government agencies have insufficient money for salaries and vehicles, and the lack of roads and trails in the forest makes it difficult to patrol wildlife areas. Under these circumstances, formal hunting regulations at the national or federal level in Mexico contribute little to wildlife management programs because they cannot be enforced.

Jorgenson (in press) proposed that officials of the national and state governments in Mexico develop wildlife management plans with the active participation of the local Maya communities. Government biologists and wildlife experts could provide technical information and advice about successful management practices used in other areas, whereas members of local Maya communities could provide their extensive knowledge of the *ejido* and the wildlife species therein, ultimately taking responsibility for the management of natural resources, especially wildlife, on these sites. Community-based conservation in Quintana Roo might very well be the key to making subsistence hunting by Maya hunters sustainable.

There are two critical aspects regarding community management of wildlife on Maya *ejidos* in Mexico (Jorgenson in press):

1. There must be common agreement between *ejido* members, by hunters as well as nonhunters, to develop and implement a wildlife management plan. This is critical for several reasons. First, all *ejido* members have an equal and collective voice in the decision-making process at the community level, regardless of the socioeconomic activities in which they participate individually. Second, different subsistence and economic activities will have an impact on wildlife and harvests in dissimilar ways. By ensuring that all *ejido* members participate in the decision-making process, there is a better chance that the management of wildlife and hunting will be undertaken within the context of other socioeconomic activities undertaken by *ejido* members. A logical consequence of this participation is that all or most *ejido* members also would stand to benefit in some way from the management of the wildlife resource.

2. There must be agreement between community members and government officials to share management responsibilities for the wildlife. This arrangement presumes active communication and the exchange of necessary management information between the two parties. Without these kinds of agreements at the local and national levels, any management plan designed to make hunting more sustainable on Maya *ejidos* likely will fail as personal interests begin to prevail over community interests.

Two approaches traditionally used in Europe and North America will probably not be successful on Mayan *ejidos*: bag limits and short hunting seasons with specific beginning and ending dates. Although these techniques are useful in areas where hunting primarily is a social activity (such as Germany or the United States), bag limits (e.g., one collared peccary/year) and short hunting seasons (e.g., 10 to 30 days/year) do not recognize the need of a subsistence hunter to harvest game essentially on a daily basis, such as in Mexico.

At least two community-based programs for wildlife conservation have been instituted recently in southern Mexico:

1. At Ejido Tres Garantías, in southern Quintana Roo, community members have been able to organize a program for sport hunting and ecotourism to attract visitors to their *ejido* (F. Quinto, personal communication). Local community members conduct regular biological surveys as a part of this program to determine the distribution and relative abundance of key wildlife species. Government officials, in turn, provide technical advice regarding specific, wildlife-management techniques (e.g., methods used to calculate the density of species in different habitats). Although it is too early to evaluate this program fully, both community members and government officials are optimistic about its success.

2. At Calakmul Biosphere Reserve, in southern Campeche, Mexico, community members and government officials are developing a wildlife management plan that will incorporate controlled levels of subsistence hunting by *ejido* members (Sanvicente 1996). In keeping with the biosphere concept (UNESCO 1984, 1985), some parts of the *ejido* will be developed as wildlife reserves (protected core area), whereas others will be open to different levels of subsistence hunting (buffer

and multiple use areas). This program is only at the beginning stage, so its success still remains to be determined.

Community-based management of wildlife might require additional mechanisms to buffer the impact of subsistence hunting on wildlife populations: wildlife refuges and no-hunting periods for specific species. The concept of wildlife refuges has been modified by the Embera Indians in the Chocó Region of western Colombia to mean an area where no hunting will occur or a species will not be hunted for a lengthy period of time, such as 1 year (Rubio-Torgler 1995). Upon completion of the no-hunting period, hunting will resume at that site while another then will be declared temporarily a wildlife refuge, or hunting will resume for that species and another species will be granted temporary protection. These concepts are promising because they can be easily understood and implemented by rural and indigenous peoples in order to promote sustainable hunting.

CLOSING COMMENTS

Hunting clearly is an important activity for the Maya at X-Hazil Sur. Although the main benefit of this activity might be the meat obtained, hunting also appears to be important in the cultural identity of the Maya. Men who have not harvested wildlife for several years still consider themselves to be hunters, and the game that they occasionally obtain reinforces a sense of being Mayan and what it means to be part of the Maya Forest. Despite the ready availability of canned meat and meat from domestic livestock, hunting will probably continue to be an important cultural and socioeconomic activity for the Maya for several decades.

Field data and anecdotal information support the idea that some wildlife species are no longer sufficiently abundant to be harvested on a regular basis. Although overhunting might be the main cause of these population declines, other human activities such as timber and chicle harvest may indirectly reduce the amount of wildlife. The *ejido* system, however, might provide an ideal opportunity to manage these species at the community level for the benefit of hunters as well as nonhunters. A community decision to manage wildlife would probably succeed, given the small size of the village and the shared perception among residents that wildlife is culturally and nutritionally important. Such a decision also would benefit the hunted wildlife species, as well as the other plant and animal species that depend on those species through a poorly known array of ecological relationships, such as pollination and seed dispersal (De Steven and Putz 1984; Terborgh 1988; Fonseca and Robinson 1990; Dirzo and Miranda 1990; Redford 1992).

ACKNOWLEDGMENTS

I thank Drs. John G. Robinson and Elizabeth L. Bennett for inviting me to contribute this chapter, and for their patience in awaiting its arrival. This study was undertaken while I was a graduate student at the University of Florida and was completed through the financial support of World Wildlife Fund-US, World Nature Association, Roger and Bernita Jorgenson, Organization of American States, Centro de Investigaciones de Quintana Roo, and Program for Studies in Tropical Conservation/Tropical Conservation and Development Program (University of Florida). The Secretaría de Desarrollo y Ecología (SEDUE) kindly granted a research permit for Quintana Roo. Marcelo Carreón Mundo and Victoria Santos Jiménez (Plan Estatal Forestal) provided maps and information about the forest in the study area. Drs. Kent H. Redford and John G. Robinson assisted in the design of the study. Amanda Barrera de Jorgenson, Armando Balam Xiu, and Rufino Ucan Chan (X-Hazil Sur) collaborated in the field work. Finally, I thank the residents of Ejido X-Hazil y Anexos for their friendship and cooperation without which this study would not have been possible.

13

The Sustainability of Subsistence Hunting by the Sirionó Indians of Bolivia

WENDY R. TOWNSEND

Many tropical forest communities in the neotropics depend on hunting to supply a good portion of their protein intake (Vickers 1984; Redford and Robinson 1987; Stearman 1989b, 1990; Alvard 1993b; Chicchón 1992; Ojasti 1993), and the Sirionó Indians of lowland Bolivia are no exception (Stearman 1987; Townsend 1996). Previously, they were semi-nomadic foragers who ranged over a large part of eastern Bolivia (Holmberg 1969), but by the late 1940s most were settled into various government camps, private ranches, and religious missionary camps (Stearman 1987). One such camp, Ibiato, persisted and is the now the center of the Sirionó Territory.

Although no longer nomadic, these people are still dependent on wildlife resources (Stearman 1987; Townsend 1995, 1997). Hunting remains a very important activity for the Sirionó, although the methods have changed. More than 75% of the wildlife is now taken by means not widely available prior to the founding of a mission in 1932, including machetes, firearms, and dogs (Townsend 1995). Central place foraging from Ibiato continues, but the capture area has certainly increased due to roads and bicycle usage. The sustainability of the Sirionó wildlife harvest depends on various factors that will be discussed in this chapter.

THE SIRIONÓ PHYSICAL ENVIRONMENT

The Sirionó Territory of Ibiato lies in the southeastern portion of the Beni River Basin, where 50% to 60% of the land floods annually (Denevan 1966). The rains, averaging 1800 mm/year (AASANA 1992), diminish in March, and by June and July the cold (to 7°C) dry southerly winds are predominant. By August and

September, surface water is scarce and only the deepest water holes remain. Most of the trees lose their leaves during the dry season and begin flowering in September and October before the onset of the November rains. Temperatures average about 26°C year round, but can reach about 37°C in the rainy season. During the 1991 to 1992 study period there was higher than normal rainfall in all of eastern Bolivia; Trinidad, the closest permanent meteorological station to Ibiato (60 km away), registered 3192 mm, almost twice the normal average for a year.

Pre-Colombians who inhabited the current Sirionó Territory before 1500 AD did major earth works which left large forest covered hills raised as much as 30 m above the surrounding savanna. These act as refuges for wildlife during floods. They are encircled by canals, parts of which serve as critical water holes during the dry season. The 340 km² of the current Sirionó Territory that is most hunted is an ecotone between the forested regions to the east and the savannas to the west (figure 13-1); 161 km² of the land is covered in forest and the remaining 179 km² is savanna (Townsend 1996).

The forested areas comprise three types: islands within surrounding savanna, long gallery forests, and larger continuous blocks generally on higher ground. Forest trees are mostly palms (*Attalea phalerata, Astrocaryum* sp., *Syagrus sancona*) (Townsend 1995), probably another indication of past human use (Balée 1987, 1988). Other common tree species include *Andenanthera macrocarpum, Gallesia integrifolia, Spondias mombin, Cordia nudosa, Guazuma ulmifolia, Triplaris boliviana,* and

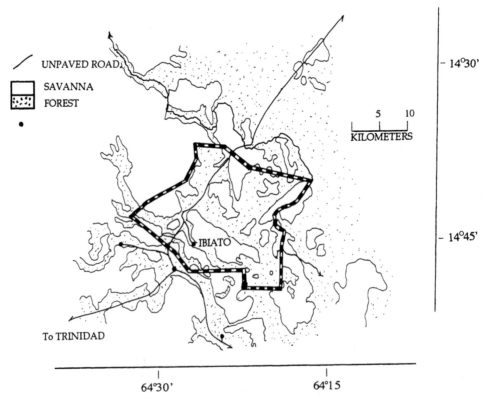

Figure 13-1. Map of the Sirionó Territory of Ibiato, Bolivia.

Chorisia spp. The Sirionó recognize a further division of the forest types according to the predominant undergrowth species, ranging from *Heliconia* sp. and *Costus* sp. to a variety of lianas and relatively open undergrowth (Townsend 1995).

The savanna vegetation consists of various communities of grasses, rushes, sedges, and broad-leafed plants distributed concentrically around lower terrain. The lowest areas are floating peat bogs that can be so thick they can support 5 m tajibo trees (*Tabebuia insignis*). The most common vascular plant species in the savanna include *Rhynchospora trisicata*, *Cyperus giganteus*, *Leersia hexandra*, *Ludwigia* sp., *Setaria* sp., *Sporobolus* sp., and, in the more humid parts, *Eleocharis elegans*, *Hydrocotyle*, and *Begonia* sp.

THE SIRIONÓ SOCIAL ENVIRONMENT

In the past 15 years the Sirionó have seen a drastic change in their living conditions and surrounding environment, due to the construction of the main highway between Trinidad and Santa Cruz, which passes within 15 km of Ibiato. Until then, Ibiato was quite isolated, its neighbors being cattle ranchers who left the natural savannas in a relatively undisturbed state (Stearman 1986). Off the highway, an auxiliary road running through the Territory has considerably changed the outlook for the future. Colonization rates have increased, and forest conversion around the Territory is extremely rapid. Several neighboring communities have begun to flourish, and more people are constantly being attracted to the area due to the improved market access for forest and agricultural products. The road also allows entry by sport and commercial hunters, both of which are prohibited by Bolivian laws, but are in actuality, uncontrolled. The subsistence hunting for consumption practiced by the Sirionó is permitted under Bolivian laws, but the commercial hunting of caiman skins previously practiced by the Sirionó (Stearman and Redford 1992) has been prohibited since 1990.

Of the approximately 500 Sirionó in the Ibiato Territory, more than 40% are under 10 years of age, and population doubling time could project to 11.1 years (Townsend 1995). The infant mortality rate has decreased considerably due to rapid access to medical attention, although tuberculosis is still a problem in some families. In-migration by family members has increased and might continue since the Bolivian government has awarded them their territorial land claims. Out-migration will probably increase in the future, and work is scarce in Ibiato. How the population will finally balance remains to be seen.

WILDLIFE HARVEST STUDY METHODS

Wildlife harvest was recorded from 46 adult male hunters, who are members of 29 households (the 192 consumers were defined as persons more than 3 years of age), during 360 days between April 1991 and May 1992. From August 1991 to May 1992, wildlife hunters were monitored each day, but from April 1991 to July 1991, the monthly sampling periods totaled 85 days (Townsend 1995). All households within the nucleated village (60% of the population) were visited twice daily by myself or trained university student volunteers and Sirionó assistants. All wildlife, fish, and honey harvested were recorded and measured. Some households farther away were visited on occasion when large animals were reported, but because the monitoring of these households was not complete, their harvest is not included in the average of wildlife biomass, but the biological data are included in the analysis of productivity.

The hunted wildlife arriving at unusual hours was brought to our attention by the hunters themselves or family members. Weights and measurements of hunted animals were recorded, and a standard set of questions was asked about location of kill, habitat, distance, weapons, animals left wounded, and other information. Reproductive condition was noted, and skulls (figure 13-2) and stomach contents were traded for soap and kerosene (US$0.15), promoting collaborative reporting

Figure 13-2. Sirionó hunter bringing in a tapir head for examination by wildlife managers in Ibiato. *Photo courtesy of Wendy Townsend.*

of information. To prevent hunting specifically to obtain these commodities, they were also exchanged for all types of agricultural products.

Wildlife harvested and consumed while on extended hunting trips was reported by hunters upon their return. Some hunters carried spring scales to weigh their hunt, which they recorded in small booklets. Others were interviewed immediately upon their return about wild meat consumed while in the forest. Although biological and other specific information obtained from the hunter were not included in analysis, identification of which species and the quantity harvested could be cross-checked with hunters participating on the same extended hunting trip. A total of 3713 records of hunted animals was registered during the study period.

The analysis of the harvest requires two sets of figures (table 13-1). First, the biomass of the wildlife harvested was calculated by multiplying the number of animals hunted by their average weights. These biomass values are converted to per capita protein estimates by dividing by the number of consumers (n = 192 over 3 years of age) in the household, monitored completely and assuming that 70% of the animal is comestible (Stearman et al. 1995) and 20% of the comestible meat is protein (Leung and Flores 1961). For calculating wildlife harvest sustainability, the number of animals killed by hunting per year is a more appropriate value. This number was obtained by extrapolating the number harvested to the entire Sirionó population (192 to 336 consumers) and from 360 days to 365 days to complete the year. Then a percentage was added to represent the wounded animals reported by the hunters. The percentage depended on how many of each species were reported as wounded in comparison with how many were taken. For most mammal species, this rate was approximately 10%.

HUNTING BY THE SIRIONÓ

The hunting methods reported by the Sirionó hunters in this study differ considerably from those that Holmberg (1969) observed in the 1940s, when he described the longest bows and arrows in the world. In 1991 and 1992, wildlife was captured by both young and old hunters of both sexes, but men brought in 80% of the biomass. Hunting occurred throughout the year. Harvests were recorded for 155 group hunts and 137 solitary hunts. Each man participated in an average of 11 group trips (SD = 7.0, range 1–29) and 8.4 solitary trips (SD = 7.2, range 0–31) during the 360 days. The modal number of participants on group hunting trips was two. Hunting trips with more than eight participants involved the driving of river turtles during the dry season.

Most animals (73%) were hunted while hunters actively and specifically searched for game, but 7% of the animals were captured by hunters while lying in wait at water holes or fruit trees. Three percent of the animals were trapped, and 17% were bagged while the hunter was engaged in activities other than hunting, such as farming or ranching.

Half of the animals were killed by firearms, with 0.22-caliber rifles being the most frequently reported (39%), then 16-gauge (8%) and 20-gauge shotguns (2%). Twenty-two percent of the harvested animals were killed with machetes,

Table 13-1. Animals Harvested by the Sirionó Indians in the Beni, Bolivia

Common Name	Scientific Name	Animals Harvested per Year	Average Weight (kg)	Biomass Harvested	Extrapolated Number of Animals Harvested for Entire Sirionó Community
(a) Mammals					
	Xenarthra				
Giant anteater	*Myrmecophaga tridactyla*	8	14.3	114.40	19
Banded-anteater	*Tamandua tetradactyla*	27	4.32	116.64	47
Nine-banded armadillo	*Dasypus novemcinctus*	642	3.79	2433.18	1146
Six-banded armadillo	*Euphractes sexcinctus*	27	3.2	86.40	49
Giant armadillo	*Priodontes maximus*	1	60	60.00	2
	Primates				
Saddle-backed tamarin	*Callithrix argentata*	7	0.38	2.66	12
Night monkey	*Aotus* sp.	12	1.15	13.80	21
Tufted capuchin	*Cebus apella*	20	2.98	59.60	40
Squirrel monkey	*Saimiri boliviensis*	1	1	1.00	2
	Carnivora				
Coatimundi	*Nasua nasua*	246	3.26	801.96	486
Kinkajou	*Potos flavus*	1	1.6	1.60	2
Ocelot	*Felis pardalis*	3	13	39.00	5
	Felis sp.	1	3.5	3.50	2
	Perissodactyla				
Tapir	*Tapirus terrestris*	5	150	750.00	21
	Artiodactyla				
Collared peccary	*Tayassu tajacu*	200	15.31	3062.00	389
White-lipped peccary	*Tayassu pecari*	139	28.53	3965.67	268
Red brocket deer	*Mazama americana*	26	29.72	772.72	51
Grey brocket deer	*Mazama gouazoubira*	21	17.28	362.88	49
Marsh deer	*Blastoceros dichotomus*	70	80	5600.00	127
	Rodentia				
Squirrel	*Sciurus spadiceus*	11	0.45	4.95	23
Porcupine	*Coendou prehensilis*	41	4.3	176.30	72
Paca	*Agouti paca*	104	7.27	756.08	207
Agouti	*Dasyprocta variegata*	167	3.31	552.77	307
Total mammals		1780		19,737.11	3347
(b) Birds					
	Rheidae				
Greater rhea	*Rhea americana* (young pets)	2	1	1	8
	Tinamidae				
Undulated tinamou	*Crypturellus undulatus*	3	0.52	1.56	8
	Phalacrocoracidae				
Neotropical cormorant	*Phalicrocorax olivaceus*	1	1.4	1.4	2
	Ardeidae				
Rufescent tiger-heron	*Tigrisoma lineatum*	44	1.05	46.2	82

Table 13-1. (*Continued*)

Common Name	Scientific Name	Animals Harvested per Year	Average Weight (kg)	Biomass Harvested	Extrapolated Number of animals Harvested for entire Sirionó Community
Ciconiidae					
Wood stork	*Mycteria americana*	8	2.46	19.68	14
Maguari stork	*Ciconia maguari*	7	3.48	24.36	14
Jabiru	*Jabiru mycteria*	5	7.96	39.8	12
Theskiornithidae					
Green ibis	*Mesembrinibis cayenensis*	1	1	1	2
Anhimidae					
Southern screamer	*Chauna torquata*	1	1.4	1.4	2
Anatidae					
Black-bellied whistling-duck	*Dendrocygna autumnalis*	14	0.57	7.98	28
Muscovy duck	*Cairina moschata*	5	2.65	13.25	11
Accipitridae					
Hawk-eagle	*Spizastur* sp. ?	1	3	3	2
Harpy eagle	*Harpia harpyja*	1	4.48	4.48	2
Hawks	Accipitridae no id.	8	0.68	5.44	14
Cracidae					
Speckled chachalaca	*Ortalis motmot*	8	0.51	4.08	21
Spix's guan	*Penelope jacquacu*	17	1.29	21.93	38
Blue-throated piping-guan	*Pipile pipile*	21	1.45	30.45	40
Bare-faced curassow	*Crax fasciolata*	3	2.65	7.95	7
	Cracidae no id.	1	1.5	1.5	2
Aramidae					
Limpkin	*Aramus guarauna*	13	1.22	15.86	23
Rallidae					
Gray-necked wood-rail	*Aramides cajanea*	1	0.45	0.45	2
Columbidae					
Pale-vented pigeon	*Columbia cayennensis*	3	0.18	0.54	5
	Columbidae no id.	4	0.5	1.5	7
Psittacidae					
Blue-and-yellow macaw	*Ara ararauna*	6	1.07	6.42	13
Red-and-green macaw	*Ara chloroptera*	4	1.07	4.28	9
Chestnut-fronted macaw	*Ara severa*	1	0.39	0.39	2
	Ara sp.	1	0.35	0.35	2
Blue-headed parrot	*Pionus menstrus*	1	0.24	0.24	2
	Psittacidae no id.	2	0.26	0.52	4
Opisthocomidae					
Hoatzin	*Opisthocomus hoazin*	3	0.85	2.55	5

Table 13-1. (*Continued*)

Common Name	Scientific Name	Animals Harvested per Year	Average Weight (kg)	Biomass Harvested	Extrapolated Number of animals Harvested for entire Sirionó Community
	Ramphastidae				
Chestnut-eared aracari	*Pteroglossus castanotis*	3	0.25	0.75	5
Toco toucan	*Rhamphastos toco*	2	0.8	1.6	4
	Ave.no id.	3	0.3	0.9	5
Total birds		198		272.81	397
(c) Reptiles					
	Pelomedusidae				
Yellow-headed sideneck	*Podocnemis unifilis*	123	2.97	365.31	215
	Chelidae				
Matamata	*Chelus fimbriatus*	57	5.88	335.16	100
Geoffry's sideneck turtle	*Phrynops geoffroanus*	23	1.35	31.05	40
Grooved sideneck turtle	*Platymys platycephala*	1	1	1	2
	Testudinidae				
Red-footed tortoise	*Geochelone carbonaria*	20	5.93	118.6	35
Yellow-footed tortoise	*Geochelone denticulata*	51	5.19	264.69	89
	Geochelone sp.	1	5.19	5.19	2
	Kinosternidae				
Scorpion mud turtle	*Kinosternon scorpoides*	40	0.45	18	70
Caiman	*Caiman yacare*	9	6.28	56.52	16
Total reptiles		325		1195.52	569

whereas 12%, mostly turtles and tortoises, were brought back alive. Dogs killed about 6% of the hunted animals, and about 1% were killed by other techniques, including clubs, axes, and bows and arrows. Only nine animals were killed with the traditional bow and arrow.

WILDLIFE HARVEST

Seventy-seven percent of the animals harvested were mammals (figure 13-3), 14% were reptiles, and only 7% were birds. Considering the total biomass harvested, 93% was from mammals, 6% from reptiles, and only about 1% from birds. The five species of artiodactyls—marsh deer (*Blastocerus dichotomus*), white-lipped peccary (*Tayassu pecari*), collared peccary (*Tayassu tajacu*), red brocket deer (*Mazama americana*), and gray brocket deer (*M. gouazoubira*) made up almost 70%

Figure 13-3. Sirionó hunter in Ibiato with coatimundi (*Nasua nasua*), paca (*Agouti paca*), and small ocelot (*Felis pardalis*). *Photo courtesy of Wendy Townsend.*

of the biomass of mammals harvested. These ungulates along with the tapir (*Tapirus terrestris*) comprised 69% of the total wildlife biomass harvested. Another 12% of the biomass of harvested mammals was from nine-banded armadillos (*Dasypus novemcinctus*), whereas three species—coatimundi (*Nasua nasua*), paca (*Agouti paca*), and agouti (*Dasyprocta variegata*)—made up an equal 12%.

Seven species of turtles from five families and one species of crocodile comprised only about 6% of the total biomass harvested (see table 13-1). Two species of river turtles, *Podocnemis unifilis* and *Chelus fimbriatus*, comprised 64% of the reptilian harvest weight (see table 13-1). Thirty-three species of birds belonging to 16 families were hunted even though they only supplied about 1% of the biomass extracted. The Cracid family supplied 24% of the avian biomass, and the tiger heron (*Tigrisoma lineatum*) made up 17% of the total avian biomass.

The total biomass of wildlife harvested for the study period was approximately 21,000 kg for the 192 consumers over 3 years of age. Of this total, 36% was harvested from the savanna and 64% from the forest. If extrapolated to the 336 consumers living in the Sirionó Territory during the study period, the total annual harvest is estimated to be 37,000 kg. Given that 30% of the biomass is nonedible (Vickers 1979; Stearman 1991; Townsend 1995), an estimated 25,900 kg of wildlife meat was exracted from the Sirionó Territory.

If the Sirionó were able to replace their wild meat with purchased beef, it would represent a value of about US$132 per person per year using the price of beef during the study period, US$1.50/kg, the same price for which the hunters exchanged wild meat within the community. Considering that the average family size in Ibiato was 6.6 members, the total cost to buy the same amount of meat they harvested would be $871/family/year, or more than 60% of the Bolivian wage earner's annual income during 1991 and 1992 (Townsend 1997).

To understand fully the importance of the wildlife harvest to the Sirionó people, it is necessary to consider the nutritional benefit they derived from their wild animal species. If we assume that the wild meat was shared equally among community members, and a mean protein content of wild meat of 20% (Leung and Flores 1961; Vickers 1979; Stearman 1991), the average daily per capita game protein extracted from the Sirionó Territory was 43 g/person/day. In addition, the 11,000 kg of fish biomass harvested by the Sirionó provides approximately 12.8 g of protein/person/day (Townsend 1995, 1996). The total protein extracted from the wild animal resources (55.6 g protein/person/day) averages only slightly less than the average protein capture of 10 other neotropical indigenous groups, 59.6 g protein/person/day (Townsend 1995). These values are above the estimated 20 g minimum daily protein intake from high-quality protein sources recommended by the FAO/WHO (1973).

THE SUSTAINABILITY OF THE SIRIONÓ WILDLIFE HARVEST

To evaluate the sustainability of the Sirionó wildlife harvest, the hunting harvest was compared with the potential production of the capture basin. First, the hunting harvest was estimated as being the number of individual animals affected by hunting per year (see table 13-1) for the ten mammal species that made up 90% of the harvested biomass. These numbers were averaged across the entire capture basin, according to the percentage of each species that hunters reported they extracted from either savanna or forest habitat (table 13-2) (Townsend 1995). Although most species were hunted in forests, the marsh deer was predominantly harvested from the savanna, and the gray brocket deer and nine-banded armadillo were routinely captured from both habitats.

Table 13-2. Comparison of Sirionó Hunting in Forest and Savanna Habitats

Species	Forest (176 km²) % of Harvest	Forest (176 km²) Number Harvested by 336 People/yr	Forest (176 km²) Hunting Pressure (no./km²)	Savanna (242 km²) % of Harvest	Savanna (242 km²) Number Harvested by 336 People/yr	Savanna (242 km²) Hunting Pressure (no./km²)
Tayassu tajacu	99	385	2.18	1	4	0.02
Tayassu pecari	97	261	1.48	3	7	0.03
Mazama americana	100	51	0.29	0	0	0
Mazama gouazoubira	62	30	0.17	38	19	0.08
Blastocerus dichotomus	3	4	0.02	97	123	0.51
Tapirus terrestris	75	16	0.09	25	5	0.02
Agouti paca	95	197	1.12	5	10	0.04
Dasyprocta variegata	100	307	1.74	0	0	0
Dasypus novemcinctus	69	791	4.49	31	355	1.47
Nasua nasua	98	475	2.7	2	11	0.05

The estimated productivity (number of young/individual/year) was observed from Sirionó hunted animals (table 13-3) and can help to explain the harvest in terms of production potential (Bodmer 1994). For example, for collared peccary, 25% of the adult females were reproductively active, having 1.5 gestations per year with an average litter size of two. This calculates as 0.75 young/female/year. Assuming a 1:1 sex ratio (as observed), the average number of young per individual per year is 0.375.

Production estimates were made by multiplying estimated productivity (see table 13-3) by population densities extracted from the literature (table 13-4). The

Table 13-3. Observed Reproductive Parameters of the Ten Most Important Games Species of the Sirionó Harvest

Species	% Females Reproductively Active	Number of Adult Females Observed	Gestations per Year	Litter Size	Estimated Productivity Young/ Female/ Year	Estimated Productivity Young/ Animal/ Year
Tayassu tajacu	25	76	1.5	1.6	0.6	0.30
Tayassu pecari	38	57	1	2	0.76	0.38
Mazama americana	83	6	1.5	1	1.245	0.62
Mazama gouazoubira	50	12	1.5	1	0.75	0.38
Blastoceros dichotomus	38	37	1	1	0.38	0.19
Tapirus terrestis	50	2	0.5	1	0.25	0.13
Agouti paca	58	55	1.5	1	0.87	0.44
Dasyprocta variegata	22	100	1.5	2	0.66	0.33
Dasypus novemcinctus	24	247	1	4	0.96	0.48
Nasua nasua	20	64	1	3	0.6	0.30

Table 13-4. Sustainability of Harvest for the Ten Most Important Game Species

Species	Average Densities (Robinson and Redford 1986a)	Average Production (No. young/animal/ km²/yr)	Harvest Rate (animal/km²)[a]	% Production Harvested	Maximum % of Production Available for Harvest (Robinson and Redford 1991b)
Tayassu tajacu	11.9	3.57	2.18	61	20
Tayassu pecari	4.9	1.86	1.48	79	20
Mazama americana	10.5	6.54	0.29	4	40
Mazama gouazoubira	10.4	3.90	0.17	4	40
Blastocerus dichotomus	0.5[b]	0.10	0.51	537	40
Tapirus terrestris	1.6	0.20	0.09	45	20
Agouti paca	27.5	11.96	1.12	9	20
Dasyprocta variegata	19.7	6.50	1.74	27	40
Dasypus novemcinctus	21.9	10.51	4.49	43	60
Nasua nasua	15.1	4.53	2.7	60	

[a]See table 13-2.
[b]Density estimate from Schaller 1983.

percentage of production that can be hunted sustainably (Robinson and Redford 1991b) varies according to the expected longevity of the species. For short-lived species such as the nine-banded armadillo, these investigators suggested that 40% of the production can probably be sustainably harvested. For long-lived species such as the tapir and the peccaries, only 20% of the potential production should be considered as a sustainable harvest (Robinson and Redford 1991b).

Comparing production potentials with harvest levels (see table 13-4) demonstrates that for the collared and white-lipped peccaries, tapir, and marsh deer, the Sirionó harvest rate was not sustainable, according to the model of Robinson and Redford (1991b). The coatimundi also might be overharvested because its life expectancy is long as defined by Robinson and Redford (1991b), so only 20% to 40% of the production could be sustainably harvested; the Sirionó harvested an estimated 60% of the production of the capture basin. The other five species seem to be harvested at sustainable rates, although locally measured density estimates are needed to improve production estimates and validate this appraisal. If the hunting practiced by the Sirionó in 1991 to 1992 is indicative of a typical yearly harvest, which might be the case (Townsend 1997), then there is good cause for the Sirionó to be concerned about the sustainability of their harvest of the peccaries, coati, and tapirs, especially with the forest conversion by colonization increasing in the region. The fact that the marsh deer harvest remains considerable may be due to a source-sink phenomenon (Pulliam 1988; Pulliam and Danielson 1991), with dispersal occurring from neighboring habitat. It is also possible that the calculations underestimate marsh deer productivity and density in the Sirionó Territory.

These evaluations of sustainability do not include wildlife harvested by persons other than Sirionó, of which there might be several groups. Although cattle ranchers do not generally hunt, colonists tend to hunt from along roads, in their farm plots, and at water holes. This increases potential hunting pressure, particularly on the brocket deer (road and farm plot hunted) and the pacas (water hole hunted). Sport hunters have been recorded to kill 10 to 15 pacas a night at a water hole, which suggests that their effect on wildlife populations might be considerable.

IMPROVING THE SUSTAINABILITY OF SIRIONÓ HUNTING

In order to produce sustainably the number of white-lipped peccaries harvested by the Sirionó during a year, it is estimated that 720 km^2 of forest is needed (table 13-5) (this assumes the density in the Sirionó Territory is the same as the average reported by Robinson and Redford [1991b]). The Sirionó Territory contains about 560 km^2 of forest, which, assuming average densities of the forest species data (Robinson and Redford 1991b), could sustainably produce the estimated Sirionó harvest for the main harvested species. However, about 400 km^2 of this forest is far away, and hunting returns are still quite acceptable closer to Ibiato. A sustainable marsh deer harvest at 1991 to 1992 levels (assuming densities reported for Brazil by Schaller [1983]) would require about 623 km^2 of savanna lands. The savanna awarded as territory by Bolivian presidential decree (No. 22609) covers less than a third of this. One method to improve sustainability would be to increase the size of the exploitation area, or the Sirionó Territory in general.

Other options for improving the sustainability of the harvest near Ibiato might be forest enrichment programs. Enriching fallow plots could improve year-round food sources for game species, particularly the collared peccary, the rodents, and the brocket deer. Reproductive rates of Sirionó animals were lower than those found by Bodmer (1994) in Peru, or those summarized from the literature by

Table 13-5. Estimated Required Area to Produce a Sustainable Harvest of the Magnitude of That Taken by the Sirionó in 1991–1992

Species	Potential Sustainable Harvest	Estimated Sirionó harvest (No. of Animals)	Estimated Required Area (km^2) to Produce a Sustainable Sirionó Harvest
Tayassu tajacu	0.71	389	545
Tayassu pecari	0.37	268	720
Mazama americana	2.61	51	20
Mazama gouazoubira	1.56	49	31
Blastocerus dichotomus	0.20	127	623
Tapirus terrestris	0.04	21	525
Agouti paca	4.79	207	43
Dasyprocta variegata	3.90	307	79
Dasypus novemcinctus	6.31	1146	182
Nasua nasua	2.72	486	179

Robinson and Redford (1986b). Habitat enrichment might increase production levels near Ibiato of those species known to feed in garden plots (Linares 1976; Jorgenson 1993).

Probably the most obvious, but perhaps not the simplest, way to achieve a sustainable harvest in the Sirionó Territory would be to limit the harvest of the ten most important game species, especially those that are overharvested from the capture basin. This is a socially difficult solution because, traditionally, only when a resource is harvested is it owned (Holmberg 1969). The concept of the community control over what cannot be owned is a difficult hurdle. In addition, although average protein consumption in Ibiato is greater than the minimum recommended protein levels (FAO/WHO 1973), the effect of heavy parasite loads observed among the Sirionó children might increase protein requirements. Furthermore, wild meat remains a socially important element as well as a food source in Ibiato, and it is also a status-gaining mechanism. If the community does not change its wildlife consumption by itself, however, the people will probably be forced to change through lack of large animals within their capture basin, for example, by switching to hunting smaller, less desirable mammals, birds, and fish. Because most of the lands surrounding the capture basin are being degraded, there is considerable risk that the local extinction of forest mammals could happen suddenly. There are indications that this has happened historically. The harvested species list includes none of the primates hunted almost exclusively by the Sirionó in the early 1940s (Holmberg 1969).

Under current socioeconomic conditions, community collaboration is needed in attaining fallow plot enrichment projects and restrictive mechanisms designed by and for themselves. Potentially, these are much more feasible than obtaining more territory for wildlife production from the Bolivian government. Bolivia's new forestry and agrarian reform laws require that all commercial use of natural resources be guided by a management plan. Participatory planning sessions in Ibiato have suggested that there is a desire to use their wildlife resources within the capture basin of the Sirionó Territory more sustainably. Community wildlife initiatives should consider ways to make the wildlife valuable as living resources, as well as how to manage the populations for sustainable harvesting. For example, the possibility of an ecotourism business spotlighting animals at protected water holes has been discussed by the Sirionó. Although considerable training in small business management would initially be necessary, ecotourism could have positive economic benefits. The many archeological sites would add to tourism potential, especially if an investigational dig could be started. However it happens, a change in the hunting practices will probably occur. That change can either be chosen and planned for by the community or forced on them as surrounding forest conversion diminishes the productivity of the system that has kept the Sirionó capture basin supplied with wildlife.

ACKNOWLEDGMENTS

I thank the Sirionó community, as well as those who provided funding: the Wildlife Conservation Society, Leakey Foundation, OAS, the PSTC and TCD Programs of

University of Florida, and the WWF-Biodiversity Support Program, a USAID-funded agreement with The Nature Conservancy, WWF-US, and the World Resources Institute. Also greatly appreciated are the efforts of Michel Pinard, Damian Rumiz, Andy Noss, John Robinson, Elizabeth Bennett, and the anonymous reviewers for their helpful comments on drafts of this manuscript.

14

Cable Snares and Nets in the Central African Republic

ANDREW NOSS

Hunting is an important economic and subsistence activity for residents of the rain forests of the Congo basin. In the Central African Republic (CAR) where firearms are still uncommon, hunters rely on nets and cable snares. Based on research from September 1993 to December 1994 in the Dzanga-Sangha Special Reserve (Noss 1995, 1997a, 1997b), this chapter describes the two methods and the hunters who use them. I analyze the economic importance of hunting for local peoples and the impacts of hunting with snares and nets on wildlife, before addressing the sustainability of the two hunting methods and more general subsistence hunting in African forests.

The Dzanga-Ndoki National Park and the Dzanga-Sangha Special Reserve are located in the southwest corner of the CAR (figure 14-1). The protected area includes 4480 km² of rain forest on the northern edge of the Congo forest basin and borders the Ndoki-Nouabalé National Park in Congo. Annual rainfall averages less than 1400 mm, but rain is not unusual during the dry season of December through February, and streams in the forest flow all year long. Seasonally flooded forests occur along the Sangha river in the southernmost portions of the Park. Forest types include evergreen, deciduous, and evergreen-deciduous transition forests. Patches of *Gilbertiodendron dewevrei* monodominant stands occur on well-drained soils, and almost pure stands of *Guibourtia demeusii*, *Gilbertiodendron dewevrei*, or raffia palms (*Raffia* spp.) grow on poorly drained soils. Herbaceous marshy forest clearings are maintained, and perhaps created, by heavy wildlife use, particularly elephants searching for minerals in the soil (Carroll 1988).

The protected area is particularly rich and diverse in large mammals: forest elephants *Loxodonta africana cyclotis*, for which it is most famous; 15 primates, including the western lowland gorilla (*Gorilla gorilla gorilla*) and chimpanzee (*Pan troglodytes*); 14 ungulates such as the forest buffalo (*Syncerus caffer nanus*) and

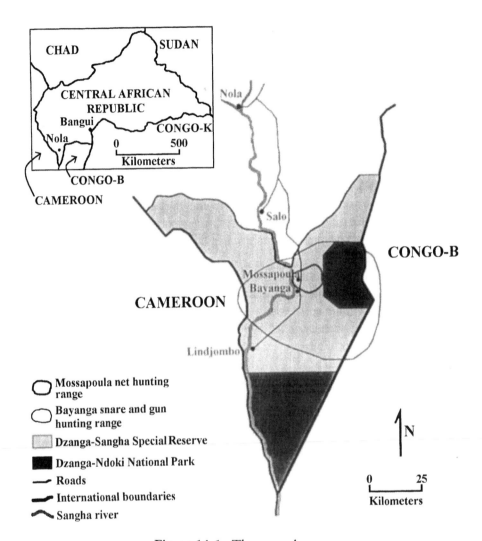

Figure 14-1. The research area.

bongo antelope (*Tragelaphus euryceros*); and 14 carnivores including the leopard (*Panthera pardus*) and golden cat (*Profelis aurata*). Elephants and gorillas are the basis for ecotourism, and bongo antelope for safari hunting; the most important game species for local hunters are the brush-tailed porcupine (*Atherurus africanus*), Peters' duiker (*Cephalophus callipygus*), bay duiker (*C. dorsalis*), and the blue duiker (*C. monticola*).

The protected area was created in 1990 and includes the two sections of the Dzanga-Ndoki National Park of 1220 km², and the Dzanga-Sangha Special Reserve of 3360 km². Only tourism and scientific research are permitted in the Park. In the Reserve a range of subsistence and economic activities are legally permitted: farming, selective logging, safari hunting, and subsistence hunting and gathering by local residents using traditional harvest methods (République Centrafricaine 1992a, 1992b). Hunting methods permitted in the Reserve include nets and legally registered firearms, whereas cable snares are illegal. Bush-meat cannot legally be exported from the area.

Roughly 4500 people inhabit the Reserve, including over 1000 BaAka Pygmies. The major settlement is Bayanga (2500 inhabitants), with its neighbor Mossapoula (370 inhabitants). Long-time local residents include the Mbimu, Ngundi, and Sangha-Sangha ethnic groups, who fished along the Sangha river; the BaAka hunted the forests east of the Sangha into what is today Congo (Demesse 1978). These populations have fluctuated through migration in response to regional economic opportunities. Colonial economic activities such as concession companies harvesting wild rubber and ivory, as well as coffee plantations, induced population shifts to and from the region. However, most of the region's current residents immigrated from elsewhere in the CAR and neighboring Congo after logging operations began in 1972, centered around the sawmill in Bayanga.

Although often distinguished by the terms *hunter-gatherers* and *farmers/fishermen*, the BaAka and other ethnic groups in the region all engage in a suite of subsistence and economic activities: hunting, fishing, gathering, agriculture, day labor, formal labor (with the logging company, local government, or the conservation project) and diamond mining. In accordance with changing opportunities and returns, individuals switch among the available activities, although the BaAka are more likely to do so from day to day, whereas others alternate from month to month. All inhabitants live in settlements surrounded by cultivated areas along the main north-south road, although individuals and family groups may spend extended periods in the forest hunting, fishing, logging, or mining diamonds. Other ethnic groups look down on the BaAka and exploit them as a cheap but unreliable source of labor and forest products (Bahuchet 1979; Berry et al. 1986; Delobeau 1989).

SNARE AND NET HUNTING

Members of all ethnic groups in the region hunt for subsistence and for sale in local markets. Although other ethnic groups historically hunted communally with nets, only the BaAka do so today. Net hunting requires community cooperation, which immigrant groups have lost. Although meat production from nets is lower per hunter than from cable snares, many other forest products are gathered during net hunting trips (e.g., honey, caterpillars, edible leaves, nuts, fruit, and construction materials). However, the BaAka too are adopting cable snares for several reasons: snares generate substantially higher returns per hunter per day than do nets (Hart this volume) and are therefore more suitable for commercial bushmeat production. In terms of time commitment, snare hunting is more compatible with formal employment, with many individuals engaging in both activities simultaneously; formal employment provides the cash necessary to buy cable for snares. At any one time snare hunting is the principal income source for about 60 men in Bayanga (15–20% of the adult male population), with 70 snares a piece. In comparison, the 21 BaAka men in Mossapoula (30% of the adult male population) who have snares set only 19 on average, and they also hunt with nets. Meat from both hunting methods is both consumed and sold, but snares produce higher individual returns and therefore a greater surplus of marketable meat.

The predominant type of cable snare is a noose set along an animal trail (figure 14-2). When an animal steps on a pressure pad, it releases a bent-over pole, which springs up to tighten the noose around the animal's leg (figure 14-3). These foot snares comprise 90% of snares set. The others are neck snares, which capture small animals as they try to pass through a cable noose that is perpendicular to the ground. A variation of the neck snare method is to build a fence of branches and leaves to direct animals to paths through the fence where as many as 30 neck snares are set 2 to 5 m apart.

Full-time snare hunters in the Dzanga-Sangha region work alone or with one or two companions, and they set a string of 55 to 100 snares a piece along a 1- to 3-km course. Most snares are set in mixed upland forest, or at the mixed/monodominant forest edge, where animal trails are more visible. Snares remain in place for a month or longer, and individual snares are moved when the hunter judges that animals are not using the trail. Hunters generally maintain a camp near

Figure 14-2. Cable noose being laid over a concealed trigger mechanism. *Photo courtesy H. Noss.*

Figure 14-3. Gbaya hunter setting a cable snare. *Photo courtesy H. Noss.*

their snare lines where they stay for a night, a week, or a month depending on yields and the distance from town. Camps are located from 3 hours to 2 days walking distance from town, and those hunting west of the Sangha river travel by dugout canoe as well. Captured animals are butchered on the spot or in camp, and the meat is then smoked over hot fires for storage and transportation to town. The resulting dried and blackened bushmeat can be stored for several weeks and is found in urban markets across central Africa.

Although they are illegal throughout the CAR and in neighboring countries, cable snares are probably the most widespread hunting method in central Africa, preferred because they are inexpensive yet effective. They are also very difficult for park guards to control, and snare hunting continues throughout the Dzanga-Ndoki National Park. Despite the economic and ecological importance of this hunting method across central Africa, few studies address cable snares (Sato 1983; Infield 1988; Wilkie 1989; Wilson and Wilson 1991; Almquist 1992; Laurent 1992; Lahm 1993a, 1993b; Hennessy 1995; WCS 1996; Dounias in press; Muchaal and Ngandjui 1999) and those that do so generally do not provide detailed and comprehensive analyses of hunting methods, economic returns, and impacts on wildlife.

In contrast, several anthropologists have described net hunting elsewhere in the CAR and in Zaire, although practices vary by area and by ethnic group (Turnbull 1961; Bahuchet 1972, 1979, 1985; Harako 1976; Tanno 1976; Hewlett 1977; Hart 1979; Tanaka 1978; Ichikawa 1983, 1986; Hart and Hart 1986; Wilkie 1988; Hudson 1990). Of particular interest is that the technique is practiced by men and women (as well as children), and BaAka net hunting groups occasionally comprise only women. The only groups that do not always include women are demonstration hunts organized for tourists. Net hunting requires a group of 5 to 100 hunters with 4 to 25 nets (average 50 hunters and 15 nets): the nets are hung on vegetation and pegged to the ground in a closed circle around a patch of dense brush (figure 14-4), for example, a treefall gap or rattan tangle. Nets measure 1 to 1.5 m in height and 5 to 40 m in length, averaging 18 m. Some of the hunters shout and beat leaves inside the circle to drive any animals they find into the nets, whereas others remain hidden near the nets to seize trapped animals, and several others with spears search inside the circle. Although women never use spears, the other roles are often traded during the course of the hunt, but men are more likely to beat while women guard the nets. Both men and women own nets.

The hunters quickly determine whether any animals inside the circle have escaped or have been caught. They then gather up the nets and move on to the next promising site, usually only a couple of hundred meters away, and begin again. The entire cycle of setting up the nets, beating, and moving to the next site

Figure 14-4. BaAka hunter setting a net. *Photo courtesy H. Noss.*

takes only 10 to 15 minutes. Thus, a net hunt comprises four to six casts per hour and averages 16 casts in 4 hours of active hunting, with another 2 to 3 hours spent walking from town and back or resting. Captured animals belong to the net owner, even if the owner is not hunting, but the person who seizes the animal in the net receives a portion, as do others who help set up and guard that particular net. Meat is also usually shared with family members participating in the hunt and again upon returning to town. On 10% of observed hunts, portions of meat were contributed at the end of the hunt by those who had killed animals to be divided among those who had nothing.

METHODS

Economic Returns to Hunting and Impacts on Wildlife

Data on hunting returns and offtakes were collected through participant observation between September 1994 and December 1995. The author accompanied snare hunters from Bayanga (of several ethnic groups including Banda, Gbaya, Kaka, and Mbimu) 1 to 2 weeks per month, and BaAka net hunters from the neighboring town of Mossapoula 1 to 2 weeks per month.

The illegal nature of snare hunting complicated and restricted research efforts. Some hunters feared that information would be passed along to park guards and that they would be arrested, others demanded payments that could have been interpreted as park management support for poaching. It was therefore impossible to monitor all snare hunting in the community. Informants willing to cooperate were identified and approached by the author's research assistant, Ngbongo Yves. Several informants could only be accompanied once: several hunters abandoned snare hunting in favor of other employment, park guards discovered a second group of three and confiscated their snares, and a third group of four made impossible demands upon returning to Bayanga. In total, the author accompanied 17 snare hunters from Bayanga to their forest camps and to check snare lines during 158 observed person days, and 18,870 snare nights. Snare nights are defined as the number of snares times the number of nights since they were last checked, excluding snares that have sprung accidentally. The following information was recorded: total snares set; injuries where the animal escaped with part of the cable or left a foot, identifying the species if possible from the foot or tracks; losses to scavengers or decomposition; biological data on captured animals; and destination of the meat. The age of duikers was determined according to the number of erupted molars. Subadult and adult duikers all have complete dentition of three molars; animals were classified as subadults if their reproductive organs were not fully developed, or if they were less than two-thirds of average adult weight.

Athough net hunting is legal within the Special Reserve and hunters were not concerned with concealing their activities, the large size and fluctuating membership of net hunting groups also posed problems. In order to monitor all net hunting in the community, the author visited Mossapoula every day during week-long periods, and accompanied the 58 hunts that took place during these 90 observation days. Data were also recorded for an additional 11 hunts on the days when two separate groups from Mossapoula hunted, with the author following the larger group. The following information was recorded: participants' names; the duration of the hunt; biological data on captured animals; and animals that were encountered but escaped. In analyzing returns from net hunting, only the net hunts during the 90 observation days (December 1993 to November 1994) were considered, but data from an additional seven net hunts prior to the observation period (September to October 1993) were included in assessing the impacts of net hunting on wildlife.

Economic returns from hunting are based on 1994 local market prices for bush meat, assuming that all marketable meat is sold. Prices paid in Mossapoula and to the BaAka net hunters are lower than those received by cable snare hunters in Bayanga. *Atherurus africanus* is sold whole for approximately US$2.60 in Bayanga and US$1.75 in Mossapoula. *Cephalophus monticola* is split lengthwise in two, with each half selling for US$1.30 and US$0.90 in Bayanga and Mossapoula, respectively. *C. callipygus* and *C. dorsalis* are quartered: forelimbs cost US$1.60/0.90 and hindlimbs US$1.80/1.30. In practice, snare hunters consume the head and organs of captured animals and sell nearly 100% of the marketable leg portions. In contrast, net hunters share captured animals among several individuals and families, and sell only 20% of marketable meat.

Two indices for comparisons among species, over time, and among hunting areas are capture rate and vulnerability. The capture rate is the frequency with which particular species are captured by snares: number of snare nights per capture. Vulnerability is defined as the proportion of animals encountered that are captured, and is relevant to both hunting methods.

Total annual hunting offtakes are extrapolated from the data. Total snare offtakes assume that approximately 60 Bayanga snare hunters maintain an average of 70 foot and neck snares a piece for 1,533,000 snare nights per year (Noss 1995). Few hunters use fence snares, and these are not included in the analysis of annual offtakes. The Bayanga snare hunting area is approximately 1000 km² (see figure 14-1). The Mossapoula net hunting area, located entirely within the Bayanga snare hunting range, is approximately 110 km². Annual net hunting harvests are extrapolated from harvests during 90 observation days.

Population Densities of Game Species

An evaluation of the sustainability of hunting requires information on population densities of game species in the Bayanga region, and on the biological productivity of these populations. However, lacking the time and resources for a large-scale census project, the author conducted very few line transects: 97.7 km of daytime line transect surveys, using another project's vegetation census trails between Bayanga and Mossapoula. The trails included four parallel and straight paths 500 m apart, each roughly 2 m wide and 2 km long, and two 1.5-km perpendicular paths connecting the ends of the four parallel trails. Two observers walked together at a pace of 1–1.5 km per hour, beginning between 1000 hr and 1200 hr, and covering 5 to 6 km per day. Because they were a secondary objective of the research, the author conducted censuses on observation days when the Mossapoula net hunters did not hunt, first visiting the community and continuing to the transect trails after 0900 hr if no hunt was organized that day and if it did not rain. Despite the late start hour, two of the four principal game species are diurnal and active throughout the day: the duikers *Cephalophus callipygus* and *Cephalophus monticola* (Dubost 1980, 1984; Feer 1989b). Night transects were not conducted because of the potential dangers in encountering the abundant elephants or gorillas, and because of night-time gun hunters active in the area. The census area was one exploited by net hunters as well, although they hunt through the forest and use trails only for traveling to and from hunting sites.

For each animal seen or heard, the observers recorded species, number, and perpendicular animal-to-path distance. Population density (D) is calculated using the following formula (Burnham et al. 1980; Whitesides et al. 1988; Buckland et al. 1993):

$$D = (N \times 1000)/2LW$$

The width of the census path W (in meters) is two times the mean animal-to-path distance for each species. L is the length of the census path in kilometers, and N is the number of animals seen.

The density estimates from line transects are presented only because they are the only such data for the Dzanga-Sangha region. However, the sample size for all species is too small for reliable density estimates: $N = 1$ to 33. Furthermore, line transects are an unsatisfactory method for counting small cryptic animals in thick forest vegetation where they are hunted: experience with net hunters revealed that duikers tend to remain hidden or to move away quietly from approaching humans.

In an attempt to exploit the available information and address sustainability issues, population densities were also estimated from total net hunt encounters: number of animals seen per area searched (Noss 1997b). Other researchers have censused duiker populations using groups of observers with nets, setting the nets randomly and counting only the nets inside the enclosed area (Dubost 1980; Hart 1985; Koster and Hart 1988). A limitation of the method with active net hunters is that they do not search randomly, but hunt principally in dry upland mixed forest and avoid monodominant *Gilbertiodendron* and marsh/swamp forests. In addition, the entire path covered by the hunters is included because animals are frequently seen outside the net circles, and nets are set to capture these animals. The search area is determined by the path length (the diameter of the net circle, the number of times the nets are set, and an estimated 200 m between sets) and the path width (two times the diameter of the net circle). The number of animals censused includes all captures and escapes during the hunt, assuming that all animals within the search area are seen, highly probable with respect to the game species when experienced hunters search a path only 200–300 m wide.

Calculating Sustainable Harvests

Given these population densities, sustainable harvest rates can be estimated using the method of Robinson and Redford (1991b). First, the maximum intrinsic rate of population increase (r) was calculated according to the following formula (Cole 1954):

$$1 = e^{-r} + be^{-r(a)} - be^{-r(w+1)}$$

where e = 2.7128 (base of natural log); r = maximum intrinsic rate of increase of a population not limited by food, space, resource competition, or predation; a = age at first reproduction; w = age at last reproduction; and b = annual birth rate of female offspring.

In captivity, all duikers reach sexual maturity (first estrus) at 11 months, and after a gestation period of 7 months they produce their first offspring at age 18 months. Conception occurs again within days of parturition, and age at last reproduction is approximately 10 years (V. Wilson, unpublished data). Therefore, assuming 1.7 young per female per year and that half of the population is female, the annual birth rate of female offspring for all three duiker species is 0.43, and r is 0.29. *Atherurus africanus* is thought to reach sexual maturity at 2 years of age, to live 8 years, and to produce one young at a time after a gestation period of 3.5 months (Rahm 1962). Assuming that females produce two young per year and that half of the population is female, the annual birth rate of female offspring is 0.50 for this species, and r is 0.28.

To estimate population growth over time, the instantaneous rate of increase is converted to a finite rate of increase λ, where $\lambda = e^r$. For the three duiker species, $\lambda = 1.34$; for the brush-tailed porcupine, $\lambda = 1.32$. Annual production of a population is then calculated by multiplying actual densities by λ and subtracting the population size at the end of the year from that at the beginning of the year. Recognizing that all production is not available to human hunters, Robinson and Redford (1991b) estimated that 20% of annual production is available to human hunters in the case of long-lived species with a life span of 10 years or more. For short-lived species with a life span of 5 to 10 years, such as *Atherurus africanus*, 40% of productivity can be sustainably harvested each year. Given population densities for the game species, it is therefore possible to compare theoretically sustainable harvests with actual offtakes per unit area by snare and net hunters in the Dzanga-Sangha region.

RESULTS

Economic Importance of Hunting to Local Peoples

Returns per person per day spent hunting averaged US$2.85 for cable snare hunters. Weekly returns from snare hunting averaged US$9.50 per hunter, based on 29 snare hunter weeks, and correspond to an average annual income from snares of US$494.00. Returns from net hunting were much lower at US$0.35 per participant in a net hunt (a single day), or US$1.75 per week given the average of five net hunts per week, dividing the value of all captured animals among all participating hunters. Total returns per net hunt averaged US$15.35 (to be divided among an average of 45 hunters), and US$76.75 per week. However, the most successful individual hunter, if he did not share the meat but sold all marketable portions, could have earned US$175 in a year of net hunting.

Impacts of Hunting on Wildlife

Game Species and Total Harvests. During 18,870 observed snare nights, the three types of cable snares (foot, neck, and fence snares) captured 124 animals: 18 species of mammals, two birds, and a reptile (table 14-1). Another reptile, the

Table 14-1. Observed Captures by Cable Snares and Nets (number of animals and percent by species)

Species	Snares		Nets	
	Number	Percent	Number	Percent
Mammals	118	95	576	98
Cephalophus monticola	38	31	440	75
Cephalophus callipygus	36	29	24	4
Cephalophus dorsalis	7	6	40	7
Cephalophus leucogaster	1	1	0	0
Cephalophus nigrifrons	1	1	0	0
Hyemoschus aquaticus	0	0	1	<1
Potamochoerus porcus	2	2	0	0
Atherurus africanus	11	9	70	12
Cricetomys emini	8	7	0	0
Rat (unidentified)	2	2	0	0
Funisciurus pyrrhopus	2	2	0	0
Protoxerus stangeri	1	1	0	0
Squirrel (unidentified)	1	1	0	0
Atilax paludinosus	1	1	0	0
Bdeogale nigripes	2	2	0	0
Herpestes naso	0	0	1	<1
Nandinia binotata	2	2	0	0
Cercocebus galeritus	1	1	0	0
Monkey (unidentified)	1	1	0	0
Manis tricuspis	1	1	0	0
Birds	5	4	2	<1
Agelastes niger	3	2	0	0
Francolinus sp.	2	2	2	<1
Reptiles	1	1	11	2
Kinixys erosa	0	0	10	2
Python sebae	0	0	1	<1
Snake (unidentified)	1	1	0	0
Total	124	100	589	100

N = 18,870 snare nights and 76 observed net hunts.

tortoise *Kinixys erosa*, was encountered and collected on four occasions by snare hunters. These animals are not included in the table because they were not captured in snares. Only three hunters used fence snares, which capture only small animals. Excluding these snares, the 105 captures from foot and neck snares during 17,803 snare nights represented 15 mammal species. The four primary game species (*A. africanus*, *C. callipygus*, *C. dorsalis*, and *C. monticola*) comprised 87% of captures (figure 14-5). Captured animals included several pregnant or lactating females whose young probably could not survive alone: three *C. callipygus*, two *C. monticola*, one *C. dorsalis*, one *C. nigrifrons*, and one *A. africanus*.

Extrapolating the snare hunting offtakes from 17,803 observed snare nights to the projected 1,533,000 snare nights per year for Bayanga cable snare hunters, annual snare offtakes are estimated to be more than 10,500 animals in the 1000-km^2 hunting range, or 10 animals per km^2. Figures include losses to scavengers and decomposition, which account for 27% of total snare captures. Total annual biomass harvested exceeds 125,000 kg and 125 kg/km^2. Table 14-2 presents annual snare and net hunting harvests for the four primary game species.

Figure 14-5. BaAka hunter placing a blue duiker (*Cephalophus monticola*) into her carrying basket. *Photo courtesy H. Noss.*

Table 14-2. Annual Snare and Net Harvests

Species	Observed Captures[a]		Total Annual Harvest[b]		Harvest Rate[c]	
	Snares	Nets	Snares	Nets	Snares	Nets
A. africanus	11	63	947	256	1.0	2.3
C. monticola	38	395	3272	1601	3.3	14.6
C. callipygus	36	21	3100	86	3.1	0.8
C. dorsalis	7	34	603	137	0.6	1.2

[a]From 17,803 snare nights (excluding fence snares), and net hunting during 90 observation days.

[b]Number of animals killed by snares and nets: assuming 1,533,000 snare nights, and extrapolating for net hunting during 365 days. Figures include snare captures lost to scavengers and decomposition.

[c]Number of animals killed per year per km^2: for Bayanga snare hunting range of 1000 km^2 and for Mossapoula net hunting range of 110 km^2.

In 76 observed net hunts, 589 animals were captured: six species of mammals, one bird, and two reptiles. The tortoises were collected rather than caught, and a python *Python sebae* was surrounded and killed with machetes. The four primary game species comprised 97% of net hunt captures, and *Cephalophus monticola* alone nearly 75%. Of the captured female duikers, seven *C. dorsalis* (25%) and 63 *C. monticola* (27%) were pregnant or lactating. Extrapolating returns from 90 observation days to probable annual offtakes, Mossapoula net hunters on day hunts harvest over 2200 animals and 13,000 kg total biomass, corresponding to 20 animals and 120 kg/km^2.

Capture Rates and Vulnerability. The most frequently captured animals by foot and neck snares were the duikers *C. monticola* and *C. callipygus*: one animal was caught per 469 and 495 snare nights, respectively (table 14-3). On average, a hunter with a line of 70 foot snares captured one *C. monticola* and one *C. callipygus* per week. One *A. africanus* was caught every 1618 snare nights, and one *C. dorsalis* every 2543 snare nights. Other species were captured infrequently by snares, only once every 8000 or more snare nights. Excluding animals encountered and captured only once (*Cephalophus leucogaster* and *C. nigrifrons*), the species most vulnerable (measured as percentage of encounters that are captured) to foot and neck snares were *A. africanus* and *C. monticola*, with 92% to 93% of encountered animals captured. Larger animals were more likely to escape by breaking the cable.

Excluding reptiles, the animal most vulnerable to net hunters is *C. monticola*, with 59% of encountered animals captured. With respect to the other three primary game species, only 39% to 44% of animals encountered were captured. Several small animals (mongooses [*Herpestes naso*], genets [*Genetta* spp.], giant rats [*Cricetomys emini*], and Bates' pygmy antelope [*Neotragus batesi*]) almost always escaped, and larger animals were rarely encountered and never caught.

Injuries. Most animals escaping from nets are not injured, although during net hunts a few animals wounded by spears do escape: four *C. monticola* in 440 captures and four *C. dorsalis* in 40 captures. In contrast, for every three animals captured in snares, two others escape with injuries after breaking the cable, corresponding to 7000 injuries and 7/km^2 inflicted each year by Bayanga snare hunters. All 64 escapes listed in table 14-3, including 12 in which the escaped animal could not be

Table 14-3. Capture Rates and Vulnerability of Mammals to Snares and Nets

Species	Snares				Nets		
	Cap	Rate	Esc	Vul	Cap	Esc	Vul
Cephalophus monticola	38	469	3.0	92.7	440	307	58.9
Cephalophus callipygus	36	495	19.8	64.5	24	37	39.3
Atherurus africanus	11	1618	1.0	91.7	70	110	38.9
Cephalophus dorsalis	7	2543	6.3	52.6	40	50	44.4
Rodents	3	5934	0.0	100.0	0	4	0.0
Small carnivores	3	5934	4.0	42.9	1	16	6.3
Monkeys	2	8902	1.0	66.7	0	—	0.0
Potamochoerus porcus	2	8902	2.0	50.0	0	0	0.0
Cephalophus leucogaster	1	17,803	0.0	100.0	0	0	0.0
Cephalophus nigrifrons[a]	1	17,803	0.0	100.0	0	0	0.0
Pangolins	1	17,803	1.0	50.0	0	0	0.0
Cephalophus sylvicultor	0	—	8.3	0.0	0	1	0.0
Tragelaphus euryceros	0	—	3.5	0.0	0	0	0.0
Gorilla gorilla[a]	0	—	1.0	0.0	0	4	0.0
Loxodonta africana[a]	0	—	1.0	0.0	0	10	0.0
Hyemoschus aquaticus[a]	0	—	0.0	0.0	1	3	33.3
Neotragus batesi	0	—	0.0	0.0	0	2	0.0
Unknown	—	—	12.0	—	0	0	0.0
All species	105	170	64	62.5	576	528	52.2

17,803 total snare nights, excluding fence snares.
[a]Fully protected species.
Cap, number captured; Rate, number of snare nights per capture; Esc, number escaped (when hunters identified the escaped animal as one of two or three possible species, 0.5 animals indicates that hunters were uncertain between two species, 0.3 animals indicates that hunters were uncertain among three species); Vul, vulnerability measured as a percentage of encounters that are captured.

identified, were cases in which all or part of the cable snare was missing, or the animal's severed foot remained. In total, one quarter of the 45 *C. dorsalis* captured by snare and net hunters during the research period had previous snare injuries, including one that was missing two feet. Another *C. dorsalis* and an *Atherurus africanus* also had two injuries each. In addition to these principal game species, safari hunters in the Dzanga-Sangha region reported cable injuries on all eight of the *Tragelaphus euryceros* killed in 1993. One informant reported capturing a *Cephalophus sylvicultor* with nine old cable injuries, on all four legs. The larger the animal, the fewer the snare captures, but the greater the proportion suffering injuries.

Sex Ratios. Both snares ($\chi^2 = 4.2, p < 0.05$) and nets ($\chi^2 = 11.82, p < 0.01$) are selective for females of all species, except for nets that are selective for male *A. africanus* (table 14-4). Although the sample size is much smaller, snares appear to take a higher proportion of females of the three duiker species than do nets. Therefore, per animal captured, snares have a greater effect on the population. However, offtakes may reflect actual sex ratios in the population: *C. dorsalis* male home ranges overlap two female home ranges (Dubost 1983; Feer 1989a, 1989b). *C. monticola* male-female pairs share territories (Dubost 1980; Hart 1985), and sex ratios in this species are more likely to approach 1:1.

Table 14-4. Sex Ratios of Animals Captured by Snares and Nets (number of animals, female/male ratio)

Species	Snares			Nets		
	Female	Male	Ratio	Female	Male	Ratio
C. monticola	20	10	2.0	234	187	1.3
C. callipygus	17	12	1.4	13	11	1.2
A. africanus	4	3	1.3	23	45	0.5
C. dorsalis	2	1	2.0	28	12	2.3

Age Structure. Neither hunting method captures many young animals (table 14-5). An important exception is net captures of *C. callipygus*, where 41% of captured animals lack complete dentition. In comparison, 15% or less of captured animals by both hunting methods of the other two duikers lack complete adult dentition. For the four main game species, subadults comprise 12% to 26% of captures, and adults 66% to 86% of captures.

Population Densities of Game Species

Table 14-6 presents estimated population densities of the four primary game species based on line transect censuses and net hunt encounters. No *A. africanus* were seen during censuses, and the sample size for the duikers is extremely small. The larger sample of net hunts can be subdivided to analyze differences in population densities in areas near and far from Mossapoula (table 14-7). A Spearman rank correlation coefficient test finds population densities increasing with distance from Mossapoula for *C. monticola* ($r_s = 0.417$, $p < 0.01$), *A. africanus* ($r_s = 0.247$, $p < 0.025$) and *C. callipygus* ($r_s = 0.246$, $p < 0.025$). There is no statistically significant relationship between distance and density for *C. dorsalis* ($r_s = 0.106$), and densities for this species are inexplicably lowest in the intermediate 5- to 10-km range.

Sustainability of Harvests

Table 14-8 compares theoretically sustainable harvest rates and actual harvests by snares and nets. Annual harvest rates per unit area by net hunters are higher than those for snare hunters for all game species except *C. callipygus*. Harvests of all species by both hunting methods exceed sustainable levels. The most extreme cases are snare harvests of *C. callipygus*, at 55 times the theoretically predicted sustainable level, and net harvests of all three duikers, at 11 to 13 times sustainable levels.

ECONOMIC IMPORTANCE OF SNARE AND NET HUNTING

At US$1.75 per week, the average individual returns from net hunting are below the range of locally available formal employment earnings, which are from US$2.00 to US$13.00 per week. At US$9.30 per week, average snare hunter

Table 14-5. Age Structure of Snare and Net Captures (number/percentage of animals by age class)

Species	Snares					Nets				
	<1 Molar	<2 Molars	<3 Molars	Subadult	Adult	<1 Molar	<2 Molars	<3 Molars	Subadult	Adult
C. monticola	0	0	2 (6%)	4 (12%)	27 (82%)	1 (0.2%)	8 (2%)	28 (7%)	71 (12%)	303 (74%)
C. callipygus	0	0	0	4 (15%)	23 (85%)	0	4 (18%)	5 (23%)	3 (14%)	10 (45%)
C. dorsalis	1 (17%)	0	0	1 (17%)	4 (66%)	1 (3%)	2 (5%)	3 (8%)	5 (13%)	29 (72%)
A. africanus	—	—	—	1 (14%)	6 (86%)	—	—	—	18 (26%)	50 (74%)

Subadult and adult duikers all have complete dentition (three molars); animals were classified as subadults if their reproductive organs were not fully developed.

Table 14-6. Estimated Population Densities from Line Transects and Net Hunt Encounters (animals per km²)

	Line Transects				Net Hunt Encounters	
Species	N	Mean	SD	Density (No./km²)	N	Density (No./km²)
C. monticola	33	9.03	7.57	18.70	659	12.45
C. callipygus	4	18.25	9.44	1.12	56	1.06
C. dorsalis	1	20.00	—	0.26	80	1.51
A. africanus	0	—	—	0.00	163	3.08

N, number of animals seen; Mean, mean animal-to-observer distance; SD, standard deviation of the mean.
Densities are derived from 97.7 km of line transect censuses, and 61 net hunts within the Mossapoula net hunting range.

Table 14-7. Population Densities and Distance from Mossapoula (animals per km²)

	< 5 km (39 Hunts)		5–10 km (22 Hunts)		10–15 km (6 Hunts)	
Species	N	Density	N	Density	N	Density
C. monticola	324	10.7	335	14.8	61	20.4
C. callipygus	28	0.9	28	1.2	3	1.0
C. dorsalis	53	1.8	27	1.2	6	2.0
A. africanus	82	2.7	81	3.6	16	5.3

Path width is two times the net circle diameter. Path length is derived from the net circle diameter, the number of times the nets were set, and a 200-m distance between sets.
N, number of animals encountered (captures and escapes) by the entire net hunting group.
Population density (animals/km²) is based on the total area searched by the net hunting group.

Table 14-8. Sustainable Harvest Rates and Actual Snare and Net Harvests (animals/km²/yr)

						Actual Harvest	
Species	λ	Density	Prod	Avail	SH	Snares	Nets
C. monticola	1.34	12.45–18.70	4.23–6.36	0.2	0.85–1.27	3.3	14.6
C. callipygus	1.34	1.06–1.12	0.36–0.38	0.2	0.07–0.08	3.1	0.8
C. dorsalis	1.34	0.26–1.51	0.0–0.51	0.2	0.02–0.10	0.6	1.2
A. africanus	1.32	3.08		0.99	0.4	0.39	1.0

λ, finite rate of increase of the population in one year. Density is in animals per km²; Prod, annual production (animals per km²) given population densities and λ: Prod = (λ × Density) – Density; Avail, the proportion of annual population growth available to human hunters [20% for long-lived species (10 years or more) and 40% for short-lived species (5–10 years) (Robinson and Redford 1991b)]; SH, sustainable harvest in animals per km² per year, given Prod and Avail; actual harvest is in animals per km² per year.

returns are within this range. Snare hunting produces eightfold higher average returns per individual hunter than does net hunting, so snare hunting is clearly a more intensive hunting method, better suited to commercial meat production. However, net hunting produces much more than meat, including numerous plant foods and other materials for domestic consumption and use or for sale.

Perhaps because of the lower economic returns, there are no "professional" or full-time net hunters. Mossapoula residents spend on average 55, and at most 158, days per year net hunting. They devote their attention on other days to a wide range of activities, for example, formal labor, day labor, agriculture, gathering, or fishing. On the other hand, snare hunting is virtually a full-time activity and supports hunters in Bayanga at an income level roughly equivalent to what they could earn from more desirable formal employment. Market prices for game meat and returns to snare hunting are not high enough to induce immigration for hunting, but those who come to Bayanga seeking employment with the logging company, or who lose their jobs when the logging company temporarily closes, can sustain themselves and their families by snare hunting.

IMPACTS ON WILDLIFE

Net hunters encounter and kill a much narrower range of species than do snare hunters. The noise of an approaching net hunting group warns away many large animals, and net hunters only exploit dry upland forest habitats. In contrast, snares are set on trails that almost all animals use and are not selective with respect to species. Therefore, net hunters are much less likely to kill or injure protected species: one protected animal (the water chevrotain *Hyemoschus aquaticus*) killed in 589 captures, versus one killed (the black-fronted duiker *Cephalophus nigrifrons*) and two possibly injured (*Gorilla gorilla* and *Loxodonta africana*) by snares in 123 captures. Some injuries may be fatal, but even nonfatal injuries may reduce an animal's ability to find a mate or to raise young. Protected animals are usually rare, and even very low offtakes may have serious consequences for the population. With respect to total population impacts from harvesting pregnant females, actual pregnancy rates may have been considerably higher than recorded because uteri were not dissected and therefore only fetuses large enough to detect by touch were recorded. Young duikers might be less vulnerable to both hunting methods because they lie motionless when danger threatens and might not be seen by net hunters, and because lighter animals are less likely to trigger snares.

POPULATION DENSITIES

Methods for estimating population densities are unsatisfactory and may underestimate densities. Indeed, net harvests of *C. monticola* and snare harvests of *C. callipygus* exceed estimated population densities. The line transects were regularly visited by net hunters from Mossapoula, as well as by residents of both Bayanga and Mossapoula seeking other forest products. Therefore, human disturbance was high, resulting perhaps in low calculated densities. Transect surveys might also underestimate actual densities at this site if animals were wary and avoided detection by census takers. The nocturnal species (*C. dorsalis* and *A. africanus*) conceal themselves by day and seldom flush.

The net hunt encounter method generated a much larger sample size and should represent a more accurate census of the four primary game species: net hunters search an area more thoroughly than do line transect census takers, and the former are therefore more likely to locate all animals in the area. *A. africanus* might again be an exception because individuals hide in holes in the ground or in hollow trees and might not flush, although net hunters surround and search these potential hiding places to drive out concealed animals. The principal game species *C. monticola* is reluctant to leave its small home range and conceals itself in thick brush when threatened. Net hunters therefore surround and search potential hiding places. *C. dorsalis* is often found lying in its resting place and is reluctant to move even when approached by hunters, and most captured animals were speared rather than netted. However, net hunt encounters may underestimate *C. callipygus*. This species is active by day, with a large home range compared with the smaller *C. monticola* (Dubost 1980; Hart 1985; Feer 1989), and is more likely to move away from the noise of approaching net hunters.

Research elsewhere in central Africa reports the following ranges of population densities in hunted and unhunted areas: *A. africanus* 30 to 78 animals per km^2, *C. monticola* 15 to 78, *C. callipygus* 1 to 16, and *C. dorsalis* 2 to 9 (Dubost 1980; Feer 1989b, 1993; Hart 1985; Lahm 1993a; Payne 1992). These densities exceed those estimated in the Dzanga-Sangha region for all game species, and suggest that the methods used in the present study underestimate actual densities. In addition, despite continuous pressure levels by snare and net hunters for over two decades, all four primary game species continue to be captured in the area surrounding Bayanga and Mossapoula.

Some hunters admit that there are fewer animals near Bayanga or Mossapoula, and that they have to go further to hunt than they did a few years ago. However, because residential and occupational mobility are high in the region, many residents cannot make long-term evaluations of resource availability in a given area. Furthermore, hunting yields are always uncertain and variable. For example, snare hunters cite numerous and contradictory reasons for poor hunting success: bad luck, God's will, angry forest spirits, too much rain, too little rain, social discord, a hunter's wife's pregnancy, and overhunting. Unsuccessful net hunts are blamed on the participation of pregnant women, sexual activity the previous day by men, death in the community, or jealousy. Most hunters mock the suggestion that they might exterminate local wildlife, saying "there will always be animals." Others blame overexploitation near larger urban areas in the CAR on the excessive use of guns (Moussa 1992), but do not believe that snares or nets can exterminate wildlife.

SUSTAINABILITY OF HUNTING

Maximum population densities recorded in the literature for the four game species are as follows: 78/km² for *A. africanus* and *C. monticola*, 16/km² for *C. callipygus*, and 9/km² for *C. dorsalis*. The theoretically predicted (from Robinson and Redford 1991b) annual sustainable offtake rates per square kilometer are 7% for the three duiker species ([1.34 − 1] × 0.2, where λ = 1.34) and 13% for *A. africanus* ([1.32 − 1] × 0.4, where λ = 1.32). Using these optimistic population density figures, annual sustainable harvests could attain the following levels: *A. africanus* 10.1/km², *C. monticola* 5.5/km², *C. callipygus* 1.5/km², and *C. dorsalis* 0.7/km². Even using these density estimates, harvests of *C. callipygus* are unsustainable, as are net offtakes of *C. monticola* and *C. dorsalis*. These results suggest that models for calculating sustainable harvests may produce very conservative estimates. Nevertheless, given the available data and models, the likelihood is great that the primary game species—the smallest and most abundant duikers—are overexploited by both snares and nets.

TOWARD SUSTAINABLE HUNTING

Much of the Park and Reserve is exploited by both snare and net hunters, in addition to gun hunters. Therefore, total offtakes exceed harvests by any single method, and are almost certainly unsustainable for these species. The regional conservation program seeks to eliminate all hunting within the Park, and restrict hunting in the Reserve to nets and legally registered firearms. Eight guns were legally registered in 1994, and probably the same number of illegal guns were in use. However, the preceding analysis suggests that this approach might not achieve sustainable use of wildlife because current net hunting pressure is already unsustainable even using traditional methods such as nets for subsistence consumption.

Management of current hunting practices to reduce offtakes is problematic. On the cultural side, neither group of hunters exhibits a conservation ethic in the exploitation of wildlife. They are opportunistic predators rather than conservationists (Alvard 1994; Hames 1991; Lahm 1993a): they capture as many animals of any species, sex, and age as they can. Any meat that is not eaten can be sold, and vice versa. Snare hunters repeatedly say "if I had more snares, I would put them here. . . ." Net hunts that are very successful are never cut short when "enough" animals have been captured; instead, hunters often push on longer because tomorrow might not be so successful.

On the technical side, snares are generally set on trails that a wide variety of species use, and cannot selectively target particular prey species. Wastage and perhaps injured escapes could be reduced by more regular surveillance, checking the snare line at least every 2 days. Injuries could also be reduced if the cable nooses are tied so that they release when the cable breaks, rather than remaining tightly attached to the animal's leg (Mossman 1989).

On the institutional side, cable snares are illegal throughout the CAR, and for a conservation organization to work with snare hunters implies that it condones their activities. Conservation organizations currently lack sufficient resources to implement their relatively straightforward mandate of stopping snare hunting. Without the cooperation of local hunters, a more complex program of regulated snare hunting requires even more resources. However, Bayanga residents strongly resent conservation measures that prohibit them from "feeding" themselves, and manifested their anger in a 3-day riot in 1993. Snare hunters continue to exploit the entire area of the Dzanga-Sangha Special Reserve and the Dzanga-Ndoki National Park.

A compromise solution might involve designating an area where cable snare hunting is legal, for example, west of the Sangha river, and encouraging hunters to operate there rather than inside the Park. Another alternative is to facilitate gun hunting: gun hunters can be more easily monitored, are less wasteful (few animals escape with injuries), and can be more selective in their prey selection than are snare hunters. Although hunting with registered firearms is permitted in the Dzanga-Sangha Special Reserve, weapons, ammunition, and the necessary permits are too expensive for most local residents. Firearms also can be much more productive than snares, and a switch to guns may intensify hunting pressure even if the number of hunters declines. When ammunition is available, Bayanga gun hunters kill several animals each day and night compared with the snare average of two animals per week. Gun hunting also may increase pressure on species previously underexploited by snares, particularly primates and possibly elephants.

As a hunting method, nets are less destructive than snares and offer a greater potential for managed hunting. Unlike snare hunters, net hunters can target particular species or even particular individuals. All animals are seen and identified before they are killed, and many of the duikers are captured in nets alive. In practice, no animal that could be caught is ever ignored, and no animal that is caught is ever released. Nevertheless, net hunters could manage net hunting offtakes: they can spare protected species, as well as females and young individuals of game species. Fully protected species are seldom killed anyway (one in over 500 captures), and this restriction would not affect hunting yields. Releasing young animals that lack complete adult dentition would also not greatly reduce yields, except in the case of *C. callipygus* (table 14-9).

However, releasing female animals would greatly reduce yields of all four game species, particularly in the case of *C. dorsalis*. Releasing only pregnant and lactating females would still greatly reduce yields for *C. dorsalis* and *C. monticola*. Enforcement of a selective hunting system would be extremely difficult, and Mossapoula net hunters would adopt such management practices that reduce yields only under duress. Restrictions on net hunting might only accelerate the rate of adoption of more productive and surreptitious cable snares. Net hunting, because it involves groups of people hunting together and makes considerable noise, is much more difficult to conceal from game guards than are snare hunting activities.

Table 14-9. Effects on Net Hunting Yields of Releasing Young, Female, or Pregnant/Lactating Animals (percentage reduction in number of animals and kilograms)

Species	Young N	Young kg	Female N	Female kg	Pregnant/Lactating N	Pregnant/Lactating kg
C. monticola	9	4	56	59	15	19
A. africanus	0	0	34	33	—	—
C. dorsalis	15	7	70	67	15	20
C. callipygus	41	26	54	51	4	7

Hunting pressure from Mossapoula is excessive because hunting is continuous surrounding a relatively large, permanent settlement. Harvests could be reduced by mandated limits on the number of hunts per week. A more cooperative alternative would be to encourage seasonal population shifts to forest hunting camps in order to extend the exploitation area and therefore reduce hunting pressure in the area immediately surrounding Mossapoula. The 400-km² Dzanga sector of the National Park since 1990 has officially prohibited net hunting more than 10 km to the east of Mossapoula, alienating the Mossapoula BaAka from their entire seasonal forest camping range. In the past, the BaAka moved for several months each year from settlements near their farming neighbors to temporary forest camps, moving every several weeks to a new location and concentrating on hunting and honey collection. Despite the hunting ban, some forest camping continues. Annual harvest rates from forest hunting camps in 1994 (extrapolating for 20 weeks of forest camping from 13 observation days) were as follows: *A. africanus* 0.7 animals/km², *C. monticola* 2.4, *C. dorsalis* 0.5, and *C. callipygus* 0.2. Although these rates also exceed sustainable harvest estimates in table 14-8, they are much lower than the harvest rates in the Mossapoula day hunting range. BaAka forest hunting camps are easy to monitor and do not threaten protected wildlife species. Forest camps and net hunting could be permitted even within the Dzanga-Ndoki National Park to the east of Mossapoula.

In the long term, conservation of wildlife in the Bayanga region cannot depend on sustainable exploitation by local hunters. Available data on hunting offtakes, population densities of game species, and theoretically sustainable harvests suggest that subsistence hunting using traditional methods is unsustainable for the Mossapoula BaAka. Mossapoula currently has 370 residents and a hunting range reduced to 110 km² by the creation of a national park and by competition with hunters from other nearby settlements such as Bayanga (2500 residents). The Mossapoula system today, with 2.7 persons/km², depends on only one-fifth the hunting territory of 1.5 to 4 km²/person exploited traditionally by Pygmy groups in the CAR and Zaire (Bahuchet 1979, 1985; Ichikawa 1978; Tanno 1976). Cable snare hunting is also likely to be unsustainable in terms of offtakes, injuries inflicted, and threats to protected or endangered species.

Current patterns of wildlife exploitation can only be expected to intensify with technology shifts from nets to snares and from snares to guns, human population growth, habitat conversion, rising economic expectations and improved transportation links to larger urban centers. In order to conserve wildlife in central African forests, methods to manage widely used hunting methods such as cable snares and nets are urgently needed. Management cannot depend solely on laws that prohibit snares or that delimit protected areas, but require cooperation with local peoples that continue to depend on hunting for food and income. In the short term, conservationists must collaborate with local hunters to manage wildlife and to identify hunting areas, methods, and practices that are less detrimental to wildlife populations. In the long term, only the development of alternative food and income sources will reduce dependence on hunting.

ACKNOWLEDGMENTS

This research was supported in part by a grant from the Joint Committee on African Studies of the Social Sciences Research Council and the American Council of Learned Societies with funds provided by the Ford, Mellon, and Rockefeller Foundations. Additional funds were provided by two Grants-in-Aid of Research from Sigma Xi, The Scientific Research Society; and by the World Wildlife Fund under the United States Department of Agriculture Agreement No. 93-G-155.

15

Saving Borneo's Bacon: The Sustainability of Hunting in Sarawak and Sabah

ELIZABETH L. BENNETT, ADRIAN J. NYAOI, AND JEPHTE SOMPUD

Humans have lived in Borneo for at least 40,000 years (Zuraina 1982) and have been hunting large mammals such as bearded pigs and orangutans for at least 35,000 years (Medway 1959; Hooijer 1960). Hunting is integral to many of Borneo's cultures. Wildlife artefacts such as hornbill feathers and clouded leopard skins are worn as personal adornment, especially for ceremonies (Bennett et al., 1997). Young men sometimes hunt to achieve status to be considered worthy to marry, and hunting is the basis of many cultural beliefs and rituals (e.g., Brosius 1986; Caldecott 1986a; Chua 1996). Traditionally, wildlife also has been one of the main sources of protein for rural peoples (Chin 1985; Strickland 1986; Caldecott 1988). This chapter examines whether hunting in northern Borneo is sustainable today. In the context of a rapidly developing economy, sustainability is critical both to ensure the physical and cultural health of the rural people and the continued survival of much of Borneo's fauna.

SARAWAK AND SABAH

Sarawak and Sabah are the two Malaysian states in northern Borneo (figure 15-1). They lie close to the Equator (from 0°50'N to 7°50'N), and the climate is warm and wet all year. Rain falls in all months, with a pronounced peak during the northeast monsoon season from November to February. Average rainfall in coastal areas is about 3000 mm, but reaches 6000 mm in parts of the interior (WWF Malaysia 1985). The natural vegetation is tropical rain forest of various types,

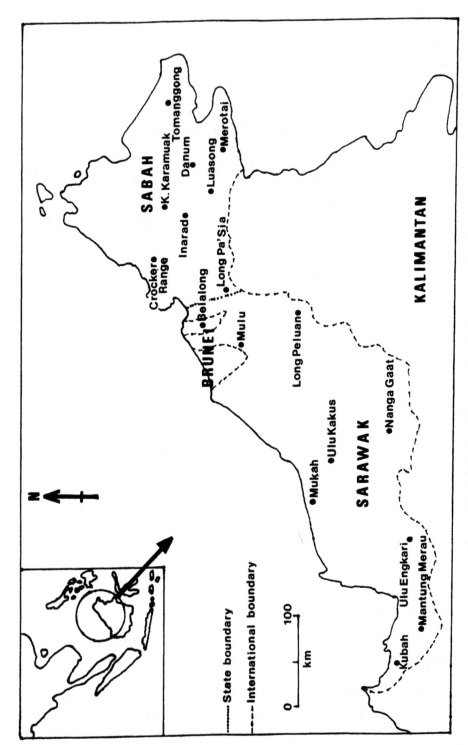

Figure 15-1. Northern Borneo, showing locations of study areas.

from mangrove and peat swamp near the coast, to mixed dipterocarp and tropical heath forests, to montane forests in parts of the interior (Whitmore 1984, 1995). Plant species diversity in the lowland forests is among the highest in the world (Whitmore 1984, 1995; Waterman et al. 1988).

Cultural diversity is also high. Traditionally, the towns mainly are composed of ethnic Chinese and Moslem Malays. Rural areas are inhabited by a wide variety of peoples. The largest groups in Sarawak and Sabah are the Ibans and Dusun-Kadazans, respectively, but between the two states, there are about 60 different ethnic groups, each with its own language, culture, and traditions. In Sarawak, almost all indigenous people in rural areas live in longhouses, which comprise a whole village under one roof. In Sabah, most people, even in the interior, live in individual family dwellings. In recent years many rural people have also moved to towns, especially those with higher education, so they are now well represented in the professions and government.

In both states, almost all rural communities practice shifting cultivation to grow hill rice, and they hunt and fish for their protein. The only significant exceptions are the Penan, who traditionally obtained their carbohydrate from wild sago (Brosius 1990), although many also now grow rice (Bugo 1995).

Sarawak and Sabah are part of the *tiger economy* of Malaysia, and are developing rapidly: from 1980 to 1990, Sarawak's economy grew by about 6.6% per year (Bugo 1995). Between improved river transport, rural flights, and logging and other roads, almost all areas of both states are now easily accessible. Everyone is now part of a cash economy to a greater or lesser extent. Thus, rural life-styles are changing rapidly. One sign of this is the spread of technology, such as outboard engines, chainsaws, electricity, and televisions. Hunting methods also have changed. Traditionally, hunting was done using traps, dogs and spears, and blow-pipes. Hunters in remote parts of the interior still use these methods, but most animals now die by gunfire. In Sarawak, there are about 60,000 legally registered shotguns (Caldecott 1988), and in Sabah almost 13,000 (Sabah Statistics Department information). Thus, modern technologies and economies are now blending with a traditional hunting life-style.

From February 1993 to June 1995, we conducted a study of hunting in Sarawak and Sabah. The aim was to investigate the current importance of wildlife and its products to rural people, especially in their diets, and also to investigate if present rates of hunting are sustainable. This chapter presents the results of the study, and derives policy recommendations that would allow us both to conserve wildlife and to sustain the diets of rural people.

METHODS

Sixteen study sites were selected, eight each in Sarawak and Sabah. Different habitats, ethnic groups, degrees of access and types of hunting were sampled (table 15-1), as were the different geographical regions of each state (see figure 15-1). An additional survey was conducted in Kuala Belalong, Brunei Darussalam. Shotguns

Table 15-1. Study Sites, Showing the Main Ethnic Groups Who Hunt There, Habitat, and Type of Hunting

Site	Ethnic Group	Habitat	Hunting Type
Kubah NP	Mixed town	Primary forest	Sport, some subsistence
Mantung Merau	Bidayuh	Old temuda	Subsistence
Ulu Engkari	Iban	Mixed age temuda	Subsistence, some trade
Nanga Gaat	Iban	Newly logged forest	Subsistence by loggers
Mukah	Iban and Melanau	Logged peat swamp	Subsistence
Ulu Kakus	Punan Ba'	Logged peat swamp and temuda	Subsistence
Mulu NP	Penan and Berawan	Primary forest	Subsistence, some trade
Long Peluan	Kelabit and Penan	Primary forest	Subsistence
Belalong	Iban	Primary forest	Some subsistence
Long Pa' Sia	Lun Dayah	Primary and logged forest	Subsistence by locals and loggers
Crocker Range NP	Kadazan and mixed town	Primary forest	Sport, subsistence, trade
Inarad	Murut	Primary, logged and burnt forest	Subsistence
Kuala Karamuak	Orang Sungai	Primary forest and temuda	Subsistence by locals, some loggers
Luasong	Immigrant and mixed town	Old logged and burnt forest, rattan plantation	Subsistence, sport, crop pest
Merotai	Immigrant and mixed town	Cocoa, oil palm, primary forest	Subsistence, sport, trade, crop pest
Tomanggong	Immigrant and mixed town	Logged forest, oil palm	Subsistence, sport, crop pest
Danum Valley	None	Primary forest	Negligible

Temuda, regenerating shifting cultivation land; NP, national park.

are illegal in Brunei Darussalam, so this survey was conducted to examine what difference the ban made to hunting patterns.

Each study site was visited two or three times, at different times of year, thereby sampling hunting patterns and animal abundance in different seasons. Through these repeated visits, people came to know and trust us, thereby increasing data accuracy on subsequent visits. Each visit lasted approximately 3 weeks. The first nine nights (eight full days) were spent living in the community, collecting the following data:

1. *General hunting interviews—to obtain an overall picture of hunting in the community, its importance to the hunters, and likely broad impact on the wildlife.* In longhouses and villages, all hunters were interviewed and results recorded on a standard form. In plantation communities of several hundred people, houses were selected at random for interviews. Questions were on the frequency of hunting, methods used, types of animals normally hunted, if hunters selected their prey according to size or sex, hunting success rate (e.g., out of every four hunting trips, how many were successful; out of every ten cartridges fired, how often would an animal be killed), reason for hunting (for subsistence, trade or killing crop pests), and if the

hunter ever sold the meat or other animal parts. A total of 391 hunters was interviewed over the course of the study. Seeking out hunters also allowed the proportion of men in the community who hunted to be determined.

2. *Interviews about individual hunts—to obtain a detailed picture of the hunting offtake (number of animals of which species per community, per hunter, and per unit area of forest) and of hunting effort (number and weight of animals killed per man hour).* All hunting trips while we were present lasted for less than 24 hours. Every time a hunter returned from a hunting trip while we were in the community, he was interviewed, and results were recorded on a standard form. This included both successful and unsuccessful hunts. Data collected included time of start and end of hunt, number of hunters participating, number of dogs, guns, spears, blowpipes and traps used, number and species of animals hunted, number of shots fired (thereby allowing us subsequently to calculate the number of cartridges used per animal killed or wounded), reason for going out (e.g., farming, hunting), use of the hunted animal, and location of hunt. A total of 345 hunting trip records were obtained.

3. *Measurements of hunted animals—to obtain as much information as possible on the importance of different species of wildlife to the community, as well as likely effects of the offtake on wildlife populations (e.g., if the hunt was strongly sex or age biased).* Any hunted animals, or parts of them, brought back to the community were measured. As many as possible of the following measurements were made: total weight, head length, total length, tail length, ear length, and hind foot length. Sex and approximate age class also were recorded.

4. *Records of diets—to assess the importance of wild meat in the diet and to cross-check with the number and species of animals being hunted in the community.* Visiting every family each day to ask about meals meant that all animals brought back to the community were detected. Each family in each community was asked what they were eating every evening while we were there, and the results were recorded on a standardized form. The composition of 4566 evening meals was recorded.

The second half of each visit to each study area was spent camping in the forest where people hunted, conducting wildlife surveys. The survey sites were determined from interviewing the hunters about where they hunted most often. Sites were usually within 1 hour by boat or on foot from the community. Two main types of survey were conducted:

1. *Line transect surveys (after Burnham et al. 1980; Marsh and Wilson 1981; Brockelman and Ali 1987; Buckland et al. 1993)—to assess the absolute abundance of readily visible large birds and mammals, i.e., primates, squirrels, ungulates, hornbills and pheasants.* In each study area, we cut one or two transects, depending on the number of observers. Transects started a minimum of 250 m from the camp, more if possible, so they were not unduly affected by noise or smoke. Transects were as straight and randomly oriented as possible. Survey walks started within 30 minutes of dawn, survey speed was approximately 500 m/hr, and a minimum of eight walks or 16 km of transect was surveyed for each visit to each site. Data were analyzed by Fourier

series analysis, using the program Transect. A total 682.1 km of survey trail was walked in the mornings. Whenever possible, 1 km of the same trail was also surveyed at night for nocturnal mammals. This was often not feasible because of heavy rain, or the known presence of hunters, which meant that the risks of being shot were high. (Night hunting accidents are common throughout Sarawak and Sabah.)

2. *Index counts for tracks and signs—to obtain measures of the relative abundance of species which are rare or shy, notably ungulates and carnivores.* Index counts were made along the transect survey trail. All tracks and signs within 2 m of the center of the trail were noted, measured, and described. Index counts were recorded along a total of 168 km of survey trail.

To assess the impact of hunting on wildlife populations, an index of hunting pressure is needed. For areas where hunting was performed by the local communities, the amount of hunting was directly sampled as noted above. In other areas, hunters from towns also hunted in the forest. Because much of their hunting was illegal, it was not possible to obtain accurate information from interviews, so their impact had to be estimated indirectly. An index of hunting pressure was devised for all areas. The number of hunters who have access to the area was estimated and divided by the square of the distance that they live from the forest (thereby recognizing that people spread out from a community across an area, not in a straight line). This assumes that hunting pressure is distributed uniformly in all directions from a community, and declines evenly with distance. For rural subsistence hunters, this is a reasonable assumption (e.g., Marsh and Gait 1988). Town hunters are more inclined to go to specific locations, but even so, the pressure on sites closer to town is far greater than on those farther away. Thus, although only an approximation, these assumptions are still valid.

Thus, hunting pressure on an area from any one community (H_A) is:

$$\frac{(\text{Population of community A}) \ (\% \text{ of community A who hunt})}{(\text{Distance of community A from study site})^2}$$

If more than one community hunts in an area, then the total hunting pressure (H) is the sum of hunting pressures ($H_A + H_B + H_C \ldots$ etc.) for all the individual communities who hunt there.

To use this index, it is necessary to know the population of each community and the proportion who hunt. For most study sites, this was obtained from interview data. For areas with outside hunters from towns, interviews were conducted by telephone. These were done by taking the first name in each column of the telephone directory, thus making it a regular permutation, and asking if the person concerned or anybody else in the household ever went hunting. Ideally, the frequency with which people hunt also should be considered, but because the accuracy of such data from telephone interviews would have been extremely doubtful, it was not included.

ROLE OF WILD MEAT IN RURAL DIETS AND ECONOMIES

In Sarawak, 28.9% of all evening meals sampled in rural communities contained wild meat, 51.5% contained wild fish, and 25.1% contained other forms of animal protein (e.g., tinned food, domestic meat, purchased dried fish). In Sabah, 13.9% of all evening meals sampled contained wild meat, 21.3% wild fish, and 58.5% other forms of animal protein. Meals were often diverse; wild meat was frequently eaten in the same meal as other forms of protein, mostly fish, and also with vegetables.

The proportion of wild meat in the diet varied greatly between sites, depending on their remoteness, habitat type, and participation in a cash economy (table 15-2). In relatively remote, traditional areas in the interior, wild meat and/or wild fish were present in most meals. This was epitomized in the highlands with little outside access and large remaining areas of tall forest. In Long Peluan, for example, 67.1% of all evening meals contained wild meat, and that increased to more than 90% in some months. By contrast, in coastal areas, wild meat was less important than marine fish. In accessible areas with alternative sources of cash, wild meat was also less important, sometimes because people had access to alternative sources of protein (such as commercially bought meat and fish), and sometimes because certain coastal habitats (such as coastal swamp forest and land under current or recent cultivation) did not support much wildlife.

At the sites where logging workers hunted, hunting was heavy and wild meat was their main source of animal protein. Loggers work in tall forest with good wildlife populations, and animals disturbed by the logging are relatively easy to hunt. The logging camp sampled was remote, so alternative fresh protein was scarce. In addition, almost all of the workers were Ibans from longhouses in nearby areas, so they were traditional hunters by background.

By contrast, workers in cocoa and oil palm plantations ate very little wild meat

Table 15-2. Proportion of Meals Containing Wild Meat, Wild Fish, and Other Forms of Animal Protein Eaten in Different Types of Community in Sarawak and Sabah

Community	% of Meals Containing Wild Meat	% of Meals Containing Wild Fish	% of Meals Containing Other Animal Protein	No. of Meals	Study Sites
Traditional interior hunter-cultivators	36.9	47.2	12.7	955	Engkari, Long Peluan, Inarad, Long Pa' Sia
Traditional coastal fishermen-cultivators	8.5	69.7	30.7	531	Mukah
Semi-traditional longhouses with easy access and other sources of cash income	15.0	57.6	35.5	656	M'tung Merau, Ulu Kakus
Logging camp workers	49.1	18.5	29.1	340	Nanga Gaat
Plantation workers	4.1	5.7	90.2	788	Luasong, Merotai, Tomanggong

and fish, despite the fact that all three plantations sampled were adjacent to large blocks of forest with good wildlife populations. There were two main reasons for this. First, plantation workers do not have much spare time or energy for hunting. Unlike the loggers, their work was not in the forest itself, so hunting required an extra journey from their work site. Second, plantation workers in all sites sampled were almost all immigrants who did not have guns and who were strongly discouraged from hunting by the plantation managers. This was largely for reasons of security and safety. With little job or home security, they generally obeyed the rules.

The dependence on wild meat throughout the interior is largely independent of ethnic group or religion. There are certain customary controls for harvesting plant products, especially among the more traditional people such as the Penan (Langub 1988), but very few for animals. Certain taboos are specific to particular families, longhouses or areas. For instance, Ibans in the Batang Ai area will not hunt orangutans, and Muruts in Inarad will not hunt hornbills. Such taboos are local and very specific, and in general, everyone will hunt and eat anything. Thus, data collected during this study showed that a group of Penan killed all of the Bornean gibbons (*Hylobates muelleri*) in a primary forest valley over a 6-month period. The trend to hunt even small species cuts across most groups of people, from town dwellers who use mist nets and catapults to catch garden birds, to the remote Kelabits and Penan who use blowpipes to hunt small squirrels and birds (figure 15-2). This also applies to whether a species is protected by law or not, or even whether the animal lives in a protected area. With a shortage of wildlife staff, and rural communities where traditional laws on animals and land are more important than government regulations, the legal status of particular species and forests is largely irrelevant to hunters.

By far the majority of the hunting was done using guns: 88% and 58.3% of all hunting trips involved guns in Sarawak and Sabah, respectively (figure 15-3). In Sarawak in 1993, 3.27 million cartridges were imported into the state (Sarawak Statistics Department Information). In the Sarawak study areas, 79.5% of all cartridges fired killed an animal. (This is similar to the figure of 75.2% recorded in 1984–1986; Caldecott 1988). In general, cartridges are treasured and few lost. The warm, moist climate means that they do not last long, so they are rarely stockpiled. Thus, if the study areas are representative of the state as a whole, then 2.6 million animals died by gunfire in Sarawak in 1993. In Sabah in 1993, 2.17 million shotgun cartridges were imported (Sabah Statistics Department information). In the Sabah study areas, 84.9% of all cartridges fired killed an animal, indicating that 1.8 million animals were shot in Sabah that year.

By far the most preferred species of wild meat throughout Sarawak and Sabah is bearded pig (*Sus barbatus*). Both the meat and fat are highly favored by people, and the animals are large so that there is plenty of food per animal hunted. Seventy-two percent of the dressed weight of animals hunted in Sarawak comprised bearded pig, and a further 10.9% deer (figure 15-4). Sambar deer are rare in Sarawakian diets, largely because the animals are so rare in the forest. This is probably due to a combination of poor soils in Sarawak compared with Sabah, and also to past hunting pressure. In Sabah, 53.7% of the dressed weight of

Figure 15-2. Penan hunter with traditional blowpipe. *Photo courtesy of Elizabeth L. Bennett.*

animals hunted comprised bearded pig, and a further 42.9% deer. Thus, ungulates comprised by far the majority of wild meat in the diet, being 82.8% and 96.6% of the dressed weight of wild meat hunted in Sarawak and Sabah, respectively.

This does not reflect accurately the number of animals hunted, however. Many animals are hunted that because of their small size, only comprise a minor proportion of the diet. Thus, ungulates only comprised 29.7% and 71.2% of animals hunted in Sarawak and Sabah, respectively (figure 15-5).

Apart from workers in plantations, people of all ethnic groups and life-styles apparently hunt animals because they are there. Thus, there is a positive correlation between the proportion of meals that contain wild meat and the abundance of bearded pigs in the forest (r_s = 0.8286; n = 6; p = 0.042). Hence, the abundance of wild meat in the diet does not reflect the lack of alternative sources of food and income so much as the fact that people will eat wild meat when it is there, and other forms of protein when it is not. This is supported by case examples. In Ulu

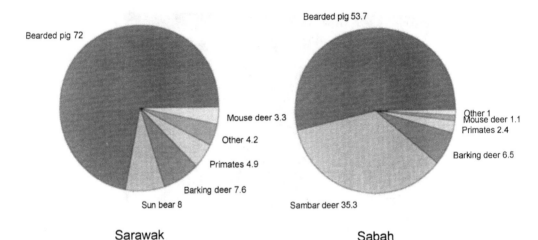

Figure 15-3. Species percentage of total weight of animals hunted: (a) by six longhouses in Sarawak (total number of animals killed = 185); and (b) by three hunting-farming communities in Sabah (total number of animals killed = 149).

Engkari, the study longhouse had fish ponds that produced large amounts of fresh fish. This made little difference to the diet of the longhouse residents, however; they sold all of the fish for cash, and hunted for their own protein. So although there were alternative sources of cash and protein, they were not used to replace wild meat. In Ulu Kakus, the longhouse was accessible, had an extremely well-stocked, large shop, and people were wealthy by local standards due to income from collecting swiftlets' nests from caves to which they had rights; their average

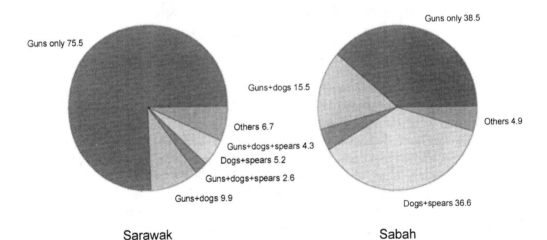

Figure 15-4. Hunting methods employed in (a) Sarawak and (b) Sabah. The figure shows the proportion of hunting trips using particular hunting methods. Sarawak, n = 192; Sabah, n = 161.

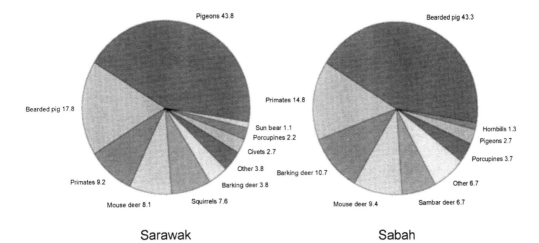

Figure 15-5. Species percentage of total number of animals hunted: (a) by six longhouses in Sarawak (total number of animals killed = 185); and (b) by three hunting-farming communities in Sabah (total number of animals killed = 149).

annual income from the nests was probably about $13,000 per family. Nevertheless, they still hunted, and wild meat was still a major source of protein.

The economic value of the wild meat consumed in rural areas is very high. In an Iban longhouse at Ulu Engkari, Sarawak (figure 15-6), the members of each family unit consume about 313 kg of wild meat per year. With approximately six people per family, that is about 1 kg per person per week. The cost of replacing that amount of meat for the whole 26 family longhouse with domestic pork at 1995 Kuching town prices would be about $26,082 per year, or about $1000 per family per year (table 15-3). In remote parts of the highlands, wild meat is more important, so at Long Peluan, each Kelabit family consumes about 396 kg of wild meat per year, and the replacement value of wild meat for the whole 17-family longhouse would be about $21,523 per year. Higher rates of consumption occur in some logging camps. In Nanga Gaat, for example, about half of the workers had shotguns, and people hunted both during the day at the logging face, and at night along the roads. The transit camp of 167 workers and their families (a total of about 500 people) consumed wild meat with a replacement value of about $93,044 per year. Consumption of wild meat varies among logging camps, depending on the type of staff (local or immigrant workers), accessibility of the forest to markets, and company policy regarding provision of other sources of fresh protein (Bennett and Gumal in press).

People in towns and coastal areas eat less wild meat than do people in the interior. Even assuming that as little as one-fourth of the population of Sarawak depends significantly on wild meat, that is about 75,000 families. If each family consumes wild meat worth about $1000 per year, that is a total of $75 million for the whole state. If these people could not hunt, many would farm their own meat rather than buy it. These figures, however, show the approximate scale by which wild meat supplements rural economies.

Figure 15-6. Iban hunters with shotgun and pig-tailed macaques (*Macaca nemestrina*), shot while raiding a swidden rice field. *Photo courtesy of Elizabeth L. Bennett.*

IMPACT OF HUNTING ON WILDLIFE POPULATIONS

The density of animals at carrying capacity (K) varies greatly between forests: logged versus unlogged, peat swamp versus hill forest, primary versus old regenerated shifting cultivation land. Nevertheless, there is a clear trend that cuts across habitats: as hunting pressure increases, the density of animals decreases. This applies to the absolute densities of primates and hornbills (figures 15-7 and 15-8), and index counts of the tracks and signs of sambar deer (*Cervus unicolor*), barking deer (*Muntiacus* spp.), and bearded pigs (figures 15-9, 15-10, and 15-11). For all except sambar deer, the correlation between animal density and hunting pressure (H) is significant (table 15-4). The pattern is so strong that hunting pressure overrides the large effects of habitat variables and is the single main determinant of

Table 15-3. Dressed Weight of Wild Meat Consumed per Year in Three Communities in Sarawak, and the Cost of Replacing It with Domestic Meat

Community	Dressed Weight of Wild Meat Consumed per Year (kg)	Replacement Cost (US$)
Ulu Engkari: 26 family Iban longhouse	8151	26,082
Long Peluan: 17 family Kelabit longhouse	6726	21,523
Nanga Gaat: logging camp of 167 workers and families	29,076	93,044

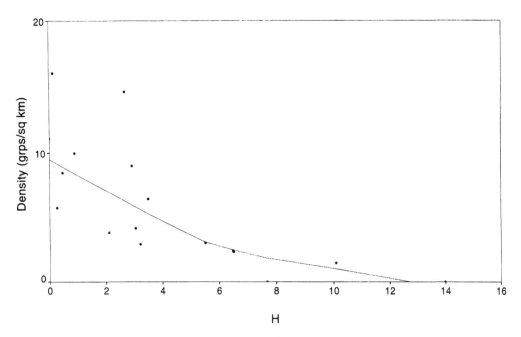

Figure 15-7. Primate density (groups per km^2) against hunting pressure index (H) in study sites throughout Sarawak and Sabah. Data from plantation monocultures have been excluded.

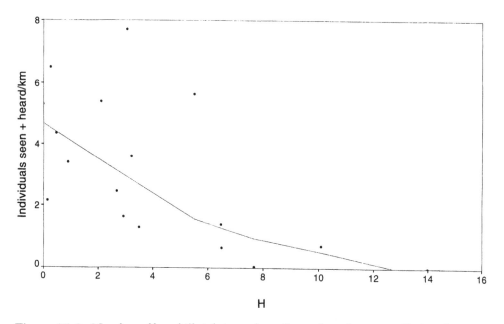

Figure 15-8. Number of hornbill sightings plus calls per km of survey walked against hunting pressure index (H) in study sites throughout Sarawak and Sabah. Data from plantation monocultures have been excluded.

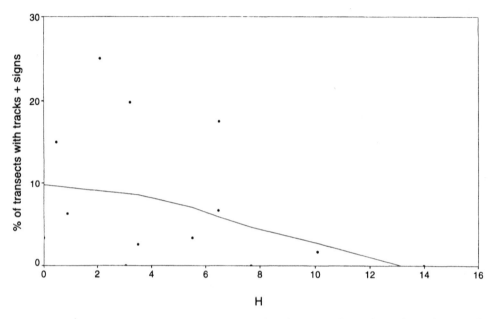

Figure 15-9. Percentage of 100-m stretches of trail (4 m wide) with tracks and signs of sambar deer against hunting pressure index (H) in study sites throughout Sarawak and Sabah. Data from plantation monocultures have been excluded.

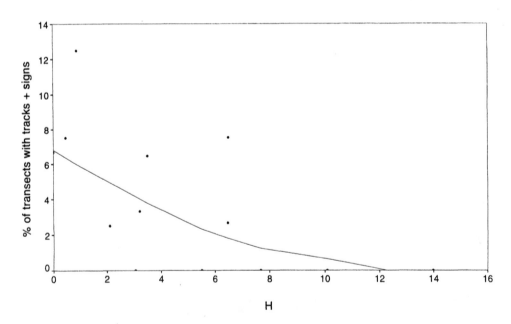

Figure 15-10. Percentage of 100-m stretches of trail (4 m wide) with tracks and signs of barking deer against hunting pressure index (H) in study sites throughout Sarawak and Sabah. Data from plantation monocultures have been excluded.

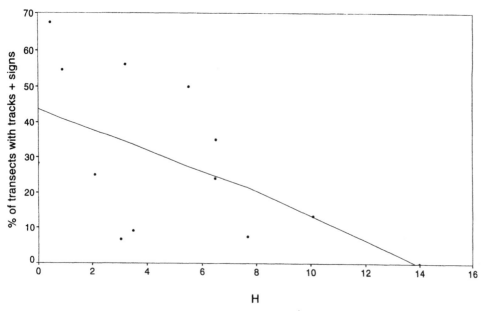

Figure 15-11. Percentage of 100-m stretches of trail (4 m wide) with tracks and signs of bearded pigs against hunting pressure index (*H*) in study sites throughout Sarawak and Sabah. Data from plantation monocultures have been excluded.

animal densities in an area. As hunting pressure increases, the density of all these animal groups decreases until, at high hunting pressures, all large animals are locally extirpated.

The fact that hunting is reducing densities of animals to below 50% of *K* in some areas, and even extirpating species in others, implies that current hunting levels are depressing wildlife populations to below levels of maximum productivity throughout much of Sarawak and Sabah. The ranges of the banteng (*Bos javanicus*), Sumatran rhinoceros (*Dicerorhinus sumatrensis*), and orangutan (*Pongo pygmaeus*) have all contracted dramatically in the past hundred years, due in large part to hunting (Medway 1977; Rabinowitz 1995; MacKinnon 1992). The increasing ease

Table 15-4. Spearman Rank Correlation Analyses of Population Densities of Large Animals Against Hunting Pressure

Species/Species Group	r_s	No. of Sites	p
Primates: groups/km²	−0.8657	17	<0.001
Hornbills: sightings and calls/km of survey walked	−0.6618	17	0.004
Sambar deer: % of 100 m × 4 m transects with tracks and signs	−0.4343	13	0.138
Barking deer: % of 100 m × 4 m transects with tracks and signs	−0.5943	13	0.032
Bearded pig: % of 100 m × 4 m transects with tracks and signs	−0.5604	13	0.046

Data from plantation monocultures have been excluded from analyses.

of access to all areas means that almost no forests remain where animals are free from hunting, including in totally protected areas. This is exemplified by the fact that, for this study, only one site (Danum Valley) could be found as a control in all of northern Borneo, where there was primary forest with negligible hunting pressure. Even in Belalong, Brunei Darussalam, where shotguns are illegal, the recent advent of hunting is affecting populations; primate densities declined by about 70% between 1992 (Bennett 1992) and the current study. Thus, throughout northern Borneo, local wildlife populations are being depressed by hunting. Local populations are becoming depleted or extinct. This applies equally to protected and unprotected species and to totally protected and unprotected areas; the only single factor offering any effective protection for wildlife is difficulty of access.

WHY IS HUNTING NOT SUSTAINABLE?

Humans have been hunting in Borneo for at least 35,000 years, yet many species still remain. There are five main reasons why hunting is a much greater threat now than it was in the past.

1. The change from traditional to modern technology. Most hunters now use shotguns. Until the mid-1980s, the number of cartridges issued was strictly controlled for security reasons, but restrictions have since been relaxed, and the total number of cartridges imported into Sarawak and Sabah per year now exceeds 5 million. This means that it is easy for less skilled people to hunt, e.g., town sports and trade hunters. For rural hunters, the easy availability of cartridges increases the chance that they will hunt extra animals to sell for cash, in addition to those for subsistence.

2. Increasing populations in some rural areas. Sarawak's population grew by about 2.6% per year throughout the 1980s. This varied considerably with area, with numbers in some rural areas decreasing due to urban migration, but in others, improved health meant that rural populations increased substantially.

3. A change from a mobile to a sedentary life. More than 50 years ago, rural communities were highly mobile, whole longhouses moving when resources became scarce (e.g., Freeman 1970; Geddes 1976). Now, they rarely move. This is due to many changes, the single greatest one in Sarawak being the Land Code of 1958, which does not recognize any new native customary claims to land after that date. This means that people cannot move to new areas of land when resources become depleted, thereby allowing areas to recover.

4. Increased access to all areas. Until about 1960, large parts of Sarawak and Sabah were inaccessible to all but occasional hunters trekking for many days on foot. Since then, the spread of logging and other roads, improved river transport, and rural air services mean that nowhere is now more than about a day's travel from the nearest settlement, and most of both states is now readily accessible from towns. This brings many changes. Some of the main ones relevant to hunting are that (a) it is easy for rural hunters to bring in cartridges; (b) it is easy to bring in

batteries for night hunting, thereby increasing hunting pressure on nocturnal species; and (c) rural communities have ready access to markets, so wild meat is sold outside the area. In addition, increased access to forests means that town hunters in four-wheel–drive vehicles or speedboats can reach most areas, and hunt for sport or trade.

5. Increasing participation in a cash economy. Until the start of the twentieth century, many rural communities were almost entirely self-sufficient and, apart from occasional barter trade, were largely outside a cash economy. This is no longer the case, and all communities are now at least partly dependent on cash. Like other people in the world, rural people in northern Borneo want to buy commodities such as radios, televisions, generators, and outboard motors. For people entering a cash economy, there is often no easy way to obtain money, and one of the easiest (sometimes the only) option is to sell forest products, including wild meat. Wild meat is sold widely in markets and restaurants in all villages and towns throughout the whole of northern Borneo, and the wild meat trade is worth an estimated US$3.75 million per year (Wildlife Conservation Society and Sarawak Forest Department 1996). Thus, the forest is no longer just supplying food to the people living nearby, but to a much larger market.

POSSIBLE SOLUTIONS

Hunting as practiced today in Sarawak and Sabah is not sustainable. If continued, this will cause local or statewide extinctions of many species of animals. Also, the forests will not be able to continue to supply the protein needs of rural people. Good planning is needed to ensure that wildlife populations are protected and that rural communities have a continuing access to animal protein. Sarawak and Sabah have separate laws and administrations on all matters pertaining to wildlife and forests, so solutions must be specific to each state.

For areas where wildlife conservation is the priority, everything possible must be done to minimize hunting, including that performed by local people. This is both to ensure the survival of the full biota and to allow reserves that would potentially act as reservoirs, supplying wild meat to rural hunters outside their boundaries. Minimizing hunting is difficult in view of the strong traditions of hunting and the major role of wild meat in rural diets. In addition, named communities have legally gazetted rights to hunt in certain national parks and wildlife sanctuaries in Sarawak.

Sarawak has recently prepared a comprehensive Master Plan for Wildlife in Sarawak (Wildlife Conservation Society and Sarawak Forest Department 1996), and all of the below measures are contained therein. The recommendations of the Master Plan have been adopted as official state policy. In Sarawak, therefore, some or all of these measures should be implemented within the next few years. Sabah also has expressed interest in conducting such a statewide plan.

Many of the measures to reduce hunting in the two states must be conducted at the state level. These include continuing the current policies of preventing the issuing of any new licenses of shotguns. In addition, cartridges should be strictly controlled by limiting the number that can be purchased per gun owner. Rural subsistence hunters still do much of their hunting using dogs and spears, or dogs and guns. If using dogs, skillful rural hunters only fire a gun when a large animal is cornered, so they are unlikely to miss. This means that they do not use many cartridges; each subsistence hunter generally uses less than 60 cartridges per year. In Long Peluan, for example, each hunter uses about 48 per year. Strictly controlling cartridges, then, would primarily affect hunting for trade and sport without causing undue hardship to subsistence hunters.

A second measure needed at state level is a moratorium on all commercial sales of meat. The wild meat trade in both states is a major drain on wildlife populations. Its scale is greater in Sarawak than in Sabah due to deer being unprotected by law in the former. In Sabah, therefore, subsistence hunting of deer continues largely unchecked, but hunting for commerce is controlled at least to some extent by the authorities. Nevertheless, wild meat coming out of newly accessible areas for trade is devastating wildlife populations in some areas and resulting in little wild meat remaining for rural communities (Gumal and Bennett 1995). Although most trade only involves large ungulates, the reduction in their numbers causes local people to hunt smaller animals; the size of these means that more are hunted, so the effect of the trade indirectly permeates the whole wildlife community (Caldecott 1988; Gumal and Bennett 1995; Bennett and Gumal in press). Banning the commercial sales of wildlife also would allow populations to recover in many areas, as occurred in North America after commercial trade in wildlife was banned (Geist 1994). In Sarawak, a moratorium on sales of all wildlife products has already been approved as state policy, so as this is implemented, monitoring at many levels is needed to examine its effectiveness in conserving wildlife populations.

A third measure urgently needed, again at the state level, is greatly increased law enforcement of all regulations pertaining to wildlife. At present, this is weak, due in part to shortages of trained staff and funds.

There also must be no ambiguity about the role of reserves. Hunting of most species even by local people in these areas is not compatible with wildlife conservation. In areas designated for biodiversity conservation, the long-term goal must be to prevent any hunting, even if for subsistence. Hunting could continue in larger, extractive reserves, provided that it is just for bearded pigs, and only for subsistence, not sale. Bearded pigs are more able to withstand hunting pressure than are primates and deer; when fruit is abundant, they breed more than once per year and produce large litters (Pfeffer and Caldecott 1986). If the excessive level of hunting bearded pigs for trade is stopped, populations ought to be able to support hunting for subsistence of local communities.

In conjunction with the aforementioned measures, a major program to provide alternative sources of protein and income for rural people must be implemented. It is impossible to prevent people from hunting wildlife meat unless they have viable alternative foods. In addition, rural communities also need other means of earning cash so that they do not suffer hardship by being prevented from selling animal products. On their own, however, provision of alternative sources of protein and income will not solve the problem. For example, around Batang Ai National Park, Sarawak, the Agriculture Department has initiated fish farming schemes that provide a major potential source of protein and income for local Ibans. In some longhouses, however, the women look after the fish ponds, and almost all of the fish are sold for cash, while the men still hunt. This is not surprising, given the result that wild meat in the diet depends more on its abundance in the forest than on life-style: if it is there, people hunt it. Another source of potential income in the area is nature tourism, and some longhouses have tourist lodges attached to them. The income derived from nature tourism is on the same order of magnitude as that from the fish farms. Unlike fish farming, however, tourism in Batang Ai occupies the men's time. As a result, it is more effective at reducing hunting (Nyaoi, personal observation). Thus, solutions must be worked out on a case by case basis.

Equally critical are conservation education and extension programs. These require the long-term presence of people on the ground to integrate people's needs and aspirations with those of the wildlife. In Batang Ai, the Sarawak Forest Department's Education Unit has been working closely with the local people for 6 years. In addition to using innovative methods such as dance and role playing to educate people about wildlife conservation, the program included teaching people how to run small businesses, helping them to establish a cooperative to maximize their returns from tourism, and English language training to enable people to obtain jobs in tourism and other enterprises. This detailed, comprehensive, and long-term extension work is the only way to reduce people's use of wildlife.

More experimental at this stage is the exploration of ways to allow rural people to benefit from reserves, thereby providing incentives to protect them, but also making the people accountable for conservation. Merely handing over control of reserves to local people will not work (e.g., Redford 1989, 1991; Wells and Brandon 1992; Redford and Stearman 1993); short-term profits are likely to supersede long-term conservation. In a region where rural populations are large, aggressively independent, and the terrain difficult for law enforcers, however, working with the local people, recognizing their needs, and giving them incentives to protect the wildlife is the only way to ensure long-term conservation. Options for areas with tourist potential include establishing boards to run individual national parks, the boards comprising local people with rights in the area, and park staff to ensure that the ultimate aim of conservation is met. Much of the revenue would return to the local people, in exchange for greatly reduced or extinguished rights to hunt inside the reserve. This has yet to be tried.

Ultimately, all of these measures must be integrated with good land use planning. National parks, wildlife sanctuaries, and wildlife reserves should be freed from all forms of hunting so that wildlife populations there are guaranteed survival. Areas outside the reserves should be for extractive forest uses, including subsistence hunting where necessary, with the totally protected reserves acting as reservoirs of potential animals. Other areas should be used for intensive production of food and cash crops, so sustainable use is at the level of the overall landscape, not within individual reserves (Robinson and Bennett, this volume).

Unless all of the above steps are taken, and innovative approaches tried, wildlife throughout the forests of northern Borneo will continue to disappear. Ultimately, this will have much wider repercussions on the diversity of the forests (Redford 1992), as well as on the life-styles and culture of the local peoples. Implementing the solutions will require major efforts from many people if they are to become effective before all of Borneo's wildlife has been eaten.

ACKNOWLEDGMENTS

The project involved work for 3 years by three people in numerous sites spread across two states, so it is impossible here to name all of the people who helped us, but we extend our warm thanks to them all. Permission for the work was granted by the then State Secretary of Sarawak, Tan Sri Datuk Haji Bujang Nor, and the Socio-Economic Research Unit of the Prime Minister's Department. The present State Secretary, Datuk Amar Haji Hamid Bugo, also has been very supportive, and his enthusiasm to conserve wildlife highly encouraging. The staffs of the Forest Department in Sarawak and Wildlife Department in Sabah were enormously helpful throughout. Particular thanks are extended to Datuk Leo Chai and Cheong Ek Choon, former and current Sarawak Directors of Forests, and Mahedi Andau, Sabah Director of Wildlife. Others of particular help included Datuk Wilfred Lingham, Datuk Chin Kui Bee, Dr. Joseph Charles, Abang Haji Kashim Abang Morshidi, Ngui Siew Kong, Sapuan Ahmad, Francis Gombek, Melvin Gumal, and Laurentius Ambu. People generous with ideas and discussion included Dr. Richard Bodmer, Barnabus Gait, Melvin Gumal, Ross Ibbotson, Dr. Jeff Jorgenson, Dato Dr. Mikaail Kavanagh, Dr. Clive Marsh, Prof. Mohd. Nordin Hj. Hasan, Dr. Junaidi Payne, Dr. Alan Rabinowitz, and Dr. Kent Redford. Special thanks go to Dr. John Robinson for his help, ideas, reprints, and suggestions from the first conception of the project until now; many of the ideas on hunting controls could not have been developed so fully without him. He also gave valuable comments on this manuscript. Finally, the fact that we have so much data is entirely due to so many of the hunters being so warm, hospitable, friendly, and informative, allowing us to share so fully in their lives. The project was funded by the Wildlife Conservation Society; many thanks to Martha Schwartz for her unfailing administrative efficiency. Additional support was provided by WWF Malaysia.

16

Agta Hunting and Sustainability of Resource Use in Northeastern Luzon, Philippines

P. BION GRIFFIN AND MARCUS B. GRIFFIN

Even within the living memory of the Agta, wild pigs and deer were plentiful throughout much of the Sierra Madre of northeastern Luzon, the Philippines (figure 16-1). Agta hunters and gatherers depended on this abundance of wild meat and fish for their sustenance, as well as for materials for exchange. Massive changes are occurring that have led to a decline in forest resources, in the quality of Agta life, and in the potential viability of the wild pig, deer, monkey and fish populations. In understanding the cause of change in Agta subsistence, we may suggest interventions that could favor a renewal of the wildlife populations and their sustainable use. This chapter aims to present an account of the Agta relationship with game animals and to suggest how hunting might be sustainable in the future. We draw on both long-term ethnographic field work and short-term visits aimed at rapid assessment of changing conditions. We conducted research from mid-1974 to early 1976 within the municipality of Palanan, Isabela, and from late 1980 to mid-1982 to the north in Cagayan Province. Our most recent visit of several months, in Palanan, ended in 1995. Graduate students C. Clark and N. Rai working under the direction of P. B. Griffin collected data during research periods of 6 months to more than 1 year. The Griffins collected data for a total of 3 years each in Agta campsites.

The Agta, although lacking a definitive ethnography, are reasonably well studied (Vanoverbergh 1937, 1938; Estioko and P. B. Griffin 1975; Peterson 1978; P. B. Griffin 1984; Griffin and Griffin 1985; Headland 1986; P. B. Griffin 1989; Rai 1990; M. Griffin 1996). Agta are best thought of at present as a marginalized

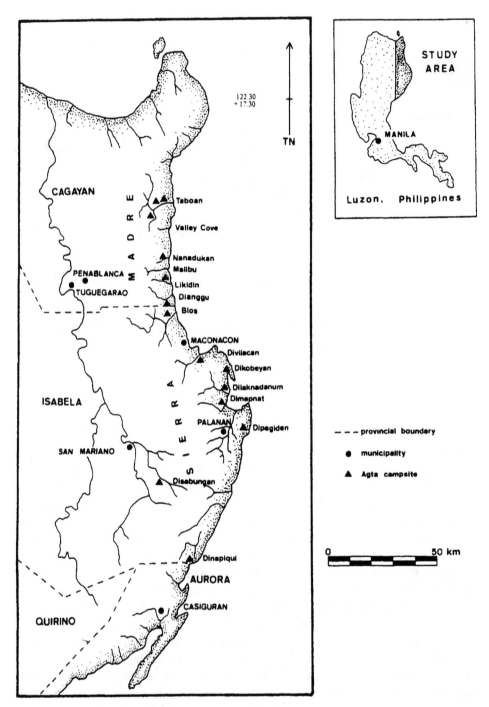

Figure 16-1. Northeastern Luzon and the study area.

people, unable to maintain the emphasis on hunting and fishing once character-istic of them. They are increasingly hovering on the bottom of the Philippine social, economic, and political hierarchy (P. B. Griffin 1991).

THE NATURAL AND SOCIAL ENVIRONMENTS

The environment of the Agta, and of the pigs and deer that they hunt, is not a uniform tropical rain forest (Allen 1985), but a highly variable montane monsoon forest, where dipterocarps are common. The area is bounded on the east by the Pacific Ocean and on the west by the descent into the rolling, arid, Cagayan Valley. The coastline is usually rocky, with scattered coral, gravel, or sand beaches backed by dense clusters of beach vegetation. The mountains are extremely broken, rugged, and eroded. Slopes are steep, some approximating cliffs. Box canyons and steep V-shaped valleys are cut by fast flowing streams and rivers that are lined on the bottom with small smooth stones as well as boulders. The land is thicketed by thorny rattans, nettles, and mazes of young saplings struggling among scattered towering and buttressed hardwoods.

The Sierra Madre is a monsoonal forest. The northeast monsoon arrives from the Pacific in about October, although timing fluctuates. By late January, the rain-fall abates, and March and April tend to be warm and dry, with scattered showers. June to September are hot and dry, except for the unpredictable yet certain typhoons. The southwest monsoons, lasting through the typhoon season, have little impact on northeastern Luzon.

The flora and fauna characteristically conform to this seasonal cycle. Diptero-carps and other trees bloom and set fruit in the dry season. Fruit can be fully mature by the wet or monsoon season. Wild pigs and deer gain body fat early in the wet season, assuming that typhoons have not destroyed the blossoms or fruit. The young pigs are in their early months of maturing at this time and benefit from the abundance of fallen fruit as well as from the wet conditions, which eases digging for roots. In contrast, during the dry season, the pigs are leanest and most difficult to hunt, while rivers and streams are low and fishable, and the fish are at their fattest.

The Agta hunt a range of mammals, birds, reptiles, and amphibians. The most important game animals are the Philippine warty pig (*Sus philippensis*) and deer (*Cervus philippinensis*). No larger prey exists. Pythons (*Python reticulatus*), a favorite food when fat, are few and difficult to capture. Crocodiles are nearly extinct locally, although they are a sporadic consumable meat. Long-tailed macaques (*Macaca fascicularis*) have in the past been avidly sought, although they are neither abundant nor easy to kill.

The Philippine warty pig is the preferred prey animal (figure 16-2), although deer are also favored. Pigs can be found nearly anywhere in the different forest zones, but increasingly they restrict themselves to higher elevations. Deer are also found in different forest types, although they apparently prefer forest some

Figure 16-2. Agta girl with Philippine warty pig (*Sus philippensis*) that she helped catch. *Photo courtesy of P. Bion Griffin.*

distance from human settlements. Neither pigs nor deer are intensive feeders in swiddens of the Agta or local farmers, although incursions are known.

Traditionally, Agta give wild pig and deer meat, fish, and forest products to their farming neighbors in exchange for cultivated starch foods such as rice and, more recently, sweet potatoes and cassava. Tobacco became a major trade item after its introduction by the Spanish, and during the tenure of our studies, liquor, sweets, and a variety of consumer goods have pervaded the exchange system. Agta depend on the cultigens and goods of the non-Agta. Going back to a pure hunter-gatherer way of life, even if the wildlife populations increased from today's levels, would be unthinkable to any Agta.

Non-Agta populations in the Sierra Madre have changed rapidly in the middle and late twentieth century (Griffin 1985). For centuries, the slopes of the Sierra Madre have supported, in addition to Agta, a variety of tribal swiddeners, simple fixed field rice cultivaters and antigovernment insurgents. Market exchanges with these people have influenced Agta hunting, and ultimately wildlife harvests until well into the twentieth century.

After World War II, the Philippines underwent a population explosion and development boom that included massive population movements, along with exploitation and degradation of natural resources. The boom and its corollaries continue to this day. By the mid-1960s, commercial logging was reaching the Pacific coast of the Provinces of Isabela and Cagayan. At this time the Philippines was the largest exporter of tropical logs in the world (Vitug 1993), and the Sierra Madre was logged. Landless farmers, wage laborers, and entrepreneurs followed the loggers wherever they led. At the same time, the New People's Army (NPA), a communist insurgent force, began to dominate much of the region, bringing sporadic armed clashes with the Armed Forces of the Philippines (AFP) and social impacts on Agta and non-Agta populations. The influx of outsiders continues, and forest resources further decline.

STUDY PERIODS AND AREAS

During 1975 and early 1976, with additional short trips from 1972 to 1979, we lived among Agta hunters (figure 16-3). They considered themselves traditional and claimed to follow the life-style and hunting practices of their ancestors. They occupied the mountain ranges south of the town of Palanan, upriver from and excluding the alluvial flats of the Palanan River. Sometimes traveling to remote rivers and coasts between Palanan and Dinipigi, and often crossing the mountain divide to the tributaries of the Cagayan River, they were judged by other, more settled Agta as *ebukid*, or "wild, mountain Agta." They were markedly more successful than other Agta in hunting wild pigs and deer, and had a meat-rich diet. They also carried meat for several hours to reach trade partners. During the dry season they tended to move far upriver, isolated from non-Agta farmers and sporadic military patrols.

In mid-1979 to 1980, Rai (1990) conducted research among Agta related to those earlier studied by the Griffins. These Agta lived primarily along the Disabungan River, which flows past the municipality of San Mariano and into the Cagayan River on the western side of the mountains. Like those Agta living at Pagsanghan, they considered themselves traditional hunters, but lived among many more lowlanders, including peasant farmers, immigrant Ifugao tribal swiddeners, and loggers (Rai 1990). We refer to all these southern Isabela Province Agta as Ihaya Agta to differentiate them from the more settled Agta downriver in Palanan and San Mariano. Disabungan Agta provide a comparison of interior groups with varying degrees of contact with non-Agta.

The Agta of Ihaya, or southern Isabela Province, in 1975 and 1980 represent the most traditional groups operating in situations of still abundant wildlife populations. Those of Pagsanghan on the Upper Palanan River were the most isolated and least disturbed by outsiders. Those on the Disabungan River on the western

Figure 16-3. The authors weighing a rusa deer (*Cervus philippensis*). *Photo courtesy of P. Bion Griffin.*

side of the mountains more frequently met loggers and encroaching farmers. The subsistence practices, especially success in hunting, reflect the availability of wild pig and deer, as well as the relative lack of pressure on Agta to secure more meat for exchange, and the lack of forest destruction.

In the Upper Palanan River in 1975, success rates on hunting trips were high: 63% of trips resulting in kills (table 16-1). More than one kill per trip was achieved, and the highly desirable wild pig was secured 38 times. The numbers of hunters on each trip were typical of traditional hunting patterns, although the numerous trips of three or more hunters sometimes saw hunters breaking up into individual units at times. Overnight hunting was popular because wildlife abundance was greatest in the deep mountain interiors. Overnight hunts nearly always resulted in multiple kills. The Ihaya Agta lived well into the mountains and had ready access to the best hunting grounds. Of special interest is their complete lack of hunting around the margins of farmers' fields; wildlife invading croplands was unimportant. Also noteworthy is the Pagsanghan hunters' lack of firearms.

A few years later, Rai (1990) reported that the Disabungan Agta often used shotguns: 44% of hunts involved shotguns, 40% bow and arrows, 3% machetes, and 12% no weapon. Agta consider wild pigs more abundant than deer, and they favor killing pigs. In both the Pagsanghan and Disabungan areas, wild pigs were killed much more often than were deer (see table 16-1).

After brief visits in the 1970s, we lived among an Agta group in southeastern Cagayan Province, mostly along the coast and the Nanadukan and Malibu Rivers. Although the group's membership included close relatives across the mountain divide, our stays on the western watershed were limited. The research was under-

Table 16-1. Hunting Success Rates and Details of Hunts for the Four Different Study Periods

Date	Location	Days of Obser- vation	No. of Hunting Trips	% of Hunting Trips Successful	Total No. of Animals Killed	Kills/ Day	Kills/ Hunt	Species Hunted	% of Animals	Hunting Method	% of Animals Killed	No. of Hunters	% of Hunts
1975	Upper Pananan River, Isabela (Ilhaya Agta)	125	87	63	100	0.80	1.15	Pig	38	Overnight trips	30	1 hunter	46
								Deer	25	Day hunt	32	2 hunters	19
								Other	37	Night hunt	6	>3 hunters	35
										With dogs	7		
										Other	25		
1980	Upper Disbungan River, Isabela (Ilhaya Agta) (from Rai 1990)	64	48	—	39	0.61	0.81	Pig	54	—	—	1 hunter	21
								Deer	2			2 hunters	33
								Monkey	23			> 3 hunters	46
1980– 1982	Nanadukan and Malibu, S.E. Cagayan	185	150	29	50	0.27	0.33	Pig	34	Forest stalk	20	1 hunter	48
								Deer	34	Tree ambush	2	2 hunters	37
								Monkey	20	With dogs	64	> 3 hunters	15
								Other	6	Night hunt	14		
1985	Dumayas Stream, S.E. Cagayan (from Clark 1990)	—	74	16	12	—	0.16	Pig	67	Forest stalk	33	1 hunter	49
								Deer	25	Trapping	58	2 hunters	24
								Other	8	With dogs	0	>3 hunters	27
										Night hunt	8		

taken intensively from late 1980 until mid-1982, then by further short visits the following years. Our most recent visit to these Agta was in 1991. Additional data on the same people come from the research of Clark (1990). For 6 months in 1985 she lived with families near the mouth of Dumayas stream, north of Nanadukan and Malibu. Her hosts were the same Agta visited by the Griffins. Interesting changes are seen in their increased loss of independence and ability to harvest wildlife populations.

HUNTING PATTERNS OF THE AGTA

Hunting strategies differ depending on the organization of hunters, general seasonal conditions, and the immediate need for meat. Tactics demand judgments concerning the number of able hunters, the specific wildlife and weather conditions of the moment, and the availability of trade partners. Strategies include driving animals with dogs, individual or team stalking, ambushing, and jacklighting (hunting at night using flashlights). Driving with dogs is a dry season strategy because the forest is noisy yet easy to move through rapidly. At this time, wildlife is lean, wary, and scent sensitive. Stalking is favored in the rainy season because dogs dislike the cold, sodden terrain, might refuse to cross swollen streams, and follow tracks with difficulty. The late dry season and early rainy season often favors ambush, usually from nearby or in a tree whose fallen fruit attracts animals, feeding either during the day or night. Other ambushes involve partners, whereby one hunter drives animals toward another along a known animal trail. This strategy is a year-round activity.

Teams of hunters cooperate in drives, teams composed of two to about 12 people, although teams that large are rare. The people and dogs travel to the forest to agreed upon locations, where dog owners, either male or female, assist their charges in jumping and following fleeing animals. Depending on luck and the general abundance of wildlife, several kills might be made. As soon as an adequate amount is secured, the hunt ceases. They might then partially butcher the carcasses (figure 16-4), rest, roast and eat the liver, feed the dogs scraps, and form backpacks of meat for the return home.

Stalking follows the same principles, except that seldom do more than two or three persons coordinate their movements. Often only one or two hunters carry out a day's search, returning when finished to an agreed rendezvous point. These tactics often include a husband and wife together, brothers, an older and younger relative, an uncle and his youthful nephew, or an aunt and her maiden niece. In the latter case, dogs are almost always companions, and occasionally an infant may be strapped on the adult's back.

The southeast Cagayan Agta in the early 1980s were traditional in many ways; they were less influenced by outside forces than Agta around the towns of Palanan, Casiguran, and San Mariano. However, they did represent an exploitation pattern that was moving from traditional to commercial. The Agta population size was similar to what it has been over the past 30 years, but the population of non-Agta

Figure 16-4. Agta hunter butchering a Philippine warty pig (*Sus philippensis*). *Photo courtesy of P. Bion Griffin.*

was markedly increased. Logging and loggers penetrated the most remote coastal areas, and miners surveyed and some established operations. Fishermen, traders, and collectors of seckry, rattan, and sea shells, as well as government agents, occasionally meet the Agta, and missionaries and anthropologists have a strong impact. Environmental destruction is great, and the Agta were encouraged to procure meat and fish for the newcomers. The introduction of desirable consumer goods enhanced the need for Agta to acquire forest products for exchange.

The Agta of southeastern Cagayan Province followed the same hunting strategies that characterized the southern Isabela Province Ihaya Agta. The major difference, and paradoxically one that the Agta themselves attribute to a more traditional or old-fashioned practice, was the frequency with which women actively hunt (Estioko-Griffin 1984; Goodman et al. 1985).

The wild pig population of Cagayan Province was much lower than that of southern Isabela. The number of farmers has steadily increased since the appearance of Acme Plywood and Veneer Company in the late 1960s. As roads extended north and south along the coast, immigrants burned the logged forest, or clear cut forest of little commercial value, and either put in swiddens or irrigated rice fields. Wild pig and deer populations declined markedly. Hunting success rates had declined rapidly by the 1980s (see table 16-1).

The data for 1980 to 1982 and 1985 show a decline in hunting success compared with the Ihaya data, and a decline among the Cagayan people themselves in only 3 years. An inadequate kill rate is evident, and by 1985, little meat was available for exchange. At Dumayas in 1985, Agta had partially abandoned both the practice of female participation in hunting and the complete reliance on bow and arrow hunting. Traps had a new emphasis, becoming the single most reliable means of securing larger animals (see table 16-1). Fishing success also sharply declined, and collection of rattan for sale escalated from a minor means of gaining food and cash to the dominant means of food acquisition. Wildlife numbers were no longer adequate to provide wild meat for farmers, loggers, fishermen, and the Agta themselves. The logging operations had finally expanded from both the north and the south, leaving no section of forest completely untouched. Agta provisioning of exchange partners was no longer a major part of their economic system; many new non-Agta needed provisioning, resulting in overharvesting the wild pig and deer. Quality of life of the Agta, as measured by protein and fat intake, had markedly fallen. They were very aware of the situation, but saw no alternative available for the acquisition of rice, tobacco, and consumer goods. The southeastern Cagayan Agta, however, still maintained a commercial exchange relationship with non-Agta, even though rattan had replaced meat and fish as the major means of payment.

SUSTAINABILITY OF HUNTING

The Agta hunting of pigs and deer is decreasingly productive. The Agta recognize the diminished faunal resources and inadequate food for larger game. During discussions in Palanan and Maconacon, Isabela, in 1994 and 1995, senior adult Agta called for a reversal in the game population declines found throughout the Sierra Madre. The Agta themselves advocate their ownership and stewardship of the forest, and they argue for inclusion in new advocacy for tropical forest preservation.

Environmental conservation, habitat management, and sustainable use of wild resources has, in the 1990s, become part of the discourse and action in the Sierra Madre. Centered around Palanan, Isabela, but extending to other municipalities, national and international nongovernmental organizations (NGOs) and funding agencies have begun committing money and expertise to reversing the destruction of the Philippine's last major forest area. Among these efforts is the creation of a wilderness preservation zone, the Northern Sierra Madre Nature Preserve (DENR 1992). M. Griffin, a consultant for Conservation International, worked with the Agta in planning their response to and participation in the NGO schemes, such as ecotourism and livelihood projects. Restoration of wildlife populations is part of the plan for the nature preserve.

Should populations be revived to former levels, the Agta might start planning how the wildlife could be managed through selective hunting. The question that dominates, however, is who should hunt: the Agta, replicating their traditional hunting system, or sport hunters guided by paid Agta? With conservation fully established, as is now being attempted through the Sierra Madre Nature Preserve in Palanan, the Agta might lose the skills and knowledge essential to replicate former hunting. Since pig and deer have been harvested for thousands of years, the forest/wildlife balance would seem to necessitate an eventual harvesting program.

Agta have over time seen a real decline in wild pig and deer populations, in their own consumption of meat and fat, and in their foraging life-style. Deforestation and excessive hunting have made the Agta way of life no longer fully viable. Timber cutting and immigration by farmers and laborers have introduced new modes of environmental destruction. Wildlife is widely trapped or shot with firearms. Few profitably arable lands remain to be cleared. In the 1990s, business and political entrepreneurs are increasingly in evidence, with proposals for roads to cross the mountains to Palanan, deep draft harbors to be built on the coast, and a lucrative wilderness park to be established.

Clearly, the sustainability of hunting is at a critical point. Only the Agta care about regaining their hunting rights and the hunting successes remembered by all adults. Other parties view the Agta as resources themselves, and of value only as a means of capital inflow and acquisition. If wildlife populations are to be restored to former levels, and if breeding populations are to be protected, one of two schemes will be necessary. One, which many Agta favor, involves the Agta as care-takers of the forest, its flora and fauna, with funding and control maintained by international NGOs. The purpose would be to preserve the Sierra Madre ecosystem by limiting use and by employing Agta to manage the resources. Agta would benefit economically from the exploitation of wildlife and would actively participate in decisions about wildlife management. The second scheme involves setting up a forest preserve to maximize levels of animal populations. In turn, hunting policy would involve decisions from outside "experts," probably profes-sionally trained in wildlife management and sport hunting.

Although business, government, and NGOs in the Philippines might favor the latter scheme as the most lucrative to their own institutional interests, the capacity for indigenous farmers and foragers to subvert the operation seems limitless. For decades, the Sierra Madre of Isabela and Cagayan have only been minimally controlled by the provincial and national governments. To be effective, any management strategy must include the Agta and consider their resource needs.

Unquestionably, an outside power and financial base is mandatory. In 1996, much commercial logging has been curtailed by the government, but illegal cutting pervades the region. Although individual voices for management are heard in the local communities, they lack power, capital and audience. Only with the appearance of international capital promoting habitat preservation has conserva-tion and sustainable management become remotely respectable.

A unification of environmental and cultural preservation which gives indige-nous people a dominant voice might be the only route to conserving wildlife populations and ensuring that hunting is sustainable. The Agta themselves have changed from abundant wildlife hunters with sustainable populations in primary forests in 1975, to being struggling foragers with inadequate wildlife resources in 1985, to landless peasants in some areas today. Building on these experiences, the Agta working with powerful outside agencies seem capable of designing a program of sustainable use of wildlife.

PART III

Institutional Capacity for Management

17

Hunting for an Answer: Is Local Hunting Compatible with Large Mammal Conservation in India?

M. D. MADHUSUDAN AND K. ULLAS KARANTH

India is a country with high levels of biodiversity, particularly with respect to its large mammal fauna (Prater 1980; Johnsingh 1986; Rodgers and Panwar 1988). This diversity is a consequence of the following factors: (a) its location at the junction of the African, Palearctic, and Oriental zoogeographic realms (Rodgers and Panwar 1988); (b) great variations in climate and geology (Mani 1974; Puri et al. 1983); (c) the establishment of government-protected forests dating back more than a century (Stebbing 1929); and (d) a traditionally tolerant conservation ethos (Gadgil and Malhotra 1985; Gadgil 1992). In the Indian social context of acute poverty, land hunger, rising aspirations, and demand for biological resources, large mammals (body mass >2 kg), as a group, are particularly difficult to conserve because of life history traits such as relatively low densities, large home ranges, slow reproductive rates, small litters, and long generation times (Eisenberg 1980, 1981). These traits, in turn, render large mammal populations potentially vulnerable to extinctions from stochastic environmental, demographic, and genetic factors (Lande 1988; Burgman et al. 1993). The fact that most large mammals attract human hunters because of their high meat yields, trophies, and commercially valuable by-products such as hides, horns, and bones further aggravates conservation problems arising from their inherent ecological vulnerability.

In the early 1970s, India embarked on a wildlife conservation strategy that focused on large mammals by introducing strong wildlife protection laws, establishing special reserves, and initiating species recovery plans such as Project Tiger (Saharia 1982; Panwar 1987). The thrust of this effort was essentially preservationist, involving curbing of poaching, forest fires, and logging and plant biomass removal, and stemming agricultural encroachments into wildlife reserves. Concurrently, a policy directed at gradually eliminating all legal hunting of wildlife, both for subsistence and recreation, has actively been implemented since 1974. As a result, India's foresters, who also manage its wildlife reserves, were quite successful in establishing several reserves where large mammals now occur at high ecological densities (Eisenberg and Seidensticker 1976; Karanth and Sunquist 1992). However, due to constraints such as scarcity of money and manpower, administrative apathy and hostility of local communities, illegal hunting of large mammals is still widely prevalent, particularly on the peripheries of and outside wildlife reserves (Karanth 1987, 1991; Sukumar 1989; WCS 1995a). Consequently, there are some areas that are rigidly closed to hunting and others that are highly prone to illegal hunting.

Indian naturalists, reserve managers, and wildlife biologists have recognized that local hunting greatly lowers large mammal populations (Dunbar-Brander 1923; Schaller 1967; Sankhala 1978; Karanth 1987, 1991; Panwar 1987; Sukumar 1989). Based on this experience, India has generally opposed the increasingly popular approach of promoting conservation by devolving management to local communities and allowing them to harvest wild species, an approach often referred to as "sustainable use." Thus, both in national and international forums, India has generally supported the ideology of preservation, rather than that of sustainable use of wildlife (Kumar et al. 1993; S. Deb Roy and M. K. Ranjithsinh, personal communication). However, quantitative data on large mammal populations, derived using rigorous estimation techniques, have been scarce in India (Karanth 1988, 1995; Karanth and Sunquist 1992). Thus, there are no reliable data on how hunting affects large mammal populations, although some conjectural accounts have been published (Malhotra et al. 1983; Malhotra and Gadgil 1988).

Thus far, quantitative studies on the impact of hunting on wildlife populations have generally been restricted to large mammal communities in the neotropics (Peres 1990; Glanz 1991; Mittermeier 1991; Bodmer et al. 1994), Africa (Marks 1973; Fa et al. 1995; FitzGibbon et al. 1995a), and North America (McCullough 1979; Irby et al. 1989). Given that population dynamics, community structure, and productivity of large mammals of Asian tropical forests are likely to be very different from those in above ecosystems (Eisenberg and Seidensticker 1976; Eisenberg 1980), data from other regions have limited relevance to South Asia.

In this chapter, we present a study examining the impacts of hunting on large mammals in two ecologically similar sites, Nalkeri and Arkeri in South India. The two sites are ecologically similar, but subject to different hunting intensities. We evaluate the impact of hunting by comparing densities of nine large mammal species in relation to hunting intensity and effectiveness of protective measures in the two sites. Finally, we consider our results in the context of two models proposed to manage hunting in India: the preservationist model and the sustainable use model. We argue that, despite shortcomings, the preservationist model appears to hold the most promise for long-term conservation of large mammals in India.

NAGARAHOLE: A CASE STUDY

Natural History

The 644-km^2 Nagarahole National Park (11°50' to 12°15' north latitude and 76°0' to 76°15' east longitute) lies in a zone where the Western Ghat mountains intergrade with the Deccan Plateau (figure 17-1). Its elevation varies between 700 and 960 m, and the soils are generally red sandy loams. Mean monthly temperatures range between 20°C and 27°C, and the annual rainfall of 1000 to 1500 mm falls mainly during the monsoonal wet season between June and September. The wet season is followed by a cool season (October to January) and a dry season (February to May). There are several streams and small rivers that drain the reserve.

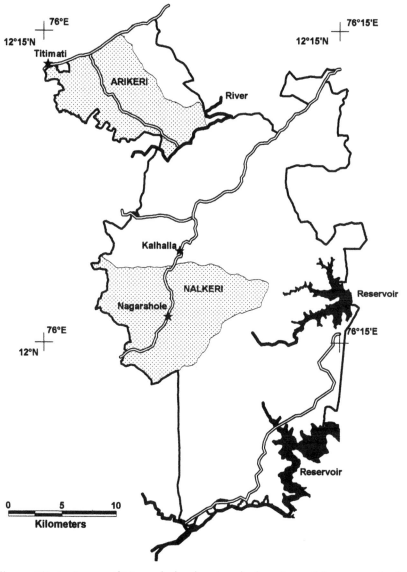

Figure 17-1. A map of Nagarahole, showing the locations of the two study sites.

The natural climax vegetation of the region consists of tropical moist deciduous forests in localities where annual rainfall exceeds about 1000 mm, and tropical dry deciduous forests where it is lower. These two forest types have been classified as *Lagerstroemia microcarpa–Tectona grandis–Dillenia pentagyna* series (moist
deciduous forests) and *Anogeissus latifolia–Tectona grandis–Terminalia tomentosa*
series (dry deciduous forests), respectively (Pascal et al. 1982).

Earlier work in Nagarahole (Karanth 1988; Karanth and Sunquist 1992) showed
that moist deciduous forests support considerably higher densities of large mammals
than do dry forests. To avoid habitat-related confounding variables, we selected both
our study sites, Nalkeri and Arkeri, in the moist forests in northern Nagarahole (see
figure 17-1). These two areas were of comparable size, being 71 km^2 and 73 km^2,
respectively. Due to reasons explained later, for over two decades Nalkeri (our less-
hunted site) has had less hunting, than Arkeri (our heavily hunted site).

The natural moist forests at both sites have a 25- to 30-m high upper tree
canopy dominated by deciduous trees such as *Terminalia tomentosa*, *T. paniculata*,
Sterospermum chelenoides, *Tectona grandis*, *Lagerstroemia microcarpa*, and the evergreen *Syzigium cuminii*. The lower story is dominated by *Kydia calycina*, *Dillenia
pentagyna*, *Cassia fistula*, *Wrightia tinctoria*, and several *Randia* species. Wet or
disturbed sites are dominated by the bamboo *Bambusa arundinacea*. The undergrowth is heavy, comprising native shrubs such as *Cipadessa baccifera*, *Helecteris isora*
and *Ardisia humilis*, *Bambusa arundinacea*, as well as exotic invaders such as *Chromo-
laena odorata* and *Lantana camara*. At both sites, as part of the official forestry policy,
substantial areas of natural forests have been clear cut to raise monocultures of
Tectona grandis (teak) between 1880 and 1980. Where such teak plantations have
failed, dense secondary moist forests occur now. The shallow, clayey valley bottoms
contain *hadlu* (swamp savannas), which are open and dominated by grasses and
sedges. Outside the park boundaries, natural forests have been totally replaced by
commercial plantations of coffee from the early 1900s onward, and some *hadlu* have
been under intermittent wet rice agriculture over several centuries.

Nagarahole and surrounding regions are recognized as one of the richest
wildlife areas in India (Nair et al. 1977; Karanth 1987). The mammalian fauna
here is particularly diverse, including an intact assemblage of seven large herbivore
species: four-horned antelope (*Tetracerus quadricornis*), muntjac (*Muntiacus
muntjao*), wild pig (*Sus scrofa*), chital (*Axis axis*), sambar (*Cervus unicolor*), gaur (*Bos
gaurus*), and Asian elephant (*Elephas maximus*), as well as their predators: tiger
(*Panthera tigris*), leopard (*P. pardus*), and dhole (*Cuon alpinus*). At least 19 other
species of mammals (>1 kg) live in Nagarahole, in addition to over 250 species of
birds. Detailed studies on large herbivores and their predators have been
published (Karanth and Sunquist 1992, 1995; Karanth 1995).

Historical Background

The presence of several derelict irrigation tanks inside Nagarahole suggests that
there were agricultural enclaves within the larger forested landscape of the region
for over 1000 years. The area was ruled by Lingayath kings of Kodagu until the
1850s, when, following a series of revolts, wars, and treaties, the British colonial

administration established hegemony over it. From the 1890s, the forests of Nagarahole, which were until then considered to be either common property or belonging to local chieftains, were gradually demarcated and notified as government-owned reserved forests (Stebbing 1929). The Forest Department was set up in the 1860s with a basic mandate to restrict forest burning and swidden agriculture by tribal groups, to control agricultural encroachments by nontribal groups, sustainably harvest valuable timber trees, and gradually replace natural moist forests with teak plantations (Stebbing 1929; Somaiah 1953). The department set about this task energetically, and had substantially accomplished its mandate by the time India gained independence in 1947. Between 1947 and 1955, the Indian government's policy was to encourage production of timber and food. This resulted in importation and settlement of tribal and nontribal groups to the *hadlu* to cultivate rice and to provide cheap labor for forestry operations. The intensity of hunting of large game mammals increased with availability of better guns, electric flashlights, motor vehicles, and a policy that promoted the liberal issue of crop-protection guns and hunting licenses. From available anecdotal accounts, wildlife populations declined rapidly. According to several informants, in the early 1950s both Arkeri and Nalkeri had a roughly comparable abundance (or scarcity) of wildlife.

LOCAL PEOPLES AND HUNTING TRADITIONS

Other than park staff, the only people who currently live inside the study areas are tribal groups known as Jenu Kuruba, Betta Kuruba (sometimes collectively called Kadu Kuruba, or simply Kuruba), and Yerava. A larger number of these tribal groups live outside the park, usually on government land, or as tenants of other land-owning castes. The tribal groups were originally hunter-gatherers, who later turned into swidden farmers and collectors of minor forest products for the commercial market. They are extremely poor, mostly landless, and socially oppressed. Their main source of employment is wage labor in coffee plantations outside Nagarahole. They also work as low-level, seasonal employees of the Forest Department. However, tribal men are also skilled hunters who deploy an ingenious array of traditional techniques. If trained, they are equally adept with shotguns. Because of their familiarity with the terrain and access trails into the interiors of the park, they are prized as guides by nontribal hunters. Tribal groups are now fully integrated into the agricultural and cash economy, have alternate sources of livelihood, and do not depend on hunting wild meat for their survival.

The people of the Kodava caste, who were warriors under their Lingayath kings, and later under the British, are the dominant local land owners. Traditionally, Kodava culture has glorified hunting prowess and, at social events like weddings, feasts of wild meat were de rigueur. Because Kodava were considered loyal to the British Crown in comparison with other seditious Indians, the colonial rulers legally exempted them from stringent gun control laws. This anachronism continues to this day, resulting in a high density of legal guns around Nagarahole. During the past two decades, however, the majority of Kodava have given up hunting because of their higher literacy and wealth. Still, a number of Kodava illegally lend their licensed firearms to their laborers (as reported by 8 of 10 Kodava interviewees), or buy illegally hunted meat. Other land-owning, meat-eating castes, although not possessing many guns, encourage illegal hunters to procure wild meat.

Most of the remaining groups of local people in the area are workers in the coffee estates. They are a heterogeneous group, comprising tribal groups, other local lower castes, and migrant labor from the neighboring district of Mysore, and the states of Kerala and Tamil Nadu. If they hunt, they either use borrowed guns in the park or use wire snares outside it. Local tradesmen who own grocery stores, liquor vends, and smithy shops also participate in local hunting. Most often, these are Moplah Muslims from Kerala state, or belong to other local Hindu castes. Their legitimate businesses occasionally cover up ancillary illegal trade in non-timber forest products, including meat or other wildlife products.

In order to determine hunting patterns and intensities, and to explore the extent to which cultural, social, and economic factors influenced them, we conducted interviews and questionnaire surveys among local hunters, informants who did not hunt, and park staff based at both sites. We approached hunters and informants by working undercover, as well as by using our extensive local contacts to gain their confidence. Our questions ascertained the cultural backgrounds of hunters, species hunted, hunting frequency, techniques employed, degrees of dependence on wild meat, and market involvement in the activity. Because we probed into an essentially illegal activity, the sample size of reliable local respondents to our survey, excluding forest department staff, was small (n = 32).

Respondents reported hunting at least 16 of the 29 species of mammals (with weights over 1 kg) present in the areas. Apart from the nine study species, other mammals hunted included chevrotain (*Tragulus memmina*), black-naped hare (*Lepus nigricollis*), large brown flying squirrel (*Petaurista petaurista*), palm civet (*Paradoxurus* spp.), small Indian civet (*Viverricula indica*), pangolin (*Manis crassicaudata*), and crested porcupine (*Hystrix indica*); several species of birds including pheasants, doves and pigeons, hornbills, barbets, parakeets, and owls; reptiles such as the monitor lizard (*Varanus bengalensis*) and two chelonians also were hunted. We documented an impressive array of hunting techniques that were in vogue (table 17-1). Shooting was the most common method of hunting, used mostly to secure larger game. A host of traditional techniques targeting smaller species was used mostly by the Kuruba tribal groups. Traditional techniques were often favored owing to their lower cost and inconspicuous nature (lower detectability). Religious taboos forbade the hunting of some species. For instance, Hindus did not kill or eat gaur, and Muslims did not eat pigs, porcupine, and pangolin. However, these taboos are increasingly being disregarded, and their value in mediating hunting impacts appears to be declining.

COMPARISON BETWEEN NALKERI AND ARKERI

Hunting Levels

In 1955, the 285-km² Nagarahole Game Sanctuary was gazetted, including both Arkeri and Nalkeri. However, on-ground protection remained poor in both areas.

Table 17-1. Currently Prevalent Hunting Techniques Used by Illegal Hunters in the Nagarahole Area

Hunting Technique	Practitioners	Target Species
Active Hunting Methods		
Walk-and-seek shotgun hunting by night using spotlights	Kodava, Vokkaliga, Moplah (all assisted by tribal guides), and Kuruba/Yerava using borrowed guns	Sambar, chital, flying squirrel, porcupine, civets, black-naped hare
Sit-and-wait shotgun hunting at fruiting trees/waterholes on moonlit nights	Same as above	Chital, pig, civets
Sit-and-wait shotgun hunting at waterholes/fruiting trees in daytime	Same as above	Chital, muntjac, langur, bonnet macaque, giant squirrel, pigeons, hornbills, pheasants
Bow and clay shots by day	Kuruba	Primates, giant squirrel, flying squirrel, monitor lizard, birds
Hunting dogs and nets by day	Kuruba	Langur, bonnet macaque
Hunting dogs and club/machete by day	Kuruba	Chevrotain, black-naped hare, langur, mongoose
Physical capture from rest-sites/nests during daytime	Kuruba	Flying squirrel, palm civet, hornbills, bird eggs/nestlings
Smoking/digging out burrows in daytime	Kuruba	Porcupine, pangolin
Catapult in daytime	Yerava	Giant squirrel, birds
Passive Hunting Methods		
Wire snare	Kuruba, Yerava	Chital, pig, black-naped hare
Baited explosives	Kodava, Vokkaliga, Moplah and others	Pig
Scissor trap	Kuruba	Small rodents, small pheasants
Deadfall trap	Kuruba	Black-naped hare, chevrotain, pig
Pitfall trap	Kuruba	Chelonians
Bird lime	Kuruba, Yerava	Frugivorous birds

New wildlife protection laws were introduced in 1974, and Nagarahole was enlarged and upgraded to a national park. Following this, considerable effort was invested in both areas to remove illegal squatters and to curb grazing by domestic stock. Throughout the park, people are prohibited from cultivating agriculture and livestock or collecting forest products. Hence, people living in both study areas earn their livelihood outside the park boundaries. However, even since the early 1970s, several factors ensured that antipoaching efforts were considerably more effective in Nalkeri than in Arkeri. Nalkeri was administered both under a wildlife range, whose separate wildlife protection staff were responsible for antipoaching vigilance and enforcement, and a territorial range concerned mainly with forestry operations. The presence of a wildlife range in Nalkeri also meant greater interest and funding from the government for wildlife protection, and greater support from local conservationists. Arkeri, however, was managed only as a territorial range, where the accent was primarily on forestry, and little attention was directed to wildlife protection. For these historical reasons, hunting levels differed significantly between the two sites when this survey was conducted in May 1996.

Although of similar sizes, Arkeri shares a much greater length of its boundary with private land than does Nalkeri (table 17-2). Fingers of cultivation extend into

Table 17-2. Susceptibility to Illegal Hunting: A Comparison of the Study Sites

	Nalkeri	*Arkeri*
Length of boundary shared with private land per unit area	0.251 km/km^2	0.426 km/km^2
History of highway access	From mid-1970s	Since early 1900s
Towns within 30 km	2 (Kutta, Srimangala)	5 (Balèle, Gonikoppa, Hunsur, Periyapatna, Virajpet)
Human habitation	6 settlements	18 settlements
Human population	c. 900	c. 2500
Mean frequency (±SE) of monthly hunting incidents (estimates made by local informants and hunters in each site)	4.3 (±1.4), $n = 11$	13.7 (±2.9), $n = 21$

Arkeri from the west (see figure 17-1), increasing the interface between the Reserve and private land. Such a spatial configuration improves access for hunters from outside and makes patrolling difficult, thus contributing to higher levels of hunting in Arkeri. Densities of potential hunters resident in the area is also higher at Arkeri (see table 17-2). We also assessed the relative capabilities of law enforcement at the two sites. Factors such as fewer staff, poor housing facilities, and lack of patrolling roads, weapons, vehicles, and communication equipment for enforcement staff at Arkeri hamper protection efforts. The differences in protection capabilities are summarized in table 17-3. In addition, poor maintenance of trenches around Arkeri causes extensive crop damage by elephants. This results in higher levels of local animosity against the park, leading to some retaliatory poaching.

Hunting at both sites is for the pot, the meat being shared by the hunters and their accomplices. Although fear of prosecution discourages the open sale of wild meat, informants at Arkeri did report a few hunters who clandestinely supplied meat to eateries in nearby towns. Sometimes hides and antlers are sold. Better access to urban centers from Arkeri (see table 17-2) predisposes the site to higher pressure from such market hunting. However, most trade in wild meat consists of bartering surplus meat for liquor, ammunition, or gun loans. Most respondents in both sites were disinclined to buy wild meat from strangers, fearing prosecution; similarly, hunters were disinclined to sell meat to strangers.

Although we made no attempt to assess the nutritional status of hunters, it is our observation that the socioeconomic and cultural profiles of local communities were similar between Arkeri and Nalkeri. Hunters at neither site considered wild meat essential for their subsistence ($n = 27$). They had access to sources of domestic animal protein: all raised livestock, even if holdings differed between communities. Local nutritional status, therefore, might not be a major influence on hunting levels at either site.

Large Mammal Densities

We surveyed both sites, applying rigorous distance sampling field methods (Burnham et al. 1980; Buckland et al. 1993) to estimate ecological densities of nine large mammal species: Indian giant squirrel (*Ratufa indica*), bonnet macaque (*Macaca radiata*), common langur (*Semnopithecus entellus*), muntjac (*Muntiacus*

Table 17-3. Comparative Law Enforcement Capabilities at the Study Sites

	Nalkeri	*Arkeri*
Area	71 km²	73 km²
Protection staff strength	37	20
Mobility/vehicles	1 jeep (4WD), 2 vans	1 van
Length of patrolling roads	c. 150 km	c. 45 km
Wireless sets	1 base set, 1 mobile set, 5 walkie-talkies	2 base sets, 1 walkie-talkie
Weapons	2 rifles, 5 shotguns	2 rifles, 1 shotgun
Patrolling frequency		
Day patrols	Daily	4–5 times/wk
Night patrols	Daily	2–3 times/wk
Number of forest antipoaching camps	3	None

muntjao), wild pig (*Sus scrofa*), chital (*Axis axis*), sambar (*Cervus unicolor*), gaur (*Bos gaurus*), and Asian elephant (*Elephas maximus*). However, we were unable to esti-mate abundances of certain other species actively targeted by hunters (e.g., black-naped hare, mouse deer, flying squirrel, palm civets) because they were not suited to line transect sampling or were nocturnal, often both.

We measured impacts of hunting using estimates of absolute abundance of the study species, rather than measures of relative abundance such as encounter rates, species richness, or species diversity (Magurran 1988). Absolute field densities of animals also reflect hunting impacts more accurately than indirect measures such as surveys of hunter bags, kitchen middens, or local markets (Caldecott 1988; Vickers 1991; Fa et al. 1995).

To estimate animal abundances, we adopted the general methodology applied in Nagarahole earlier and described by Karanth (1988) and Karanth and Sunquist (1992). During May 1996, we performed systematic line transect sampling (Burnham et al. 1980; Buckland et al. 1993) simultaneously at both study sites to estimate absolute densities of study species. We cut four permanent transect lines in Arkeri (total length = 10.4 km) and five lines in Nalkeri (total length = 12.6 km). Two trained observers walked each transect during the two time blocks of the day (0600–0900 hr and 1600–1900 hr) when animals were most active. The species, group size, sighting distance (measured with a rangefinder), and sighting angle (measured using a compass) for each detection were recorded. The total sampling effort was 271.4 km for Arkeri and 327.6 km for Nalkeri. We then estimated mean animal densities and their related parameters using the computer program DISTANCE version 2.0 (Laake et al. 1993). We chose appropriate detection probability estimators using the half-normal model and the fitment procedures based on the Akaike Information Criterion, and goodness-of-fit tests generated by DISTANCE (Buckland et al. 1993; Laake et al. 1993). We adopted a 5% right truncation of outlier distance data to improve the precision of our estimates.

Results (table 17-4) clearly show that all small-bodied species (giant squirrel, bonnet macaque, and langur), two medium-bodied species (wild pig and chital), and one large-bodied species (gaur), occurred at significantly ($p < 0.05$) lower densities at Arkeri, the more hunted site. The giant squirrel, bonnet macaque, and

Table 17-4. Density Estimates and Comparisons at the Two Study Sites

Common Name	Arkeri (Animals/km²)			Nalkeri (Animals/km²)			z Test Results	
	N	D	SE	N	D	SE	z	p
Small prey (<10 kg)								
Giant squirrel	83	5.5	1.1	139	8.1	0.9	1.81	0.035
Bonnet macaque	11	4.5	2.2	60	15.3	3.1	2.78	0.003
Langur	12	4.1	1.8	196	32.6	3.7	6.90	0
Medium prey (10–100 kg)								
Muntjac	80	7.7	1.1	53	6.2	1.1	−1.01	0.156
Pig	9	0.5	0.2	55	6.2	1.5	3.76	0
Chital	83	9.1	1.5	365	66.2	5.9	9.31	0
Large prey (> 100 kg)								
Sambar	39	4.4	1.0	66	3.9	0.8	−0.40	0.350
Gaur	21	1.7	0.5	64	5.8	1.2	3.05	0
Elephant	34	2.2	0.6	26	0.8	0.2	−2.27	0.01

N, number of detections; D, mean density of animals; SE, standard error; z, z test statistic; p, significance level.

z Test: H_0: $D_{Arkeri} = D_{Nalkeri}$; H_A: $D_{Arkeri} < D_{Nalkeri}$.

langur are arboreal mammals susceptible to similar hunting techniques. Daytime shooting, which targets these diurnal species, is perceived as risky by the hunters, given that some amount of daytime vigilance from the forest department staff exists even in the more hunted site. However, these species are also hunted vigorously by tribal hunters using bows with clay shots. Langurs are also hunted by driving them into nets placed in trees, using trained dogs. Although langurs and giant squirrels are killed and eaten by all communities, bonnet macaques (whose meat is said to be distasteful) are only rarely hunted and eaten by tribal groups. However, the macaques are shot outside the park for damaging crops. Active, traditional, nonshotgun hunting, which is less conspicuous than daytime shotgun hunting, appears to take a heavy toll on these arboreal species.

Twenty-three of the 27 local hunters named chital (figure 17-2) and wild pig as the species that they hunted most commonly. Both species are vulnerable to snaring because they frequent areas close to human habitation, where snares are typically set. In addition, the chital's habits of traveling in groups, bedding in open areas, and frequenting fruiting trees make them a favored quarry of night-time shotgun hunters. None of the local communities observe taboos against hunting or eating chital. Wild pig, unlike chital, are not easily shot at night by spotlighting. However, they are more effectively targeted by snaring and by using the wait-and-shoot technique at fruiting trees and water holes inside Arkeri. Their tendency to raid crops also exposes them to shooting, snaring, and baited explosive devices. Local tribal groups also capture whole litters of piglets from their nests. Only the Moplah Muslims observe taboos against eating pork; all other communities keenly seek and hunt pigs. Comparatively depressed densities of both chital and pigs at Arkeri appear to result primarily from the higher hunting pressure.

Although gaur are found at significantly lower densities in Arkeri, we are unable to explain this satisfactorily as a function of hunting alone. Young gaur are sometimes accidentally caught in snares set for chital and pigs. Adults are too large

Figure 17-2. Chital (*Axis axis*). *Photo courtesy of H. N. A. Prasad.*

to be snared and can only be hunted with shotguns. However, religious taboos among the majority of Hindus forbid them from killing gaur. Even when Hindus lend their guns, they forbid the borrowers from killing gaur. This taboo ensures that gaur are rarely hunted even by the minority non-Hindus. Only two of 27 hunters interviewed admitted to having shot gaur. Moreover, the large body size of gaur increases the risk of detection for the hunter and is a further deterrent to its being hunted.

Neither muntjac nor sambar show significant differences in abundance between Arkeri and Nalkeri, although they are hunted and eaten by all hunting communities. Muntjac are reportedly hard to snare using the technique of local hunters. The only effective way of hunting them is daytime shooting under fruiting trees, a technique infrequently used because of high detection risks. Sambar are secretive and keep to dense cover, visit fruiting trees infrequently, and do not bed in the open. This makes nighttime spotlight hunting of this species on foot difficult. Adults are normally not snared because of their large size, but young sambar are occasionally caught. The solitary nature of muntjac and sambar, their affinity for dense cover, and logistical difficulties in hunting ensure that hunting pressures on these species are low at both sites. These factors may explain the absence of significant differences in densities of muntjac and sambar between sites.

Our results indicated significantly higher elephant densities at Arkeri. Elephants are too large to be hunted without detection, regardless of the technique employed. Our questionnaire survey revealed cultural restrictions among all local communities against killing elephants for their meat. Elephants are occasionally killed while raiding crops and, rarely, for their ivory. Considering, however, that elephants undertake large-scale seasonal movements, density differences between the two study sites are more likely to be an artifact of the spatial scale on which their abundance was sampled, rather than indicative of any real difference.

Overall, hunters at Arkeri target small and medium-sized prey that are easy to kill using specific hunting techniques. Although the level of protection at Arkeri was lower that that at Nalkeri, it sufficiently deterred daytime shotgun hunting and the hunting of large species. A breakdown in current protection levels will almost certainly expose even the largest species, such as gaur and sambar, to overhunting, which has depressed the densities of smaller species so strikingly. A range of highly specialized traditional techniques (see table 17-1) also target smaller species, whose abundances we did not estimate (figure 17-3). The impact of hunting on these species, most of which are vulnerable tree cavity and burrow dwellers, might be even more serious.

ECOLOGICAL IMPLICATIONS OF LARGE MAMMAL DECLINES

It is well recognized that persistent overhunting lowers animal densities and subsequently leads to local, regional and overall extinctions of species (Diamond 1989; Nitecki 1984; Rabinowitz 1995). Past experience has shown how hunting works synergistically with forces of habitat degradation and fragmentation to produce local and regional extinctions of large mammals in Asia (e.g, rhinoceros [*Rhinoceros unicornis, Dicerorhinus sumatrensis, D. javanicus*], banteng [*Bos javanicus*], and Nilgiri tahr [*Hemitragus hylocrius*]). Large ungulates are known structurally to modify habitats and affect plant regeneration, thereby influencing forest composition in the long term. Lowering their densities could trigger changes in the habitat accompanied by a cascading effect on smaller herbivores. The focused killing of keystone frugivores and seed dispersers by hunters in Nagarahole (see table 17-1) could set off such long-term changes. Furthermore, lowered densities of prey such as ungulates, resulting from overhunting, can significantly reduce the carrying capacity, recruitment, and survival rates of large predatory carnivores such as the tiger, leopard, and dhole. Karanth and Stith (unpublished data) have used demographic models of tiger populations to suggest that prey depletion is likely to be a more critical determinant of the viability of tiger populations than is direct hunting of tigers. Hence, at Arkeri, where the density of medium-bodied prey is greatly reduced by hunting (86% for chital, 92% for pigs; see table 17-4), populations of large predators are almost certain to be affected.

CONSERVING LARGE MAMMALS IN INDIA: PRESERVATION OR SUSTAINABLE USE?

The preservationist model holds that it is virtually impossible to manage hunting under the social conditions that prevail in India. In precolonial India, hunting for

Figure 17-3. Bow and clay shot, locally known as Thattä Billu, a traditional method of hunting small and medium sized animals. *Photo courtesy of M. D. Madhusudan.*

subsistence and for feudal sport was already a widespread activity, relying on native techniques and skills. After the advent of colonial powers, introduction of firearms and systematic trophy hunting aggravated the kill (Gadgil 1992; Rangarajan 1996). State-sponsored bounty hunting of large predators was also introduced in some regions (Rangarajan 1996). With increasing economic development and social mobility, hunting patterns of colonialists were assimilated by upper class Indians, intensifying impacts. Finally, in the 1960s, recognizing the severity of the wildlife declines (Schaller 1967), Indian conservationists, backed by high-level political support, succeeded in establishing antihunting laws and wildlife preserves. The presence of a regimented forest service capable of implementing the new antihunting laws on ground was a key ingredient of this endeavor. After 1974, legalized hunting was successively restricted and was finally banned in 1991 in most of India.

The preservationist model thus advocates total banning of hunting as the only practical, clear-cut, enforceable policy. It relies on the use of force as a deterrent against illegal hunting. Prima facie, this system has been reasonably successful in areas where adequate manpower and money were invested in protective efforts. As our results show (see table 17-4), higher densities of most large mammals at Nalkeri are strongly correlated with the strict protection of the area against hunting. All local respondents (n = 32) in our survey noted a decline in hunting rates over the past 5 years, and 91% attributed this decline to intensified enforcement. Another successful institutionalized example of the Indian preservationist model is the recovery of tiger and prey populations in several reserves under Project Tiger (Panwar 1987). On the other hand, in the hill states of northeast India, it has not been possible to implement the preservationist model because of lack of government control over community-owned forest land and due to political and ethnic factors. As a result, the impact of overhunting on wildlife has been severe (Choudhury 1995, S. Deb Roy and M. K. Ranjithsinh, personal communication).

Critiques of the Preservationist Model

A preservationist approach potentially antagonizes a section of the local community denied access to forest resources. Consequently, it has come under criticism for ". . . treating conservation as a matter of keeping people out of a few large nature reserves and preventing them from killing most larger species of wild reptiles, birds and mammals." Its style of ". . . imperialistic . . . [operation] . . . with an impersonal centralized bureaucracy . . . [working by] . . . alienating local people from control of, and access to resources" has been lambasted (Gadgil 1991, 1992). The preservationist model also has been criticized for ignoring the potential for using wildlife resources sustainably for the benefit of local people, and making them stakeholders in wildlife conservation (IUCN/UNEP/WWF 1980; IUCN/UNEP/WWF 1991; Allen and Edwards 1995). Whether originating from an Indian traditionalist perspective (Gadgil and Malhotra 1985; Gadgil 1992), or from a Western technocratic perspective (Eltringham 1984; Robinson and Redford 1991b, 1994a), critiques of preservation argue that hunting can be reconciled with wildlife conservation if annual harvests are maintained at levels at or below annual productivity. Although Indian traditionalists argue that such regulation can be achieved subtly by reinforcing folk traditions, the technocrats favor a deliberate, rationalized management approach. The point, however, is this: Can this happy state of affairs be achieved in real-world India?

Critiques of the Sustainable Use Model

The sustainable use model of hunting management is founded on the premise that it is essential to make hunting compatible with large mammal conservation. This premise is itself questionable in the Indian context, where most viable populations of large mammals occur within wildlife reserves that comprise less than 5% of the land area (ICFRE 1995). Furthermore, tropical forest mammals are not candidate taxa for unproven wildlife production systems because of their intrinsically low densities and productivities (Eisenberg 1980, 1981; Robinson and Bennett this volume.) Any attempt to institute regulated hunting will contribute to pushing hunting levels above sustainable limits. The situation would be exacerbated by the high market values for illegal wildlife by-products (bones, hides, horns) and existing hunting pressures in defense of crops or livestock. Moreover, market linkages between the direct beneficiaries and potential consumers mean that demand for meat and other more valuable wildlife by-products is likely to be far higher than the productivity of the biological system.

In addition to the above biological considerations, sustainable use of large mammals is exceedingly difficult to define and implement in the Indian context for several social reasons. First, there is no true subsistence economy based on hunting large mammals, with exceptions, perhaps, of a few remote sites in northeastern India. On the other hand, India is characterized by high human population densities of potential consumers. Furthermore, the prevailing social divisions in terms of caste and class make it impossible to define who the beneficiaries in any wildlife harvest scheme would be. Whatever the defining yardstick, the number of beneficiaries would be so large that per capita returns from any truly sustainable harvesting of wildlife are likely to be economically insignificant. Lastly, systemic abuse, which plagues all other natural resource extraction enterprises in India (forestry, fisheries, mining), leading to well-recognized failures, is almost certain to affect wildlife harvesting systems. The precarious status of India's wildlife reserves allows no room for such failures.

Proponents of sustainable use (Gadgil et al. 1983; Gadgil 1990) have documented the spectacular, century-old failure of foresters in India to harvest trees sustainably under the technocratic model and urged more community control. But if the sustainable harvest of tropical trees, with their long life spans and relatively predictable life histories cannot be achieved without disrupting biological communities, we fear that sustainable harvest of large mammal communities, with their poorly understood dynamics, will be an impossible goal.

There are no examples (supported by reliable data) of successful self-imposed, community-based, traditional regulation of large mammal harvest that demonstrate such sustainability. In the absence of data, we believe that mere reinforcement of native "conservation" traditions is a naive management strategy, fraught with grave risks to large mammals. The rare Indian examples of traditional wildlife preservationism (e.g., the Bishnois of Rajasthan who protect antelopes) (Gadgil and Malhotra 1985), are completely overwhelmed by numerous examples of rampant native wildlife consumerism (tradition-driven hunting for meat, hides, horns, and other wildlife products). Moreover, wildlife managers in India experimented with regulation of hunting between 1955 and 1972 and failed. Learning from this, they progressively closed down all legal hunting between 1972 and 1991.

Our data show high densities of large mammals in Nalkeri, which has benefited from exemplary protection, and distinctly lower densities in Arkeri, even with some degree of hunting control. On the other hand, observations from north-eastern India (Choudhury 1995, S. Deb Roy and M. K. Ranjithsinh, personal communication) and southeast Asia (Rabinowitz 1991, 1995; WCS 1995a; Bennett et al. this volume) demonstrate the virtual elimination of large mammal fauna in the absence of effective on-ground hunting controls, despite the existence of strong regulations on paper.

CONCLUSION

Although the Indian preservationist model of wildlife conservation undoubtedly suffers from some weaknesses that need to be addressed (Gadgil 1992; Kothari et al. 1995), the model does recognize the biological and social problems of applying the sustainable wildlife use approach in tropical forests. The preservationist model's efficacy for conserving large mammals is clearly demonstrable through data-driven examples, unlike the other alternative models. Ensuring the survival of large Asian mammals into the twenty-first century is a challenge that the conservation community must face realistically. Our data support Robinson (1993), who argued that "sustainable use is a powerful approach to conservation, but it is not the only one." Although conservationists continue the hunt for a biologically sound and socially acceptable answer, hunting, clearly, is not that answer.

ACKNOWLEDGMENTS

We are indebted to Wildlife Conservation Society, New York, for funding the study and to Karnataka State Forest Department for facilitating our work. We are also grateful to K. M. Chinnappa, Neelamma Chittiappa, K. K. Lakshman, Ranjan Poovaiah, and V. Ramakantha for their valuable support. We acknowledge the assistance provided by Forest Department Staff of Nagarahole, the volunteer naturalists who helped with line transect surveys, and all the informants who participated in our questionnaire surveys.

Appendix 17-1. Scientific Names of Animals Referred to in Text

Common Name	Scientific Name
Mammals	
Indian pangolin	*Manis crassicaudata*
Indian porcupine	*Hystrix indica*
Large brown flying squirrel	*Petaurista petaurista*
Indian giant squirrel	*Ratufa indica*
Black-naped hare	*Lepus nigricollis*
Indian chevrotain	*Tragulus meminna*
Chital	*Axis axis*
Sambar	*Cervus unicolor*
Muntjac	*Muntiacus muntjao*
Wild pig	*Sus scrofa*
Four-horned antelope	*Tetracerus quadricornis*
Gaur	*Bos gaurus*
Asian elephant	*Elephas maximus*
Mongoose	*Herpestes* spp.
Common palm civet	*Paradoxurus hermaphorditus*
Small Indian civet	*Viverricula indica*
Tiger	*Panthera tigris*
Leopard	*Panthera pardus*
Dhole	*Cuon alpinus*
Bonnet macaque	*Macaca radiata*
Hanuman langur	*Presbytis entellus*
Birds	
Pheasants	*Galloperdix spadicea*
	Gallus sonneratii
	Pavo cristatus
Pigeons	*Family Columbidae*
Parakeets	*Psittacula* ssp.
Owls	Family Strigidae
Hornbills	*Tockus griseus*
	Anthracoceros coronatus
Barbets	*Megalaima* spp.
Woodpeckers	Family Picidae
Reptiles	
Indian monitor lizard	*Varanus bengalensis*
Chelonians	*Lissemys punctata*
	Melanochelys trijunga

Nomenclature follows Daniel (1993) for reptiles, Ali and Ripley (1983) for birds, and Prater (1980) for mammals.

18

Enhancing the Sustainability of Duiker Hunting Through Community Participation and Controlled Access in the Lobéké Region of Southeastern Cameroon

CHERYL FIMBEL, BRYAN CURRAN,
AND LEONARD USONGO

There is increasing evidence that mammal populations of forests in east, central, and west Africa are being overhunted (Fa et al. 1995; Fitzgibbon et al. 1995a; Muchaal and Ngandjui 1995; Wilkie et al. 1997). Even mammals in relatively large intact forest blocks such as the Lobéké region (> 5000 km²) of southeastern Cameroon and the Okapi Reserve (13,000 km²) in the Ituri forest of northeastern Zaire, areas traditionally supporting low human population densities (fewer than two individuals per km²), are showing signs of overhunting (Wilkie et al. 1997; Hart this volume).

Residents of the Lobéké region of southeastern Cameroon, like most of Africa's rural communities, make extensive use of natural resources, including wild meat and aquatic resources such as fish and shrimp, for subsistence and barter trade. Hunting activities in the region are increasing due to deteriorating economic conditions and the depletion of wildlife elsewhere in Cameroon and the neighboring Republic of Congo and Central African Republic (CAR). In addition to the traditional residents of the Lobéké region, a growing number of nonindigenous peoples are migrating to Lobéké in search of employment with timber and safari hunting companies, and to hunt wild meat illegally for urban commercial markets.

Despite the many levels of forest resource exploitation, including commercial hunting of birds and mammals, and industrial logging, little of the revenue generated is returned to the area. Restricted finances have resulted in a limited presence by the government institution responsible for managing and conserving natural resources in this relatively isolated region. Consequently, the forest resources have been viewed as common-property resources by all stakeholders, including traditional residents, recent immigrants, safari hunters, parrot trappers, and logging company employees. Historically, the abundant resource base in the Lobéké region was sufficient to meet most demands, with limited competition between different users. Now there are signs that some species are being extracted at levels in excess of renewal rates. Thus, future competition will become more severe as favored species such as duikers become increasingly scarce, unless wildlife management programs are rapidly instituted to control the harvest.

Our analysis begins by addressing the question of whether duiker hunting as currently practiced in the Lobéké region of southeastern Cameroon can be sustained in the long-term. The analysis is a treatment of data obtained from several studies conducted concurrently in the forest and villages of the Lobéké region of southeastern Cameroon, complemented by information derived from published literature. Using Lobéké as a model, we then review options for managing hunting in African forests where finances, legal and institutional mechanisms, and the political will for enacting wildlife management are limited. The importance of creating designated protected and hunting zones, and adapting management to local social conditions, are discussed.

THE LOBÉKÉ REGION

The Lobéké forest region, an area of approximately 5000 km², is in the northwestern corner of the greater Congo Basin and is characterized by lowland (300–700 m elevation) evergreen and swamp semi-deciduous forests (CTFT 1985; Letouzey 1985 [figure 18-1]). The forests of the region, although not as floristically diverse as evergreen forests in other areas of the Guineo-Congolean or Congolean forest blocks, still support a highly diverse plant community minimally disturbed by recent human activities (Hall 1994; WCS 1996). In addition, they support extraordinarily high densities of large mammals, including elephants, gorillas, and bongo (Stromayer and Ekobo 1991).

A central protected zone for the Lobéké forest was proposed in 1995 to prohibit all hunting and timber exploitation activities (figure 18-1). There are no permanent human habitations within the proposed reserve, although hunting is widely practiced there. The ramifications of establishing this reserve are currently under review by the Cameroonian government.

The traditional inhabitants of the Lobéké region comprise the Bangando and Bakwele ethnic groups of Oubangui origin, and the semi-sedentary forager-farmer

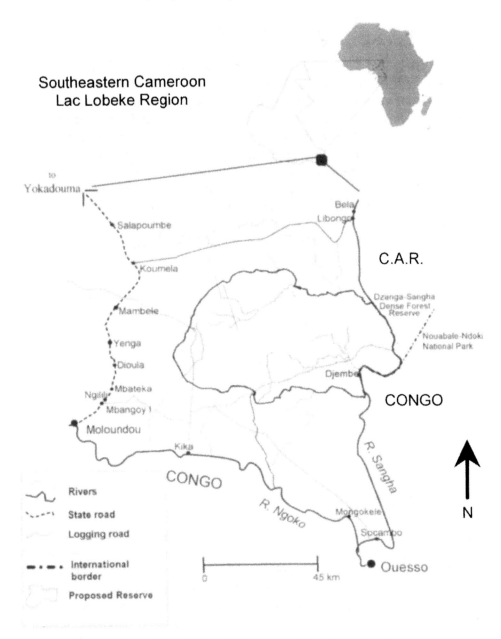

**Southeastern Cameroon
Lac Lobeke Region**

to
Yokadouma

Salapoumbe

Koumela

Mambele

Yenga

Dioula

Ngilili· Mbateka

Mbangoy 1

Moloundou

Kika

CONGO

Bela
Libongo

C.A.R.

Dzanga-Sangha
Dense Forest
Reserve

Nouabale-Ndoki
National Park

Djembe

CONGO

R. Sangha

R. Ngoko

Mongokele

Socampo

Ouesso

N

Rivers

State road

Logging road

International
border

Proposed Reserve

0   45 km

Figure 18-1. Map of the Lobéké region of southeastern Cameroon.

Baka pygmies. There is a smaller nonindigenous sector contributing roughly 10% to the region's population, made up of Moslem traders and Cameroonians from other areas of the country, plus individuals from the Congo and the CAR that work for the logging companies and commercial bush-meat operators. Population numbers are low on a regional level (fewer than two individuals/km²), with the majority concentrated along the main road running north to south between Yokadouma and Moloundou, and at the sawmill/logging sites throughout the region (see figure 18-1).

Hunting and the consumption of wild meat, especially duikers, is very much a part of the culture and economy of the traditional residents, particularly the Baka, who grow few agricultural products other than plantains, some root crops, and leafy greens (WCS 1996). Domestic meat such as goats and chickens is not considered normal dietary fare and is consumed only during festive occasions. At present, game meat is still relatively easy and inexpensive to obtain compared with domestic animal meats and agricultural or imported products, and local residents have few incentives to change their dietary habits. Local subsistence hunters also use their catch to barter for goods among neighbors and to sell to external markets to gain money for social and subsistence needs (soap, cooking oil, salt, clothing, health care, and school fees). Hunting was found to be very lucrative in villages adjoining the nearby Dja Reserve of southern Cameroon, providing up to four to five times more income than with the sale of cocoa, one of two major commercial agricultural crops of the region (Muchaal and Ngandjui 1995). In summary, hunting plays a major role in the economy of southeastern Cameroon, and the greater region as a whole, including Kabo and Ouesso in the neighboring Republic of Congo (Blake 1994).

Hunters

Nearly all residents of the region engaged in some aspect of the hunting process, if only to help fashion weapons or occasionally tend traps of another family member. The primary hunters were young males between the ages of 16 and 25, although men estimated to be 73 still engaged in hunting. Hunter interviews revealed that the hunting population, as with the general population, was composed nearly equally of Baka and Bangando, although a greater proportion of the young Baka hunted, whereas their Bangando age-mates were generally occupied with attending school.

Documentation of Hunter Harvest

Estimates of wildlife harvest by 46 subsistence hunters were calculated from hunting data gathered over 17 months (November 1994 to March 1996) by seven resident assistants in villages located along the main Yokadouma-Moloundou road. A reference list of animal names was generated in Baka and Bangando languages for use by the assistants. The seven local assistants, representing both Bangando and Baka ethnic groups, were trained in the use of standardized data forms and spring balance scales. Each assistant worked with a fixed group of approximately four to eight individual hunters crosscutting ethnic groups, so that each assistant collected data from Bangando and Baka hunters alike. Hunters were selected by the assistants based on the hunter's willingness to participate in the study and the proximity of the hunter's residence to the assistant's residence to facilitate communication and data collection. An effort was made to include representative hunters from the region by avoiding especially avid or unskilled hunters. The assistant collected biological data (species, weight, age, sex) on the animals killed, as well as hunting technique and location, and destination of meat (local consumption or sale). The local assistants remained in their villages to record kills brought back to the village and did not accompany the hunters on their trap lines. They were paid minimal monthly sums for this part-time work, and the hunters received monthly gifts of tobacco and cola nuts for their cooperation. There were no rewards based on the number of animals captured to avoid providing an incentive for additional kills. These methods resulted in a conservative estimate of bushmeat offtake by villagers because the total harvest of animals taken during lengthy hunting expeditions in the forest were not recorded, nor is it likely that all animals killed (animals lost to spoilage, etc.) were reported despite our efforts to encourage reporting of all kills and assurances of confidentiality. It is worth noting that both (a) illegal kills such as gorillas, sitatunga, and bush-pig and (b) spoiled meat were recorded by the assistants, but we suspect they were under-represented.

HUNTING IN THE REGION

Hunting Methods

The most common hunting method employed wire snares (84% of captures), which captured virtually the entire range of exploited wildlife. Other technologies were specialized for particular game, including guns (6% of captures) used for arboreal species and gorillas, and occasionally during night hunts with spotlights for forest antelopes. Crossbows (3%) were used exclusively for arboreal species such as palm civets, birds, and especially primates. Spears (4%) and machetes (3%) were typically employed fortuitously, the former often in conjunction with dogs for antelopes. Machetes were used for small ground-dwelling species such as pangolin, turtles, rats, and porcupines.

Figure 18-2 shows the frequency of captures according to study zone and distance from the main Yokadouma-Moloundou road. Although distances were estimated, the local residents were familiar with measuring distances in kilometers in the forest because the timber companies conducted measured surveys in these forests with the help of local residents and also left behind measurement markers when their work was finished.

Harvest Composition

Thirty-six different animal species were captured at least once during 354 hunter/months of data collection, yielding a total of 1024 captures. Table 18-1 summarizes the frequencies and biomass of the 17 species that were trapped at least ten times during the course of this study. Blue duikers (*Cephalophus monticola*) were trapped most often (325 animals, 32% of all captures), yielding 14% (1393 kg) of total captured biomass. By comparison, the larger Peter's duiker (*C. callipygus*) and bay duiker (*C. dorsalis*) were trapped less often, but their contribution to the harvest biomass was greater (186 animals = 26% [~2600 kg], and 97 animals = 14% [~1400 kg], respectively). The red duikers as a group

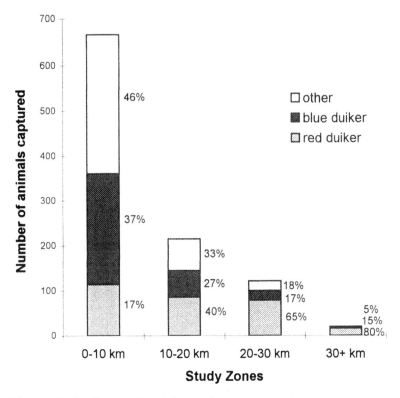

Figure 18-2. Composition of game harvest in southeastern Cameroon, 1995, according to four study zones at different distances from the main road. The "other" category includes all non-*Cephalophus* species, including brush-tailed porcupines, primates, and giant rats.

(*C. callipygus, C. dorsalis, C. nigrifrons,* and *C. leucogaster*) were clearly important to hunters in southeastern Cameroon, constituting 29% of all animals recorded captured and 44% of the harvested biomass. All duikers considered together (including *C. monticola*) accounted for 61% of the captures and 58% (5713 kg) of the harvested biomass recorded for the region.

Primates, especially white-nosed monkeys (*Cercopithecus nictitans*) and gorillas (*Gorilla gorilla*), were hunted regularly. Chimpanzee captures were not recorded by hunters during this study, nor are they known to be taken on a regular basis in the region. Among prosimians, the potto (*Perodicticus potto*) is sometimes captured near villages.

Calculation of Duiker Harvest Rates

A total of 325 blue duikers and 296 red duikers was taken over the course of 354 hunter months in the Lobéké region. Annual offtake rates for duikers were calculated using per hunter kill rates per study zone, in conjunction with the total number of hunters and the size of the study zones. Based on a combination of

Table 18-1. Total Number of Captures and Total Biomass of Species Captured at Least 10 Times During 354 Hunter Months in the Lobéké Region of Southeastern Cameroon, 1995

Common Name	Scientific Name	Total Frequency	Total Biomass (kg)
Blue duiker	*Cephalophus monticola*	325	1393
Red duiker (group)[a]	*C. callipygus/dorsalis/leucogaster/nigrifrons*	296	4162
Peter's duiker (red duiker)	*Cephalophus callipygus*	186	2600
Bay duiker (red duiker)	*Cephalophus dorsalis*	97	1403
Brush-tailed porcupine	*Atherurus africanus*	83	270
Primates (excluding apes)		61	381
Giant rat	*Cricetomys* spp.	33	53
Long-tailed pangolin	*Manis tetradactyla*	31	64
Bate's pygmy antelope	*Neotragus batesi*	30	107
Tree hyrax	*Dendrohyrax arboreus*	16	52
Sitatunga	*Tragelaphus spekei*	16	821
Bush-pig	*Potamochoerus porcus*	16	472
Water chevrotain	*Hyemochus aquaticus*	14	125
Tortoise	*Chelonia*	14	21
Gorilla	*Gorilla gorilla*	13	1550
African linsang	*Poiana richardsoni*	11	5
Palm civet	*Nandinia binotata*	10	18
Marsh mongoose	*Atilax paludinosus*	10	34

[a]Includes all red duikers captured, including those recorded separately if the species was known.

formal and informal village surveys, we estimated that 80% of 1330 adult males over the age of 15 in the local population actively engaged in hunting, yielding a total of 1064 hunters (unpublished census data, WCS 1996). This yields an estimated total annual harvest of 11,619 blue duikers per year and 10,597 red duikers per year by local hunters of the region. To calculate annual harvest estimates for the three study zones within 30 km of the villages, the percentage of captures in each zone was multiplied by the regional totals to determine the number of captures per zone. This was then divided by the area of each zone, which was approximately 1003 km² per zone, to yield a density figure (table 18-2). The Reserve study zone, defined as more than 30 km from the main road, was unbounded, and therefore harvest rates on a per km² basis could not be determined.

There are several potential errors in these figures. The first is that we do not know to what extent the 46 hunters cooperating in this study provided representative harvest figures. Therefore, we should consider our values to be maximum estimates. In contrast, we doubt that 100% of all captures were reported, for example, if the hunter was too tired to inform the resident technician, or the animal was rotten. In this way, the harvest figures become minimal estimates. In summary, these two unknown factors probably cancel one another out, in which case the offtake estimates are representative. Finally, this study concentrated on village hunters and did not incorporate the harvest by nonindigenous commercial hunters. The latter removed 29 animals/hunter/month from the forest, a rate ten times higher than a typical village hunter (WCS 1996). Thus, although these hunters are relatively few in number (45 were interviewed in 1995; WCS 1996), their high animal capture rates resulted in a considerable harvest. The harvest

Table 18-2. Estimated Harvest Rates of Blue and Red Duikers According to Study Zones[a] in the Lobéké Region of Cameroon, 1995

	Harvest Rate (Duikers/km²)			
Common Name	*Zone A (0–10 km)[b]*	*Zone B (10–20 km)[b]*	*Zone C (20–30 km)[b]*	*Regional Average*
Blue duikers	8.20	2.07	0.71	3.2
Red duiker group[c]	3.79	2.98	2.79	3.7

[a]Harvest estimates per km² for the fourth study zone. Reserve could not be calculated because the distance for this zone was defined as 30+ km from the main road, and therefore was an unbounded area, precluding calculation on a density basis.

[b]Distance from main road.

[c]Includes *C. callipygus, C. dorsalis, C. nigrifrons,* and *C. leucogaster.*

figures of the commercial hunters were not incorporated into this analysis due to a different study design. In summary, our estimates of harvest by local villagers are minimum estimates of duiker removal for the region.

Estimating Duiker Densities

The current synthesis centers on a treatment of the small-bodied (~4.5 kg) blue duiker (*C. monticola*) and the larger-bodied (~17.5 kg) red duiker group. The abundance of duikers was estimated using census walks to record abundance of signs, primarily dung, along straight line transects. Because the dung pellets of the red duikers (*Cephalophus callipygus, C. dorsalis, C. nigrifrons,* and *C. leucogaster*) were of a similar size and shape, it was difficult to distinguish the dung of these four species. Thus, these duikers were combined into a common category of "red duikers" (figure 18-3) to distinguish them from the smaller blue duiker.

Figure 18-3. White-bellied duiker (*Cephalophus leucogaster*), one of the four red duikers in the Lobéké region. *Photo courtesy of William Karesh.*

The forest east of the main north-south road was stratified into four zones: A, B, C, and the Reserve (0–10 km, 10–20 km, 20–30 km, and >30 km from the main settlement road, respectively), representing progressively decreasing levels of human hunting. Fifteen transects of 5-km length were distributed among the four study zones, and each was sampled three times during the study period. Transects were normally oriented on east-west bearings and were minimally cut to avoid bias in subsequent samplings due to use of the path by animals and to prevent their use by hunters. Transects were walked at an average speed of 1 km/hr by a team of three persons, normally between the hours of 0700 and 1300, and adhering to the general guidelines for line transect sampling described by Burnham et al. (1980) and Barnes and Jensen (1987). All animal signs were recorded, and estimated ages and perpendicular distances were included for direct sightings and dung. Transect widths and then density estimates were calculated for blue and red duiker dung using the computer software program DISTANCE (Laake et al. 1993). Dung density was converted to animal density using the following equation (Barnes and Jensen 1987):

Animal Density = dung pile density / (defecation rate/day) × (dung decay period)

Values taken from the literature (Koster and Hart 1988) gave defecation rates of 4.9 and 4.4 pellet groups/day and decay periods of 18 days and 21 days for blue and red duiker pellets, respectively. We recognize that these estimates are imprecise, and their applicability to species and conditions in the Lobéké forest are questionable because dung defecation and decay rates vary with climate, habitat, and diet. However, in the absence of data from our site, they allow calculation of density estimates, bearing in mind the constraints on accuracy. The nonparametric Kruskall-Wallis procedure for analysis of variance was used to test for differences in the frequency of duiker sign among study zones.

Duiker Densities in Relation to Human Settlement Patterns

Estimates of blue duiker densities were similar throughout the four study zones (table 18-3), with no significant differences detected between study zones. Lahm (1993a) also found blue duikers at relatively similar densities in zones that were close to settled human populations and in remote areas of Gabon. Despite the high harvest rates, blue duiker density data obtained in this study do not reveal a significantly lower population in the most heavily hunted zone, zone A. At this time, the heavy hunting pressure has not produced a measurable effect on the blue duiker population.

In contrast, red duiker density estimates were significantly lower in zone A than in the Reserve zone ($H = 14.476$, 3 df, $p = 0.002$; see table 18-3). The red duikers appear to be showing a response to the high hunting pressure in zone A, the region of highest human activity, and to a lesser extent in the adjacent zones B and C. The higher densities in the relatively remote Reserve zone are presumably the result of less hunting pressure or reduced habitat disturbance from logging or agriculture.

Thus, blue and red duikers appear to respond to hunting in the Lobéké forest

Table 18-3. Densities of Blue and Red Duikers According to Study Zone in the Lobéké Region of Cameroon, 1995

	Animals/km²				
Common Name	Zone A (0–10 km)[a]	Zone B (10–20 km)[a]	Zone C (20–30 km)[a]	Reserve (>30 km)[a]	Regional Average
Blue duiker					
Density	3.61*	2.26*	1.41*	3.82*	2.78
Lower and upper 95% confidence interval limits[b]	2.15–6.06	0.97–5.26	0.63–3.19	0.42–35.05	
Red duiker group					
Density	2.45*	5.50*†	6.31*†	15.12†	7.35
Lower and upper 95% confidence interval limits[b]	1.21–4.95	0.68–44.58	2.39–16.67	5.38–42.52	

[a]Distance from main road.

[b]Confidence intervals shown are not comprehensive, including only the variance from the dung density estimates, ignoring the variability associated with the other components of the duiker density equation such as the defecation and decay rates for duiker dung, which are not known. Therefore, these confidence intervals should be considered as the minimum ranges.

[c]Where densities of a species or species group are followed by the same symbol, they are not significantly different at $p \le 0.05$.

in different ways. Blue duiker populations appear to be more resilient to hunting pressure than red duiker populations. The reasons for this difference are not known, but it is likely due to the shorter gestation periods and greater compensatory reproduction in blue duikers compared with red duikers (Hart this volume). Other contributing factors include higher red duiker mortality rates due to selection by leopards (Hart et al. 1996), and higher rates of movement by red duikers, with consequent greater potential of being caught in a snare (Hart this volume).

Struhsaker (1997) determined that selective logging had long-lasting negative effects on duiker populations in the Kibale Forest of Uganda, with blue duikers more adversely affected than red duikers. Wilkie and Finn (1990) reported that roadside agriculture had little effect on the densities of blue and red duikers (although dung counts revealed more red duikers in climax forest than secondary forest). However, where there are high levels of hunting in conjunction with habitat disturbance such as logging or agriculture, hunting is usually determined to have the greater impact on wildlife populations (Wilkie et al. 1992, 1997; Bennett and Gumal in press). In the Lobéké region, relatively low levels of logging and clearing for agriculture are widely distributed, such that the impacts from the heavy hunting pressure are expected to play a greater role in influencing the densities of preferred game species such as duikers.

SUSTAINABILITY OF HARVESTS

Maximum Possible Sustainable Harvest Rates

Robinson and Redford (1991b) developed a model for calculating species-specific maximum sustainable harvest levels based on the inherent ability of the species to replace losses. This model quantifies the number of animals expected to join the target population through births and immigration, labeled the annual maximum production (P_{max}). Lacking data on the actual rate of increase of a population, a theoretical value may be calculated based on reproductive parameters of the species (Robinson and Redford 1991b). Table 18-4 shows values used to calculate P_{max} for red and blue duikers.

Redford and Robinson (1991b) provide recommendations for potential harvest levels according to a species' typical longevity because this also serves as an index of the natural rate of replacement for a species. These investigators proposed harvest levels of 60% of annual production for very short-lived species (<5 years), but harvest levels of 40% for short-lived species (5–10 years) and 20% for long-lived species (>10 years). The duikers under consideration in this study probably fall into the "short-lived" category because they are likely to live more than 5 years, but not more than 10 years (Ralls 1973; Payne 1992). In accordance with this model, 40% of their annual P_{max} would be considered a sustainable hunting level over the long term.

Comparisons of Actual and Theoretically Sustainable Rates

Figure 18-4 shows the harvest rates observed in this study in relation to the theoretical maximum sustainable harvest according to study zone. Blue and red duikers are being harvested at rates far in excess of the theoretical maximum sustainable harvests. Thus, according to the model, the observed harvest rates are not likely to be sustainable over a long period of time. Our study is only a snapshot of time, but it is our assumption that the data reflect an increase in hunting pressure compared with historical levels, and subsequent response by duiker populations, rather than a relatively steady-state (dynamic equilibrium) phenomenon. This is supported by (a) claims of local people that they need to go farther into the forest to catch duikers compared with recent memory, and (b) the recent immigration of

Table 18-4. Dietary and Biological Characteristics of Blue and Red Duikers According to Published Literature

Common Name	Dietary Category	Body Mass	Age at First Reproduction	Age at Last Reproduction	Annual Female Production Rate	r_{max}	λ_{max}
Blue duiker	FH[a]	4.7 kg[b]	1.07 yr[c]	7 yr[d]	0.68/yr.[c]	0.49	1.63
Red duikers[e]	FH	18.8 kg[b]	1.67 yr[f]	8 yr[d]	0.35/yr[f]	0.21	1.24

[a]Dietary category FH is frugivore/herbivore.
[b]Koster and Hart 1988.
[c]Von Ketelhodt 1977, in Payne 1992.
[d]Payne 1992: estimates based on consideration of captive longevity.
[e]Refers to the combination C. callipygus, C. dorsalis, C. nigrifrons, and C. leucogaster, and characteristics for this group were calculated using weighted average data from C. callipygus and C. dorsalis according to their approximate respective contributions to the harvest data (67% and 33%).
[f]Feer 1988, in Payne 1992.

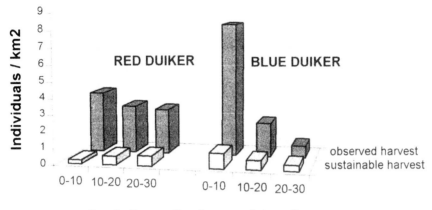

Figure 18-4. Relationship between the theoretical maximum sustainable harvest and the observed harvest of red and blue duikers in the Lobéké region, Cameroon, 1995, according to study zones.

Cameroonian and Congolese hunters from areas where wild meat has been depleted in recent years. Additionally, the logging industry is fairly new in southeastern Cameroon, having commenced large scale operations in the early 1970s. This industry brought many changes to the Lobéké region, including (a) increased access to the forest interior through the creation of logging roads; (b) the introduction of a nonindigenous labor force; and (c) a mechanism for transporting wild meat to external markets on logging trucks. Finally, the recent economic crisis has caused more people in the central African region to return to natural resources to earn a living (Lahm 1993b). All of these factors translate to increased activity in the wild meat trade (figure 18-5), with heavier pressure on areas that contain large ungulates, such as the Lobéké region. Therefore, it is evident that pressures on wildlife have been increasing in the recent past, and we suggest that the resulting high harvest rates observed cannot be sustained into the future.

OPTIONS FOR MANAGING HUNTING IN LOBÉKÉ

Spatial Management of Harvest

In the more developed nations, the harvest of game animals is generally regulated through a numerical, or quota, system in which limits are placed on the number of animals harvested. Alternatively, or in conjunction, game harvest may be regulated via a spatial system, where the areas open to hunting are limited, perhaps incorporating a temporal element as well, such as a restricted hunting season. These systems require relatively sophisticated knowledge of the biology and harvest statistics of the game species being regulated, along with large inputs of effort and infrastructure by state regulating agencies. Both of these requirements are typically lacking for game species and wildlife agencies in Africa. Thus, a suitable

Figure 18-5. Guenons being sold along the road. *Photo courtesy of John G. Robinson.*

alternative to regulating harvest of game, especially duikers, is needed to ensure the continued survival of hunted species and, thus, future hunting opportunities over the long term. A hunting management system suitable for use in remote areas must require little in the way of (a) knowledge of the density, ecology, behavior, and demographics of the species being harvested, as well as (b) financial inputs by the state agency charged with managing the country's fauna. Furthermore, a harvest system should be relatively foolproof, such that it is resistant to unintended overharvest of preferred species (McCullough 1996a). A system based on spatial controls of harvest would best meet these challenges for continuously distributed animal populations (McCullough 1996a) in the Lobéké region.

Spatial Harvest Theory and the Importance of Protected Areas

Spatial harvest theory (McCullough 1996a) is based on the concept that the area under management is divided into hunted and unhunted (protected) zones and that animals are free to move among these zones. A further caveat is that the protected area, or refuge, must be large enough to support a minimum viable population of the species being hunted. Thus, the primary limiting factor to over-hunting is a defined protected zone. As long as animal populations in protection zones remain unexploited, animals may be hunted to extirpation in adjacent hunting zones without eliminating the population as a whole (although such intense exploitation is obviously not recommended as a harvest goal). In this way, hunting will be self-regulating because hunting effort will probably decrease as the catch/unit effort decreases below a threshold at which hunting is considered not worth the effort. These thresholds are imprecise because they will vary with hunter and species. Ideally, spatial harvest systems are meant to manage hunting by restricting the offtake of animals to the annual production in designated hunting zones. This limits the catch to the annual production of animals within those hunting zones, as well as the animals migrating to these zones. If both annual production and reproducing stocks are depleted within hunted zones, causing a decrease in the population density, dispersal into the hunted zones is expected to continue as long as the protected areas serve as centers of reproduction, thereby providing stocks for replenishing adjacent depleted populations.

The size of the area open to hunting in relation to the protected area is a critical consideration for effective management of hunting. The protected area must be large enough to sustain minimum viable populations of the hunted species, allowing for annual movement patterns according to species' needs within the protected zone. However, this consideration needs to be balanced with the dispersal characteristics of the species being hunted, such that the protected area allows for dispersal of some animals out of the protected zone and into the open hunting area to replenish depleted stocks. Thus, the size and arrangement of the hunted and protected areas will influence the efficiency of the harvest (McCullough 1996a).

The most efficient system in terms of hunter harvest would consist of a large-scale patchwork of hunted and protected zones, where the number and configuration of hunting blocks were manipulated to achieve the highest sustained yield. However, given real world situations, this system of trial and error might prove too complicated for conditions of social and political instability coupled with little government presence, circumstances that exist throughout much of Africa today. Furthermore, although species with small ranges would be expected to fare well under this patchwork arrangement, wide-ranging species, such as elephants, which regularly move between hunted and protected blocks would be susceptible to overexploitation. Finally, it is important to keep in mind that the goal of maximum sustained yield is not truly obtainable due to natural background fluctuation in animal numbers. Instead, managers should strive to ensure that harvest levels are sustainable through time.

Limitations of Spatial Harvest Theory

Despite recommendations based on theory, real world scenarios would be expected to unfold quite differently. For example, managers will be limited by the size of the area open to management and by the ability to manipulate restrictions for different zones. Although size is less of a problem in large forest blocks such as Lobéké and Ituri, compliance with and enforcement of any land use restrictions that change through time would probably pose a problem for rural communities. Even where hunters are convinced to restrict their hunting to specific zones, it is not likely that they will cooperate in the face of repeatedly changing restrictions, unless those changes mean an increase in the area open to harvest. Thus, a trial-and-error system would have to move very cautiously, to avoid surpassing the peak efficient yield. Given that this peak would be difficult to determine in advance, it would be best to err on the conservative side of the area open to harvest. Furthermore, the fewer boundaries to monitor, and the less complicated the system, the more likely it is to succeed. Thus, the reserve design that includes a single central core protected area, surrounded by multiple use zones that include hunting blocks, is likely to provide the best solution in large areas such as the Lobéké and Ituri forests.

In the interest of gaining local support, it is possible that selected sites within the core protected area could be opened to hunting on a seasonal basis. For example, traditional fishing spots within the core area may be opened to hunting of those animals with relatively low dispersal distances. This strategy should be employed only if the core protected area is very large, the temporary hunting areas are clearly defined in space and time, and access to those areas within the reserve can be monitored.

Community Hunting Zones in Wildlife Management: Opportunities and Constraints

Traditional approaches to conserving wild lands around the world often fall short of their target objectives due to a lack of support by the local communities (Newmark et al. 1994; Western et al. 1994). Rasker et al. (1992) proposed that solutions to wildlife management problems are likely to be found somewhere between the extremes of public ownership and private commercialization of faunal resources. Heinen (1996) outlined a number of conservation strategies for conserving wild lands, emphasizing the need for consideration of local human behavior patterns, combined with social and economic incentives and education programs. A marriage of the concepts of community participation in conservation and the application of spatial harvest theory might prove to be the best option available for the conservation and sustained yield of faunal resources in areas such as Lobéké for the near future.

In the Lobéké forest region, there is little evidence of governmental or social controls, and there are no powerful traditional structures in place, such as exist in western and northern Cameroon. Thus, local communities and individual households are used to freedom of access to natural resources and there is relatively little interference from neighbors or governing authorities. At the same time, this laissez-faire system has allowed access to the forest by many nonindigenous peoples, and local residents have observed that the additional pressures are having a strong negative impact on their wildlife resources. In light of this, local peoples recognize the importance of gaining control over a community hunting zone in which access is restricted. The controlled access would limit the number of nonindigenous hunters that come to hunt for commercial purposes, and would go a long way toward satisfying the local residents, whose primary complaint is the unrestricted access by "outsiders." Thus, controlled access to community lands could be offered to resident villagers as an incentive for compliance with protected area regulations.

Many residents of the Lobéké region embraced the concept of controlled access to community lands in return for the creation of a central protected area. However, all residents were unanimous in their concern that if a protected area were created, outsiders, including big game safari hunters, not be allowed to hunt in this zone. To this end, several village leaders requested assistance in the development of a social structure that would be recognized by civil authorities. Aided by a conservation nongovernmental organization (NGO), five villages organized conservation committees. Their purpose was to define conservation issues important to the residents and to provide a unified voice for discussions with civil and government authorities, safari hunters, and logging companies, in matters concerning their adjoining lands. With continued technical assistance, these village committees could develop wildlife management plans and apply to the state for the creation of community hunting zones in the forest adjoining their village lands. If authorized by the state, these community hunting zones would grant wildlife management authority to the community in accordance with their approved management plan, although the plan would remain subordinate to the existing wildlife laws.

Monitoring of harvest rates and hunting efficiency trends is crucial to the long-term wise management of community lands. Thus, community hunting management plans should include provisions for a basic monitoring program. As part of our study, Lobéké villagers gained experience in monitoring hunting offtake. With continued technical assistance by NGOs and long-term maintenance of this monitoring system, community members could review trends in harvest offtake and make sound decisions to manage this activity. Marks (1996) has shown this to be a workable goal in Zambia.

In conjunction with community management efforts, effective guardianship of the designated protected zones would be essential to ensure that the reproductive stocks remained intact. This protection would either need to come from a collective of communities who agree to undertake the responsibility of patrolling the protected zones, or from the state agency responsible for management of the protected zone, or, preferably, via a joint effort by the two groups. In Lobéké, some villagers were beginning to organize a local vigilance program whereby hunting activities in their vicinity were reported to village authorities. If transportation to the protected reserve area were to be provided, along with food rations, it is likely that many villagers would join state agency personnel in patrols of the protected area for the purpose of excluding outsiders.

The sale of wild meat also should be controlled. Although commerce in wild meat is illegal without proper permits, both live and dead animals were openly sold along the road throughout the region. Most importantly, the sale of large quantities of meat destined for external markets, whether by indigenous or nonindigenous hunters, should be discouraged by enforcing the law.

These initiatives, which require cooperation and group effort by the community, should be undertaken in conjunction with regular environmental education programs to reinforce support by local residents for conservation initiatives. Finally, measures to control immigration to the area should be considered because immigration is often a problem when communities near protected areas gain special privileges (Hofer et al. 1996). For example, logging companies could be required to hire only from the local workforce rather than importing workers from prior concessions, except in the case of exceptionally skilled labor needs.

CONCLUSIONS

Community Involvement

The first step in guarding against overexploitation of heavily hunted game species is to establish a central protected area where hunting is not allowed. Following the creation of a protected zone, designated hunting blocks could be incorporated into multiple-use buffer zones surrounding the central protected area. The establishment of community hunting zones would constitute a concrete example of support for indigenous land rights, which is widely recognized as one of the best options for the conservation and rational use of tropical forests outside protected areas (Redford and Stearman 1993). The creation and maintenance of adjoining hunting and protected blocks should provide a relatively risk-free opportunity for management of the fauna by local people, while minimizing the threat to biodiversity.

The granting of a community hunting zone should be contingent upon acceptance of a wildlife management plan for the zone by the government. However, the plan should be prepared by the community with technical assistance from someone familiar with local conditions, to ensure participation of the hunters and forest users in the management of their resources. Within these community hunting zones, the communities could be granted full management capacity over their faunal resources, subordinate to the existing national wildlife laws, including the option to refuse access to noncommunity members. In return for exclusive management of the community hunting zones under a state-approved plan, community members should be required to agree to restrict hunting to their mandated area and to agree collectively to refrain from hunting within the protection zone. The community also should be required to monitor hunting offtake and hunter efficiency within their designated hunting sector.

Other conservation measures might include discouraging the establishment of snare lines in remote areas to avoid spoilage and waste of meat (Muchaal and Ngandjui 1995). A good rule of thumb would be to limit snare lines to within 10 to 15 km of the village, or the distance a hunter can achieve within a half day's walk to allow visits to and from snare lines in a single day. Other methods of hunting, including guns and crossbows, could be employed in the less accessible areas (20 to 30 km from the village) of the hunting zone. Additionally, hunters should be required to visit their snare lines at least three times per week to avoid spoilage of animals dying in traps. If particular community members are consistent violators of the community conservation codes, fines or penalty measures could be imposed, with the money to be managed by a conservation committee. This would encourage neighbors to monitor each other for the common good, at least in the case of flagrant violators. Given that the majority of captures probably occur within a day's walk of the village for most hunters in Africa, this objective of limiting snare lines might not be hard to achieve.

All of the above community management options should include a mechanism for oversight by a state or parastatal agency, or NGO, to ensure compliance with management plans and wildlife laws, and ensure that the resources are not being degraded.

Adapt Management Strategies to Local Conditions

Lessons learned to date (Strum 1994; Becker and Ostrom 1995) highlight the need to (a) tailor management strategies to incorporate consideration of local social conditions and (b) use adaptive management techniques, making changes to include lessons learned on site. With this in mind, special exceptions to restrictions and privileges should be considered for both hunting and nonhunting blocks. High faunal use habitats such as forest clearings or mineral licks should be protected from hunting, even when they occur within community hunting zones. Thus, in the interest of maintaining high levels of animal reproduction, hunting should be prohibited within at least a 500-m radius of such high animal use areas. In return, specially designated areas of the protected zone could be opened to hunting to provide an incentive for hunter compliance. Local peoples of the Congo Basin have traditionally made extensive forays deep into the forest during the fishing and wild mango (*Irvingia* spp.) harvest seasons, and it would be impractical to prohibit hunting during these extended periods of a month or more in the forest camps. Thus, in the interest of gaining cooperation from local communities, strictly defined foraging (e.g., fishing or mango collecting) areas in the protected area could be opened to hunting on a seasonal basis. These specially designated hunting zones could be subject to special restrictions in hunting methods (e.g., wire snares prohibited, select species only). Research among the local residents would be necessary to identify areas of traditional use and to develop a system that addresses their needs but that also could be effectively implemented to achieve conservation objectives without being abused.

The alternative to making changes such as those outlined above would be to allow continued unrestricted access until the preferred game populations were eliminated in the region. This scenario has been repeated often enough throughout the world to know that it is a consequence of unrestricted exploitation of wildlife populations (Hofer et al. 1996) and could occur even in relatively large and remote wild-lands such as those of Africa's Congo Basin. Therefore, where government institutions are limited in their capacity to enforce management regulations, our main hope lies in creating innovative programs for controlling exploiters by providing incentives for them to adopt conservation measures (Hilborne et al. 1995).

ACKNOWLEDGMENTS

Funds for this study were provided by the Wildlife Conservation Society, the Biodiversity Support Program, and the Liz Claiborne Art Ortenberg Foundation. We are grateful to the Government of Cameroon, specifically the Ministry of Scientific Research and the Ministry of the Environment and Forestry, for permission to work in southeastern Cameroon. We sincerely thank the many residents of the Lobéké region for their participation and assistance in these studies. We also thank J. Oates, D. Wilkie, and the editors for their many helpful comments on an earlier draft of this manuscript.

19

Traditional Management of Hunting by a Xavante Community in Central Brazil: The Search for Sustainability

FRANS J. LEEUWENBERG AND JOHN G. ROBINSON

Recent studies of hunting by indigenous peoples in South America have indicated that the present-day harvest rates of many game species in many areas might not be sustainable (Vickers 1988; Stearman 1990; Redford 1991; Alvard 1993a; Robinson and Redford 1994a; Leeuwenberg 1994). Yet indigenous peoples have lived in neotropical forests for at least 12,000 years (Roosevelt et al. 1996) and have probably exploited wildlife populations throughout that period. Across the landscape, and for extant species, hunting may be assumed to have been sustainable. Where indigenous hunting is now no longer sustainable, the harvest (the number of animals taken by hunters) exceeds production (the addition to the wildlife population through births and immigration, accounting for animals lost through deaths or emigration). This can come about through either a decline in the production from wildlife populations, an increase in the harvest, or both (Robinson and Redford 1991b).

Declines in production commonly result from declines in wildlife densities (Peres this volume). In many studies of indigenous hunting, density declines have been ascribed to increased fragmentation of natural habitats, restricting movements of wildlife species across the landscape (Stearman 1990); hunting by colonists or commercial hunters of European extraction in areas bordering indigenous land (Beckerman 1978; Peres 1993c; Leeuwenberg 1995); and overhunting by the indigenous peoples themselves (Yost and Kelley 1983; Hill and Hawkes

1983; Peres 1993c). Increases in harvests over the short term are commonly the consequence of increased indigenous participation in economic markets, with their increased demand for game meat (Saffirio and Scaglion 1982), although commercialization does not always have this result (Stearman and Redford 1992).

Declines in wildife densities in localized areas such as around villages is associated with the increased sedentarization of indigenous people in the neotropics (Gross et al. 1979; Stearman this volume). Indigenous communities thus continue to hunt in areas with depleted game densities (Vickers 1980). Another socioeconomic change is the abandoning of customs that act to disperse or relocate hunting areas, customs such as hunting zone rotation, outlier camps, and trekking (Redford and Robinson 1987).

The loss of hunting sustainability is of concern because indigenous groups living in neotropical forests rely extensively on wild meat (Vickers 1984; Redford and Robinson 1987; this volume). Preliminary evidence indicates that diminishing wildlife harvests over the long term might be more the rule than the exception for indigenous peoples living in neotropical forests. In informal discussions with one of the authors (F.L.) during 1993 and 1994, representatives from a variety of indigenous groups in Brazil (the Tereno in Mato Grosso do Sul; the Bororo in Mato Grosso; the Surui in Rondonia; the Tukano, the Tikuna, and the Sateré-Maué in Amazonas; and the Ashaninca, Kaxinawá, Katukina, Yawanawá, and Poyanáwa in Acre) indicated that harvests appeared to have declined or are in decline in many indigenous reserves in Brazil.

This chapter reports on an initiative taken by the Etenhiritipá community (also known as Pimentel Barbosa) of the Xavante, an indigenous group of the *cerrado* (a dry forest-grassland mosaic) of Brazil, who were concerned about the apparent decline of game harvests around their village. The Association Xavante Pimentel Barbosa requested funds, initially from the World Wildlife Fund, to evaluate game populations and harvest rates. The Xavante then hired F.L., a wildlife biologist associated with the Indigenous Peoples Research Center (Centro de Pesquisa Indigena), to evaluate the sustainability of their subsistence hunting and to advise them on appropriate management of their wildlife resources.

Here we report on the first 36 months of this project, which made a preliminary evaluation of the sustainability of hunting by (a) examining total harvests and hunting success of the Xavante, (b) comparing the demographic characteristics of harvested wildlife populations over 3 years, and (c) applying a model that evaluates hunting sustainability for neotropical forest mammals.

Throughout the study, the Xavante sought to use preliminary results and information to change their ongoing hunting activities. This arrangement introduced experimental variation into the design of the sustainability study and on occasion made interpretation of results somewhat problematic. However, this study documented an important attempt (Stearman and Redford 1995) to integrate the wildlife management approach of indigenous peoples and that of scientifically trained biologists. It also examined the approaches that an indigenous community itself took to manage its wildlife resources.

THE XAVANTE OF ETENHIRITIPÁ

The Xavante inhabit the *cerrado* of Brazil, a dry forest-grassland mosaic (Eiten 1972, 1990). The dry forest is seasonally deciduous and has a variable canopy cover. Trees and shrubs are characterized by a tortuous physiognomy and thick barks. The gallery forests lining the rivers and streams in the area, however, are dense and evergreen and contain many species that are typically Amazonian. Rainfall is seasonally distributed, with a dry season lasting from May to September (peaking in July) and a wet season from October to April.

The Xavante were traditionally hunter-gatherers (Flowers et al. 1982), practicing extensive hunting, some fishing, and traditional small scale agriculture. The first effort to bring this indigenous group into the mainstream Brazilian society started in the 1930s, was deferred in 1941 following the massacre of the Pimentel Barbosa expedition, and was finally achieved by Chico Meirelles in 1946 (Maybury-Lewis 1984; Giaccaria and Heide 1984). Following the creation of the new capital Brasilia in 1960, colonists of European extraction moved into the interior of the country, increasingly coming into conflict with the Xavante. After a long period of negotiations with the Brazilian government to legalize their territory, and their forceful expulsion of cattlemen in 1979 and 1980, the Xavante were given title to six reserves in the early 1980s (CEDI 1990).

The Rio das Mortas Reserve is one of these six and is situated 350 km north of Barra dos Garças, on the west slope of the Rio das Mortas watershed (figure 19-1). The reservation was demarcated in 1980 and comprises a total of 3290 km². The Reserve is the northernmost of the six and is considered to be the land of origin of the Xavante. It is politically divided between the community of Etenhiritipá in the north and the villages of Caçula, Agua Branca, and Tanguro in the south. The community of Etenhiritipá was the focus of this study, and its area of influence comprises some 2200 km². In 1992, the project census of the Etenhiritipá community documented a population of 298 (53% children under the age of 15), a population density within the Reserve of 0.13 persons/km². At the time there were 57 active hunters (figure 19-2). The principal village is Etenhiritipá, which resulted from the coalescing of several of the villages within the Rio das Mortas region.

The Xavante still derive most of their animal protein from wildlife. They still practice small-scale swidden agriculture, growing corn, beans, and potatoes in fields close to the village. But in recent decades, the Fundaçao Nacional do Indio Brasília (FUNAI, the Brazilian institute for Indian affairs) has encouraged a greater emphasis on more mechanized agriculture, and there has been a decrease in the diversity of traditional crops. The traditional collecting of plant products from the *cerrado* has decreased. Cattle raising is increasing, and the Xavante of Etenhiritipá presently have some 60 head.

The Rio das Mortas Reserve is increasingly surrounded by lands converted for agricultural production, primarily large-scale soy beans, rice, and pasture for cattle. Examination of LANDSAT images taken in August 1990 shows that about 80% of the *cerrado* habitat around the Reserve has been converted, and the Reserve itself is

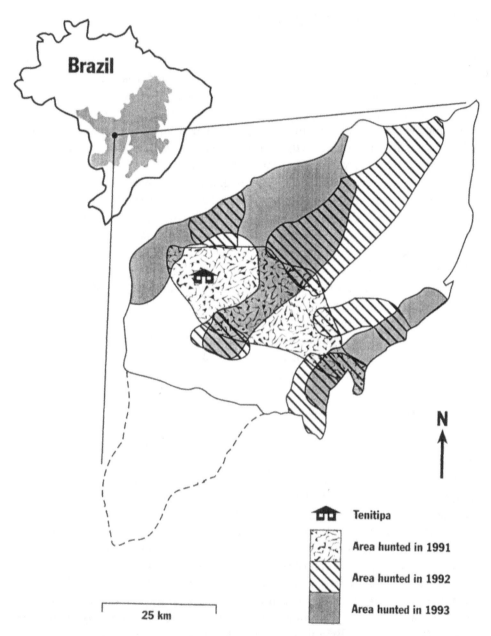

Figure 19-1. Map of the Pimentel Barbosa Reserve, showing its location in the *cerrado* (*shaded portion*) of Brazil, the village of Etenhiritipá, and the areas hunted in 1991, 1992, and 1993. The Etenhiritipá community occupies the northern part of the Reserve, divided politically from the southern part (*dashed lines*) and the Xavante villages of Caçula and Tanguro.

Figure 19-2. A Xavante hunter. *Photo courtesy of Frans Leeuwenberg.*

the largest demarcated area of relatively undisturbed *cerrado* vegetation in Brazil. The Reserve itself has lost about 10% to illegal pasture conversion by neighboring colonists in the past.

MEASURING INDICES OF HUNTING SUSTAINABILITY

In order to evaluate the sustainability of hunting over the course of the project, data were collected on harvest rates, hunting success, area hunted, and the age structure of hunted animal populations. F.L. visited Etenhiritipá for 2 to 3 weeks every 2 months between February 1991 and January 1994 to design and modify the research methods, discuss results with the Xavante, and monitor hunting patterns. During this period, hunting data were collected over the 3 complete years. Day-to-day data collection on hunting was handled by the Xavante themselves. From the beginning of the study in February 1991, hunting records were recorded on standardized questionnaires, information being taken shortly after the return of hunters. To document hunting methods and hunting success, F.L. was present on 14 hunts over the course of the study.

Wildlife Harvests

Trends in harvest numbers over time provide an initial indication of hunting sustainability. Declining harvests over time raise a question about the sustainability of hunting. To record game harvests, information was recorded on questionnaires after every hunt from every house in the community on (a) the number of each species killed during each hunt, (b) the type of hunt (individual, wedding, family, or community), and (c) the type of weapons used. All hunters were inter-viewed, but not all kills were registered. For smaller bodied species, data are not accurate. During hunts, such species were directly consumed in the field and frequently not mentioned during interviews. Birds, with the exception of rheas (*Rhea americana*), were rarely reported. Opportunistic kills, especially when people were working in the fields or collecting fruit in the *cerrado*, were frequently not mentioned. We consider data for the larger bodied species to be reliable.

Hunting Success

Harvest data by themselves are not a strong indicator of sustainability because they do not take into account annual variation in hunting effort. Did hunters have to spend more time to harvest a similar quantity of wildlife? To measure hunting success, information on (a) the number of hunters in each hunting party and (b) the duration of the hunt was collected by both accompanied hunts and by detailed questioning of hunters on their return. Hunting success was described as harvest per 100 hunter days. Harvest per 100 hunter days was calculated as the number of kills for each game species × 100 / number of hunters × number of days. Days were used as the time measurement because the Xavante do not subdivide the day into hours; therefore "days" was considered to be the more appropriate and reliable measure of time spent hunting.

Age Structures of Hunted Populations

Comparisons of age structures of hunted and unhunted populations can be used to evaluate sustainability, but the approach is somewhat problematic because the population age structures of neotropical forest species are poorly known. The most notable exceptions are those of Collett (1981), who described the demog-raphy of hunted and unhunted agouti in Colombia, and Bodmer et al. (1994), who described demography of ungulate populations at locations in Peru and Brazil. In addition, how hunting affects age structures is unclear and much debated (McCul-lough 1994). The age structure of a population does not vary with hunting inten-sity, unless hunting selectively has an impact on certain age classes (Caughley 1977). Nevertheless, a number of studies have noted differences in age structures of populations between hunted and unhunted (or lightly hunted) sites. Hunting tends to shift the age structure of the population so that (a) the proportion of juve-niles in the population increases (Robinson and Redford 1994b), and (b) among adult animals, the distribution of animals in a hunted population is more skewed toward younger age categories (McCullough 1974; Collett 1981; Bodmer 1994, 1995b).

The reasons for these possible shifts are unclear. They could arise through changes in the way the population is sampled without concomitant changes in age structure (e.g., older animals might become more wary, and thus be less likely to be killed by hunters), or they could arise through actual changes in the age distri-bution of the population. Actual changes might result for the following reasons:

1. Hunting selects for certain age categories of animals, and this shifts the population age structure. If hunters select for larger or older animals, for instance, then this might have the effect of increasing the proportion of animals in younger age classes.
2. Hunting, by decreasing population density, might increase the population birth rate. If birth rate is density dependent, increasing at lower densities, this would also have the effect of increasing the proportion of animals in younger age classes (this demographic response has not as yet been demonstrated for any neotropical game species).
3. Hunting might lower the abundance of large predators, and the resulting decrease in predation on game species might result in a lower mortality rate for young and very old age classes. The result would be greater proportions of animals in these age classes.

To collect data on the age structure of hunted populations, lower jaws from the larger game species were collected throughout the study (method of Bodmer 1995b). Collection of the entire cranium was not possible because the Xavante eat the brains of game animals. The age of each specimen was estimated from tooth wear on molars and premolars.

To calculate the age structure of peccaries and deer, we defined three broad categories of tooth wear, assuming that wear relates to the age of the animal and is constant throughout the adult life of the animal. The first included all juvenile animals that were close to adult size and young adults with little or no wear on their teeth. Juvenile teeth were replaced by 2.5 years of age in peccaries and 1.5 years for deer (Nowak and Paradiso 1983; Sowls 1984). The second included adults with moderate wear, and the third included adults with extensive wear. Each wear category therefore is roughly equivalent to 3 to 4 years for peccaries, and to 2 to 3 years for deer (Nowak and Paradiso 1983; Sowls 1984). Survivorship curves (l_x)were constructed with the first age category (l_1) defined as equal to 1000. The last adult category (l_4) was defined as equal to 0 (all animals were dead). This method of calculating species age structure is plagued by assumptions (kill samples are biased by hunter techniques and differential vulnerability of different age-sex classes, and the harvest itself can reshape the standing population; see discussions in Caughley 1977; McCullough 1977; Downing 1981) hence we only use it here to give a general indication of the impact of hunting on the populations of these ungulate species.

Sustainability Models

Sustainability of hunting was also evaluated by comparing actual harvest per km² to the maximum potential harvest per km² calculated using Robinson and Redford's (1991b) population growth model. This model generates potential harvests of wildlife species assuming maximum wildlife densities and maximal reproduction. As such, this model can indicate when hunting is clearly not sustainable but does not indicate when hunting is sustainable. With lower than maximum game densities, or less than maximal reproduction, the number of animals that could be harvested sustainably would be lower. Actual densities and reproductive rates at Pimentel Barbosa, which are unknown, might be and probably are lower than these maxima.

The catchment area used by hunters from Etenhiritipá during each of the 3 years was estimated by having hunters note the location of all hunts, which were subsequently mapped with the aid of a global positioning system by F.L. This area was then used to estimate harvest per km² for each of the 3 years for each important game species.

XAVANTE HUNTING OVER 3 YEARS

Wildlife Harvests

Xavante prefer to hunt in the more open habitats because it is easier to move, track, and spot prey. The more dense habitats are rarely entered, except when pursuing prey. The only clear exception to this is when the Xavante hunt white-lipped peccaries. The highest hunting intensity occurs in the dry season (Leeuwenberg 1994) because all habitats are more accessible, most game is forced to water spots, tracking is easier, and hunters make little noise as they follow animals.

The Xavante harvest of specific wildlife species in part reflects the habitat they prefer for hunting. Species restricted to dense habitats, such as agouti and red brocket deer, are less likely to be hunted. Prey selection also arises from the Xavante preference for certain species. The giant anteater is highly appreciated (figure 19-3), a gustatory preference shared by other *cerrado* indigenous groups but

Figure 19-3. Giant anteater (*Myrmecophaga tridactyla*), with young, asleep during the day. *Photo courtesy of John G. Robinson.*

not by Amazonian groups. In contrast, the capybara *Hydrochaeris hydrochaeris* and all the primate species are ignored by the Xavante. The Xavante also have some taboo restrictions on game (e.g., armadillos, brocket deer, and peccaries are prohibited for 6 months after the birth of a child). Prey selection is also an individual characteristic. Some hunters (especially the younger ones) focus on deer, whereas others (almost always the elder ones) specialize in tracking giant anteaters. Nevertheless, there are some generalities. Given a choice, hunters state that they will invariably choose to track larger bodied game, and when given a choice, they take the larger animals first. They evince no preference for the sex of the animal.

Annually between February 1991 and January 1994, hunts took place on a total of 101, 123, and 123 days. For each of these years, totals of 1523, 1261, and 2898 hunter-days were recorded, yielding a total harvest of 506, 473, and 761 individual animals, respectively. Table 19-1 presents the harvest of major game species during these periods. Smaller bodied species also were harvested but incompletely registered on the questionnaires, so they are not included here (figure 19-4). In the comparison across years, this simple count does not indicate a decline in the availability of any game species. Indeed, in 1993, overall harvest of most species was greater.

Hunting Success

Hunting success data, however, alter our interpretation of the encouraging harvest data. Comparisons across years show variation in hunting success measured as the number of kills per 100 hunter-days. In 1991, overall hunting success was 33.2 animals (of the major game species) per 100 hunter-days. This increased to 37.5 animals in 1992 and decreased to 26.3 animals in 1993. In other words, the average hunter had the greatest success in 1992.

This variation reflects, in part, different hunting areas. In 1991, the first year of

Table 19-1. Harvest of Major Game Species Over 3 Full Years

Common Name	Number of Animals Harvested		
	1991	*1992*	*1993*
Marsh deer (*Blastocerus dichotomus*)	17	22	**37(+)**
Pampas deer (*Ozotoceros bezoarticus*)	34	40	**73(+)**
Red brocket deer (*Mazama americana*)	6	1	5
Grey brocket deer (*Mazama gouazoubira*)	11	3	**4 (−)**
Tapir (*Tapirus terrestris*)	26	18	**43(+)**
White-lipped peccary (*Tayassu pecari*)	114	**62(−)**	**204(+)**
Collared peccary (*Tayassu tajacu*)	140	136	**230(+)**
Giant anteater (*Mymecophaga tridactyla*)	93	122	**155(+)**
Tamandua (*Tamandua tetradactyla*)	5	**9(+)**	5
Giant armadillo (*Priodontes giganteus*)	8	5	7
Six-banded armadillo (*Euphractes sexcinctus*)	52	55	38
Overall	506	473	761

Boldface numbers indicate 50% increases (+) or decreases (−) in harvest relative to 1991.

Figure 19-4. Xavante children carrying harvested coatis (*Nasua nasua*), one of the many small-bodied mammals that are hunted. *Photo courtesy of Frans Leeuwenberg.*

the study, the Xavante hunting range extended out to a maximum of 41 km from the village of Etenhiritipá, but was concentrated within the first 20 km. After examining the first year's harvest data, and taking into account the perception of the Xavante themselves that hunting was increasingly more difficult, F.L. recommended to the Men's Council that hunters cease hunting in this area. This recommendation was followed, and in 1992 hunts were made in more peripheral areas of the Reserve (see figure 19-1). As measured by the number of kills for the average hunter, species-specific hunting success increased by more than 50% for 3 of the 11 most important game species (marsh deer, giant anteaters, and tamandua), but decreased by more than 50% for the two brocket deer species (table 19-2). The more extended treks away from the village were not generally popular, and in 1993, hunters returned to areas close to the village, hunting over an area partially overlapping the 1991 hunting range. In comparison with 1991, overall hunting success decreased by more than 50% for 5 of the 11 species (the two brocket deer, tamandua, giant armadillo, and six-banded armadillo), although none of these species are among the most important sources of wild meat (see table 19-2). These

Table 19-2. Hunting Success (Harvest per 100 Hunter-Days) of Major Game Species Over 3 Years

Common Name	Number of Animals Harvested		
	1991	*1992*	*1993*
Marsh deer	1.12	**1.75(+)**	1.27
Pampas deer	2.23	3.17	2.52
Red brocket deer	0.39	**0.08(−)**	**0.17(−)**
Grey brocket deer	0.72	**0.24(−)**	**0.14(−)**
Tapir	1.71	1.43	1.48
White-lipped peccary	7.49	4.92	7.04
Collared peccary	9.19	10.79	7.94
Giant anteater	6.12	**9.68(+)**	5.35
Tamandua	0.33	**0.71(+)**	**0.17(−)**
Giant armadillo	0.53	0.40	**0.24(−)**
Six-banded armadillo	3.41	4.36	**1.28(−)**
Overall	33.22	37.51	26.26

Boldface numbers indicate 50% decreases (−) and increases (+) in hunting success relative to 1991.

data suggest that in their preferred hunting area close to the village, densities of some important game species declined over the course of the study.

Age Structure of Harvested Populations

There is some evidence that the Xavante do select for adult over juvenile animals, and this could have effects on age structure. When hunting peccaries, the hunters take animals from all age-sex classes, but when given a choice, they will tend to track and harvest the larger animals. When hunting deer, hunters tend not to take young animals. Once juveniles begin moving with the group, their vulnerability to hunters is equal to that of adults because hunters will frequently kill the entire group. For both peccaries and deer, therefore, while there is probably little selection among adult-sized animals, there is apparently selection for adult over juvenile animals.

From examination of lower jaws, information on age structure is available for white-lipped peccaries (n = 130 skulls), collared peccaries (n = 283), pampas deer (n = 79), and marsh deer (n = 29). Juvenile skulls were excluded from this sample because they were incompletely sampled. We assumed that even though the Xavante hunters were hunting in different areas in the 3 years, they were probably exploiting the same source population; thus, the age structure of game populations would not differ significantly across the landscape. Accordingly, we lumped together age-structure data from the 3 years.

To examine whether the age structure of the white-lipped peccary population at Rio das Mortas (figure 19-5) exhibits a clear response to hunting, we compared the age structure to that of white-lipped peccary populations at two well-studied sites in Peru: Tahuayo and Yavari Miri. Tahuayo is characterized by persistent hunting pressure, whereas Yavari Miri has only light hunting pressure (Bodmer et

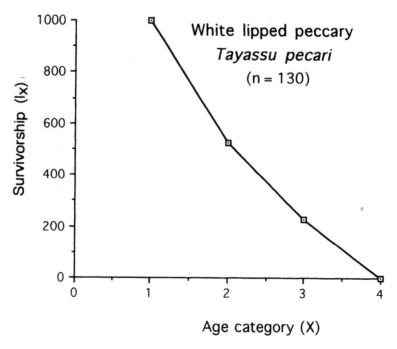

Figure 19-5. Survivorship curve for white-lipped peccary, based on animals harvested by the Xavante.

al. 1994; Bodmer 1995b). The age categories used in Peru are the same as those used in this study. The age structure at Rio das Mortas was not significantly different from that at Yavari Miri ($D = 0.10$, $0.15 > p > 0.10$, Kolmogorov Smirnov test), but is was different from that at the Tahuayo site ($D = 0.13$, $p < 0.05$, Kolmogorov Smirnov test). We also compared the age structure of the collared peccary at Rio das Mortas (figure 19-6) to the collared peccary populations at Tahuayo and Yavari Miri. Age structure was not significantly different from populations at Yavari Miri ($D = 0.04$, $p > 0.20$, Kolmogorov Smirnov test) and Tahuayo ($D = 0.08$, $0.10 > p > 0.05$, Kolmogorov Smirnov test), although the probability was less at the latter site. For both white-lipped and collared peccary populations, age structure at Rio das Mortas is not significantly different from that at Yavari Miri, a lightly hunted site.

In contrast, the distribution of hunted animals among age classes in pampas deer suggests that hunting is having an impact. The age structure of the population is highly skewed toward the youngest age class (figure 19-7). The age classes of marsh deer show a similar pattern (figure 19-8). Although no age distribution from unhunted or lightly hunted populations of pampas or marsh deer are available for statistical comparison, the pronounced skew suggests that hunting might be having a significant impact on the deer populations.

Harvests per Square Kilometer

Actual harvests per km² over the course of each year were calculated recognizing annual differences in known catchment areas. The area over which the Xavante

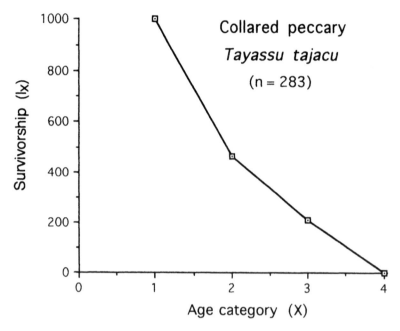

Figure 19-6. Survivorship curve for collared peccary, based on animals harvested by the Xavante.

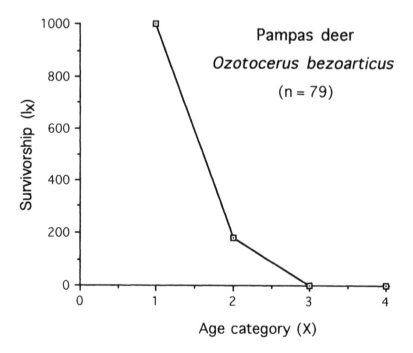

Figure 19-7. Survivorship curve for pampas deer, based on animals harvested by the Xavante.

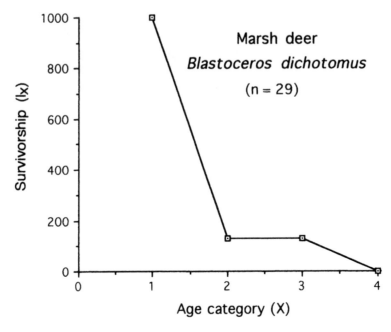

Figure 19-8. Survivorship curve for marsh deer, based on animals harvested by the Xavante.

hunted varied across the 3 years. In 1991, the Xavante were hunting over an area of approximately 650 km² (see figure 19-1). In 1992, following F.L.'s suggestions, the hunting range was moved to the northeastern part of the Reserve, and covered an area of 850 km². In 1993, the Xavante were hunting in the west and northwest, and their hunting extended over an area of 1150 km². Compared with 1991, the harvest in 1992, as measured by number of kills per km², decreased by more than 50% for 5 of the 11 major game species (the two brocket deer species, tapir, white-lipped peccary, and giant armadillo) and increased by more than 50% for none (table 19-3). The same comparison shows that the harvest per km² in 1993 decreased by more than 50% for 5 of the 11 major game species (the two brocket deer, tamandua, six-banded armadillo, and giant armadillo).

In the absence of density and production estimates for the hunted populations, the sustainability of these harvest rates per km² is difficult to evaluate. However, use of Robinson and Redford's (1991b) population growth model, which generates a maximum potential harvest for each species assuming maximum wildlife densities and maximum reproduction, can clearly identify which species are not being harvested sustainably (if rates are higher than this absolute maximum). Comparisons of maximum potential harvest per km² (table 19-4) with actual harvest per km² for eight important and large-bodied game species (see table 19-3) indicate that the hunting of tapirs was clearly not sustainable in 2 of the 3 years, and hunting of giant anteaters was not sustainable in any year—harvests exceeded the absolute maximum. The harvest of marsh deer was slightly below the maximum potential level in 2 years and slightly above it in one. Unless conditions are

Table 19-3. Harvest per km² for Major Game Species Over 3 Years

	Number of Animals Harvested		
Common Name	*1991*	*1992*	*1993*
Marsh deer	0.026	0.026	0.032
Pampas deer	0.052	0.047	0.063
Red brocket deer	0.009	**0.001(–)**	**0.004(–)**
Grey brocket deer	0.017	**0.004(–)**	**0.003(–)**
Tapir	0.040	**0.021(–)**	0.037
White-lipped peccary	0.175	**0.073(–)**	0.177
Collared peccary	0.215	0.160	0.200
Giant anteater	0.143	0.144	0.135
Tamandua	0.008	0.011	**0.004(–)**
Giant armadillo	0.012	**0.006(–)**	**0.006(–)**
Six-banded armadillo	0.080	0.065	**0.033(–)**
Overall	0.778	0.556	0.662

Boldface numbers indicate 50% decreases (–) in harvest relative to 1991.

maximal, it is likely that hunting is also not sustainable for this species. For the other species, harvests are less than theoretically possible, but the model by itself does not indicate whether the hunting of these other species is sustainable.

EVALUATING HUNTING SUSTAINABILITY

Accurately assessing the sustainability of hunting requires information on harvest rates, as well as densities and productivities of hunted species (Robinson and Redford 1994b). However, this is rarely feasible, and in the absence of such complete information, indices can be used to suggest management options. In this project, sustainability was first evaluated by comparing overall hunting success

Table 19-4. Calculation of Maximum Potential Harvest for Some of the Most Important Game Species

Common Name	*Maximum Population Density (No./km²)*	*Maximum Potential Production (No./km²)*	*Maximum Potential Harvest (No./km²)*
Marsh deer[a]	0.6	0.15	0.03
Pampas deer[a]	4.6	1.9	0.38
Red brocket deer	10.5	1.7	0.67
Grey brocket deer	10.4	3.1	1.2
Tapir	1.6	0.16	0.03
White-lipped peccary	4.9	4.2	0.83
Collared peccary	11.8	12.0	2.41
Giant anteater[b]	2.0	0.44	0.09

For actual harvests of these species during the 3 years of the study, see table 19-3.

[a]Density from Leeuwenberg et al. (1997), production calculated using method of Banse and Mosher (1980).

[b]Density from Redford and Eisenberg (1992), production calculated using method of Banse and Mosher (1980).

All other data from Robinson and Redford (1991b).

across 3 years to get some indication of trends in the availability of game. These data suggest that game densities, especially close to the village of Etenhiritipá, were generally declining. Following what they considered to be a disappointing hunting year in 1991, on the advice of F.L., the Xavante shifted their hunting to areas away from the village in 1992. Overall hunting success (as measured by harvest per hunter-day) increased, but hunting success on individual species was mixed. In addition, the hunters were traveling farther from the village and hunting over a larger area, and their perception was that they were having to work harder to get game. In 1993, the Xavante decided to shift the location of their hunting again, partially returning to areas hunted in 1991. In comparison with 1991 and 1992, overall hunting success measured by harvest per hunter-day declined. Hunting success for five important game species also was more than 50% lower compared with 1991 (the two brocket deer species, tamandua, giant armadillo, and six-banded armadillo). Taken in aggregate, these trends in hunting success suggest that (a) by 1993, densities of a number of species in areas closer to the village were lower than in 1991, possibly because they had not recovered from hunting in previous years, and (b) hunting in more peripheral areas involved greater travel costs and ranged over a wider area, and hunting success, and possibly wildlife densities, were not clearly better than in more central areas.

Sustainability also was evaluated by considering the age structure of some of the more important game species. Although there was little to suggest that hunting of peccaries was not sustainable, the age structures of both marsh and pampas deer populations were highly skewed toward the younger age classes. The great preponderance of juveniles and young adults in these species suggests (a) significant hunting pressure in the past or (b) a direct impact of the hunting on the standing crop population. Hunting sustainability also was evaluated with reference to Robinson and Redford's (1991b) population growth model. This model suggests that it was highly unlikely that the hunting of tapirs and giant anteaters was sustainable, and unlikely that is was sustainable for marsh deer.

None of these indices are conclusive, but taken in aggregate, they suggest that a number of wildlife species important to the Xavante might be either at low densities as a result of past hunting or their densities are declining. Nine of the most important game species are flagged by one or more of these indices. The only two species for which there is no indication of low or declining densities are the two species of peccaries.

SOCIOECONOMIC AND CULTURAL CHANGES IN THE XAVANTE

A number of factors might have contributed to the presumed decline in hunting sustainability in the Etenhiritipá region. The most important is that the Xavante have changed from a semi-nomadic to a sedentary existence. Prior to their contact with Brazilian society, the Xavante regularly moved their villages and cropfields.

This pattern of movement is likely to have allowed hunted wildlife populations time to recover. In addition, the Etenhiritipá community has coalesced into the single village of Etenhiritipá. This resulted in hunting becoming more localized to an area immediately adjacent to the village. This centralization was an explicit policy of FUNAI, who provided the Xavante with gifts of clothing, medicine, pottery, a tractor, guns, and farming implements at a central location. Another contributing factor has been deforestation around the Reserve and the illegal hunting by colonists from neighboring areas. This hunting pressure could contribute to lower wildlife population densities, especially in more peripheral areas. Xavante elders informed F.L. that immediately following the establishment of the Reserve, wildlife populations did not appear to decline (possibly because of immigration from adjoining areas), but in recent years, densities appeared to decline significantly.

Other cultural changes potentially affecting hunting sustainability include the intensified use of fire by the Xavante as a hunting method, as well as the abandonment of family hunts. Traditionally the Xavante would burn the *cerrado* during the dry season (normally July through September), a hunting technique practiced by all *cerrado* indigenous groups (Flowers et al. 1982). In this collective hunt, an area of perhaps 50 to 150 km^2 was set on fire and partially surrounded by hunters. In addition to increasing hunting success immediately, this method stimulated vegetative growth, especially in herbs and grasses. The Xavante return to hunt in previously burned areas in succeeding months. Traditionally, use of fire was limited to a period from the middle to the end of the dry season, and the frequency of burning was determined by habitat type. For instance, more open areas were burned every 1 to 2 years, *cerrado* vegetation every 3 to 4 years. In recent years, the timing and frequency of burning has changed. Burning has been much more frequent, and has extended throughout the dry season. This intensified use of fire has apparently degraded the habitat in some areas, presumably with concomitant effects on wildlife densities.

Another change affecting the sustainability of hunting was the erosion in the popularity of traditional family hunts. Family hunts are extended hunting and collecting trips involving family social units that last up to 2 months into distant regions of the Xavante territory, and effectively served to disperse hunting. In addition, these hunts reinforced family bonds and trained youngsters to hunt. At the end of the 1970s, Flowers (1983) reported that half of the village still engaged in family hunts, but with the centralized activities at Etenhiritipá; however, family hunts have become much less common than in the past.

A further change adopted by the Xavante is the use of 0.22-caliber rifles. Their use was encouraged by FUNAI, and now only about 15% to 20% of hunters still use bow and arrow (figure 19-9) even occasionally. The use of firearms probably does not increase the size of the hunter's bag. Observation of hunts indicated that hunters continue hunting until they have accumulated about 50 kg of game, at which point they return to the village (Leeuwenberg 1994). But rifles do increase the wounding of game. Larger animals in particular tend to escape after being hit,

Figure 19-9. Bow and arrow are still used by some Xavante hunters. *Photo courtesy of Frans Leeuwenberg.*

and they presumably are more likely to die later if wounded. Such wastage might be as high as 25% (Leeuwenberg 1994). A similar wastage level using 0.22-caliber rifles also was noted by Townsend (1995) for the Sirionó in Bolivia.

MANAGING HUNTING: A RETURN TO THE TRADITIONAL

During the course of this project, preliminary data were discussed with the Wara, the tribal council, during most visits by F.L., and the Xavante responded by changing their way of hunting. This process was a gradual one and relied on first identifying the reason for declines in hunting success, and then determining possible solutions to the problem. Initially, the Xavante even expressed considerable skepticism on the usefulness of studying hunting patterns and of collecting lower jaws and crania. However, when convinced that such information would be useful in determining the reasons for wildlife declines, they were enthusiastic participants in data collection. After an initial appraisal of the data, the Wara determined that hunting declines resulted from the concentration of hunting around the village. They drew an analogy with the ranging behavior of large predators such as the jaguar, which the Xavante describe as rotating the focus of its hunting behavior over its range, thus allowing prey populations to recover. The Wara then identified the social change responsible for this problem: traditional hunting practices were not being followed, and knowledge of these practices was not being transmitted to the younger generation. Hunting knowledge is typically transmitted formally by the older hunters during family hunts and during the 5-year training period for boys between the ages of 9 and 17, known as the Wapté. However, many younger members of the community were increasingly being drawn to fishing, sports, and activities in local urban centers. As a result, the young men in the community were not as knowledgeable about hunting and traditional practices as their elders had been at the same age.

Once the Xavante Men's Council had recognized the social changes responsible for the declines in hunting success, they were able to identify approaches to alleviate the problem. The Wara resolved, during a meeting in March 1992, to focus on intensifying traditional family hunts and to use distant hunting grounds. In addition, they decided to use fire in a more traditional way. They expressed little interest in western management techniques, such as imposing bans, seasonal limits, or bag limits on any species. They rejected the suggestion of a hunting ban on pampas and marsh deer. In other words, novel ways to manage game were rejected in favor of more traditional approaches.

Rotating the location of hunting away from areas close to the village was the most significant change adopted by the Xavante community. As well as giving hunters access to areas not recently exploited, the shifts in the area hunted in 1992 and 1993 allowed the Xavante to patrol their boundaries and thus more efficiently prevent the invasion of illegal gold-diggers and farmers. Hunting pressure on wildlife populations near to the village was relieved, and the hunting success of some species increased.

An increase in family hunts was also strongly encouraged by the Wara. These hunts would serve to disperse hunting over the landscape. The tribal council was also enthusiastic about these hunts because the present unpopularity of family hunts meant that the younger generation was not being trained to hunt and collect forest products. As a result of this encouragement, the number of family hunts increased during the study from 6 (4% of hunts) in 1992 to 16 (10% of hunts) in 1993 (Leeuwenberg 1994). The Wara also argued that hunting was a community responsibility, not an individual right, and the study observed that the cases of a single person hunting decreased from 27% of all hunts to 19% in the same period.

"Excessive" use of fire in hunting was also discouraged by the Wara as part of the effort to return to more traditional hunting patterns. This use of fire for hunting is now diminishing. In 1991, burning for the purposes of hunting began in June and continued through the end of the dry season in August. In 1992, burning did not begin until August. In 1993, there were fires in May and August, but most were set by colonists living to the south and northeast of the Reserve.

Reestablishing a management system that allows for sustainable use of game species clearly depends, in this case, on reestablishing the traditional hunting patterns. This is being accomplished through information transfer between scientifically trained wildlife biologists and indigenous people. The case of the Xavante illustrates that a concern of indigenous peoples for declining wildlife harvests can lead to an adaptive and considered response to this problem.

ACKNOWLEDGMENTS

Greatest appreciation should go to F.L.'s wife Susana de Lara Resende for her unlimited patience and suffering during the research. The whole Xavante community, but especially Serebura, Suptó Bupréwen Wairi, and Simao Wapsa-wawé, provided great help and hospitality throughout the study. Understanding the effects of acculturation was greatly aided by discussions with Ailton Krenak (Centro Pesquisa Indigena) and Cipassé Xavante (Association Xavantes of Pimentel Barbosa). Funding in 1991–1993 was provided by World Wildlife Fund US, and in 1993 by the Wildlife Conservation Society. Elizabeth Bennett, Richard Bodmer, Mac Chapin, Dale McCullough, Kent Redford, and Fred Wagner provided useful comments on earlier drafts.

20

Community-Based Comanagement of Wildlife in the Peruvian Amazon

RICHARD BODMER AND PABLO E. PUERTAS

Community-based conservation approaches the conservation of species and ecosystems by appreciating the fundamental role of rural communities in managing and using wildlands (Little 1994). Local communities living around coastal estuaries, in tropical forests, savannahs, and mountain ranges have begun to take on responsibilities for conserving and managing natural resources in their vicinities (Maltby et al. 1992; Bodmer 1994; Poffenberger 1994; Wells 1994). Community-based conservation has arisen from the realization that rural people not only dwell in the vast wildlands, but have a meaningful long-term stake in their surroundings and an interest in the well-being and production of these environments (Western and Wright 1994).

Wildlife management is an important component of many community-based conservation strategies because game hunting is both economically important for local people and directly affects species conservation. Wildlife management cannot function without the input of hunters in the development of regulations. Regulations should make sense to most hunters if the management system is to be successful. However, it usually takes more than grass-roots community initiatives to attain wildlife management that concurs with conservation goals (Rettig et al. 1989). It would be unrealistic to assume that hunters could manage wildlife by themselves in the complex political, economic, and natural systems.

Thus, community-based wildlife management is likely to function best if comanaged (Pinkerton 1989). Comanagement is the division of management responsibilities between local communities and other parties through formal and informal partnerships. It strengthens technical reasons for actions, the legality of the system, and social sanctions brought against violators, among other important aspects of community-based management (Rettig et al. 1989). Comanaged systems will result in conservation of species only if the partners agree on conservation objectives.

This chapter analyzes information and events of the Reserva Comunal Tamshiyacu-Tahuayo (RCTT) located in northeastern Peru to see if community-based comanagement of wildlife can work as a true solution for conservation in Amazonia. First, we review the structure of community-based comanagement in the RCTT and then examine the events that led to its creation and continuance. We examined harvests of game mammals before and after communities set up wildlife management programs to evaluate the impact of community-based comanagement. Lastly, using the case of the RCTT as an example, we discuss the need for training conservationists in community-based comanagement so that they understand how to integrate biological information with the needs of local people.

THE RESERVA COMUNAL TAMSHIYACU-TAHUAYO: A CASE STUDY

Community-Based Comanagement

Wildlife management in the RCTT involves a combination of community-based and comanagement strategies. The community-based side recognizes that communities are responsible for performing wildlife management. The comanagement side involves stakeholders who have a meaningful interest in the appropriate management of the Reserve and includes local communities, government agencies, NGO extension workers, and researchers.

Communities of the RCTT make decisions on how to manage resources of the Reserve. Community members vote on resource use and management issues democratically during community meetings. Government officials, extension workers, and researchers are not usually present when communities vote on management and resource use issues. However, government programs, extension activities and research results influence the management and resource use decisions. Wildlife extension workers link government regulations and results from wildlife research back to the communities.

Community-based comanagement relies on acquisition and communication of information. For example, local people perform management that affects game populations. Biological studies on game populations generate information on the impact of hunting and effectiveness of management. Extension activities convey results from biological studies to local people. The feedback loop linking game populations to local people can only be completed if it contains a research and extension component (figure 20-1). In other words, the impact of management can only be determined through research on game species. Therefore, research and extension link the realities of game populations back to community-based management.

Wildlife research and extension in the RCTT use participatory methods that

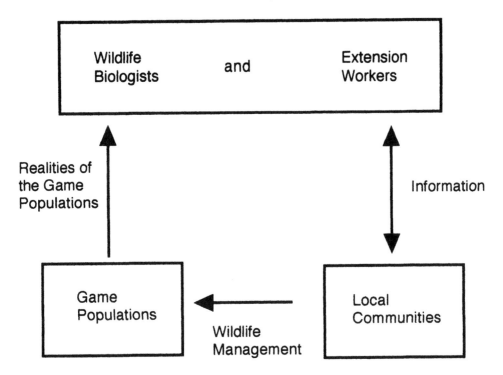

Figure 20-1. Schematic flow chart of the community-based comanagement in the Reserva Comunal Tamshiyacu-Tahuayo.

involve local people. This entails building interest in community-based wildlife management by researchers working with hunters when evaluating the impact of harvests. For example, one of these methods uses skulls from animals that hunters have shot. By collecting skulls, hunters and their families become involved in data collection. Women actively participate because they usually cook and clean skulls (figure 20-2), and often help their husbands or sons label and store them.

This participatory method helps researchers, extension workers, and hunters find common ground to discuss wildlife issues. This common ground, in this case, is the animal skull. When a researcher or extension worker discusses the sex, age, and species of an animal's skull with the hunters, they also discuss such things as the interest in community-based wildlife management or more technical issues such as registering the numbers of animals hunted. It is also possible to get hunters' opinions on issues such as hunting by people not belonging to the community. This participatory technique helps hunters to think about different aspects of wildlife management and to learn about game registries.

Pinkerton (1989), using examples from North American fisheries, presented a set of variables that favors the development and maintenance of comanagement. We found that many of the same variables applied to developing and maintaining community-based comanagement of wildlife in the RCTT specifically.

Comanagement of wildlife functions in the RCTT because there is a dedicated core group that applies consistent pressure to advance the process. Currently this

Figure 20-2. Community woman cleaning an animal skull as part of the participatory wildlife program. *Photo courtesy of Pablo Puertas.*

core group consists of the wildlife extension-research component, which includes dedicated professionals who work with communities, and community representatives. This core does not always comprise the same people. Before establishing the Reserve, for example, it consisted of a few researchers, a few government officials, and community representatives.

The preconditions that applied to the initial phases of community-based comanagement in the RCTT were a real or imagined crisis in game depletion and an opportunity for experimental management of wildlife.
Comanagement of wildlife in the RCTT operates because:

- There are formal, legal, and multi-year agreements.
- There is the assumption that long-term economic benefits will be realized through management.
- There is a mechanism for conserving wildlife in a manner compatible with the cultural system.
- There is external support from universities, research institutions, and research NGOs that provide technical information.

- The area is not too large (i.e., a watershed system).
- The number of community members is not too large for effective communication, and communities already have cohesive social systems and can effectively define their boundaries and membership.
- Technical decisions are separate from allocation decisions.

New relationships have been created because of community-based comanagement in the RCTT:

- Cooperation of individuals in planning the use and conservation of wildlife
- Commitment among local hunters to share the costs and benefits of their efforts
- Increased motivation to negotiate sharing of access to the resources
- Creation of a more equal negotiating relationship between hunters and other users
- Willingness among all stakeholders to share information and reach a more complete understanding of the resource
- Creation of greater trust between stakeholders and a greater sense of control by hunters, thus reducing the motivation to overhunt
- Greater trust between stakeholders that has led to more appropriate enforcement regimens.

DEVELOPMENT AND MAINTENANCE OF MANAGEMENT

The RCTT is in the northeastern Peruvian Amazon, in the state of Loreto, and covers an area of 322,500 hectares (ha). The Reserve is in the upland forests, which divide the Amazon valley from the Yavari valley (figure 20-3). The closest city to the RCTT is Iquitos, which has approximately 300,000 inhabitants and is around 100 km northwest of the Reserve.

The RCTT is a community reserve decreed regionally on June 19, 1991 (Resolución Ejecutiva Regional No. 080–91-CR-GRA-P). Community reserves in Peru legally give the responsibility of managing resources to local communities. They are conservation areas, and communities are responsible for managing resources in a manner consistent with biodiversity conservation.

To realize the conservation objectives, the RCTT is divided into three land use zones: (a) a buffer zone for subsistence use of approximately 160,000 ha, (b) a fully protected core area of approximately 160,000 ha, and (c) an area of permanent settlement that lacks definite boundaries. The fully protected and subsistence areas fall within the official limits of the Reserve and have no human settlements. The fully protected zone does not usually have extractive activities and is far from any human settlements. This zone acts as a refuge and source area for species. Local residents of the permanent settlement zone use the subsistence zone for extraction of natural resources. Residents cannot set up permanent settlements or clear land for agriculture within the boundaries of the subsistence use or fully

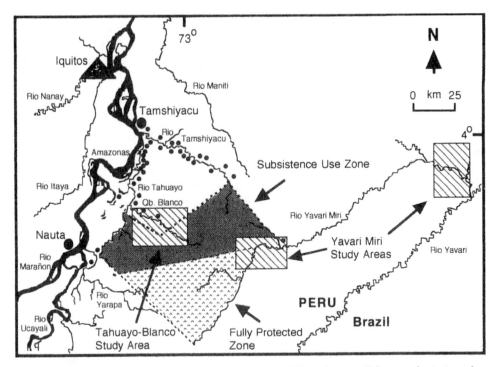

Figure 20-3. Boundaries of the Reserva Comunal Tamshiyacu-Tahuayo depicting the fully protected and subsistence zones. The small black dots are villages of the permanent settlement zone. Also shown are the persistently hunted Tahuayo-Blanco study area and the slightly hunted Yavari Miri study areas.

protected zones. The zone of permanent settlements along the Tamshiyacu, Tahuayo, Yarapa, and Yavari Miri rivers is next to the Reserve. This area encompasses the villages and is for intensive land-use activities, such as agriculture. The permanent settlement zone was not officially incorporated into the Reserve to avoid conflict over land uses, but it is an important part of the RCTT management plans (Bodmer et al. 1997b).

Nontribal people, known in Loreto as *ribereños*, inhabit the settlement zone of the RCTT. *Ribereños* have diverse origins and include detribalized Indians and varied mixtures of Indians, Europeans, and Africans (Lima 1991). They commonly practice fishing, agricultural production, small-scale lumber extraction, collection of minor forest products (such as fruits, nuts, and fibers), and hunting. Hunters in the RCTT obtain economic benefits from market sales and subsistence consumption of mammals. Hunters receive cash for the meat of peccaries (*Tayassu* spp.), deer, lowland tapir (*Tapirus terrestris*), capybara (*Hydrochaeris hydrochareris*), and paca (*Agouti paca*) in city markets. Peccary hides are also legally sold by hunters. Mammals not sold in markets have value as subsistence food and substitute for purchases of animal protein and include primates, small rodents, edentates, marsupials, and carnivores. The most frequently hunted mammals in the RCTT include collared (*T. tajacu*) and white-lipped peccaries (*T. pecari*), red brocket deer (*Mazama americana*), paca (*Agouti paca*), agouti (*Dasyprocta* spp.) (figure 20-4), and large-bodied primates (Bodmer et al. 1994).

Figure 20-4. Agouti (*Dasyprocta* sp.). *Photo courtesy of Andrew Taber.*

There are 32 villages in the Tahuayo, Tamshiyacu, Yarapa, and upper Yavari Miri river basins, with a population of approximately 6000 inhabitants. They use resources of the RCTT to varying degrees. For example, only about 40 households from the Tahuayo River regularly use the Reserve for extraction, and only 9 villages consider themselves close enough to the Reserve to participate in management programs. Almost 100% of 541 households surveyed in the Tahuayo basin practice some type of agricultural production, whereas 42% fish as a major financial activity, 19% hunt wildlife, 23% commercially extract nontimber plants, and 6% extract timber (calculated from Coomes 1992).

Communities of *ribereños* in the RCTT organize themselves around political units, often with an elementary school and several health officials. Inhabitants within each community decide by concensus on rules for land use and extraction of natural resources. These rules govern titled land owned by community members and land officially recognized as part of the community reserve.

The first *ribereños* entered the Tahuayo River basin shortly after the construction of a naval base in Iquitos in 1862 (Coomes 1992), but it was the rubber boom of 1880 to 1920 that brought a large influx of people to the area. With the crash of the rubber boom, the area experienced a net emigration. Communities of *ribereños* consolidated during the recession of the 1930s saw an influx of people of Cocama/Cocamilla Indian origin. With the increase of market-oriented agriculture and an increase in extraction of forest resources after 1940, the population of the Tahuayo River basin increased and continued to do so until the end of the 1980s.

The abolition of estates after the enactment of the agrarian law of 1969 produced an open access system that initiated the uncontrolled extraction of natural resources. Natural resources were rapidly declining through the 1970s and early 1980s, and many were scarce by the mid-1980s. These resources fulfilled both financial and subsistence needs of local inhabitants. The communities were particularly unhappy about the exploitation of fish by freezer vessels, the extraction of timber by city-based operators, and the hunting of meat by merchants from Iquitos. As a result, communities organized a system of controls that began to prohibit the extraction of natural resources by nonresidents.

The environmental actions taken by the communities of the upper Tahuayo during the 1980s were the major influence promoting the legal creation of the RCTT. During the 1980s people living closest to the proposed reserve were seriously discussing the issue of fair natural resource use. They began to take community initiatives to protect natural resources by setting community regulations among themselves.

Comanagement also began during the early 1980s, because community representatives approached the Ministry of Agriculture and scientists working in the area to gain support for their community conservation initiatives. The Ministry of Agriculture and the scientists worked with the communities to begin the legal actions required to gazette a reserve. Fortunately, the Peruvian government had recently created the protected area category of "community reserve." This coincided nicely with the communities' requirements and the conservation ambitions of the Regional Ministry of Agriculture.

Government agencies and nongovernment groups took particular interest in the area and in comanaging the Reserve because of its unique biodiversity. For example, the RCTT is the only protected area in Peru that includes the red uakari (*Cacajao calvus*). This species is rare in Peru and considered vulnerable to extinction (IUCN 1996).

After the creation of the RCTT, much debate began over who had access to the Reserve and how much could be taken out. The four groups involved with settling this included (a) the local communities, (b) government agencies, (c) NGO extension workers, and (d) researchers. These groups coordinated many activities, but often had different approaches to comanaging the Reserve depending on their interests and the resource of concern.

For example, after establishing the Reserve, NGO extension workers collaborated with the regional government and local inhabitants to ensure that the number of people using the RCTT did not increase. Extension workers held many informal and formal meetings in villages of the upper Tahuayo to discuss the concept of the Reserve and resource extraction. Extension workers encouraged agreements whereby communities could have authority to manage resource extraction through reasonable accords. This process has continued since 1989. It has evolved from a consciousness-raising exercise to detailed discussions about the legalities of the Reserve and the biology of sustainable resource use.

Currently, decisions on resource use and management in the RCTT are voted upon democratically during community meetings. This allows communities to experiment with different types of management and to find management systems that are compatible with their culture. In addition, the size of communities is not too large for effective communication, and they can easily define their boundaries and membership. The four villages of the upper Tahuayo River and Blanco Creek are closest to the Reserve, and they have been the most active in managing it. They regularly meet to discuss resource issues and amend intracommunity agreements.

Community-based comanagement in the RCTT has resulted in management actions. For example, communities restrict access to hunting grounds, with access permitted only to people who live in nearby villages. Professional hunters from urban centers, such as Iquitos, cannot enter hunting areas. The communities have also established a game register and appointed game inspectors who are responsible for noting the number of animals harvested by each family. In addition, the communities have experimented with a game tax system, a quota system, and a male-directed harvesting program (Bodmer 1994).

Comanagement in the RCTT has external support from universities, research institutions, and research NGOs that provide technical information. Although researchers sometimes differ in their approach to the RCTT, most have interest in local participation and develop research that aids local people with community-based management.

Under the comanagement, technical decisions are separate from allocation decisions. Technical decisions use information from researchers, which is communicated to the hunters by extension workers. In contrast, community members make allocation decisions during community meetings. Shortly after the creation of the RCTT, researchers thought that NGOs and government agencies were not adequately communicating research results to local communities and that community-based management required input from objective studies to attain the goals of more sustainable resource use, including socioeconomic analyses. This input is now achieved through extension workers.

Because the communities already had an interest in managing their resources, it was easy for them to register the amount of resources used and to stop people from outside the boundaries from using resources. However, converting nonsustainable use of resources to more sustainable use by community members is more difficult because it often entails short-term economic costs.

A financial cost/benefit analysis showed that over the short term (0–5 years) there would be economic costs for local inhabitants if they used wildlife more sustainably in the RCTT (Bodmer et al. 1997b). These costs are around 25% of the annual financial income that would be earned by maintaining the current unsustainable system. Over the long term (6–30 years), there would be financial benefits for local people if they establish a more sustainable system in the RCTT. These benefits are around 66% above the annual income that would be earned by continuing unsustainable practices. Local inhabitants will only use resources more sustainably if short-term costs can be overcome. Lowering revenues would not be acceptable to many families because of their poverty and would only increase their discontent. However, if people do not set up a more sustainable system, poverty will eventually worsen once resources become depleted.

Communities are staggering management programs as a way to spread the economic costs over a longer period. This enables local people to accept economic costs more readily than bearing them all at once. They are doing this by setting up management programs for a given resource only when they receive the economic benefits of a previous management program. For example, the increased access to game animals that resulted from restricting access to outsiders has enabled local residents to consider additional game management programs.

IS COMMUNITY-BASED COMANAGEMENT CONSERVING WILDLIFE POPULATIONS?

Studies were conducted on the harvests and populations of game mammals in the RCTT to evaluate the effectiveness of community-based comanagement. Studies on animal populations included comparative density analysis, age structure analysis, harvest models and vulnerability models. These studies clearly showed that before establishing the Reserve, people overhunted primates and tapirs, but did not overhunt artiodactyls and large rodents.

Densities of mammals in the persistently hunted areas of Tahuayo-Blanco were compared with the lightly hunted areas of Yavari Miri. Results showed that collared peccary, white-lipped peccary, red brocket deer, grey brocket deer, and agouti densities were similar between these areas. However, densities of lowland tapirs and large primates were considerably less in the persistently hunted areas (Bodmer et al. 1994).

The harvest model calculated the percentage of production taken by hunters. Again, results indicated no overhunting of peccaries and deer in the Tahuayo-Blanco area, but severe overhunting of lowland tapirs (Bodmer 1994).

An age-structure analysis compared the demography of ungulate populations in persistently hunted and slightly hunted sites. Results concurred with the other analyses by showing no significant difference in age distributions of peccaries and deer with hunting pressure. In contrast, there was a significant depression in the age distributions of lowland tapirs in the persistently hunted site, again suggesting overhunting (Bodmer 1995b).

Vulnerability models suggest that lowland tapir and large primates are vulnerable to overhunting because of their low rates of reproduction and slow intrinsic rates of population increase. In contrast, deer, peccaries, and large rodents are less vulnerable to overhunting because they have faster rates of reproduction and intrinsic rates of population increase (Bodmer et al. 1997a).

Research played an important role in determining the thrust of extension programs. Extension workers conveyed the information on overharvesting of primates and lowland tapir to the communities. They stressed the need to decrease harvesting of these species and maintain current harvest levels on artiodactyls and large rodents.

Harvests were evaluated in 1991 before communities set up wildlife management programs. They were reevaluated in 1994 and 1995, 3 and 4 years after the onset of the community-based programs. The impact of management was examined by comparing harvests of artiodactyls, large rodents, primates, and tapirs pre- and postmanagement (Bodmer et al. 1997b). The number of animals hunted was determined from skulls collected by hunters and by interviewing hunters about skulls not collected. An error margin was added to the hunting pressure to account for animals hunted but not recorded either by skulls or interviews. This was calculated by determining which local hunters were not participating in the project. The error margin varied between 10% and 20%.

Over the 4-year period spanning premanagement and postmanagement, harvests of artiodactyls showed a slight, but not significant, difference between 1991, 1994, and 1995, measured as the number of animals harvested per 100 km² per year (p = not significant [NS]) (figure 20-5). Similarly, harvests of large rodents and tapirs showed no significant difference between 1991, 1994, and 1995 (p = NS) (figure 20-6). However, hunters harvested significantly fewer primates between 1991 and 1994, and again between 1994 and 1995 (f ratio = 9.823, p = 0.002) (figure 20-7).

To examine whether harvest patterns were related to changes in the wild populations or a real decrease in hunting effort, we looked at the abundance of game mammals in Tahuayo-Blanco between 1986 and 1994. The year 1986 corresponds to the premanagement census and 1994 to the postmanagement census. Transects were used to calculate abundance of mammalian game species. Trails of 3 to 5 km were cut in the forest and censused in the morning and evening, and records were made of the number of groups sighted of each species, the number of animals in each group, and the perpendicular distance of the first sighting to the trail. Totals of 120 km and 626 km were surveyed in 1986 and 1994, respectively.

The wild populations of artiodactyls did not change significantly between 1986 and 1994 (p = NS) (see figure 20-5). Similarly, the abundance of large primates in the forest did not change significantly between 1986 and 1994 (p = NS) (see figure 20-7). People hunt primates in the RCTT mainly as a source of subsistence food because they have little market value, so it is unlikely that a change in market demand caused the change in primate harvests. Hence, the decrease in the harvests of primates between 1991 and 1995 was ascribed to community-based comanagement.

Lowland tapir harvests should also be decreased to prevent overhunting. However, the first several years of community-based comanagement did not result in a decrease in such harvests. Tapirs are the largest terrestrial mammal in the area and represent substantial cash income for hunters. Extension activities have focused on finding solutions to tapir overhunting. Communities of the upper Tahuayo have recently stated that community members will not be allowed to sell tapir meat to city markets. Setting up this policy might be difficult, however, because of economic demands. The best management strategy for the species might be total protection of the fully protected zone, allowing it to act as a source area.

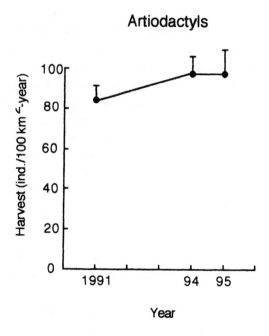

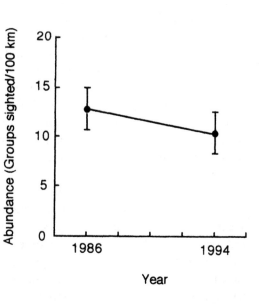

Figure 20-5. Harvests and abundances of artiodactyls in the Tahuayo-Blanco area of the RCTT. The error bars of the harvests were determined through interviews with local people and represent the error in calculating actual harvests. The error bars of the abundance data represent the standard deviation calculated using the coefficient of variation of relative densities for foot transects with an average transect length of 7 km (Seber 1982).

Large Rodents

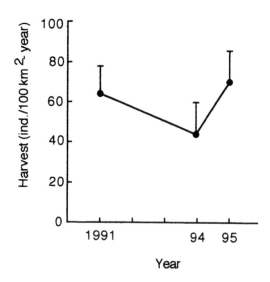

Tapir

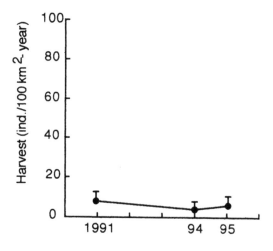

Figure 20-6. Harvests of large rodents and lowland tapirs in the Tahuayo-Blanco area of the RCTT. The error bars of the harvests were determined through interviews with local people and represent the error in calculating actual harvests. Large rodents have a greater error because hunters often sell pacas with the skull attached, thus making the harvest more difficult to calculate.

Primates

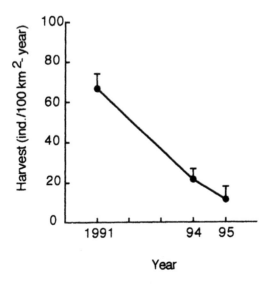

Large Primates

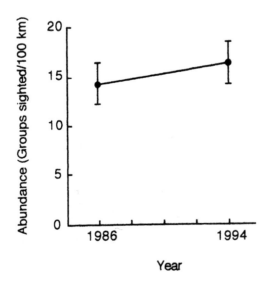

Figure 20-7. Harvests and abundances of large primates in the Tahuayo-Blanco area of the RCTT. The error bars of the harvests were determined through interviews with local people and represent the error in calculating actual harvests. The error bars of the abundance data represent the standard deviation calculated using the coefficient of variation of relative densities for foot transects with an average transect length of 7 km (Seber 1982).

CONCLUSION

Community-based efforts will only lead to successful conservation if the socioeconomic realities of local people operate within the biological limits of the ecosystems. Recently, prerequisites for sustainable development have focused on the socioeconomic conditions of local people and their need to attain an improved life (Robinson 1993). There is no doubt that this is of the utmost importance, especially in light of community-based conservation. Of equal importance, however, is the recognition that better information on the biology of ecosystems and species must be incorporated into community-based conservation efforts. Community-based conservation will undoubtedly fail if these biological attributes are not adequately considered (Robinson and Redford 1994a).

Throughout the tropics, rural communities have taken initiatives to set up their own resource management. These communities frequently seek assistance on the technical aspects of resource use; for example, communities often want to know the sustainable level of resource extraction. Currently, there is a lack of adequately trained professionals who can give these local people a workable answer. Failure to provide technical support is a tragedy for community-based conservation because it discourages the efforts of rural people. On the other hand, by providing adequate technical assistance, communities will be encouraged and community-based conservation can endure.

Recently there has been a rapid increase in the number of communities who have taken their own initiatives to manage and conserve their neighboring habitats (Western and Wright 1994). Unfortunately, many of these initiatives become distorted when it becomes evident that there is insufficient knowledge on managing many natural resources in a way that is compatible with the socioeconomic abilities and aspirations of the local people. One of the greatest challenges of conservation today is whether professionals can be trained fast enough to address resource use and simultaneously respect local people's needs.

Community-based comanagement appears to be working as a conservation strategy in the RCTT. Our experience is that combining research biologists with extension responsibilities is necessary for comanagement because these professionals must assist local people on the biological limitations of resource use while fully respecting the socioeconomic realities of the communities. Community-based resource specialists are in a new field of expertise. Through their efforts, local communities will assume the responsibility of designing, implementing, and monitoring the management of their natural areas.

PART IV

Economic Influences on Sustainability

21

Wildlife Use in Northern Congo: Hunting in a Commercial Logging Concession

PHILIPPE AUZEL AND DAVID S. WILKIE

Throughout much of central Africa, people still rely on wild animals for most of their daily protein intake (Aisbey 1974). Both rural and recently urbanized populations provide a ready market for bushmeat in a region where livestock is viewed more as a form of savings than as a dietary item (Wilkie et al. 1997). Forest foragers (pygmies) throughout the region have traditionally traded bushmeat for forest farmers' agricultural crops that provide over 60% of their annual calories. More recently, as the rural infrastructure has fallen into disrepair and markets for agricultural products have collapsed, both foragers and farmers have often relied more heavily on hunting and selling bushmeat as a source of income.

Selective logging is, at present, the dominant industrial economic activity in moist forest regions of Central Africa. The industry generates considerable revenue for national governments—US$506 million per annum for Cameroon (ITTO 1988)—and provides undocumented but apparently sufficient profits for companies involved in logging. Concerns about the impacts of logging on forest structure and function have focused primarily on the direct removal of biomass and changes in canopy closure, as well as on the indirect consequences of these changes on forest-dependent plant and animal species (TELESIS 1991; Stoll 1992). Study results suggest that one-time, highly selective logging that removes less than one tree per hectare (ha), although catastrophic to size class distribution of the selected tree species, has only limited long-term impacts on forest structure and function; rather than reducing animal populations, this type of logging might increase densities of herbivores that benefit from the lush and accessible regrowth vegetation (Skorupa 1986; Thiollay 1992; Struhsaker 1997). Although timber removal itself might not have severe adverse environmental impacts, recent research has begun to show that the process of timber extraction can have severe adverse impacts on wildlife populations by indirectly or directly facilitating commercial hunting to supply the bushmeat trade (Wilkie et al. 1992; Chardonnet et al. 1995; Auzel 1996a; Auzel 1996b; Wilkie et al. 1997) and by raising household income and thus the demand for bushmeat.

This chapter examines the impact of selective logging on the involvement in and scale of commercial bushmeat hunting by rural communities. Data are presented on domestic and commercial bushmeat hunting in rural communities that vary in their access to and level of economic integration with a logging company in the northern Republic of Congo.

BACKGROUND

Forest of Northern Congo

Approximately 60% of the country's 34,200 km^2 is covered with tropical moist forest, of which 60% is considered productive for logging (Hecketsweiler 1990). By the late 1980s, only about 5000 km^2 of forest, mainly in the northern part of the country between the Sangha and Oubangui rivers, was considered untouched by logging operations (FAO 1991).

Rainfall in northern Congo ranges from 1400 mm to more than 2100 mm per annum, with an average of 1715 mm in Ouesso and 1846 mm in Impfondo (CIEH 1990). The forest is a mixture of four principal types:

1. Moist semi-deciduous forest with high densities of commercially valuable timber species: *sipo (Entandrophragma utile)*, *sapelli (E. cylindricum)*, *tiama (E. angolense)*, *bossé (Guarea cedrata)*, *ayous (Triplochiton scleroxylon)*, *iroko (Chlorophora excelsa)*, and *wengue (Millettia laurenti)*
2. Monodominant forest (*Gilbertiodendron dewevrei*)
3. Open canopy forest including raphia and swamp forest
4. Secondary forest patches found in logged areas and near actual or former human settlements.

These areas still contain some of the highest densities of forest wildlife in the world with at least eight species of *Cercopithecus*, *Cercocebus*, and *Colobus* monkeys, six species of duikers (*Cephalophus* spp.), as well as Bongo (*Boocercus euryceros*), giant forest hog (*Hylochoerus meinertzhageni*), forest buffalo (*Syncerus caffer nanus*), red forest hog (*Potamochoerus porcus*), sitatunga (*Tragelaphus spekei*), and many threatened species, such as African forest elephant (*Loxodonta africana cyclotis*), lowland gorilla (*Gorilla gorilla gorilla*), and chimpanzees (*Pan troglodytes*) (Fay et al. 1990).

Concession Description

In 1963, a 445,000-ha forest reserve was set aside for further development of logging in northern Congo, and logging companies were granted 25,000-ha concessions. In 1974, the forest reserve was divided into forest management units (FMUs) created to regulate timber production and ensure sustainable exploitation of commercially valuable tree species. Some 15 FMUs subdivided northern Congo at the present time (figure 21-1). It was not until 1970, after much of the

Figure 21-1. The Congo, showing the location of the Pokola concession and other forest concessions in the northern part of the country.

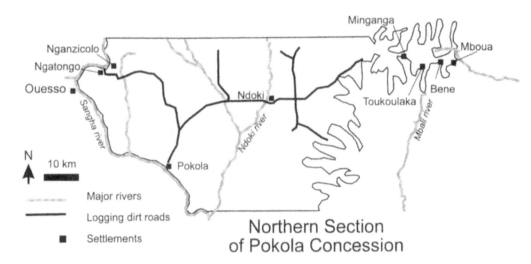

Figure 21-2. Northern section of the Pokola concession.

country's southern forest had been logged, that a foreign company, Congolaise Industrielle des Bois (CIB), requested and received a permit to log 25,000 ha in the Pokola FMU (figure 21-2). In 1980, CIB signed a timber transformation contract with the Congolese Government and began sawmill operations in Pokola, south of Ouesso (the regional administrative center). Since then, the logging and sawmill industry (figure 21-3) in Pokola has grown steadily, as has the size of the concession, now 480,000 ha. Today it employs approximately 700 Congolese workers and extracts annually 150,000 m³ from 22 tree species, although 90% is from two *Entandrophragma* spp. (*sipo* and *sapelli*).

Figure 21-3. Logging trucks. *Photo courtesy of John G. Robinson.*

Human Population

The northern part of the Republic of Congo remains relatively unpopulated. More than 80% of Congo's population of 2.5 million people live in the southern 30% of the territory (Lopez Escartin 1991), with the majority living in or near the two major cities of Brazzaville and Pointe Noire. Estimated rural population densities in the northern Sangha and Likouala provinces are between 0.7 to 0.8 people per km[2] (Bahuchet 1993). Except for Ouesso and a few logging towns on the Sangha River such as Pokola and Kabo, most of the population inhabits small villages or temporary camps scattered through the forest, most often along rivers.

Although a complete census of the human population has not been undertaken, the local (Bantu, Oubangian, and Pygmy) and immigrant (Emboshi, Teke, and other Lingala-speaking citizens of the Congo, Central African Republic, Cameroon, or Democratic Republic of Congo) population within the CIB concession is estimated to exceed 5000. CIB's main operations are run from Pokola, and in 1992 the company developed a second base camp along the Ndoki River for 100 to 120 workers. Selective logging is relatively labor intensive, at least during the survey phase, and can result in the transient relocation of skilled workers and their families (Wilkie et al. 1992; Auzel 1996b). Local workers are usually hired for their knowledge of the forest and are usually paid less than the immigrant skilled workers. Preconcession residents of Pokola often complain about the arrival of "southerners" who outcompete them for higher paying logging jobs and can outspend them to obtain valued forest or trade goods. Although irrevocable changes in the indigenous community structure might be occurring, baseline information about such phenomena are not yet widely available (Hardin 1997).

Traditional Hunting

Bushmeat has traditionally been and remains an important source of protein and fat for rural families in central Africa (Hladik et al. 1990). Historically, Pygmy hunters used crossbows (introduced by the Portuguese in the 14th century), bows, nets, and spears to capture forest animals. Forest farmers also set snares made from *Raphia vinifera* fiber (Wilkie 1996). The establishment and arming of village militias in border zones of the Republic of Congo dramatically increased hunters' access to firearms after independence in 1960 and saw a shift to the use of guns for hunting. This trend has been continued with the manufacture of shotgun cartridges in Pointe Noire, and the steering rods of Land-Rover 110s are the exact caliber for local manufacture of 12-gauge shotguns (Pearce and Ammann 1995). As a result, acquiring firearms is not a problem in the northern Congo, even among Pygmy hunters, some of whom possess their own firearms. The shift from traditional hunting methods to jack (spot) lighting with shotguns constitutes a dramatic change in the efficiency of hunting (Hames 1979) and has resulted in hunters having to travel farther afield to find wildlife to hunt (Auzel 1996b). Although Congolese law prohibits jack lighting, and the use of nontraditional material (wire-cable) in the construction of snares, both techniques are openly used and are acknowledged as the most common, preferred, and effective techniques for obtaining game (Wilkie et al. 1992).

Study Villages

The purpose of this study was to examine the impact of the logging concession practices on the scale of commercial bushmeat hunting. Hence, settlements were selected to represent a range of labor involvement in the concession and a range of degrees of access to concession infrastructure and transportation. Four of the eight settlements in the northern section of the Pokola concession were selected for study (figure 21-2). Settlement characteristics are described in table 21-1. Spatial segregation of ethnic groups made it impossible to select settlements with comparable ethnic composition.

METHODS

Data for this chapter were collected by P.A. and Congolese field assistants between June and August of 1995 and 1996. The 1995 field season was used primarily for qualitative data collection, to train local field assistants, and to habituate human subjects to the presence of the researchers.

Table 21-1. Characteristics of the Study Settlements

Name	Toukoulaka	Nganzicolo	Ngatongo	Ndoki (CIB)
Location	Village near Likouala swamps	Sangha riverside village		Logging camp
Families per village	35	19	28	101
Total population	197	144	153	648
Average family size	5.6	6	5.4	6.4
Ethnic groups	Kabonga (99%)	Yeswa (38.6%) Ngundi (19.3%) Pomo (18.4%) Bengélé (14.9%) and 5 ethnic groups (8.8%)	Pomo (53%) Bengélé (7.7%) Bomassa (7%) Gbaya (7%) and 18 ethnic groups (25.3%)	Embosi (13.7%) Bengélé (13.1%) Téké (10.3%) Kouyou (9.7%) and 26 ethnic groups (53.2%)
Settlement type	Long term	Long term	Long term	Established in 1992
Market access	Limited access, although a dirt logging road was going to reach the area in 1997	Easy access by road with CIB trucks or by pirogue along the Sangha river		Easy access with CIB trucks
Time to dirt road	8 hr walk	0.25 hr paddling		On the road
Time to Pokola	+ 2 hr truck	1.5 hr truck		>1 hr truck
Time to Ouesso	+ 3 hr motor pirogue	2.5 hr paddling		>3 hr motor pirogue
Main economic activity	Agriculture, swamp fishing, and hunting	Agriculture, river fishing, palm wine, and hunting		Logging, agriculture, and hunting
Sample size in days	53	38	43	58

Demography

A complete household census was undertaken in 1996 in all study settlements other than Ndoki, where 73% of all households were surveyed.

Household Bushmeat Consumption

The qualitative importance of wild game to rural populations was determined using irregular interviews of the female head of household in a sample of five households within in each village during the period of time that a researcher was resident. Women were asked to recall the composition of the primary meal of the day that they had cooked or were planning to prepare for their household. The frequency that animal protein and that fresh and smoked bushmeat were components of the household's meals was then collated from the interview data.

Hunting Returns

The community level consumption of bushmeat was estimated using methods developed in Cameroon (Koppert and Hladik 1990; Dounias et al. 1996). The species, gender, capture method (shotgun or snare), source, destination (local consumption or trade to market), and extent of processing (none or smoked) were recorded for all bushmeat entering a village during each day that a researcher was resident. An attempt was made to validate destination data by monitoring all bushmeat leaving the village, although this approach fails to measure meat that was sold and consumed within the settlement.

The biomass of bushmeat captured was estimated using literature values for the mean weight for each species and an average carcass butchering weight loss of 40%. Consumption estimations obtained with these methods were comparable with those obtained by nutritionists who systematically weighed all food consumed by households (Koppert 1996). The price of each species of bushmeat that was exported from each settlement was determined by interviewing traders.

RESULTS

Variance in Hunting Strategies Among Pokola Households

In all study sites, shotguns and snares were the only two methods used to capture forest animals (table 21-2). Use of traditional methods such as crossbows, bows, and nets was never observed. Guns were used to kill the majority of all game brought into the settlements (51–94%; mean 76%). Jack lighting at night was the preferred hunting method in heavily hunted areas, such as those near long-term settlements (these included villages along the Sangha River and long-term CIB forest camps). In newly opened forests, hunters (CIB workers, their relatives, or pygmies) attracted wildlife during the day using duiker or sometimes monkey distress calls. The only site where wire snares accounted for a large fraction of kills

Table 21-2. Number and Biomass of Animals Consumed Directly and Traded by Each Settlement

Sites	Number Captured	Relative Frequency	Total Weight (kg)	Relative Biomass	Biomass Captured per capita/ day (kg)	Total Butchered Weight (kg)[a]	% Biomass Traded	Bushmeat Consumed per capita/ day (kg)
Toukoulaka								
Duikers	7	5%	99	8%	0.01	59	0	0.01
Primates	52	39%	430	35%	0.04	258	0	0.02
Suidae	6	4%	300	24%	0.03	180	0	0.02
Others	70	52%	411	33%	0.04	247	15	0.02
	135	100%	1240	100%	0.12	744	5	0.07
Nganzicolo								
Duikers	75	18%	976	57%	0.23	586	59	0.06
Primates	27	6%	259	15%	0.06	156	31	0.02
Suidae	8	2%	400	23%	0.09	240	88	0.01
Others	311	74%	86	5%	0.02	52	62	0.00
	421	100%	1721	100%	0.40	1033	62	0.09
Ngatongo								
Duikers	242.8	78%	3186	77%	0.48	1912	76	0.07
Primates	51.3	16%	388	9%	0.06	233	50	0.02
Suidae	9	3%	450	11%	0.07	270	89	0.00
Others	10	3%	123	3%	0.02	74	88	0.00
	313.03	100%	4147	100%	0.63	2488	75	0.09
Ndoki								
Duikers	800	82%	10,938	75%	0.29	6563	36	0.11
Primates	110	11%	896	6%	0.02	538	17	0.01
Suidae	44	5%	2200	15%	0.06	1320	50	0.02
Others	18	2%	512	4%	0.01	307	40	0.00
	972	100%	14,546	100%	0.39	8728	37	0.15

[a]A 40% weight loss is assumed to result from butchering.

(49%) was the flooded forest settlement of Toukoulaka, where they were used to trap dwarf crocodiles. However, all settlements did set trap lines. Ethnic differences might have accounted for Ngatongo households using snares more than households in Ngazicolo (28% versus 6%) because they shared all other settlement characteristics. In Toukoulaka and Ngazicolo, households restricted their hunting to their own territory, whereas in Ndoki hunters exploited areas of forest being prospected or felled by the logging company. Hunters in Ngatongo exploited the broadest range of forest, using home territory, logging areas, sections of forest along the Sangha River, and even forest in Cameroon. Hunting occurred both directly from the settlements and during lulls in agricultural activity from hunting camps that were often more than 5 km from the settlement. Hunters in Toukoulaka, Ngatongo, and Nganzicolo walked to their hunting areas and transported bushmeat on their backs. Ndoki hunters used logging vehicles under contract to CIB to transport bushmeat along the Ngatongo-Pokola road, which allowed them access to transport down the Sangha River.

In Ndoki, Pygmy hunters were often given a shotgun and two or three 12-gauge cartridges by company employees. Logging trucks then transported them to and from the forest in the process of taking employees to the areas being prospected and felled. The hunters were usually paid with a portion of the catch. Employees were able to buy shotgun cartridges with what remained from their monthly wages after payment of debts and necessary expenses, and traded, shared, or consumed the bushmeat to augment their salaries or improve their living conditions.

Many teenage boys set snare lines when they returned to their villages during school vacations. These snares were often made of nylon from old grain sacks and were set in old agricultural fields to catch porcupines (*Atherurus africanus*) and the few nocturnal duikers (*Cepalophus dorsalis*) still remaining in these heavily disturbed areas.

Variance in Hunting Returns, Species Harvested, and Quantity Traded Among Pokola Households

To account for ecological and socioeconomic factors that influence the quantity of bushmeat brought in by hunters, it would be necessary to stratify the monitoring of captures throughout the year and to monitor captures over more than 1 year. Because data for this study were collected only during a few months in the year, we must restrict our analysis to comparisons across settlements rather than attempt to extrapolate values for annual consumption of bushmeat. In addition, we must interpret these results with care because seasonal demands on labor may vary significantly among settlements; thus, investments in hunting during the study period may not be representative of annual investments.

To facilitate comparisons across settlements, all bushmeat species captured were tabulated into four categories (small antelope, primates, pigs, and others; see table 21-2). Somewhat surprisingly, settlements showed considerable variation in the relative contribution of different species groups captured. Small antelope contributed over 75% of all individuals and biomass captured by hunters in the Ndoki logging camp, as well as in Ngatongo; indeed the profile of bushmeat hunted by these two settlements was very similar. In contrast, fruit bats (*Pteropus indeterminata*) accounted for over 70% of all animals brought into Nganzicolo, whereas small antelope and pigs contributed 70% of the biomass. In Toukoulaka, primates and dwarf crocodiles (*Osteolaemus tetraspis*) contributed over 82% of animals captured by hunters, and primates, pigs, and crocodiles contributed 89% of the biomass. Toukoulaka has very little terra firma forest, and hunters paddle and pole dugouts along rivers to hunt. This probably explains the low incidence of antelope and the high proportion of primates and crocodiles captured by hunters in Toukoulaka.

Most bushmeat entering settlements was fresh (65–95%). Smoked bushmeat only accounted for 5% of carcasses brought into Nganzicolo, 14% in the Ndoki camp, 24% in Toukoulaka, and 35% in Ngatongo.

Per capita daily captures of bushmeat were highest in Ngatongo (0.63 kg per capita/day), almost double that of Nganzicolo and the Ndoki camp (0.40 and 0.39 kg per capita/day) and over five times that of Toukoulaka (0.12 kg per capita/day).

All settlements were involved in bushmeat trading, with 5% to 75% of all biomass brought into settlements entering the market. Ngatongo was the most involved in bushmeat marketing and Toukoulaka the least. Although Ndoki camp had the greatest access to transportation, hunters only traded 37% of the meat that they captured outside the camp, partly because of the extensive trading opportunities within logging communities. Primates entered the market less frequently than all other species groups. In Ngatongo, 28% of carcasses brought into town were consumed locally directly or after exchange or trade on the domestic market. Only 3.5% of all the game recorded comes from Ngatongo forest. Most came from temporary hunting camps along the Ngatongo-Ndoki dirt road (41%) or from hunting camps upstream, on the Cameroonian side of the border (37%). Ten percent of small antelopes brought into the settlement were destined for the bushmeat market in Pokola, whereas 64% were destined for the market in Ouesso. Of all carcasses brought into Ngatongo, 63% were sold in Ouesso.

Assuming a 40% weight loss during butchering, direct per capita consumption of bushmeat ranges from 0.06 to 0.20 kg meat per capita/day, with Ndoki camp residents consuming more than twice the amount daily compared to all other settlements.

Variance in Bushmeat Eaten by Pokola Households

Forest families in the Pokola FMU ate animal protein (bushmeat, fish, insects, reptiles) as a component of their main daily meal 73% to 97% of the time, with bushmeat occurring in 39% to 76% of meals (table 21-3). Although only a small fraction of bushmeat entering the village is smoked, 50% of meat consumed by households is smoked. Only the Ndoki camp residents eat primarily fresh bushmeat.

Variance in Bushmeat Prices Within and Outside the Pokola FMU

Average price estimates for bushmeat within and outside the logging concession show (a) an almost tenfold increase as bushmeat moves from the hunting area (US$0.24/kg) within the concession to Brazzaville (US$2.28/kg), the capital city, and (b) that consumers are not making a price distinction between duiker and primate meat. Although it is unclear why the price per kilogram of the larger Peter's

Table 21-3. Frequency of Animal Protein and Bushmeat as a Component of the Meals of Forest Families

Site Description	Villages	N	% of Meals Containing Animal Protein	% of Meals Containing Bushmeat	Fresh Meat	Smoked Meat
Swamp villages	Toukoulaka ($n = 41$) Bene ($n = 22$) Mboua ($n = 14$)	77	72.7	39	50%	50%
Sangha riverside	Ngatongo ($n = 76$) Nganzicolo ($n = 47$)	123	95.1	48.8	50%	50%
Logging camp	Ndoki ($n = 124$)	124	96..8	75.8	85.1%	17%

duiker (*C. callipygus*) is often lower than that of the blue duiker, one trader noted that this was to keep the price of a whole or quarter carcass affordable to customers. When we combine this information with the quantity that enters the market in each settlement, we find that the daily average value of bushmeat ranges from US$0.17 to US$28.42, and the average income generated ranges from US$0.004 to US$0.06 per capita/day. The rank order importance of bushmeat as a source of household income is Ngatongo, Ngazicolo, Ndoki, and Toukoulaka.

DISCUSSION

Impact of Logging on Hunting Patterns in Northern Congo

Results of this study corroborate and expand on the findings of previous research by demonstrating that logging increases hunting (a) directly by increasing demand through immigration and by raising household income of logging employees and (b) indirectly by increasing hunters' access to the farthest reaches of the forest and to markets by providing access via logging roads and logging vehicles.

Residents of the most logging-integrated settlement, the Ndoki camp, ate the most bushmeat, and captured, in absolute terms, the most bushmeat (figure 21-4). However, per capita bushmeat exploitation was not unlike other settlements with access to logging roads and vehicles; Ndoki camp residents only traded 37% of all bushmeat captured. Most interesting is the contrast between the isolated settlement of Toukoulaka and the settlements of Nganzicolo and Ngatongo that have progressively more access to roads and markets. Toukoulaka eats about as much bushmeat

Figure 21-4. Sorting and processing wood at a logging camp. *Photo courtesy of John G. Robinson.*

per capita as the other two settlements, but trades only 5% of the animals it captures and hunts only one quarter as intensively. Nganzicolo, a mere 15 minutes by dugout to the logging road, hunts less and trades less in absolute and relative terms than Ngatongo that lies directly on the logging road. It appears that the focus of Ngatongo households on bushmeat hunting and marketing of wild meat is based on a combination of access to markets and absence of logging-based salaries.

In summary, logging company employees eat more bushmeat than their village counterparts, but they hunt less and sell less than those households that have equal access to markets. Households not employed by the logging company but that have access to market eat no more bushmeat than the residents of isolated villages, but they hunt more intensively and market a much higher proportion of all game captured.

Managing Wildlife Populations in Logging Concessions

Results of this study show that logging companies affect the intensity of hunting in two ways: (a) by increasing demand through immigration and through increases in household income of company employees, and (b) by increasing hunter access to the forest and reducing bushmeat transportation costs by constructing roads and allowing hunters access to company vehicles.

If the Congo, the international conservation community, and logging companies themselves are concerned about hunting in logging concessions, then two approaches must be taken to reduce logging-facilitated bushmeat hunting. The first addresses transportation of hunters to the forest and of bushmeat to markets; the second addresses the issue of demand. Both approaches assume willingness to enact new policies or enforce old ones, as well as willingness to pay for implementing, monitoring, and enforcing such management policies. Neither assumption is necessarily likely to be true in Congo at present.

Curbing the Transportation of Hunters and Bushmeat

Reducing access by hunters to the forest and to markets can be achieved quite simply by (a) banning the use of logging trucks to transport hunters and bushmeat, and enforcing the ban using road blocks and spot checks of logging vehicles; and (b) destroying bridges and gouging the surface of roads to make them impassable by vehicles, in sections of forest that have already been logged. This directly reduces the profitability of market hunting, which is largely determined by access to and cost of transportation. When CIB started transporting logs to Douala from the Sangha River port at Sucambo (near Ouesso), bushmeat from Cameroon soon comprised over 13% of the game sold in Ouesso markets (Bennett Hennessey 1995). However, a dispute between the trucking company and the concession halted traffic from Congo through Cameroon, resulting in the temporary collapse of the bushmeat market and the closure of hunting camps that border the roads during August 1995 (Pearce and Ammann 1995). Bushmeat marketing is clearly a risky business. Thus, raising the level of risk by making access to transportation more unreliable and costly is an appropriate way to reduce the viability and hence intensity of hunting within concessions.

Enforcers of hunter and bushmeat transportation prohibitions should be trained wildlife officers from the Ministry of Water and Forests. Wildlife officers could be paid indirectly by the logging concessions, who should be required to post a conservation bond (Karsenty and Maître 1994) each year, paid to the appropriate government ministry, for an amount indexed to the area of forest to be exploited that year. These monies would be earmarked for natural resource conservation within logging concessions, and thus could only be used to support forestry and wildlife officers, and plant and animal surveyors, stationed in logging concessions. Repayment of the bond to the logging concession could be indexed to the ratio of pre- and postlogging wildlife survey figures, with the highest rebates occurring at parity. If the bond was set high enough, logging companies might comply with recommendations that wildlife and firearms laws of the country be respected by personnel of logging companies, and that vehicles, roads, facilities, and company time should not be used in support of poaching. A conservation bond would also help strengthen Congo's capacity and institutions to enforce wildlife protection, as Verschuren (1989) has urged. This approach will, of course, only work (a) if the logging companies do not attempt to bribe forestry and wildlife officers, and (b) if the forestry ministry establishes and enforces wildlife conservation bond legislation and uses the earmarked funds appropriately.

Reducing Demand for Bushmeat

Altering the demand for bushmeat is much more challenging. For logging companies to reduce the level of immigration into concession areas, they would have to be willing to train local residents to undertake the tasks of the skilled workers that they presently hire from outside the area. Although this would minimize the impact of immigrant consumers, it would be costly. Rather than attempting to reduce the number of consumers, it is probably more feasible to reduce the per capita demand for bushmeat by concession workers. This could be achieved in two ways:

1. Importing meat of domestic animals for sale to concession workers at a price equal to that of bushmeat. This will undoubtedly require a subsidy by the logging company.
2. Establishing a small livestock-raising program within the concession. Techniques for raising rabbits or cane rats are well established (Jori et al. 1995; HPI 1996). It would not be difficult for the logging concession to raise these animals at both Pokola and Ndoki and sell them to employees at a price below that of bushmeat.

Assuming that the demand for bushmeat is elastic, then providing consumers with an abundant, acceptable, and competitively priced substitute should drive bushmeat traders out of business, and should reduce the incentive for employees to invest their labor and capital in hunting. One concern about either importing domestic animal meat, or raising domestic livestock is the introduction of zoonoses that could result in high mortality in previously unexposed wildlife populations.

To encourage logging concessions to establish small livestock-raising programs, repayment of the conservation bond could also be indexed to this activity.

Impacts of Regulating the Bushmeat Trade on Local Economies

This study has demonstrated that in communities where residents are not directly employed by the logging concession but have access to logging roads and vehicles, bushmeat trading becomes an important part of the local economy. Any attempt to curb bushmeat marketing will have an adverse impact on the living standards of hunters and their families. One possible solution is for the logging concession to contract the raising of small livestock for its employees to nonemployee residents of concession villages. This would require providing the families with technical assistance and the materials and capital to start livestock raising. To ensure that domestic meat prices are competitive with bushmeat prices, it might initially be necessary for the livestock farmers to sell their animals to the logging concession which then resells it to its employees at a loss. However, once roadblocks increase the scarcity of (and thus the price of) bushmeat, livestock farmers should be allowed to sell their meat directly to consumers. To promote livestock raising by local residents further, and to help reduce bushmeat demand in regional population centers, the logging company should allow farmers to transport their livestock carcasses to market on company vehicles.

Commercial logging in the forest of northern Congo has a clear impact on the demand for bushmeat and on the scale of bushmeat marketing. Though mitigation is possible, the management of hunting in logging concessions is going to take political will and money, both of which are often in short supply in the region.

ACKNOWLEDGMENTS

The research reported here was funded by The Wildlife Conservation Society and GEF Congo. Special thanks are extended to the Ministry of Agriculture, Water and Forests of the Republic of Congo who gave permission for this study that was conducted under the auspices of the Nouabalé-Ndoki project. We thank J. M. Fay, NNP Project Director, and R. G. Ruggiero, the project Scientific Advisor, for their help and support in this research, as well as the CIB company, who allowed us to conduct the study in the Pokola FMU and provided assistance on numerous occasions. We also thank Rebecca Hardin for her comments and editorial assistance.

22

Socioeconomics and the Sustainability of Hunting in the Forests of Northern Congo (Brazzaville)

HEATHER E. EVES AND RICHARD G. RUGGIERO

Hunting plays an important and sometimes essential role in the diet and local economy of the inhabitants of northern Republic of Congo and its surroundings (Bahuchet 1991; Wilkie et al. 1992; Blake 1994; Eves 1995, 1996). The advent of foreign-owned logging companies has resulted in vicissitudes in the local economy and human demography, including increased employment opportunities and the immigration of workers and their families. Activity by international conservation organizations has begun in the region within the past 10 years, which also provides economic opportunities for some communities. As the area's economic and demographic conditions evolve, so do activities that may impact the diversity of the forest; one of these is hunting. In addition to the continued practice of subsistence hunting, a commercial bushmeat trade has grown to supply logging communities and markets in urban centers.

Successful management of protected areas might depend on the development of economic alternatives to dependence on wildlife resources, and the alleviation of hunting pressures to reduce unsustainable exploitation of wildlife (McNeely 1988; Brandon and Wells 1992; Gibson and Marks 1995). This necessitates, however, a comprehensive understanding of household decision making and how this is linked to proposed conservation activities (Ferraro and Kramer 1995). Although integrated conservation and development projects (ICDPs) and community-based conservation (CBC) are frequently promoted as a means to engage and empower local communities to conserve biodiversity (Munashinghe and Wells 1992; IIED 1994; Western and Wright 1994; Gibson and Marks 1995; Barrett and Arcese 1995), their potential in northern Congo has not yet been examined. To investigate the effects of increased employment opportunities and economic development on wildlife use, we conducted a socioeconomic study from January to April 1996 among villages surrounding the Nouabalé-Ndoki National Park (figure 22-1). This chapter focuses on the nutritional and economic importance of bushmeat and hunting to inhabitants of villages that are influenced by logging activities at Kabo (logging villages), conservation activities near the National Park bases at Bomassa and Makao (conservation villages), and those villages having no locally available cash employment opportunities (no-industry villages). Particular emphasis is placed on an examination of the factors that underlie wildlife use patterns and that may affect the sustainability of wildlife use. A complementary study focusing on hunting economics was conducted from May 1995 to May 1996 in two villages associated with the Wildlife Conservation Society (WCS) forest conservation project in the National Park. Results of the latter study are presented to provide a detailed economic evaluation of subsistence hunting in this region.

BACKGROUND

The Forest

Tropical dry forests form 70% to 80% of forested regions in Africa (Murphy and Lugo 1986). The Republic of Congo is covered in large part by forest (19,865,000 hectares [ha]; 58.2%), much of which is typical of the dry forest inner equatorial regions whose distinctive patterns of rainfall result in regular annual dry seasons (TFAP 1994). The nation's largest area of intact forest is found in the north (Wilkie and Sidle 1990), where human population density (3 individuals/km²) is among the lowest of nonarid areas in Africa (Wilkie et al. 1992).

The forest in this area is described as Sterculiaceae-Ulmaceae semi-deciduous forest and represents the northern fringes of the Guineo-Congolean forest block, which stretches from west Africa to the Democratic Republic of Congo (formerly Zaire). The Park and surrounding areas contain both primary and secondary forest, which can be grouped into three basic types: mixed-species forest, the largest and most diverse vegetation type dominated by species of Meliaceae and Leguminosae families; swamp forest, with a permanently flooded floor, found along streams and rich in trees of the genera *Alstonia*, *Mitragyna*, and *Xylopia*; and *Gilbertiodendron dewevrei* (monodominant forest), a riverine vegetation type with a relatively clear understory.

Gaps of various origins are present in the primary and secondary mixed forests, and terrestrial herbaceous vegetation may be dense in some places and characterized by species of the Commelinaceae and Marantaceae families (Moutsambote et

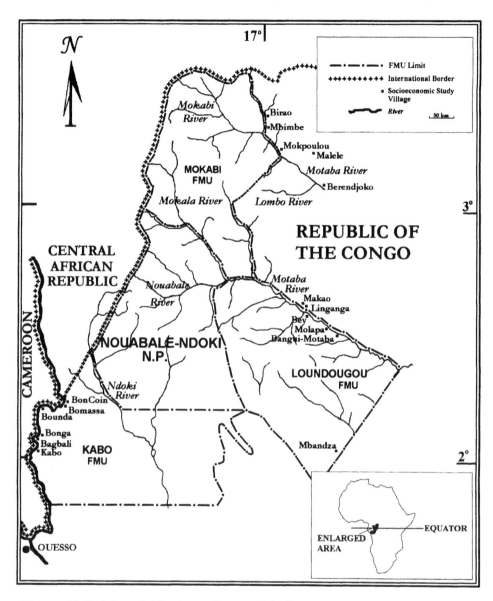

Figure 22-1. Map of Northern Congo with forestry management units (FMUs) surrounding the Nouabalé-Ndoki National Park and villages surveyed in the socioeconomic study of January to April 1996.

al. 1994). An important subgroup of the swamp forest is a type of marshy grassland called *bai* in the native Baka language (Moutsambote et al. 1994). These forest openings are usually covered by herbaceous species from the Araceae, Commelinaceae, Cyperaceae, and Melastomataceae families. Because large, open places and palatable herbaceous vegetation are relatively rare in the forest, these *bais* regularly attract a variety of large mammals and birds and have particular conservation importance (Turkalo and Fay 1994; Ruggiero and Eves 1998). The

Table 22-1. Wildlife Species Hunted in the Northern Republic of Congo with Level of Exploitation in 1996

Latin	English	Bomassa	Exploitation Level
Loxodonta africana cyclotis	Forest elephant	Iya	Common
Cephalophus monticola	Blue duiker	Dengbe	Common
C. callipygus	Peter's duiker	Ngandi	Common
C. dorsalis	Bay duiker	Mbom	Common
C. leucogaster	White-bellied duiker	Senge	Common
C. nigrifrons	Black-fronted duiker	Monjombe	Common
C. sylvicultor	Yellow-backed duiker	Bemba	Common
Tragelaphus spekei	Sitatunga	Mbilya	Common
T. euryceros	Bongo	Mbongo	Occasional
Syncerus caffer nanus	Forest buffalo	Mbo'o	Occasional
Potamochoerus porcus	Bush pig	Pame	Common
Hylochoerus meinertzhageni	Giant forest hog	Bea	Occasional
Hyemoschus aquaticus	Water chevrotain	Mbegene	Occasional
Viverra civetta	African civet	Liabo	Common
Gorilla gorilla gorilla	Western lowland gorilla	Bobo	Common
Pan troglodytes	Chimpanzee	Se'o	Occasional
Cercopithecus cephus	Moustached monkey	Moutenge	Common
C. nictitans	White-nosed monkey	Koi	Common
C. pogonias	Crowned guenon	Mambe	Common
C. neglectus	Brazza's monkey	Mossela	Common
Cercocebus albigena	Gray-cheeked mangabey	Ngada	Common
C. galeritus	Crested mangabey	Tamba	Common
Colobus guereza	Black and white colobus	Kalou	Common
Piliocolobus badius	Red colobus	Niao	Occasional
Atherurus africana	Brush-tailed porcupine	Mbo'e	Common
Manis spp.	Pangolins	Kandzono/Kalepa	Common
Osteolaemus tetraspis	Dwarf crocodile	Mokouakele	Common
Varanus niloticus	Monitor lizard	Dzama	Common

National Park and its surroundings are known for the numbers and diversity of large mammals (table 22-1). Two hundred seventy-three bird species have been identified in the National Park and its potential buffer zone (Dowsett-Lemaire, unpublished data).

The People

The area is populated by Bambenjele and Bangombe Pygmies, collectively referred to as Aka (Sato 1992), who traditionally hunt in the forest for subsistence in addition to performing agricultural activities. There are also numerous distinct Bantu-speaking groups who variously practice hunting, fishing, and subsistence agriculture. These two groups live in a complex, interdependent economic and social relationship (Bailey et al. 1992). Although a considerable body of literature exists on Pygmy populations in the area (Bahuchet 1985; Cavalli-Sforza 1986), there is a paucity of information on other groups in the Sangha Region (Wilkie and Sidle 1990). Archeological and linguistic data indicate that as long ago as 1500 BC, Bantu-speaking inhabitants hunted elephants and other large game, gathered forest products, and practiced farming and animal husbandry (Giles-Vernick

1996). Agriculturists and hunter-gatherer groups have had contact with each other for at least 2000 years (Bailey et al. 1992).

Extraction of Forest Products

Historical accounts suggest that the Sangha River was important for transporting slaves, ivory, rubber, and other products even before the arrival of Europeans in the area (Froment 1887; Glave 1889; M'bokolo 1992). Large rivers also provided the primary means of communication and trade between communities on the Sangha River and Brazzaville (Cana 1911). The installation of the French concessionary system greatly affected exploitation of forest resources. These companies derived much of their profits from ivory and wild rubber (Coquery-Vidrovitch 1972). A report by the Ngoko-Sangha Company (1911) described the area around Bomassa as containing many villages on navigable rivers and with hinterlands where elephants lived in peace. This region was administered from Ouesso, which was established at the confluence of the Sangha and Ngoko Rivers in 1891 and formed part of the Middle Congo Colony of French Equatorial Africa (Cana 1911).

Local populations continue to be dependent on forest products for subsistence (Bahuchet 1992; Kitanishi 1995a, 1995b; Noss 1995, 1997b). Relationships between agriculturists and hunter-gatherer communities in the region have traditionally involved patterns of exchange, with forest-products being traded for agricultural products (Bailey et al. 1989). Currently, villages in northern Congo are composed of households with subsistence production based on agriculture and forest products. The relationships between these cultural groups have continued to evolve and are affected by changing settlement patterns resulting from commercial exploitation of forest resources (Hart 1978; Sato 1992; Wilkie and Curran 1993; Kitanishi 1995b), of which wildlife is an important component (table 22-1).

Commercial economic activity in northern Congo is based on the exploitation of selected hardwoods such as *sapelli* (*Entandrophragma cylindricum*) and *sipo* (*E. utile*) (CITES 1994) by large, mechanized logging companies. Commercial hunting has developed to supply markets in large urban centers and logging towns (Wilkie and Sidle 1990; Blake 1993; Colchester 1993; Auzel 1995). Commercial trade in wildlife products emerging from northern Congo's forests has been documented in selected sites in the region (Wilkie et al. 1992; Hennessey 1995). Where few or no controls exist, there is intensive hunting activity, particularly in settlements associated with logging concessions (Blake 1994; Auzel 1995). Reports indicate that large volumes of bushmeat are produced for consumption in large settlement areas and to provide for basic subsistence needs of villagers providing the hunting labor (Hart 1978; Noss 1995). Duiker antelopes (*Cephalophus* spp.) comprise over 60% of the biomass of the commercial bushmeat trade in many areas (Lahm 1993a; Auzel 1995; Noss 1995; WCS 1996) and are a focus of this study.

Legal Status of Hunting

Hunting in the Republic of Congo is subject to governmental regulation, mostly contained in Law 48/83 of April 21, 1983, which defines the conditions of the conservation and exploitation of wildlife. Certain wildlife species, such as forest elephants (*Loxodonta africana cyclotis*), western lowland gorillas (*Gorilla gorilla gorilla*), and chimpanzees (*Pan troglodytes*), are completely protected. Others may be hunted subject to certain restrictions, and some are not protected. Subsistence hunting for species that are not completely protected can be performed using traditional methods and equipment such as nets, spears, and crossbows made of indigenous materials. Use of modern means such as shotguns is subject to a permit system, payment of a species-specific tax, and a closed season from November to May. Use of snares made of modern materials, such as steel wire cable, hunting with flashlights (jack-lighting), and poisons is prohibited. Many other restrictions apply, particularly in reference to commercial hunting or hunting within an attributed forestry concession.

NOUABALÉ-NDOKI NATIONAL PARK AND SURROUNDINGS

The National Park consists of 386,592 ha of forest in the northern Republic of Congo and is bordered by the Central African Republic and the Republic of Cameroon. The park was gazetted by Presidential Decree No. 93–727 of December 31, 1993. This decree prohibits all human use and disturbance of the park. It also mandates preparation of a management plan and recommendations for a buffer zone in areas adjacent to the park.

Human settlements are found only outside the National Park. They are located in three forestry management units (FMUs): Kabo FMU (280,000 ha), Loundougou FMU (390,816 ha), and Mokabi FMU (370,500 ha) (see figure 22-1). The Kabo FMU has been attributed to the Société Nouvelle des Bois de la Sangha (SNBS), which performed limited timber exploitation during the study period. The other FMUs were not attributed to a logging concessionary at the time of the study, but the Loundougou FMU is likely to be awarded in the near future. In effect, however, the attribution of an FMU to a forestry concessionary negates the implicit land tenure rights of local inhabitants (Wilkie et al. 1992) without which sustainable use of wildlife resources is unlikely to occur (IIED 1994).

The villages located within each FMU are characterized by subsistence farming, fishing, and forest-related activities (Sato 1992; Kitanishi 1995a). Conservation villages (Bomassa, Bon Coin, Makao, and Linganga) are located on the east and west borders of the National Park (see figure 22-1). Limited commercial opportunity for agricultural products exists within these villages; a cash economy exists based primarily on the salaries of Park employees. Logging villages are those within the Kabo FMU where people are employed in the logging industry. Commercial opportunities for agricultural and forest products exist in the larger population centers surrounding the SNBS logging operation. No-industry villages are those where the people are not employed by either the logging industry or the National Park. Residents of both conservation and no-industry villages are largely self-employed; those of logging villages have a considerably

Table 22-2. Bushmeat Consumption (Mean Number of Days/Week/Household) According to Village Type and Household Wealth

Bushmeat	Conservation (n = 32)	Logging (n = 40)	No Industry (n = 101)	Poor (n = 42)	Average (n = 73)	Wealthy (n = 32)
Mean no. of days that meat was consumed/wk	1.05 (1.71)	3.40 (2.15)	1.40 (1.73)	1.27 (1.77)	1.80 (1.83)	2.58 (2.57)
Mean no. of estimated days that meat was consumed/wk	3.30 (1.93)	3.47 (1.57)	3.39 (1.74)	2.88 (1.37)	3.62 (1.89)	3.41 (1.81)
Difference between observed and estimated	2.18 (1.97)	0.08 (2.61)	1.93 (1.95)	1.61 (1.79)	1.82 (2.10)	0.82 (3.13)
p > (t = 0)	0.0001	0.8471	0.0001	0.0001	0.0001	0.1457

Statistically significant difference between actual and estimated number of days per week bushmeat is consumed based on paired data *t* test at *p* = 0.05.

Values in parentheses are standard deviations.

Data were collected in 22 villages in northern Congo, January to April 1996.

Sample sizes represent number of households.

higher formal employment rate (table 22-2). Cultivation, hunting, and fishing are commonly practiced by the majority of adults in conservation and no-industry villages but are relatively limited in number in logging villages. Many residents of logging villages are immigrants from other areas, whereas most conservation and no-industry village households comprise original residents.

METHODS

An initial census of 4000 people showed that approximately 1500 (37.5%) lived in six villages associated with logging operations at Kabo, 835 (20.9%) in four conservation villages, and 1635 (28.4%) in 12 no-industry villages. With the exception of a few villages and camps containing approximately 100 people, this represents a total census of the area likely to be directly affected by National Park buffer zone management recommendations. To provide a portion of the information base toward a management plan for the National Park and buffer zone, two major studies were conducted. The first was a socioeconomic study involving 22 villages in the vicinity of the Park's potential buffer zone (see figure 22-1). The other was a 1-year study of hunting in the conservation villages of Bomassa and Bon Coin. Twenty-two percent (*n* = 173) of all households were sampled for the socioeconomic study. Households in no-industry villages received more intensive sampling (58%) in relation to their relative frequency in the population due to a general lack of information about this village type, and the potentially greater impact caused by changes in land use regulations in the proposed buffer zone. The population of both the Likouala and Sangha Regions within which this study area is located is estimated to be between 100,000 and 150,000 (Sato 1992; MEFP 1994).

Socioeconomics of Wildlife Use

Basic Design and Sampling Units. The socioeconomic study was conducted from January to April 1996 and was based on a pilot study among villages on the Sangha River conducted between April and October 1995 (Eves 1995). This dry season study period was selected because many villages in the region are inaccessible during the wet season and a maximum number of villagers was expected to be near the villages (Kitanishi 1995a). Data collection lasted 7 to 12 days in each village, depending on village size. Data were collected by three separate teams of four Congolese researchers who were trained together by Heather Eves and Louis Bodjo. The training required 2 weeks, using Bomassa and Bon Coin as training sites. As a precaution against misreporting of information deemed sensitive due to the illegal nature of some hunting activities, we using cross-checks both within the questionnaires and by interviewer. By hiring local residents as key informants and assistants, we were also able to verify certain key elements of the information base.

The socioeconomic analysis was structured as follows:

1. A complete census and detailed map of each village
2. An extensive questionnaire detailing domestic economics, food consumption, demographics, and attitudes toward wildlife (after House 1991; Mordi 1991)
3. A concomitant analysis of daily activity, household food consumption, and the sources and cost of food procurement
4. Subjective information concerning hunting and commercial wildlife trade using unstructured and semi-structured interviews with key informants (after Mitchell and Slim 1991; Chambers 1994; and Raval 1994).

All of these studies were conducted in all 22 villages surveyed. Simultaneously, in selected villages a study was conducted to estimate the relative abundance of duikers in hunting zones around villages using an imitated duiker distress call (Fay et al. 1994).

Preliminary meetings were held in each village among researchers, the village chief, and other members of the community to describe the work and to receive permission to work in the village. After permission was obtained, research teams visited each household to conduct the census and identify households for the questionnaire and daily activity portions of the study. Five consecutive days were devoted to data collection for each of the eight households. The final day was spent visiting with each family and preparing a farewell party.

The economic sampling unit was the household (after Ferraro and Kramer 1995). Based on initial visits and following the census of the entire village, eight households in each village were selected with the following stratified sample objectives: two wealthy Bantu, two poor Bantu, two wealthy Pygmy, and two poor Pygmy. This sampling scheme was used to permit analysis according to village type, ethnicity, and level of wealth. In villages of more than 40 households, 16 households were selected for sampling, with four households per analysis category (wealth/ethnicity). The stratified design was preferred to assure sufficient data according to ethnicity and income level within village types (Kramer et al. 1994).

Village Census. The initial research activity in a village was to locate and map all salient features including each household. The census for each village was conducted following map making. The census form included the following components: date, interviewer name, village, number, house number, surname, name, sex, age, relationship to proprietor of household, ethnic group, number of children living and deceased, year of arrival in village, arrival from where, education level, and primary income-generating occupation. Histories for each village and information about hunting were solicited from villagers during semi-structured and informal interviews (Mitchell and Slim 1991).

Questionnaire. A questionnaire, which required approximately 1.5 hours to complete, was administered to each sampled household. Usually the male and female heads of household were present during interviewing. The questionnaire contained four general sections: demographics and movement patterns; income generating activities/household items; estimated production and consumption of agricultural/forest products; and attitudes/valuation of wildlife and the forest. The following results are reported here: household wealth and estimated bushmeat consumption. Household wealth was estimated by an evaluation of the quantity, condition, and source for 21 common household items, such as cooking pots, spoons, chairs, beds, and shoes (after House 1991). The estimated value of each item as new on the local market was multiplied by (a) the quantity; (b) condition, which accounted for depreciation (very good = 1.0; good = 0.80, average = 0.60, poor = 0.40, very poor = 0.20); and (c) source (paid cash = 1.00, given = 0.50, homemade = 0.75). Total values for each item were summed to give overall household wealth. Household wealth categories were assigned based on total estimated value of accumulated household goods as follows: Wealthy >US$300; average <US$300 and >US$50; and poor <US$50.

Researchers asked each interviewee to estimate the number of days per week that each of 15 commonly consumed food items is consumed in the household. The data on estimated weekly bushmeat consumption derived from these interviews were compared with those obtained by our observations during the daily activity study to examine the accuracy of estimations by household members.

Daily Activity and Food Consumption. Daily activity patterns of adults and daily food consumption were assessed through interviews of members of each household over a 5-day period (n = 865 house-days). Each sample household was visited at the end of each day, and members were interviewed regarding each adult's (over 16 years of age) activity during the morning (0600–1000 hr), mid-day (1000–1400 hr), and afternoon (1400–1800 hr). Household food consumption was also observed directly by researchers. Interviews were conducted as a group activity to aid in cross-checking facts, and information was verified by the researcher by random visits to the sample households.

Duiker Call Survey. To estimate effects of relative hunting pressure around villages, we employed an experimental system based on a traditional hunting method (Fay et al. 1994). This method uses the number of responses (animal approaching caller) to imitated duiker distress calls as an indicator of prey avail-ability. Duikers were selected as an indicator because they usually make up most of the bushmeat hunted for subsistence and commercial uses (Lahm 1993b; Auzel 1995; Noss 1995). Teams of three individuals (one trained research assistant and two experienced duiker callers/hunters) traveled 25 km to and from each village along normally used hunting trails in 10-km sections each day over a 5-day period. The team stopped every kilometer, and the caller imitated a duiker distress call for 5 minutes. At every kilometer, the following data were recorded: distance from village, time of day, habitat type, and weather. Data on animals entering the field of vision of the team were recorded, including species, distance, type of approach (e.g., slow, fast), and whether the hunter/caller considered the subject close enough to kill.

We used data on the number and type of animals observed in the duiker call survey and the estimated weight of each animal then to determine a potential rate of return (PRR). This is an indicator of relative duiker abundance vulnerable to hunting by calculating the kg/hr observed along each 5-km segment of the transect. Although several species responded to imitated duiker distress calls, only responding duikers were used in the analysis.

Hunting Study in Bomassa/Bon Coin

The primary objective of this study was to examine the dynamics and economics of hunting in two conservation villages associated with the National Park. The work was conducted from May 1995 to May 1996. Following a training period of 1 month, hunting observations were made by the same research assistant (RA) throughout the study. Use of local hunters in research and data collection has been effective in other areas (Marks 1994). The RA is a resident of Bomassa as well as an experienced hunter. Weekly meetings were held with the research assistant to review data sheets. Observations were randomly verified using additional inter-views by one of the authors (R.G.R.) of hunters and other villagers. Data were collected for hunters living in the villages of Bomassa and Bon Coin, hereafter referred to collectively as Bomassa. The households of all known hunters were visited once or twice each morning and once in the evening, and each hunter was interviewed about the day's activities. Data collected included hunter's name, gun owner's name, location of hunt, transport used, number of persons participating, type of hunt (e.g., snare, shotgun, or other), time of departure, time of return, number of cartridges shot, number of shots missed, species of prey, age class, sex, weight, total body length, horn length and circumference, general condition, and pregnancy status. The economic details of each hunt were gathered, including the payment (usually in the form of bushmeat) that the hunter received for his work, portion of the carcass consumed by the gun owner, portion sold and price, and purchases made with the money earned. All carcass parts for each species were valued according to local Bomassa market rates and were incorporated into the economic analysis to determine profits to gun owners and total economic value of hunting in these villages.

The total weight of animals killed and hours spent on each hunt were used to determine an estimated rate of return (ERR) in kg/hr (described by Blake 1994). Measures of hunting yield (kilograms harvested per unit hunting effort) have been used previously in studies of hunting sustainability (Vickers 1991). This figure was used to compare efficiency of hunting methods, hunter types, ethnicity of hunters, and transport.

HUNTED SPECIES AND HUNTING PATTERNS

The large mammal community of this area is characterized by forest elephants and 12 species of ungulates. Primates form a numerous and diverse group represented by nine diurnal species (Ruggiero and Eves 1998). All ungulates and primates are hunted to some extent (see table 22-1). Smaller mammals (e.g., brush-tailed porcupines [*Atherurus africana*]) and a wide variety of birds and reptiles are also hunted, mostly for subsistence.

Hunting in the villages outside the Park is carried out using two principal means: 12-gauge shotguns and steel-cable snares. Some shotgun hunters hunt at night using flashlights (jack-lighting), and some hunters use an imitated duiker distress call to attract prey within range. Net hunting was reported from only a small portion of the study area; it does not contribute significantly to commercial bushmeat production, in contrast to other regions (Kitanishi 1995a, 1995b; Noss 1997a). Although a considerable portion of hunts were performed by Pygmies, all gun owners were Bantus.

In the villages surveyed during the socioeconomic study, regular hunting territories extended to a distance of 15 to 30 km from the village. This varies depending on prey densities, season, and hunting technique. Data collected during the socioeconomic study suggest that where hunting pressure is continually high (e.g., if hunting for commerce), hunters have to walk for several days from the village to find sufficient prey. For example, in the Kabo logging concession, prey densities are depleted up to 40 km from villages, thereby necessitating transport by logging trucks (Blake 1994). Although we did not measure the proportions of bushmeat sold commercially or consumed as subsistence, Lahm (1993b) found in a similar region in Gabon that one third to two thirds of bushmeat produced is marketed commercially. In southeast Cameroon, another study found that Bantu households sold nearly one third of the bushmeat captured, whereas most Pygmy households consumed the majority of bushmeat procured (WCS 1996).

In the Bomassa hunting study, 61% ($n = 188$) of all hunts analyzed ($n = 309$) were performed by hunters who were hired by the shotgun or ammunition owner, and 97% ($n = 183$) of these hunts were by Pygmy hunters. Of the hunts performed by owners of the shotgun/ammunition, 70% ($n = 83$) were by Bantus. The majority of hired (91%) and gun owner (65%) hunts were performed during the day using 12-gauge shotguns loaded with buckshot. Less than 6% of hunts were recorded as performed using snares, but this may be an underestimate due to under-reporting of snare hunting. A slightly higher percentage of hunts were performed on foot, whereas a considerable number of hunts also involved canoe river transport. Duikers composed 70% (n = 217) of prey items; monkeys (17%; n = 52) and bush pigs (13%; n = 42) made up the remainder.

CONSUMPTION AND NUTRITIONAL IMPORTANCE OF BUSHMEAT

Manioc root (*Manihot esculenta*) is the food most frequently consumed in all villages. Other commonly consumed foods include manioc leaves, *Gnetum* spp. leaves, oil palm nuts (*Elaies guineensis*), and wild mushrooms. Bushmeat consumption among sampled households was analyzed using the estimated number of days per week from the questionnaires and the daily activity survey of recorded meat consumption (see table 22-2). The highest reported bushmeat consumption (mean number of days/wk) was among logging villagers (3.40 days/wk), followed by no-industry (1.40 days/wk) and conservation (1.05 days/wk) villages. Consumption was highest among wealthy households (2.58 days/wk), followed by average (1.80 days/wk), and poor (1.27 days/wk). These results are consistent with expectations of increased demands for bushmeat where standards of living are improved (Barrett and Arcese 1995).

The estimated number of days per week when bushmeat is consumed was similar among all village types and household wealth categories. Differences between reported and estimated weekly bushmeat consumption were significant for all categories except for logging and wealthy households. This might be due to the commercial or less variable supplies of bushmeat available to these household types. It might also be due to the associated higher education levels (table 22-3). Increasing wealth increases a wealthy household's ability to employ hunters and, therefore, increases potential for acquiring bushmeat (figure 22-2).

To quantify the nutritional importance of bushmeat in the diet, we examined data from the Bomassa hunting study. Protein requirements for adults are estimated to be 40 g/person/day (Bailey and Peacock 1988). Each 100 g of meat contains an estimated 15 g of protein, with 60% of a carcass being edible meat (Noss 1995). Approximately 7100 kg of hunted wildlife were recorded during the 1-year study at Bomassa. The estimated edible bushmeat therefore was nearly 4260 kg, which translates to approximately 638,800 g of protein per year to the 159 members of the Bomassa community. Assuming equal distribution of meat to each person, this is equivalent to 0.07 kg/person/day. This equals approximately 11 g of protein/person/day for every day of the year, representing roughly 25% of daily protein requirements. Protein is also provided by fish, invertebrates, leaves of *Gnetum* spp., manioc leaves, and other sources.

The Bomassa estimate of 0.07 kg/person/day corresponds closely with calculations based on data for the area around Bayanga, Central African Republic (Noss 1995), where residents (n = 2500) received an estimated 0.06 kg/person/day from bushmeat obtained by snare hunting. Households in a large logging camp on the Ndoki River in the Pokola FMU consumed an estimated 0.16 to 0.29 kg of

Table 22-3. Results of Village Census in the Region Surrounding the Nouabalé-Ndoki National Park, January to April 1996

Analysis variable	Conservation (n = 167)	Logging (n = 300)	No industry (n = 327)	Poor (n = 47)	Average (n = 63)	Wealthy (n = 29)
Household members						
Wife	**1.20**	0.88	1.11	**1.28**	0.98	1.03
Husband	**1.12**	0.99	0.94	**1.07**	0.95	0.67
Child	3.32	2.91	**3.52**	3.39	3.24	**3.52**
Family	0.52	**1.48**	0.52	0.55	0.84	**1.14**
Employment status						
Employed	0.20	**0.77**	0.03	0.21	0.18	**0.35**
Unemployed	0.02	**0.16**	0.05	**0.12**	0.03	0.09
Self-employed	2.52	1.56	**2.53**	2.67	2.33	1.82
Self-employment categories						
Hunter	0.27	0.11	**0.67**	**0.61**	0.44	0.11
Fisherman	**0.21**	0.02	0.04	0.14	0.09	**0.47**
Cultivator	**1.31**	0.06	0.81	**0.78**	0.74	0.65
Wine maker	0.05	0.07	**0.13**	0.12	0.10	**0.14**
Resident status						
Resident	**3.51**	2.37	3.43	2.80	**3.26**	2.47
Immigrant	1.45	**2.35**	1.43	1.97	1.70	**2.54**
Literacy						
Literate	0.24	**1.83**	0.18	0.07	0.34	**1.05**
Illiterate	4.42	3.02	**4.50**	**4.81**	4.32	3.54
Student	0.32	**1.12**	0.24	0.12	0.40	**0.88**

Values are in terms of number of people per household based on 5 people per household. **Bold** indicate highest value for a variable within a category (village type or household wealth). Sample sizes indicate number of households.

Figure 22-2. Bushmeat consumption among poor households in northern Congo is much less than among wealthy households. *Photo courtesy of Richard Ruggiero.*

bushmeat/person/day (Auzel 1996, in Wilkie et al. 1997) which is two to three times more than was observed in Bomassa. Chardonnet et al. (1995) estimated bushmeat consumption in sub-Saharan Africa communities in forested areas to be 0.104 kg/person/day for foragers and 0.043 kg/person/day for farmers; our estimates for Bomassa fall within this range.

Due to the unequal distribution of carcass parts, these figures do not reflect actual meat consumption in every household. The hunting study in Bomassa, and results of semi-structured interviews in the region surrounding the Nouabalé-Ndoki National Park, indicated that distribution of meat from subsistence hunting follows a pattern wherein hired hunters and their families consume lower quantities than gun owners. This practice follows the tradition whereby a hunted animal is owned by the owner of the weapon rather than by the hunter (Kitanishi 1994; Noss 1995).

In other hunter-gatherer cultures, nutritional benefits from hunting are influenced by wealth, age, gender, and ethnic group, with certain groups having predetermined rules for distribution of meat (O'Brien 1994). Most hunts in Bomassa were conducted by Pygmy hunters hired by Bantu gun/ammunition owners. Most of these hunts resulted in the hired hunter receiving the head and entrails as payment for the hunt. The majority of hunts where hired hunter payment included a leg or the filet resulted in the hunter trading the meat for some other item, such as manioc (40.3%), other food saved (36.4%), alcohol (10.5%), cigarettes (10.5%), or shotgun cartridges (2.2%). Thus, the nutritional value to the hunter was contained primarily in the head and entrails of the animal, which contain less edible meat by weight than the other portions of the animal retained by the gun owners. The gun owners, who are wealthier individuals, also receive the highest quality portion of the animal with the greatest potential nutritional value. Because most hired hunts were performed by Pygmy hunters (97%), it is likely that hunting supplies less of the daily nutritional and protein requirements for Pygmy households than Bantu households.

ECONOMIC IMPORTANCE

Bushmeat

Bushmeat is sold throughout the region, in local villages, logging towns, and the regional capital (table 22-4 and figure 22-3). Here we consider the economics of the trade at the local level. Economic consideration of production and consumption are confounded by cultural practices involving trade and gift-giving of food items (Kitanishi 1994). Such trade in food items may be a means toward ensuring necessary caloric intake if gift receivers reciprocate on days when a family is providing for themselves (Cashdan 1985). Alternatively, food sharing might be a desirable activity if there are higher costs to not sharing (Hawkes 1993). Bushmeat can be considered a public good, which is nonexcludable, having a total value worth more to the community at large when shared than if consumed by single

Table 22-4. Average Bushmeat Prices (US$) at Village Level (Bomassa), Logging Town (Kabo), and Prefectoral Capital (Ouesso), Republic of Congo, 1996

		Bomassa		Kabo		Ouesso	
Species	*Common Name*	*Small*	*Large*	*Small*	*Large*	*Small*	*Large*
Potamochoerus porcus	Bush pig	16.00	20.00	14.00	20.00	110.00	210.00
Cephalophus monticola	Blue duiker	2.40	3.00	2.40	3.00	4.00	5.00
Cephalophus dorsalis	Bay duiker	8.00	10.00	8.00	10.00	10.00	14.00
Cephalophus callipygus	Peter's duiker	8.00	10.00	8.00	10.00	10.00	14.00
Cephalophus leucogaster	White-bellied duiker	8.00	10.00	8.00	10.00	10.00	14.00
Cephalophus nigrifrons	Black-fronted duiker	8.00	10.00	7.00	8.00	8.00	12.00
Tragelaphus spekii	Sitatunga	14.00	110.00	20.00	114.00	40.00	310.00
Syncerus caffer nanus	Forest buffalo	60.00	80.00	70.00	120.00	80.00	140.00
Loxodonta africana cyclotis	Forest elephant	NA	NA	50.00 per sack		75.00 per sack	
Manis tricuspis	White-bellied pangolin		2.00		2.00	2.00	3.00
Manis tetradactyla	Long-tailed pangolin		2.00		2.00	2.00	3.00
Manis gigantea	Giant pangolin	10.00	10.00	10.00	10.00	15.00	15.00
Atherurus africanus	Brush-tailed porcupine	2.00	4.00	3.00	3.00	2.60	3.00
Cercocebus galeritus	Crested mangabey	3.00	4.00	5.00	6.00	6.00	8.00
Cercocebus albigena	Gray-cheeked mangabey	3.00	4.00	5.00	6.00	6.00	8.00
Cercopithecus cephus	Moustached monkey	2.00	4.00	3.00	4.00	4.00	5.00
Cercopithecus nictitans	Greater white-nosed monkey	2.00	3.00	4.00	5.00	6.00	7.00
Cercopithecus pogonias	Crowned guenon	2.00	3.00	3.00	4.00	4.00	5.00
Cercopithecus neglectus	Brazza's monkey	2.00	3.00	4.00	5.00	6.00	7.00
Colobus guereza	Black and white colobus	4.00	5.00	5.00	6.00	7.00	9.00
Piliocolobus badius	Red colobus	4.00	5.00	5.00	6.00	7.00	9.00

NA, estimated price not available.

households (Gibson and Marks 1995; Hawkes 1995). This system of trade and gift giving can be affected by the development of commercial markets for bushmeat (Hart 1978; Sato 1992; Kitanishi 1995a).

The single greatest source of meat in both conservation and no-industry villages was receiving it as a gift (figure 22-5). Conversely, in logging villages, over 60% of bushmeat was acquired through cash purchase. Meat acquisition by cash purchase was highest among wealthy households (figure 22-6). Poor households received over 80% of their meat as gifts or from husbands' hunting activities. This suggests that distribution patterns change in areas of altered employment and demographics (i.e., logging villages). Where employment opportunities exist but demographics have remained relatively stable (i.e., conservation villages), patterns appear to follow more traditional patterns of distribution.

Figure 22-3. Truck with bushmeat for the trade. *Photo courtesy of Richard Ruggiero.*

In Bomassa and Bon Coin, two conservation villages, the total economic value and economic value per hunt were analyzed according to hunter type (tables 22-5 and 22-6). Most hunts were performed by hired hunters (57.8%) compared with those by gun owners (37.2%) or snare hunts (4.9%). Because the ban on snare hunting is enforced in Bomassa, low reporting may under-represent the true figure. Hired hunters earned less per hunt (US$2.25) than gun owners (US$3.04) did in net profits. Gun owners hunting for themselves decreased their profits per hunt (US$2.50) due to their higher opportunity costs for time spent hunting. The highest per hunt net earnings (US$11.80) were from snare hunting.

The total time spent hunting in hired hunts was 1881.5 hours. Using the average hourly rate for unskilled laborers in this area (US$0.39), the total value of time spent hunting is approximately US$730. The hired hunters, however, were

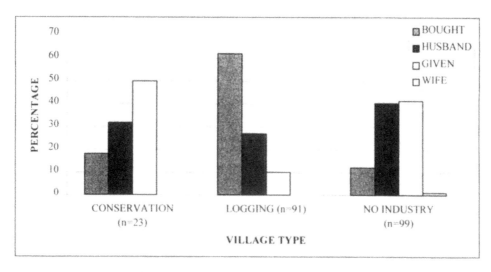

Figure 22-4. Household sources of bushmeat according to type of village in northern Republic of Congo. Sample sizes indicate number of days in which bushmeat was reported consumed. Data were collected between January and April 1996.

paid US$405.56, only 55% of the potential local earning rate. Gun owners received economic benefits in two ways: using hired hunters and hunts they conducted themselves. The combined benefits were US$874, equivalent to the annual salary of an unskilled laborer at National Park wage rates.

Assuming that the study year is typical, the annual net total economic value from both hunting types (including both hunter and gun owner profits) to the local communities of Bomassa and Bon Coin is estimated to be US$1297 or

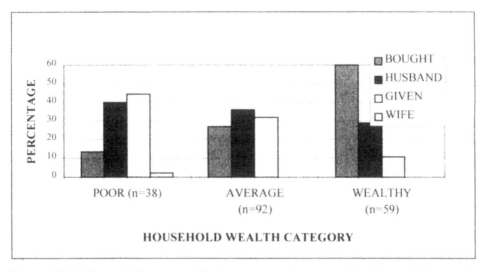

Figure 22-5. Household sources of bushmeat according to wealth of household. Sample sizes indicate number of days in which bushmeat was reported consumed.

Table 22-5. Total Economic and Per Hunt Values for Hunting in Villages of Bomassa and Bon Coin, Northern Republic of Congo, May 1995 to May 1996

Description	Hired Hunters (n = 188)	Gun Owners as Hunters (n = 121)	Snare Hunting (n = 16)
Labor cost (US$): Value of meat earned by hired hunter as payment for hunting or opportunity cost of gun owner hunting for himself	405.56	376.74	23.30
Capital cost (US$): Cost of hunt to gun owner including payment of hunter and cost of ammunition used	601.68	144.66	23.30
Game meat consumed (US$): Value of game meat consumed by the gun owner	494.47	284.47	35.63
Game meat sold (US$): Value of game meat sold for cash by the gun owner	588.64	394.76	190.29
Gun owner profit (US$): Net profit to gun owner equal to total benefits (meat consumed and sold) minus total cost	481.43	157.83	200.68
Mean per hunt hunter profit (US$): Mean value of game meat or payment earned by hired hunters per hunt including successful and unsuccessful hunts	2.25	0	1.29
Mean per hunt gun owner profit (US$): Mean value of net profit to gun owner per hunt including successful and unsuccessful hunts	3.04	2.50	11.80
Hours hunting: Total hours spent in each hunting type	1881.5	966	215

Table 22-6. Mean Economic Gain per Hunt (US$) for Hunts Taking Place During May 1995 to May 1996 in Bomassa and Bon Coin, Republic of Congo

Hired Hunters		Gun Owners as Hunters	
All hunts (n = 188)	3.04 (3.67)	All hunts (n = 121)	2.50 (5.07)
Transport			
Foot (n = 145)	2.85 (3.51)	Foot (n = 68)	2.08 (4.64)
Canoe/foot (n = 43)	3.69 (4.12)	Canoe/foot (n = 53)	3.02 (5.56)
Hunt type			
Shotgun (n = 172)	2.79 (3.29)	Shotgun (n = 79)	1.70 (4.67)
Shotgun at night (n = 8)	2.2 (2.59)	Shotgun at night (n = 34)	2.51 (3.67)
Snares (n = 8)	9.5 (6.26)	Snares (n = 8)	10.25 (7.61)
Ethnic group			
Bantu (n = 5)	0.70 (3.23)	Bantu (n = 83)	2.22 (4.48)
Pygmy (n = 183)	3.12 (3.66)	Pygmy (n = 36)	3.02 (6.24)

Economic gain is to gun/ammunition owners who either performed the hunts themselves or hired hunters. Economic gain = total benefits – total costs.

Total benefits = value of meat consumed by gun owner + value of meat sold.

Total costs = value of meat "paid" to hunter or gun owner opportunity cost + cost of ammunition.

Values in parenthesis are standard deviations.

US$108 per month. The economic input from the National Park project averages US$2912 per month to the local community through employees' (n = 30) salaries. Hence, subsistence hunting in Bomassa is worth 3.7% of the combined cash economy in the two villages. This figure only reflects hunting and salaried income opportunities and does not consider the economies of alcohol production, fishing, or agriculture.

There are 39 households in these two conservation villages, most of which have at least one National Park employee. Averaging income of employees across households, each household receives US$76.58 per month. If hunting is considered in terms of benefits equally divided among households in Bomassa, its estimated value is US$2.77 per month to each household. Based on the census conducted for this study, there are approximately 5.1 people per household. The per capita economic benefit from the National Park employment is approximately US$15/month compared with US$0.54/month from hunting.

Bomassa and Bon Coin are atypical northern Congolese communities. Hunting is managed by agreements made between the National Park management and the village residents through their village committee. These agreements require compliance with Congolese laws that regulate or prohibit hunting protected species and hunting using wire snares. In addition, these two villages have agreed to refrain from market hunting. In return, bona fide residents receive preferential hiring status with the National Park.

This system has resulted in a relatively low percentage of income from hunting as compared with villages where hunting is uncontrolled and where few income-generating opportunities exist, or when compared with logging communities where hunting levels are considerably higher and hunters have the potential to earn an income similar to that of employees of the logging company. Comparable information on local economies in this region is very limited, but Noss (1995) estimated that for communities in the Central African Republic, per capita benefits from snare hunting were approximately US$600/yr, net hunting contributed US$200/yr, and formal employment yielded US$400 annually.

Elephants: A Special Case. It is difficult to determine the overall economic importance of hunting small animals (nonelephant) in the other study villages without extended studies such as conducted in Bomassa. To compare our results with other areas surrounding the National Park, we examined known benefits, based on semi-structured interviews conducted as part of the socioeconomic study. Elephant hunting has significant economic potential for villagers. Information from more than 50 interviews in 24 villages situated in the three FMUs surrounding the National Park is summarized in table 22-7. These numbers are conservative, and only elephants reported killed by hunts originating from the village from October 1995 to January 1996 are included. Elephant hunts that were organized outside the village (e.g., Impfondo, Dongou, Ouesso) are not included.

Elephants were usually hunted unselectively, thereby resulting in highly variable size and potential economic return both from tusks and meat. Distance from the nearest village also affects return from an elephant hunt due to the difficulty

Table 22-7. Summary of Elephant Hunting in the Regions Surrounding the Nouabalé-Ndoki National Park, Northern Congo

FMA	No. of Villages Surveyed	12-Gauge Shotguns	Large-Bore Rifles	No. of Elephants	Estimated Value per Month	Estimated Value/ Village/ Month	Destination Market
Mokabi	8	60	40	186	US$18,600	US$2325	Central African Republic
Loundougou	11	123	19	74	US$7400	US$672	Impfondo/Pokola
Kabo	5	92	7	13	US$1300	US$260	Kabo/Ouesso
Total	24	275	66	273			

Data represent elephants reported killed over a 4-month period (October 1995 to January 1996). Estimated returns based on $400 per elephant value. Data for number of 12-gauge shotguns, large-bore rifles, and number of elephants killed were based on semi-structured interviews with hunters in villages.

(expense) of meat transport. On average, about US$400 profit was reported per elephant, including both meat and tusks as sold from the village. It was evident throughout this study that the commercialization of meat has become an important incentive to hunt elephants, despite the decreasing tusk size that results from overexploitation of the population (figure 22-6). Elephant meat is readily available in Impfondo, Ouesso, and Pokola (Blake 1993; Auzel 1995; Malonga 1995).

Estimated monthly revenue from elephant hunting varies widely among villages in each FMU. The estimated monthly economic returns per no-industry village from elephant hunting in the Mokabi FMU is US$2325, similar to the US$2912 provided by the Nouabalé-Ndoki project to Bomassa. In the Loundougou's two conservation and three no-industry villages, the monthly benefits to residents from elephant hunting was only US$672. In the conservation villages (Linganga and Makao), elephant hunting has recently diminished as a result of the installation of a National Park project base, and it is no longer a major contributor to the local economy (Kitanishi, personal communication). This trend is also seen among villages in the Kabo FMU, where returns were estimated to be US$260 per village per month; this is attributable to the two conservation villages (Bomassa and Bon Coin) where elephant hunting is also controlled. These estimates reflect income solely from elephants and do not include the commercial hunting of duikers, bush pigs (*Potamochoerus porcus*), and monkeys (*Cercopithecus, Cercocebus,* and *Colobus* spp.), which generates large revenues for both no-industry and logging villages.

CONCLUSIONS

The economic importance of bushmeat to rural communities varies according to the combination of alternative employment opportunities and hunting controls. Where they exist together (i.e., conservation villages), hunting appears to

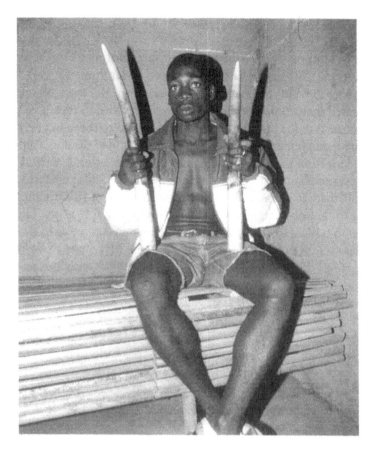

Figure 22-6. Economic returns from ivory are decreasing as populations comprise younger animals with smaller tusks. *Photo courtesy of Richard Ruggiei.*

contribute relatively little to the overall cash economy. Where few alternatives for income exist, and hunting is not controlled (i.e., no-industry villages), bushmeat is a primary contributor to the local economy. Where alternatives exist (i.e., logging villages) and no control is exercised, income from bushmeat is a major contributor. This is true despite the fact that several economic alternatives are available in logging towns (e.g. wage labor, selling agricultural products or other goods, fishing); hunting and meat consumption are higher than in most other areas studied. Hunters in nonlogging areas cited limitations on hunting levels being primarily due to insufficient guns and ammunition, whereas transport and access to the forest were thought to enable the high levels of hunting seen in logging concessions. This is consistent with the results described by Noss (1995) for a similar area in the region in the Central African Republic. Blake (1994) and Auzel (1995) emphasized the logistical advantages of hunting in a forestry concession, where little or no effective control on hunting is exercised.

IMPACT OF HUNTING ON WILD SPECIES

The impact of hunting as described by most key informants in these 24 villages followed expected patterns. More experienced older hunters were selected as informants ($n = 52$). With the exception of those from Bomassa, all hunters interviewed indicated that wildlife populations have decreased in recent years. In some areas the success rate for hunting has dropped to two to three in every ten hunts, and then only if one hunts at night with jack-lights and at distances at least a day's walk (20–30 km) from the village. In Bomassa, opposite trends were reported. All interviewees stated that hunting conditions in the late 1980s and early 1990s were more difficult than at present. At that time, Bomassa and Bon Coin were both contributing to the commercial meat market in Kabo and Ouesso. They also sold smoked meat to traders from Ouesso or on river barges traveling to Brazzaville. In addition, Bomassa, like most villages on the Sangha River, was a staging point for elephant hunting. This situation changed in late 1991 with the advent of the Nouabalé-Ndoki project. Subsistence hunting continues all year, and at present, hunting success rates are as high as 82% in the forest surrounding Bomassa.

The duiker call surveys provided a relative index of hunting pressure in the different study areas. The PRR in kg/hr was calculated for each 5-km segment, and results were pooled by region, notably: Mokabi FMU, north of the National Park (all no-industry villages); Loundougou FMU, east of the National Park (two conservation and three no-industry); and Bomassa (two conservation villages), the area closest to the park on the southwest (figure 22-7). No PRRs were estimated for the logging villages near Kabo. Blake (1994) reported limited hunting potential as measured by response to duiker calls within this area. The Loundougou

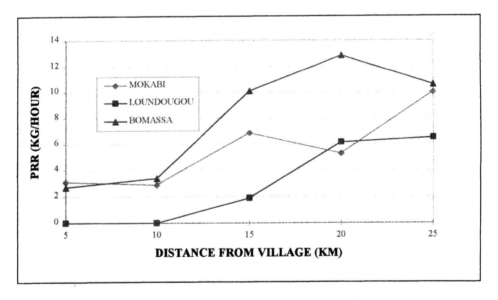

Figure 22-7 Potential rate of return (PRR), in kg/hr, based on duiker call surveys in regions surrounding the Nouabalé-Ndoki National Park, showing relative abundance of duikers vulnerable to hunting. Data were collected in December 1995 (Blake, unpublished data).

FMU, which has supported a high level of commercial hunting for many years, showed consistently lower PRRs (0.00–6.48 kg/hr; $n = 200$) than the Mokabi FMU (2.86–9.99 kg/hr; $n = 200$), where commercial hunting is in an earlier stage. The highest PRR values were found in the area around Bomassa (2.71–12.85 kg/hr; $n = 148$) and close to the National Park (Blake, unpublished data). Although habitat differs somewhat among these areas, all areas where both commercial and snare hunting occur showed lower PRRs, particularly within 10 km of the village, although they increased farther than 10 km from the village.

The ERR was calculated for actual hunting returns during 309 hunts in Bomassa (table 22-8). The mean ERR for hired hunters was 2.79 kg/hr. The mean ERR for gun owners hunting for themselves showed a slightly higher value at 3.29 kg/hr. The mean ERR for hunting during the day with 12-gauge shotguns of 2.67 kg/hr was lower than the ERR for hunting at night (3.57 kg/hr). Snare hunting had an ERR of 6.59 kg/hr. Differences in mean ERRs were statistically significant according to hunt type and mode of transport but not for either location or type of hunter (table 22-8). The difference due to mode of transport was primarily due to increased weight of meat hunted rather than a decrease in hours spent hunting, which suggests that hunters harvest more game when a canoe is used.

Blake's (1994) ERRs for day hunts in logging villages, for an animal population that has been commercially exploited for up to 25 years, were lower than those determined for Bomassa (1.3 kg/hr for day hunts within 40 km and 1.9 kg/hr for day hunts over 40 km from Kabo). Auzel (1995), working in the Pokola logging concession south of Kabo, reported a range of ERRs from 1.86 kg/hr in an area exploited by loggers in 1981 and 1982 to 4.41 kg/hr in a zone exploited in 1986. These data suggest that in areas of long-term or uncontrolled hunting activity, the net return per unit effort is lower, which may be suggestive of a diminishing resource base.

Table 22-8. Mean per Hunt Estimated Rate of Return (ERR) in kg/hr of Animals Killed for Hunts Conducted from Bomassa, Northern Republic of Congo May 1995 to May 1996

Analysis Category	Mean ERR in kg/hr per Hunt	$p > F$
Hired hunter ($n = 187$)	2.79 (3.67)	0.23
Gun/ammunition owner as hunter ($n = 121$)	3.29 (3.65)	
Transport		
Foot ($n = 212$)	2.68 (3.51)	0.03[a]
Canoe/foot ($n = 96$)	3.65 (3.91)	
Hunt type		
12-gauge shotgun ($n = 251$)	2.67 (3.46)	0.001[a]
12-gauge shotgun at night ($n = 42$)	3.57 (3.96)	
Snares ($n = 15$)	6.59 (4.26)	
Ethnic group		
Bantu ($n = 88$)	3.34 (3.59)	0.19
Pygmy ($n = 218$)	2.75 (3.58)	

Values in parentheses are standard deviations.

[a]Statistically significant differences in ERR at $p = 0.05$.

SUSTAINABILITY OF HUNTING PRACTICES

Ecological Sustainability

Ecologically sustainable activities can be defined as those that do not degrade the natural resource (Robinson 1993). To determine ecological sustainability, it is necessary to have information concerning natural replacement rates, harvest rates, and an estimation of the population density of harvested wildlife (Robinson and Redford 1994b). Many studies of the sustainability of hunting rely on data from transects to estimate prey density, but the use of transect techniques in central African forests has proven problematic, and accuracy is often questionable (Payne 1992; Noss 1995). In certain cases, an extremely large sample (up to 800 repetitions per site) is necessary to obtain significant results (White 1992). This poses practical problems for wildlife managers who must make decisions based on reliable information. As our study of the region surrounding the National Park was primarily focused on socioeconomic activities, time was limited to perform extensive studies. Hence, PRRs were used to evaluate trends and impacts of hunting.

An analysis of variance indicated that the mean PRRs for Bomassa were significantly higher than for Mokabi or Loundougou ($p < 0.03$). ERRs for Bomassa did not exceed the PRRs for the area, except for snare hunting. PRRs include only returns for duikers, whereas ERRs estimate total returns of all species, including monkeys and bushpigs, for actual hunting. Thus, PRRs are a conservative estimate of bushmeat. These two parameters are useful indicators of hunting pressure and potential sustainability of hunting activities. Without detailed studies, however, it is difficult to determine if hunting is ecologically sustainable. Based on qualitative data from interviews and duiker call surveys, however, it seems that villages in the Mokabi and Loundougou are hunting from a declining prey base, which might imply lack of sustainability.

An estimated annual harvest of at least 7097.6 kg from the approximately 300 km^2 hunting area around Bomassa yielded an estimated rate of 23.7 kg/ km^2. Noss (1995) determined annual bushmeat production for all hunting methods to be 185 kg/km^2 in the Bayanga area. This might be influenced by the potentially higher productivity of the forest around Bayanga but was deemed to be unsustainable (Noss 1995). Without data extending over many years, it is difficult to assess whether these hunting rates are sustainable (Vickers 1991). However, based on observed trends and comparative analysis, we suggest that current hunting practices and rates in the Bomassa area might be sustainable.

However, in large portions of all three FMUs, as elsewhere in the region where commercial hunting is practiced, the evidence suggests that hunting levels are at risk of being unsustainable and require the development of appropriate management recommendations (Moussa 1992).

Socioeconomic Sustainability

Socioeconomically sustainable activities are those that meet the economic needs and aspirations of the users (Robinson 1993). Long-term sustainability implies that not only natural resource use but also the economy and culture of local inhabitants should be sustainable (Kitanishi 1994; Robinson and Redford 1994b). The question of socioeconomic sustainability of wildlife use in northern Republic of Congo is one that cannot be easily answered. Sustainability of any kind requires resource use in a manner that enables the continued availability of that resource for future generations. When sustainability is viewed in its socioeconomic context, it implies that the activity under investigation favors the continued cultural and economic integrity of the society practicing the activity.

The most obvious conflict between current hunting practices and socioeconomic sustainability may be seen in the case of a hunter-gatherer society such as the central Africa Pygmies. Traditionally, when hunters agreed that game was scarce, groups would migrate to find better areas (Bahuchet 1991, 1992). This is likely to become increasingly difficult as game populations are exhausted and forest area declines. Serious threats to their culture are posed by overhunting, changing demographics, and resulting effects on traditional hunting patterns (Sato 1992; Kitanishi 1995a).

Increasing pressure from commercial hunting is now evident in most areas outside the National Park where game remains. Several traders indicated that the scarcity of game in bordering forests of the Central African Republic causes them to enter northern Republic of Congo in search of bushmeat for export. As the distances grow longer, costs of transport and risks increase, and wildlife populations decrease over large areas, the industry becomes less economically viable.

The threat to the cultural tradition of hunting, particularly in Pygmy communities, implies that current practices are sociologically unsustainable. Economically, hunting is likely to be unsustainable in the long term due to the unrealistically low opportunity costs for hunting in some areas resulting from acceptance by Pygmy hunters of underpayment and by the rapidly increasing marginal costs to be borne by future generations. In addition, economic sustainability is dependent on the future presence of the resource. Where hunting is ecologically unsustainable, it is also, by definition, economically unsustainable. The socioeconomic effects of commercial hunting threaten severely to compromise the integrity and therefore sustainability of the resource and life-styles of people depending on the forest. In logging and no-industry villages, it appears that wildlife populations are threatened by unsustainable levels of hunting. The socioeconomic structure of hunting activities is geared toward unsustainable levels of commercial exploitation. In conservation villages, where employment opportunities exist, human populations are relatively low, and exploitation is limited to subsistence, the socioeconomic sustainability of the system appears to be promising.

ENHANCING THE SUSTAINABILITY OF HUNTING IN A CHANGING WORLD

Why Is Hunting Not Sustainable?

The sustainability of hunting is governed by a complex set of socioeconomic and ecological factors. In northern Congo, these factors are dynamic and vary from place to place, but all are affected to some extent by demands of the marketplace and the concomitant impacts of economic development in the region (Wilkie and Sidle 1990; TELESIS 1991). As transport infrastructure improves, employment opportunities increase, human populations grow, and new or larger commercial markets appear outside of production areas, pressure on wildlife populations rises dramatically (Wilkie and Sidle 1990). Commercial hunting in the forests of northern Congo under the current uncontrolled conditions is more than likely to be unsustainable in the long term.

Restriction of hunting is an essential remedial step because the demand of urban centers will continue to stimulate unsustainable levels of hunting. Although this appears to be an obvious part of the solution, its realization depends on enforceable legislation that responds to changing demographic patterns and economic conditions and does not overburden or alienate local communities (Colchester 1993; Western and Wright 1994). The advent of both logging and conservation activities in this region has resulted in vastly different outcomes in terms of the integrity of the forest, its resident wildlife populations, and the culture of those dependent on them. These adjacent yet opposing scenarios, although initially borne of similar histories, can be attributed primarily to the large increases in human disturbance and decreases in management authority over wildlife populations. In areas where no economic alternatives are readily available, exploitation levels are comparable with those found among commercially based communities. Here the demographics of distant population centers and the availability of transport enable the demand for commercial level production of bushmeat to be met.

Approaches Toward Sustainable Hunting

If effective conservation policy is to be realized in this region, it clearly must incorporate, at the very least, an understanding of local community needs, authority relationships, and resource management systems, including perceptions of land ownership and associated use (Colchester 1993; IIED 1994). Likewise, such policies must include the perspectives of national and local government agencies, international development and conservation support organizations, commercial logging and mining companies, indigenous NGOs, and traders (Murphree 1994). Each of these user groups has a role in the exploitation of products from these forests and therefore must be incorporated into decision making (Ascher and Healy 1990). Without a clear understanding of these perceptions and where they are aligned or in opposition, it is unlikely that effective wildlife policy and legislation will be effective in this region (Dale and Lane 1994).

Due to current political and economic constraints, it is unlikely that former models of legislation and enforcement will be successful in achieving sustainable use of the wildlife resources in this region. Throughout Africa, an emergent model of wildlife management, community-based conservation (CBC), which incorporates and empowers local communities in the decision-making process, is being tested (IIED 1994; Western and Wright 1994). Although there are undoubtedly numerous challenges to the effective implementation of such programs, and it is clear that no single strategy could be applicable in all situations, it is recommended that an investigation of the potential for CBC in northern Congo should be made. It is essential that each stakeholder group involved with wildlife hunting and trade be incorporated into the process of evaluation and decision-making (Murphree 1994). Alienation or perceptions of power loss of any one group could undermine attempts at wildlife management (Gibson and Marks 1995; Ndinga 1996). Incentives to villages that maintain control of hunting in legally authorized zones should be formalized (IIED 1994; WCS 1996).

Because the financial means to support conventional regulation and enforcement are limited, and incentives to support them are restrained, it is unlikely that such systems will succeed. Funding for enforcement of game laws within logging concessions should be provided by the logging companies as part of their formal agreement with the government. As enforcement increases, so will the price of bushmeat in these communities, enabling the development of markets for alternative sources of protein. Additionally, in remote areas where high hunting levels and commercial trade occur across international borders, installation of effective border control and international cooperation are essential. Development of indigenous environmental groups responsible for monitoring, and supported by international NGOs responsible for maintaining communication linkages among stakeholder groups, is encouraged to facilitate this process (Western and Wright 1994).

These links between behavior patterns, economics, and tenure authority relationships are too complex for simple substitutions to be effective (Ferraro and Kramer 1995). If unsustainable hunting is to decrease, employment options must be combined with increased costs of hunting through effective detection and penalties, formal authority over resource use, and labor and capital investment (Ferraro and Kramer 1995; WCS 1996). These, combined with empowering local communities in the decision-making process involving management of local wildlife populations, have a higher potential for success than existing systems (IIED 1994). Economic opportunities and infrastructure development attract members of this highly mobile society into developing areas, thereby increasing pressure on the forest. Once this starts, it is unlikely that it will decrease even when the commercial operations that initially attracted them no longer exist (Loudiyi 1995). If human population densities exceed the system's ability to provide acceptable benefits to individuals, a management strategy dependent solely on compensation for loss of access to resources will fail (Barrett and Arcese 1995).

Sustainable conservation strategies are initiated by in-depth studies of the role of wildlife in local communities (Gibson and Marks 1995). Hunting has been shown in this study to decline where integrated incentive-disincentive systems and controls on immigration are implemented. The model established in villages associated with the National Park, although unique in many ways, may provide support for the level of commitment and action that is required among all stakeholders if effective conservation and sustainable levels of use are to be realized. Without such commitment and careful management planning, it is unlikely that conservation efforts will succeed. In this Ndoki model, government agencies, international conservation organizations, and local communities have worked cooperatively, resulting in a vastly improved situation where the sustainability of wildlife exploitation is possible. With the establishment of a long-term funding solution, the foundation upon which this model has been created could be assured.

ACKNOWLEDGMENTS

We extend our gratitude to the Government of the Republic of Congo for their support of this research. The Wildlife Conservation Society, Global Environment Facility, and USAID provided logistical and financial support without which this project would not have been possible. We are extremely grateful to Dr. J. Michael Fay and Mr. Stephen Blake, who made essential contributions to the design and implementation of this project. Our thanks are also extended to our dedicated research team, including our principal research assistants: Louis Bodjo, Guy Ebeke, André Lombe, Yves Constant Madzou, Gabriel Mobolambi, Antoine Moukassa, Marcel Ngangoué, and Valentin Yako. Our sincere appreciation is extended to the residents of the Likouala and Sangha Regions, who welcomed us into their villages and shared their hospitality and knowledge.

23

Impact of Subsistence Hunting in North Sulawesi, Indonesia, and Conservation Options

ROB J. LEE

Of all the animal species known to have become extinct world-wide since 1660, and whose causes of extinction were known, 38% became extinct due to the introduction of exotic species and diseases, 33% due to loss of habitat, and 27% due to hunting (Smith et al. 1995). How species go extinct depends on the scale on which these forces occur in a region. In certain regions where hunting is intense, threats from habitat destruction might be secondary to hunting (Mittermeier 1991; Oates 1996a). Compared with habitat disturbance, hunting tends to have a direct and immediate effect on wildlife populations. If hunting pressure exceeds the reproductive output of a given population, the outcome will range from a steady decline to local extinction.

Hunting of wildlife for subsistence presents a difficult set of problems for wildlife conservation because it is often intrinsically tied to the nutritional and economic needs of rural people. Therefore, the tasks of the conservation researcher are to identify the segment of the human population that is harvesting wildlife, determine their level of dependence on wild meat as a source of protein, understand the nature and scale of harvesting and the cultural basis for it, understand the impact of hunting on local wildlife populations, and seek ways to offset these effects.

This chapter focuses on hunting by the people of North Sulawesi in Indonesia, where wild animals are hunted and consumed on a regular basis. I focus specifically on subsistence hunting by communities surrounding protected areas. This research was conducted from June 1994 to September 1995. Data collected in the field came mainly from two sources: wildlife population surveys and household surveys by myself and field assistants.

NORTH SULAWESI

The 159,000-km^2 island of Sulawesi (figure 23-1) is the largest island in the biogeographical subregion of Wallacea. This region encompasses adjacent portions of the Oriental and Australasian biogeographical regions and consists of a central land mass and four peninsulas: northern, eastern, southern, and southeastern. The complex geological events of Sulawesi's history and its long isolation are apparent in its mix of Asian and Australasian fauna and the extraordinary endemicity. Of the 127 mammal species, 328 bird species, and 104 reptile species, 79 (62%), 88 (27%), and 29 (28%) species, respectively, are endemic, and mammal species endemism rises to 98% if nonvolant mammals are excluded (FAO 1982; Whitten et al. 1987a).

The present study took place in Minahasa and Bolaang Mongondow, the two eastern districts within the province of North Sulawesi. The topography is characterized by steep mountains that were formed by recent volcanic events, with elevations reaching over 2200 m. The climate is generally characterized by a long wet period and a short dry period, with the eastern part of North Sulawesi having a longer wet period than the western part (Whitmore 1984a, 1984b). Annual rainfall in North Sulawesi varies, although it averages between 1500 and 2400 mm (Whitten et al. 1987a).

The fauna and flora of North Sulawesi have been well documented, with

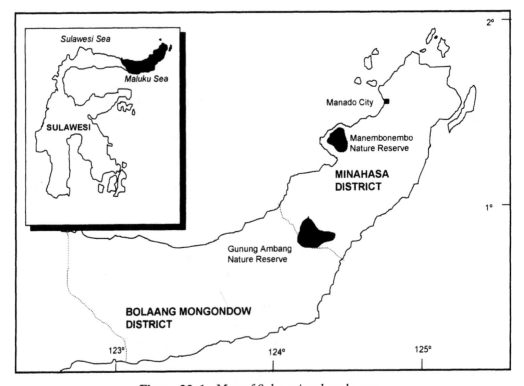

Figure 23-1. Map of Sulawesi and study area.

studies dating back to the nineteenth century by naturalists such as Wallace (1869), Guillemard (1886), and Hickson (1889). Whitten et al. (1987a) compiled a list of recent studies. The natural vegetation dominating North Sulawesi is evergreen rain forest and includes lowland, submontane, and elfin/moss forests. Coastal forests include beach and mangrove forests. Floral diversity is variable, with more than 240 species at Tangkoko-Duasudara-Batuangus (Tangkoko) Nature Reserve (WCS 1995b) and approximately 130 species at both Manembonembo Nature Reserve (Lee 1997) and Dumoga-Bone National Park (Kohlhaas 1993). Among the more conspicuous animal species are anoa (*Bubalus depressicornis*), babirusa (*Babyrousa babyrussa*), crested- black macaque (*Macaca nigra*), Gorontalo macaque (*Macaca nigrescens*), bear cuscus (*Ailurops* [formerly *Phalanger*] *ursinus*, dwarf cuscus (*Strigocuscus* [formerly *Phalanger*] *celebensis*, redknobbed hornbill (*Rhyticeros cassidix*), and maleo bird (*Macrocephalon maleo*). For simplification, I will refer to the general study area as North Sulawesi, and will use *Minahasa* and *Bolaang Mongondow* when specifically referring to these two districts.

Although a majority of the Indonesian people (86.9%), including the people of Sulawesi, are Muslim, approximately 85% of the people in Minahasa are Christian (KSPSU 1994). This stems from a long history of mission work in Minahasa, first by Portuguese missionaries in the 1560s followed by the Dutch for nearly 300 years. In addition to Christian influences, the Dutch brought to Minahasa, an area directly governed by the Dutch, a number of institutions and practices typical of western European society. Among these were plantation agriculture, the education system, and road and irrigation systems. As a result, Minahasa has historically been and is today one of the most western-educated, wealthy, and developed provinces in Indonesia.

Modernization in recent decades also brought to North Sulawesi increased life expectancy (Soemantri 1983), higher nutritional intake (BPS 1984a), and lower infant mortality (Soemantri 1983). Unlike other outer island provinces whose population growth is partly due to the central government's transmigration program, North Sulawesi's population has grown in size with little effect from transmigration.

These three characteristics of North Sulawesi—religion, modernization, and increasing human population—have an important bearing on hunting patterns. First, unlike Islam, there are no religious laws in Christianity governing the eating of wildlife. Second, modernization brings with it centralized economies in which consumption of wild meat might increase because of the development of more sophisticated technology for capturing and transporting wild animals, or might decrease because of a centralized market system in which domesticated meat is distributed widely. Third, the increasing severity of environmental problems in the region is a reflection of an ever-increasing human population. Minahasa is the most densely populated district within North Sulawesi, with approximately 300 people per km^2 (KSPSU 1994). As a culture that consumes wildlife on a regular basis, efforts must be made to harvest sustainably so that animal populations in the surrounding regions are not adversely affected. However, it might be too late. Already, wildlife might have been hunted out of accessible parts of Minahasa, and suppliers for local markets seek animals outside Minahasa (Clayton and Milner-Gulland this volume).

PROTECTED AREAS AND SURROUNDING COMMUNITIES

Two protected areas were selected for this study: Manembonembo Nature Reserve (6500 hectares [ha]) and Gunung Ambang Nature Reserve (8368 ha). Prior to this study, except for a brief ecological assessment by MacKinnon and Artha (FAO 1982) and a macaque population survey (Sugardjito et al. 1989), no studies had been conducted at Manembonembo and Gunung Ambang.

Manembonembo is located adjacent to the northwest coast in Minahasa. The Trans-Sulawesi Highway runs along the south side of the Reserve and a small asphalt road lines the north and west boundaries. An extensive coral reef outside the Reserve, part of the Bunaken Marine National Park, borders the coastal side of the Reserve and serves the fishing needs of the local communities. The elevation extends from 50 m above sea level to the peak of Gunung Tanuwantik at 665 m. Central reserve areas can be divided into two major distinguishing vegetative structures. At lower elevations (50 to 300 m), palms such as the fan palm (*Livistona rotundifolia*) are abundant, whereas at higher elevations (>300 m) various species of rattan such as *Calamus* spp. are found. Abundant canopy and mid-canopy trees with large buttresses can be found at all elevations. The topography is characterized by steep mountains with clayish soils.

Gunung Ambang is on the border of Minahasa and Bolaang Mongondow districts. The elevation ranges from 250 m to 1780 m. The area consists of lowland rain forest with abundant palms, including *L. rotundifolia* and the Wallacean palm (*Pigafetta filaris*), and large mid-canopy to canopy-sized trees. The submontane forest and montane forests reach to the peak of the active volcano Gunung Ambang. The forests occur on steep terrain, are interspersed by rocky cliffs, and are rich in ferns and orchids (FAO 1982).

Both areas are fraught with a host of illegal activities such as hunting, slash and burn agriculture, cultivation of coconuts and cloves, and wood collection for commercial (e.g., rattan) and noncommercial (e.g., firewood) use. Sulfur mining also has been performed in Gunung Ambang. During the study in Manembonembo, four to six park guards were responsible for management of the Reserve. Except when officially required to accompany me or perform a special patrol, they rarely entered the Reserves.

The inhabitants surrounding Manembonembo and Gunung Ambang make their living in a variety of ways. There are a few educated professionals such as civil servants and teachers who work in the local district offices and schools. Most families make a living farming and fishing, and running small shops and services. Agriculture is mostly restricted to small-scale cultivation, mainly in two forms: *kebun* (orchards) and *ladang* (farms). *Kebun* is the main producer of income. Found near settlements, this comprises coconuts, cloves, and bananas. *Ladang* is maintained mostly for subsistence and consists of rice, cassava, maize, tomato, and various fruit trees. Although a majority of inhabitants of these communities are farmers, others derive a significant portion of their income from family members who are employed in Manado, the provincial capital.

METHODS

Forest Surveys

Population surveys of wildlife were conducted from June 1994 to December 1995. Forest surveys were conducted at Manembonembo Nature Reserve using the line-transect sampling method (Burnham et al. 1980) for eight mammal species (table 23-1). The species surveyed were known or thought to exist in the Reserve from previous surveys and/or anecdotal evidence, and were those included in the household surveys. I measured abundance for each species as the number of animals sighted per kilometer censused. Four trails each 5 km long (total 20 km) were established, and they were surveyed 2 weeks after being cut. Censuses were conducted by myself and an experienced research assistant. Transects were censused in the morning from 0600 to 1100 hr, at the rate of 1 km per hour, once a month throughout the year. Data collected included weather, location, time, path to animal distance, height of animal, mode of detection, number of animals, age and sex of animals, reproductive status of adult females (i.e., whether adult females were visibly pregnant, carrying young, or nursing), and gross habitat type. In addition, to obtain presence/absence data, we searched for animal tracks and signs along transects and in habitats where certain animal species were likely to be found.

Table 23-1. Survey Results for Mammal Species Surveyed at Manembonembo Nature Reserve, June 1994 to November 1995

Species		Density (No./km²) June 1994	Abundance June 1994	Density (No./km²) November 1995	Abundance November 1995	% Change June 1994 to November 1995
Ailurops ursinus	Bear cuscus	2.5	162.5	1.5	99.5	−38.8
Macaca nigra	Crested black macaque	22.8	1480	14.7	955.5	−35.8
Macrogalidia musschenbrockii	Sulawesi palm civet	Rare	Rare	Rare	Rare	Unknown
Viverra tangalunga	Malay palm civet	Rare	Rare	Rare	Rare	Unknown
Bubalus depressicornis	Anoa	Rare	Rare	Rare	Rare	Unknown
Cervus timorensis	Rusa deer	0.8	51.4	0.2	15.6	−69.9
Babyrousa babyrussa	Babirusa	Rare	Rare	Rare	Rare	Unknown
Sus celebensis	Sulawesi wild pig	3.7	239.2	5.1	331.1	+38.4

Household Surveys

I collected data on consumption rates of wild meat per household. In villages surrounding Manembonembo and Gunung Ambang with at least 200 residents, households were selected and questioned. The first household within a village was randomly selected for interviews followed by a household five houses away. This continued until the sample size represented approximately 5% to 10% of the total number of households in the village. Subjects were asked a set of standardized questions to determine the extent to which rural people depend on wildlife as a source of food (table 23-2). In addition to questions specifically pertaining to wild meat consumption, all information concerning age and sex, religion, education level, income level, profession, length of residence in the village, and family size were obtained.

In nearly all of the interviews, the women, who prepare all meals, were present and responded to the questions. Each interview took between 15 and 30 minutes. Local names of animals were used to avoid confusion in nomenclature. If the subject was unclear about the species in question, its description and photographs or illustrations were presented. Voluntary information concerning other animals was noted. However, I did not incorporate interview techniques to cross-check systematically the accuracy of responses. In addition, villages were surveyed just once. Therefore, assessment of temporal variation in harvest rates cannot be made.

Species for household surveys (table 23-3) were originally chosen on the basis of their conservation importance and their known role in rural diets. In addition, they possess unique physical features, and therefore could not easily be mistaken for other species. However, during the survey, subjects frequently volunteered information about additional taxa (e.g., bats, rodents). This presented a problem in analyzing the data because, for certain taxa such as bats and rodents, the local classification of animals did not comply with scientific nomenclature. In addition, because small animals are caught in large numbers and are immediately prepared

---

Table 23-2. List of Questions Asked During Household Surveys

1. How many people live in the house?
2. How much meat (not including fish) do you consume in a month?
3. What percentage of this meat is obtained from the forest?
4. How much meat of the following species do you consume in a month?[a]
5. How is the meat obtained?
    a. Family member is a hunter.
    b. Trade or buy from hunter directly.
    c. Buy from middlemen.
    d. Buy from central market.
    e. Elsewhere.
6. If you buy the meat, how much is the meat for the following species?[a]
7. What animals don't you consume, and why not?
8. Are animal less or more difficult to obtain in the past 5 years? Why?

All interviews included information on the village, date, age, sex, education, occupation, and ethnicity.
[a]See table 23-3 for animal list.

Table 23-3. Commonly Harvested Animals and Total Annual Harvest Rates

Scientific Name	Common Name	Average Weights	Manembonembo Total Harvest (kg) per Year	Gunung Ambang Total Harvest (kg) per Year
Varanus indicus	Indian monitor lizard	31.0	840.00	0.0
Python reticulatus	Reticulated python	37.0	567.00	188.9
Rhyticeros cassidix	Red knobbed hornbill	2.5	147.0	15.7
Strigocuscus celebensis	Dwarf cuscus	0.5	315.0	1409.8
Ailurops ursinus	Bear cuscus	3.5	168.0	554.4
Family Sciuridae	Squirrels	NK	NK	NK
Order Chiroptera	Bats	0.2	11,739.0	7242.2
Family Muridae	Rats	0.2	26,502.0	24,072.0
Macaca nigra	Crested black macaque	4.1	1025.0	471.5
Tarsius spectrum	Spectral tarsier	0.1	15.0	31.4
Macrogalidia musschenbrockii	Sulawesi palm civet	4.9	63.0	0.0
Viverra tangalunga	Malay palm civet	7.8	0.0	31.4
Bubalus depressicornis	Anoa	225.0	31.4	1243.6
Cervus timorensis	Rusa deer	82.5	323.8	834.4
Babyrousa babyrussa	Babirusa	71.5	124.4	61.8
Sus celebensis	Sulawesi wild pig	32.5	5315.0	2865.4
			47,175.6	39,022.3

NK, not known.

All animals listed are consumed, but absence of figures on harvest denotes that they were consumed on a rare basis. Weights are based on carcasses weighed at markets by author and also obtained from other sources: Grzimek (1975), Stuhan (1983), Wemmer et al. (1983), Caldecott (1986), Whitten et al. (1987), Novak (1991), Watanabe (personal communication), Napier and Napier (1967), Kinnaird and O'Brien (personal communication).

for consumption (often charred), the local people did not attach much significance to their species, and identification at the species level was very difficult. Thus, the data for certain taxa are grouped.

Hunters in two areas were known to hunt on a regular basis as a means of income. They were contacted opportunistically and asked to provide information concerning their catches. The sample size from these interviews is too small (n = 13) to make definite conclusions about harvest rates and catchment areas. However, the responses do have an important bearing on the findings from the community surveys.

The figures on household consumption presented here are conservative estimates. For example, responses were discarded if the estimate of the amount or frequency with which a particular animal was consumed was considered inaccurate. Responses concerning consumption of larger animals were often expressed in gross weight, whereas responses concerning consumption of smaller animals were often expressed as numbers of animals or parts of animals (e.g., one bat, a leg of a monkey). I obtained weights of animals by weighing whole carcasses at markets. For species that could not be weighed at markets, I used the recorded weight of that species or a similar species from other sources (e.g., anoa [Grzimek 1975]; babirusa [Whitten et al. 1987a]; spectral tarsier [*Tarsius spectrum*] [Napier and Napier 1967]; red knobbed hornbill [Kinnaird and O'Brien personal communication]).

FOREST WILDLIFE POPULATION

Population density and abundance data were compiled from a total of 340 km of transects. Results are presented in table 23-1 and figure 23-2. Although reports from villagers indicate that the two civet species might exist in the Reserve, they were not detected during the surveys, and therefore are likely to be extremely rare. Sightings of anoas and babirusas (see figure 24-2) were also rare. During the entire study period, babirusas were only seen four times and anoas once. Three species—crested black macaque, bear cuscus, and rusa deer (*Cervus timorensis*)—experienced rapid declines, showing a 36% to 70% decrease in densities during the study. The crested black macaque was surveyed in 1988 (Sugardjito et al. 1989) and its population had declined by 63.6% between then and the end of this study (November 1996).

The pattern of wildlife depletion at Manembonembo is in keeping with recent trends in North Sulawesi. Kinnaird and O'Brien (WCS 1995b) reported the extirpation of the anoa and babirusa from Tangkoko Nature Reserve over 15 years. Similarly, there was a sharp decrease in densities of the crested black macaque and Gorontalo macaque from 1978 (MacKinnon 1983) to 1988 (Sugardjito et al. 1989). As for other large mammal species in Manembonembo, because no baseline data existed before this study, it is unclear whether densities of these species are naturally low or whether they are being reduced by hunting. Solely based on the

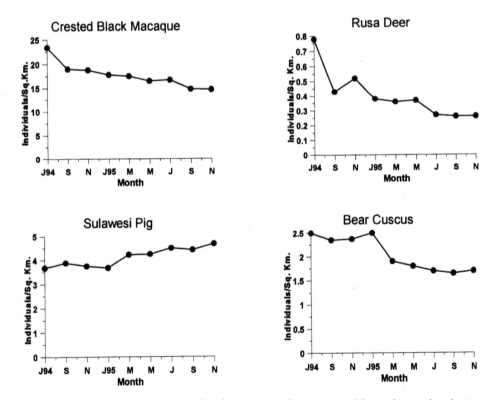

Figure 23-2. Population trends for four mammal species at Manembonembo during study period.

population census data, at the present rate of decline, seven of the eight mammal species censused may face extirpation at Manembonembo in 1 to 5 years.

Not all species are decreasing, however. Populations of the Sulawesi wild pig (*Sus celebensis*) appear not to be affected by the current level of hunting. In fact, the species increased in density by 38.4% during the study period. The Sulawesi wild pig is reported to produce litters of two to eight piglets with a mean of five (Anonymous 1983) and, therefore, is relatively resilient to hunting pressure. At Tangkoko Nature Reserve in North Sulawesi, the wild pig population was reported to have doubled in 2 years (WWF 1980b).

HUNTING SUSTAINABILITY

Three measures of sustainability are used here. First, animal populations at Manembonembo were monitored to see if there were changes in population density through time. This provides a simple sustainability index; if hunting is known to occur in a catchment area, a steady decline in population numbers indicates that harvest is not sustainable (Robinson and Redford 1994b). Although sustainability indices provide clues as to whether hunting is sustainable, they cannot be considered conclusive. Monitoring of population numbers does not evaluate sustainability directly. Rather, it merely indicates whether a population is declining or not; therefore, inferences are weak.

Another way to assess sustainability is to use models. Models have been proposed by Robinson and Redford (1991b, 1994b) and Bodmer (1994). Both use demographic characteristics of the species and, therefore, allow for stronger inferences.

The model put forth by Robinson and Redford (1991b) evaluates whether an actual harvest might be sustainable under conditions of maximum production (P_{max}). To calculate production (P_{max}), I used the following formula:

$$P_{max} = (D \times \lambda_{max}) - D,$$

where D is the observed density. Use of the observed density is appropriate when it can be assumed that populations are not at carrying capacity. I used Cole's (1954) equation to calculate the natural rate of increase (r_{max})

$$1 = e^{-r_{max}} + be^{-r_{max}(a)} - be^{-r_{max}(w+1)}$$

From P_{max} values, potentially available harvest values are calculated based on the longevity of a species: 60% of the production in very short-lived species, 40% in short-lived species, and 20% in long-lived species (after Robinson and Redford 1991b). Table 23-4 shows P_{max} values, and potential harvest levels.

Current hunting levels for the bear cuscus (see figure 9-2) and the crested black macaque (figure 23-3) are clearly unsustainable, the model indicates, at both Manembonembo and Gunung Ambang. This is also consistent with Bodmer's

Table 23-4. Sustainability Estimates for Censused Species Based on Robinson and Redford's Model (1991b, 1994b)

Common Name	Density (No./km²)	r_{max}	P_{max}	Manembonembo				Gunung Ambang			
				PH/km²	PH	AH	Sustainable	PH/km²	PH	AH	Sustainable
Bear cuscus	2.53	0.293	0.86	0.34	22.36	48	No	0.34	30.3	158.4	No
Crested black macaque	22.84	0.145	3.42	0.68	44.21	250	No	0.68	60.5	115.0	No
Anoa	0.18[a]	0.312	0.04	0.02	1.04	0.14	Yes	0.02	1.8	5.53	No
Rusa deer	0.79	0.412	0.39	0.16	10.27	47.38	No	0.16	14.2	10.1	Yes
Sulawesi wild pig	3.68	0.62	3.16	1.27	82.55	163.38	No	1.27	113.1	88.1	Yes

r_{max}, intrinsic rate of increase; P_{max}, maximum production; PH/km², potential number of animals for harvest per km² per year; PH, potential harvest within reserve per year; AH, actual number of animals harvested per year.

[a]Density estimate based on WWF (1980).

Figure 23-3. Sulawesi crested black macaque (*Macaca nigra*).
Photo courtesy of Margaret Kinnaird.

(1995c) finding on the strong relationship between a species' vulnerability to hunting and r_{max}. For the other three species, hunting sustainability varied between the two sites. For the rusa deer and Sulawesi pig, current harvest rates were apparently sustainable at Gunung Ambang, but not at Manembonembo; conversely, the harvest rate for the anoa was apparently sustainable at Manembonembo, but not at Gunung Ambang. These results are difficult to explain. Particularly problematic is the fact that Sulawesi pigs showed a significant population increase, but current hunting levels were considered unsustainable. The conflicting findings might be due to small sample sizes in both the forest and household surveys.

Bodmer's (1994) model of sustainability compares production (*P*) with the harvest rate (*H*) in a catchment area. If *H* exceeds *P*, then the harvest rate is considered unsustainable. *P* is derived by multiplying the total productivity, *S* (average number of young/female/year ÷ 2) by the density estimate, *D* (individuals/km²). *S* can be obtained from the proportion of adult females that are reproductively active, the average litter size, the number of gestations per year, and the percentage of reproductively active females in the censused population. This value is divided by 2, assuming a 1:1 population sex ratio. The proportion of adult females that were reproductively active was estimated by noting adult females who were carrying young, nursing, or visibly pregnant.

We were only able to collect data on the age structure and reproductive conditions of the crested black macaque, so Bodmer's model was only useful for this species. With 40% of the adult females reproductively active, an average litter size of one, and 0.43 gestations per year, the average number of young/individual/year was 0.086. The density estimate at the beginning of the study was 22.8 individuals per km^2. P, then, is 1.96 individuals per km^2 per year. An annual harvest of 250 macaques or 3.85 individuals per km^2 at Manembonembo (table 23-4) is nearly twice the recruitment. Therefore, the present harvest rate of macaques is clearly not sustainable according to Bodmer's model.

CONSUMPTION OF WILDLIFE AND PATTERNS OF HUNTING

The household survey included interviews with 175 people from nine villages around Manembonembo and 164 respondents from eight villages around Gunung Ambang. All respondents but one were native to North Sulawesi. Nearly all of the respondents (92% at Manembonembo and 96% at Gunung Ambang) ate wild meat either on occasion or on a regular basis. The rest did not eat any meat from the forest due to religious dietary restrictions. On the average, families obtained 35% to 37% of their meat from the forest. The most frequently eaten wild animals in both communities were rodents, bats, and Sulawesi pig, comprising 56% to 62%, 19% to 25%, 11% to 14% respectively, of the total number of individuals (see figure 23-4 and table 23-3).

The extent to which a species is exploited in North Sulawesi is determined by biological and cultural factors. Important biological factors are animal availability (distribution and abundance) and mass per carcass. First, frequently taken species are habitat generalists. Three taxa—the Sulawesi wild pig, rodents, and bats—are

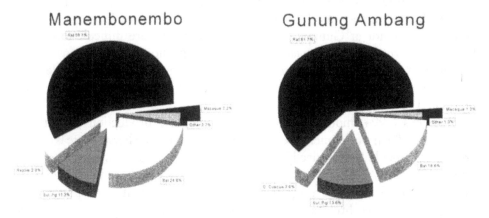

Figure 23-4. Percentages of wild animals harvested and consumed at Manembonembo and Gunung Ambang.

known to occupy diverse habitats (Whitten et al. 1987a). In particular, rodents and certain fruit bats are commensal with humans (Marshall 1985; Whitten et al. 1987a; Fujita and Tuttle 1991). Because these animals are more commonly found near human settlements, they are easy to capture. Second, smaller mammal species were more frequently harvested than larger ones. This is consistent with Redford and Robinson's (1987) findings on hunting patterns in the neotropics, where small-bodied prey species, which are more common than larger species, were hunted more frequently by Indians, and smaller mammal species were more frequently harvested in North Sulawesi. Third, in addition to availability, large-bodied species such as the Sulawesi wild pig and rusa deer are highly favored due to the amount of meat they yield. The crested black macaque, the fourth most hunted species at Manembonembo, is also a highly preferred animal. In addition to the large body size, because of the large number of individuals in a social group, a single hunting effort can yield a large amount of meat.

Prey selection is also determined by the type of hunting that takes place in North Sulawesi. Hunting can be broadly divided into two categories: active and passive. Active hunting is labor intensive and time consuming and requires that the hunter actively pursue the animal. Active hunting methods include the use of guns, dogs, and slingshots. Shotguns and ammunition are not readily available, so air rifles are often used. Although this method is relatively inexpensive—a box of 100 pellets costs US$0.50—it is labor intensive and quite ineffective for hunting most forest animals, so its use is relatively infrequent.

Hunters in North Sulawesi generally use passive techniques that require intensive effort at first (e.g., building a trap) but do not require active pursuit of the animal. Passive techniques consist of traps and snares that can be built from materials found in the forest (e.g., bamboo, lianas, and palm leaves). Fruit found in the forest is often used as bait. Sometimes nails, nylon rope, or metal wire are used to assemble the trap; these materials are inexpensive and can be reused, so the cost is not prohibitive.

The fact that passive hunting is not expensive or time consuming is a major reason for hunting in North Sulawesi being so prevalent. Although hunting is part of a work routine, being incorporated into a schedule based on farming chores, it is of secondary importance and is often done opportunistically. Hunting is tied into farming duties. It requires little or no money to build traps and snares and little time to set them while taking care of farming chores. This is important in considering the long-term dynamics of hunting. Previous discussions have concluded that if the harvest of a hunted population does not meet the socioeconomic needs of the hunters, then harvesting is not socioeconomically sustainable (Robinson 1993; Robinson and Redford 1994b). In North Sulawesi, however, hunting is only a secondary activity to farming, and hunting is performed as a convenience. Hence, profitability (hunting success per unit effort) is not a major consideration when examining hunting dynamics. If the animal population reaches a point where most animals have been hunted out, hunting effort will obviously decrease significantly. The major prey species (e.g., Sulawesi wild pigs and rodents) appear to withstand current hunting pressures; thus, it is unlikely that hunting effort will decrease.

Although the focus of this chapter is on subsistence hunting, the impact of hunting to meet market demands or for special occasions is also major (O'Brien and Kinnaird this volume; Clayton and Milner-Gulland this volume). Subsistence hunters, when they are not able to consume, trade, or sell all of their catches at the village level, will sell to central market vendors. Fujita and Tuttle (1991) found 16 species of bats available at local markets in North Sulawesi. In conducting a market survey, at two locations on a single day, I encountered 22 dwarf cuscus, four crested black macaques, one babirusa, and approximately 175 bats.

ECONOMIC AND NUTRITIONAL CONDITIONS

In many cases, a rural community depends on wild meat for nutrition or income. This raises the following questions: (a) to what degree do people in North Sulawesi depend on wild meat for nutrition or income; and (b) can North Sulawesi rural communities manage without exploiting wildlife?

A review of basic economic and nutritional indicators provides a convenient starting point to examine wild meat consumption in the context of rural poverty. Various indicators such as the literacy rate, availability of roads, annual income, gross domestic product, and calorie and protein intake are used to identify regions of poverty within Indonesia. Much of the information used here is from official government reports, either from national or provincial censuses or technical reports. They might or might not reflect actual rural conditions.

Most evidence indicates that North Sulawesi is not impoverished. Socioeconomic indicators show that North Sulawesi has a higher standard of living than most other provinces of Indonesia. North Sulawesi is one of four provinces that has consistently been in the high-growth and high-income bracket (BPS 1986a). The education and literacy rates are very high (BPS 1980), and living standards are higher than in most other provinces (BPS 1984b, 1984c, 1986b, 1986c). In addition, North Sulawesi had the eighth lowest incidence of rural poverty among all Indonesian provinces, with only 11.4% of its population below the poverty line (Booth 1988).

Rural economies are notoriously complex to decipher. It is often unclear from what sectors incomes are derived and how much of a person's gross income comes from different sectors. Accordingly, the extent to which wildlife contributes directly to the income of rural people is unclear. From informal observations of local markets and interviews with villagers and meat sellers, income earned from these sales is probably minor. At most village markets, for every kilogram of wild meat that is sold, 20 to 30 kg of fish is sold. Even if there is much money to be made from the sale of wild meat, it probably benefits such a small group of people that it serves no significant role in the local economy. There is also little commercial use of nonedible wildlife products. Feathers from hornbills, reptile skins, deer antlers, and babirusa tusks are sometimes traded among villagers, but for the most part, there is little market value for these items.

Related to this generally higher standard of living, several lines of evidence indicate that North Sulawesi is not deficient in food and its people are not nutritionally deprived. The Food and Agriculture Organization recommends 2100 kilocalories and 46 g of protein per person per day (Thorbecke and van der Pluijm 1993). Compared with national figures, North Sulawesi has a consistently higher caloric and protein intake and lower incidence of associated diseases (Tarwotjo et al. 1978; BPS 1984a). Whereas the national average caloric intake for 1980 and 1984 were 1794 and 1798, respectively, the averages for North Sulawesi were 2007 and 1898. The national average protein intake for 1980 and 1984 were 42.7 g and 43.3 g, but for North Sulawesi were 47.8 g and 48.64 g (BPS 1984a; KSPSU 1988). Indeed, per capita protein intake in North Sulawesi has steadily increased since 1982 (KSPSU 1988, 1990, 1992) (figure 23-5). Much of the protein is from fish and vegetables (KSPSU 1988, 1990, 1992).

These findings demonstrate that North Sulawesi rural populations are neither nutritionally nor economically dependent on wild meat. So, if not strictly economics or nutrition, what is driving the present desire for wild meat?

Hunting serves an important purpose. While providing meat for one's family and neighbors, it kills agricultural pests. Macaques, pigs, deer, and rodents are considered serious pests, and most respondents reported that although obtaining wild meat is important, it was more important to protect crops that are the source of most rural household income.

Protection of crops through hunting has been part of what is known as garden hunting (Linnares 1976). Although garden hunting has been reported mostly in the neotropics (Linnares 1976; Nations and Nigh 1980; Murphy 1990; Jorgenson 1993), it has only occasionally been observed in Asia (King 1978; Morris 1978; Caldecott 1986a). This is probably the most significant component of hunting in

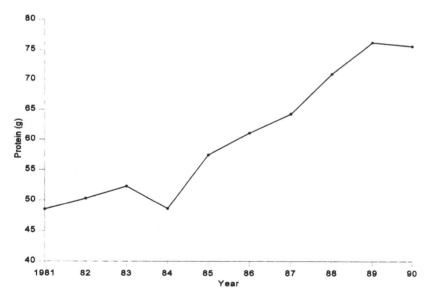

Figure 23-5. Protein intake per capita per day in North Sulawesi 1981–1990. *Source of data: KSPSU 1988, 1990, 1992.*

North Sulawesi. Most people in the community possess garden forests. Because upland gardens are located several kilometers from the village, farmers want to ensure protection against crop predators. Additionally, in order to obtain as much wild meat as possible with the relatively infrequent outings into the forest, garden hunters set up as many snares and traps as possible near gardens and in the forest. As a result, most animals are caught this way.

The evidence presented here demonstrates that the North Sulawesi rural population is not merely eking out a living, and that they are not exploiting wildlife to survive. One of the most revealing pieces of data from the household surveys is that family size is not correlated to the amount of wild meat eaten. This seems to suggest that a large family will not expend the extra effort in obtaining wild meat as a source of protein ($r^2 = 0.15$). Normally, as long as there are tasks to be done in the garden, snares and traps will be set. In the absence of farming chores, snaring and trapping will be infrequent.

CONSERVATION ISSUES AND OPTIONS

Findings from this study indicate that current subsistence hunting levels in North Sulawesi for many animal species are not sustainable. The changes needed to ensure the continued survival of wildlife populations in North Sulawesi will require long-term thinking and planning in collaboration between government agencies and research institutions. Immediate conservation efforts and social programs need to be planned, and research undertaken.

Research

The present study used gross demographic and economic figures from national and provincial censuses. However, social, religious, and economic differences between subdistricts and even between villages can be wide. Therefore, to make better informed recommendations, detailed data will be needed from economic, nutritional, and ecological studies to understand the dynamics between rural economies, health and nutritional conditions of rural communities, and the biology and status of wildlife populations.

Law Enforcement

Effective management policies should reflect particular concerns for the given region. Illegal hunting has been recognized as a serious problem in Latin America (Ojasti 1984) and has been given proper attention in formulating natural resource management policies and rural development (Ponce 1987). With some exceptions, however (e.g., Sarawak [Caldecott 1986a; Bennett and Dahaban 1995]), it has not been a serious consideration in Asia. In Indonesia, habitat destruction has been the central concern in management policies, and rightly so, because forest conversion into agricultural land is the primary conservation problem in most parts of

Indonesia. However, for North Sulawesi, such policies are less relevant to conservation problems than consideration of hunting.

At present, conservation problems in North Sulawesi stem mainly from hunting, so management should revolve around deterring such activities through effective law enforcement. Regular patrolling by guards from PHPA (Directorate General of Forest Protection and Nature Conservation) should be given the greatest priority.

Patrolling should take place in three forms. First, regular patrols should be performed along existing trails to detect traps and snares and catch hunters. Second, to detect traps and snares, special sweeps should be conducted through gardens within protected areas. Because traps and snares invariably belong to the land holder, identification and apprehension of poachers should be relatively simple. Third, PHPA with assistance from the local police should target wild meat sellers at village and town markets, making the selling of illegal meat difficult. Although patrolling alone will not ensure that wildlife populations will not be depleted, it is a precondition.

An effective legal system of deterrence not only means patrolling and apprehending those who violate wildlife laws, it also means having laws that can be enforced. Fines should be set to deter violators from committing future violations. As decreed by the President, the penalty for hunting protected species holds a maximum fine of US$45,454 and/or up to 5 years in prison. Considering that the average per capita annual income is approximately US$325 for a North Sulawesian (KSPSU 1993), a more realistic set of fines and penalties should be drawn up.

Nutritional Program

If population sizes of vulnerable species can be stabilized through patrolling, further deterrence of hunting will come from a program to offer less expensive meat than currently available. The most economically feasible program would be to assist in forming village cooperatives for small-scale farming of domesticated animals such as pigs and chickens.

Village and township cooperatives serve practical and ideological functions in Indonesia. The Indonesian word *kooperasi* (cooperative) is found in both the Republic of Indonesia Constitution, in which the formation of cooperative organizations is encouraged, and the National Ideology *Pancacila* (five principles). In practical terms, the situation in North Sulawesi lends itself well to the use of cooperatives for rural programs because most economic activities of rural communities take place at these levels (Thorbecke and van der Pluijm 1993).

Local cooperatives might be a suitable means to increase the availability of meat through increased livestock production and increased harvests of vertebrate pests. Much of the damage to agricultural products in Sulawesi can be attributed to vertebrate pests (Whitten et al. 1987a). Rodents, in particular, damage crops greatly. They build nests in coconut trees and feed on live plant material such as leaves and nuts as well as copra. Rodents also cause losses to other crops such as maize, tomatoes, cassava, and yams. It seems feasible that a program to increase wild rodent harvests would reduce the amount of crop damage and might also reduce the need to exploit other wildlife. In Central Java, rodents have been harvested on a sustainable basis and used as animal feed (Whitten et al. 1987a). Increased production of pigs and rodents through a subsidized program would provide a steady and inexpensive way to reduce the present rate of wildlife harvesting.

Education

Education about the impact of intense animal harvesting will be critical to the survival of wildlife populations. The smattering of information currently available in the form of pamphlets, posters, public presentations, and discussions are inadequate. Because religion plays such an important role in the lives of North Sulawesians, efforts at working with churches and community groups must play a pivotal role in the establishment of ecologically sustainable practices.

Although the people of North Sulawesi consume wildlife, they are apparently not dependent on it. However, much needs to be done to study how and why people exploit wildlife and to develop strategies to conserve it. Finding a balance between preserving wildlife and winning the support of local people for sustainable practices will be fraught with a host of difficulties. Even though the rationale for recommendations might be understood by government agencies and communities, adoption of new ideas and techniques will require much time, effort, and thought. Particularly problematic is the issue of time. Unless programs are designed and implemented soon, efforts to save imperiled populations will be futile because the animals will already have been extirpated in many places.

ACKNOWLEDGMENTS

This work was supported by the Fulbright Foundation, Jersey Wildlife Preservation Trust, and International Primate Protection League. I thank the Indonesian Academic of Sciences (LIPI) and Perlindungan Hutan dan Pelestarian Alam (PHPA) for granting permission for me to work in North Sulawesi. Special gratitude goes to Dr. Jatna Supriatna (Universitas Indonesia), Mr. Ramon Palete (PHPA), and Mr. Endang Mashudi (Sub-Balai of North Sulawesi KSDA). Helpful advice by Drs. Margaret Kinnaird and Tim O'Brien, Mr. Graham Usher, and comments and suggestions by Dr. John Robinson, Dr. Elizabeth Bennett, Mr. Scott Byram, Dr. Aletta Biersack, and anonymous reviewers are appreciated. Finally, I thank Piter Sombowadilu for assisting with data collection.

24

The Trade in Wildlife in North Sulawesi, Indonesia

LYNN CLAYTON AND E. J. MILNER-GULLAND

A key conservation issue throughout Asia is the effect of hunting on wildlife populations. Sulawesi, Indonesia, is the largest and most central island of Wallacea, the biogeographical transition zone between the Asian and Australasian faunas. The fauna of Sulawesi has very high levels of endemism: 62% of mammal species (98% excluding bats) and 27% of bird species are endemic (Whitten et al. 1987a). Endemic species include the babirusa (*Babyrousa babyrussa*), anoa (*Bubalus depressicornis, B. quarlesi*), and seven species of Sulawesi macaques. Indonesian wildlife law gives full protection to these species (Republik Indonesia, 1990), yet our research has shown that the local wildlife trade is driving them toward extinction in North Sulawesi province. The problem facing national and local authorities and conservationists is therefore how to bring about effective enforcement of existing wildlife protection laws. In this chapter we describe the structure of trade in these species and present data on market sales and records obtained from meat dealers. We describe the current laws protecting these species and known prosecutions for offenses against the wildlife laws, and consider how these laws might be more effectively enforced to protect Sulawesi's endemic fauna from extinction.

BACKGROUND TO THE STUDY

The province of North Sulawesi is a long, narrow arm of land (25,000 km²) comprising three districts (figure 24-1). Trade in wildlife is closely related to religious differences within the province; in Minahasa district, 95% of the population are Christian, whereas the two remaining districts, Bolong Mongondow and Gorontalo, are almost entirely inhabited by Muslim peoples. The meats of babirusa, Sulawesi wild pig (*Sus celebensis*), anoa, macaque, bat, snake, and dog are extremely popular with Minahasan people but are not consumed by residents of

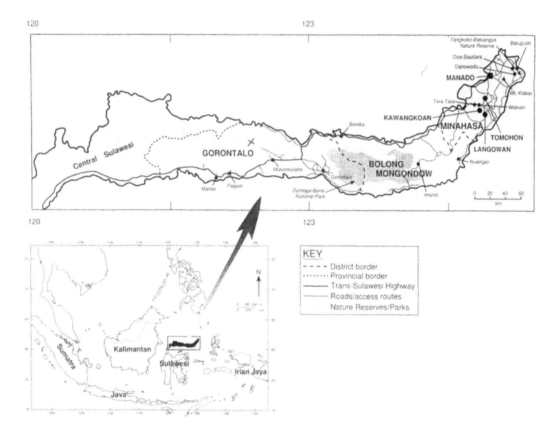

Figure 24-1. A map of North Sulawesi, showing the places mentioned in the text. The cross indicates the position of L.C.'s field site in the Paguyaman forest.

Bolong Mongondow and Gorontalo. Minahasa has lost most of its populations of these species, so wild meat dealers must drive from Minahasa to Bolong Mongondow and Gorontalo, where wildlife is still relatively plentiful.

About 30 dealers, all Minahasan, currently sell wild meat in the markets of Minahasa, chiefly Langowan, Tomohon, and Kawangkoan. Dealers from these towns drive out every week in small pickup trucks to purchase meat from hunters. Today they mostly purchase in the Gorontalo district and at the North/Central Sulawesi provincial border, a round trip of up to 1200 km. They purchase this meat at the forest edge or at collection points and carry it back smoked or alive for sale in the weekly market. Dealers traditionally have their own hunters who regularly supply meat to them. Hunters are typically Minahasans, and they hunt singly or in pairs. The dealer transports them from Minahasa to a hunting area where they remain for 2 to 6 months at a time to trap wild pigs before moving to a different forest area. There are currently about 1000 hunters trapping in North Sulawesi province.

The number of dealers and hunters working in the wild meat trade has increased dramatically in recent years. From 1948 to 1970 there was only one wild pig meat dealer operating in North Sulawesi; two more dealers became active from 1970 to 1984, and 12 full-time dealers were active in 1993. In 1996 there were 30 active dealers. The commercial traders have benefited enormously from recent road improvements, allowing them to travel faster and farther to buy meat. The road system of North Sulawesi is shown in figure 24-1. Just one road links Minahasa with the Gorontalo region. This road first became passable in 1980, although then it was necessary to raft vehicles across major rivers; it was first fully tarmacked in 1992.

Hunters are paid in cash by dealers at the forest edge or at a collection point. The usual method by which hunters trap wild pigs and anoa is using nylon string leg snares set on paths used regularly by the animals, as well as around wallows, salt-licks, and fruiting trees. Hunters each set 15 to 50 snares, which are checked usually every third day. Once the hunter is at the forest site, string is only available to him via the dealer, with the cost deducted from the payment for his meat. Either leg snares or baited nets are used to trap macaques. The eggs of the mound-building maleo bird, five times the size of a chicken's egg, are also on sale in North Sulawesi markets, but are not carried by the wildlife dealers. Forest rats are popular with Minahasans, but they are still common in Minahasa and are sold in the markets by local women rather than by the wild meat dealers.

This chapter draws on 5 years of field work performed by one of the authors (L.C.) in North Sulawesi between 1989 and 1995. Data were collected on the ecology and behavior of the babirusa, as well as on the wildlife trade in North Sulawesi. The trade data come from several sources: monitoring of the end markets; monitoring of dealer vehicles as they travel from the forest to market; records of an individual dealer; discussions with many hunters and dealers; and personal observations by L.C.

The ecology of the babirusa was studied at the Paguyaman Forest in Gorontalo district. The three main markets in Minahasa where wild pig meat is sold (Langowan, Tomohon, and Kawangkoan) were monitored between February 1993 and July 1995, with visits once a month or every 2 weeks to each market at 0730 hr on a Saturday, the main market day. The price of wild pig meat and the quantity on sale in each market were recorded. It was possible to distinguish between babirusa (figure 24-2) and Sulawesi wild pig meat because babirusa skin appears hairless and Sulawesi wild pig skin is conspicuously hairy. Skin is present on all pieces of meat. The market monitor was a local woman who posed as a customer.

Detailed records were also obtained of the business transactions of one wild pig meat dealer (dealer X) over the period December 1991 to June 1994. The records do not distinguish between babirusa and Sulawesi wild pig meat because trading in babirusa meat is illegal, whereas trade in Sulawesi wild pig meat is not. The records give the number of pigs supplied by individual hunters each week, the price paid for them, and the age and sex class of each pig purchased. Dealer X was the first full-time wild pig trader in North Sulawesi. He has traded continuously in wild pigs since 1948. Until 1979, he purchased wild pigs and dogs at Nuangan, 100 km from Manado (see figure 24-1). With the opening of the Trans-Sulawesi Highway in 1980 he began purchasing in Boroko district and extended his journey

Figure 24-2. Babyrusa (*Babyrousa babyrussa*) at a waterhole. *Photo courtesy of Margaret Kinnaird.*

westward to Molombulahe (1988), Paguat (1989), and Marisa (1991). Thus, it is possible to trace the development of the wildlife trade through the experiences of this particular dealer. His movements west are especially important in tracing the depletion of wild pig populations along the Trans-Sulawesi Highway: as the wild pigs in an area become hunted out, the dealer has had to drive farther in order to buy meat. At the same time, the road improvements have allowed him to drive farther and still get back to the market before the quality of the meat deteriorates enough to make it unsaleable.

In North Sulawesi, wild pig hunting is purely commercial and is performed almost entirely by Minahasan people rather than by local people in the areas where the animals occur (Bolong Mongondow and Gorontalo). Gorontalonese and Bolong Mongondow people do not hunt wild pigs for religious reasons. The wild pig hunters also trade commercially in species that are hunted by local people such as anoas, thus leading to a substantially higher hunting mortality rate for those species as well.

THE VENDING STRUCTURE FOR WILD MEAT

Data on the major wildlife species sold in Minahasan meat markets are presented in table 24-1. Some species, such as forest rats, are caught and sold by local people, but the majority of the wild meat sold in the markets is traded by dealers. Babirusa, Sulawesi wild pig, macaques, anoas, bats, and domestic dogs are traded together by Minahasan dealers, although only the babirusa, anoas, and macaques are legally

Table 24-1. Traded Wildlife Species of North Sulawesi

Common Name	Scientific Name	Endemic?	Legal Status[a]	Where Eaten	Price per Individual in Market (Rp)	CITES status[b]	1993 Red List Status[c]	1994 Red List Status[d]
Babirusa	*Babyrousa babyrussa*	Yes	Protected	M	50,000	I	V	EN
Sulawesi wild pig	*Sus celebensis*	Yes	No	M	50,000	—	—	VU
Anoa	*Bubalus depressicornis, B. quarlesi*	Yes	Protected	M,L	50,000–75,000	I	E	EN
Bear cuscus	*Ailurops ursinus*	No	Protected	M	8000	—	—	DD
Dwarf cuscus	*Strigocuscus celebensis*	No	Protected	M	5000	—	—	DD
Tarsier	*Tarsius spectrum*	No	Protected	M	5000	II	K	LR
Macaques	*Macaca nigra*	Yes	Protected	M	15,000	II	I	EN[e]
	Macaca becki	Yes	Protected	M	15,000	II	-	LR(nt)[e]
	Macaca nigrescens	Yes	Protected	M	15,000	II	—	LR(cd)[e]
Rusa deer	*Cervus timorensis*	No	Protected	L	?	—	—	LR
Bats	Various, including Pteropidae	No	No	M	500–2500	—	—	
Rats	Muridae	No	No	L, M	400–1750	—	—	
Maleo bird (eggs)	*Macrocephalon maleo*	Yes	Protected	L	750–1500	I	V	EN[f]
Monitor lizard	*Varanus salvator*	No	No	M	6000	II	—	LR(lc)
Reticulated python	*Python reticulatus*	No	No	M	6000+[g]	II	—	?
Rock python	*P. molurus*	No	Protected	M	6000+[g]	II	V	?

Where eaten: M, Minahasa; L, locally.

Market price: US$1 = Rp 2200.

CITES status: I, listed in Appendix 1; II, listed in Appendix 2; —, not listed.

IUCN 1993 Red List categories: E, endangered; V, vulnerable; R, rare; I, indeterminate; K, insufficiently known; —, not listed.

IUCN 1994 categories: CR, critically endangered; EN, endangered; VU, vulnerable; LR, lower risk (cd, conservation dependent; nt, near threatened; lc, least concern); DD, data deficient.

[a]Republik Indonesia 1990.

[b]World Conservation Monitoring Centre 1996.

[c]Groombridge 1993.

[d]IUCN 1994.

[e]IUCN 1996.

[f]Dekker and McGowan 1995.

[g]Python prices depend on length.

protected. Most of the dealers' meat is sold in Langowan, Tomohon, and Kawangkoan markets, although wild meat is sold in all Minahasan markets. Examination of 15 dealers' vehicles revealed that a typical vehicle might contain about 10 bats, 15 Sulawesi wild pigs, 15 domestic dogs, five babirusa, one macaque, and one anoa. Because the wild pig meat trade is particularly lucrative and is the backbone of the dealers' business, it is examined in detail later. First we discuss the trade in other wildlife species.

Legally Protected Species

Anoa. The meat of this endemic dwarf buffalo is eaten both by Muslims and Christians, making it particularly vulnerable to local extinction from hunting. It is usually consumed locally rather than sold, occasionally reaching village markets in Bolong Mongondow and Gorontalo. For example, at Imandi market, Bolong Mongondow, one was observed on January 27, 1990, cut into 30 pieces and sold at Rp250 per piece (US$1 = Rp2200 at the time of writing). Anoa are also trapped by Minahasan wild pig hunters and sold to dealers for Rp25,000 each. Dealers resell the meat in the Minahasan markets, smoked and cut into pieces, for Rp75,000, although anoa is rarely observed in these markets.

Macaques. Monkey is eaten only by some Minahasan people, others regarding this as tantamount to eating human flesh. *Macaca hecki* are caught by wild pig hunters in Bolong Mongondow and Gorontalo districts and sold to wild pig dealers. They are usually transported to Minahasa alive and smoked on market day prior to sale. The forest hunters receive Rp3000 per monkey, and dealers resell them for Rp15,000. *M. nigra* (see figure 23-3) are caught in the Klabat area of Minahasa by local villagers. The hunters smoke them and take them to the market town to sell to a market trader. Trade is focused on the hill-town markets of Tomohon and Langowan, although the number sold is low; typically 5 to 15 monkeys per month in each town. Numbers rise before Christmas and Twelfth Night; 20 monkeys were observed for sale by one dealer on December 23, 1989, and 16 on January 6, 1990. Monkey meat is considered to be medicinal, including curing skin diseases.

Maleo Eggs. Maleo eggs are available in villages close to their nesting grounds throughout the province. These enormous eggs are highly valued for human consumption (Dekker and Wattel 1987). The maleo's communal egg-laying habit, on open beaches or volcanically heated soils, makes them an easy target. The status of the maleo has been well described (Dekker and Wattel 1987; Argeloo 1994). Opportunistic exploitation of maleo eggs has resulted in a severe population decline (Argeloo and Dekker 1996). Eggs are collected for local consumption and sale to householders, small restaurant owners, or market traders. Eggs were available at Marisa for Rp1500 each; at Tambarana, Poso, in Central Sulawesi province (Rp750–1000); and 16 eggs were observed in Palu market (Rp1000 each) on March 10, 1989. Collection by villagers in North Sulawesi has a take-as-much-as-you-can attitude (Argeloo and Dekker 1996). Eggs may be ornamentally wrapped in palm leaves and given as gifts.

Other. Both species of cuscus (*Ailurops ursinus* and *Strigocuscus celebensis*) and tarsiers (*Tarsius spectrum*) have been observed occasionally for sale in Minahasan markets but are not usually carried by dealers.

Unprotected Species

Rats. Forest rats (e.g., the introduced *Rattus argentiventer*) are a common and popular item in Minahasan markets. Typically about 1000 to 1500 rats were observed in Tomohon market each Saturday (the main market day). Rat sellers are numerous, usually women from surrounding villages (e.g., Woloan, Tara-Tara) who sell approximately 30 to 50 rats each. These are trapped by their husbands around agricultural gardens, smoked, and carried to market. There is no dealer structure.

Bats. Bats, especially large fruit bats (Pteropidae), are popular among Minahasan consumers, purchased particularly as a delicacy for festivals, and are carried by wild pig meat dealers. From 1989 to 1994, 25 to 50 bats per week were typically sold at Tomohon, but this increased to 300 per week early in 1997. Prices depend on size. Hunters devastate whole roosts, using sling shots and/or nets. In some areas, the bats are caught using kites with hooks on the strings to catch the animals as they leave the roosts.

Other. Monitor lizard (*Varanus salvator*) and python (*Python reticulatus, P. molurus*) meat is popular and occasionally available in the markets, although there is no structured trade.

TRADE IN BABIRUSA AND SULAWESI WILD PIG: A CASE STUDY

The Ecology of Wild Pigs in Sulawesi

The babirusa (figure 24-2) is an extraordinary curly-tusked pig, found only in rain forest areas, with a total wild population of about 5000. It has a very limited distribution, being found in only three of Sulawesi's four provinces (and the small or neighboring islands of Sula, Buru, and Togian). Densities of the babirusa are believed to be highest in the northern province (MacKinnon 1981), although they are still relatively low there; observations at L.C.'s study site in Gorontalo suggest a babirusa density in the area of about 1.25 animals/km². One babirusa weighs 50 to 100 kg (Macdonald 1993a).

The babirusa has a rather low reproductive rate, producing one, two, or occasionally three piglets at one time. Interbirth interval and age of first reproduction are unknown for wild babirusa. Babirusa social organization comprises solitary adult males or matriarchal groups of one or a few adult females and their immature young. Groups usually comprise five or fewer animals. Short-term data collected over a maximum period of 5 weeks on eight radio-tracked babirusa indicated that areas used were 0.8 to 12.8 km² (using minimum convex polygon methods). Data from six animals were compatible with nomadism, although data from two animals suggested the use of small ranges (<1 km²) from which excursions were made (Clayton 1997).

The babirusa is officially protected by Indonesian law (Republik Indonesia, 1990), which states that it is illegal to hunt, kill, or trade babirusa. The species is undergoing rapid range contraction; nineteenth century naturalists recorded babirusa at locations in the northern tip of Sulawesi (Wallace 1869; Guillemard 1886; Hikson 1889), whereas today most babirusa in North Sulawesi are found at the western end of the province. We demonstrate in this chapter that this range contraction is ongoing and can be entirely explained by the wild pig meat trade.

The Sulawesi wild pig is as yet unprotected and is sympatric with the babirusa. It is smaller than the babirusa, weighing 30 to 60 kg (Macdonald 1993b). It is also found in disturbed areas, such as secondary forest, as well as in the rain forest. The carrying capacity of the Sulawesi wild pig is approximately 10/km²; a density of 12 animals/km² was found at Tangkoko Dua-Saudara Reserve, North Sulawesi, a reserve with a rather higher density of Sulawesi wild pig than is usual (WWF 1980b; O'Brien and Kinnaird 1996). The Sulawesi wild pig produces four to eight piglets at one time (Macdonald 1993b). The ecology of this species has not been studied in the wild, and interbirth intervals, age of first reproduction, group sizes, and ranging behavior are not well known. Although the Sulawesi wild pig is heavily hunted, it is more resilient to hunting than the babirusa, due to its higher density and higher reproductive rate.

Both species are almost entirely hunted using snares, with occasional use of dogs and spears. The two pig species differ in their susceptibility to capture by snares. Experienced hunters indicate that 90 snares are needed to catch two babirusa and one Sulawesi wild pig per week. The hunters report that of the two species, babirusa is easier to catch, Sulawesi wild pig being more able to detect and avoid snares. A comparison of data from dealer X's records, and data from field observations, shows that leg snares are relatively unselective by age or sex. Leg snares catch wild pigs in approximately the same proportions as were observed in the wild, although juveniles are underrepresented in the hunted sample (table 24-2).

Market Sales of Wild Pig Meat

Long-term market monitoring (figure 24-3) has given us a detailed picture of wild pig sales from the period 1991 to 1995. Figure 24-4 shows that Langowan is the major market for wild pig sales. It is also the market where a significant proportion of the wild pig meat sold is babirusa (figure 24-5), whereas babirusa meat was rarely observed at the other two markets. Langowan market has been important for wild pig sales for some time. Occasional surveys in 1990 and 1991 revealed that about 50 to 60 pigs were sold there every week, of which a third to a half were babirusa. This is a similar proportion of babirusa to that found in the market in 1995, although the average number of wild pigs sold each week in 1993 to 1995 had increased to 74. Data from Tomohon market suggest that although the number of pigs sold showed no significant changes in the period 1991 to 1995, wild pig meat prices increased from about Rp2750/kg in 1989 to about Rp3000/kg

**Table 24-2. Mean and Standard Deviation for the Age:
Sex Ratio of Wild Pigs Sold to Dealer X, 1991 to 1994**

	Mean Proportion Sold	Standard Deviation	Observed in the Wild
Adult male	0.33	0.116	0.23
Adult female	0.45	0.141	0.36
Juvenile	0.17	0.091	0.26

The remaining 5% of individuals sold were classified as of unknown age and sex. The proportion of each class observed at a salt lick in the wild over the same period is also shown.

in 1991, Rp3600/kg in 1993 to 1995, and Rp5000/kg in 1997. Because there are no quantitative data available on the local economy, it is unclear whether this price increase is inflationary or a real increase driven by changes in factors such as supply, consumer income, or human population size. The analysis presented here concentrates on Langowan market in 1993 to 1995 because it is the most important market for wild pig meat, and there are more data available for it than for the other markets.

Figure 24-5 shows a sudden increase in the proportion of Sulawesi wild pig sold in Langowan market from an average of 0.59 in February to May 1993, to 0.88 in June 1993 to August 1994. There was a nonsignificant decline in the proportion of Sulawesi wild pigs sold over the period June 1993 to August 1994, but when monitoring recommenced in January 1995, the proportion had

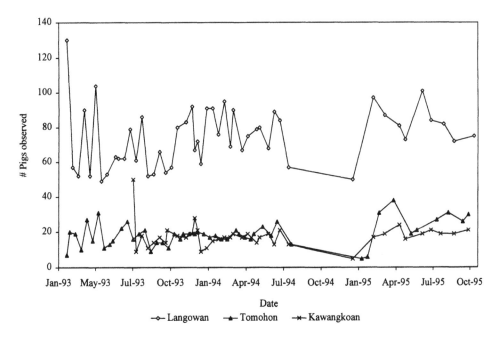

Figure 24-3. The number of wild pigs observed for sale in the three major wild meat markets of Minahasa, January 1993 to October 1995. Observations were made at 0730 hr on a Saturday, when the weekly maximum number of pigs was on sale.

Figure 24-4. Market sales of pig. *Photo courtesy of John G. Robinson.*

decreased to 0.6, a similar mean to that observed in February to May 1993. The proportion of Sulawesi wild pig sold then remained at this level until monitoring ceased at the end of October 1995. The total number of wild pigs sold did not show a similar trend (see figure 24-3), nor did the price charged per kilogram of pig meat (figure 24-6a), suggesting that consumers were equally happy to buy either species of wild pig.

The change in the proportion of babirusa sold was caused by the notification

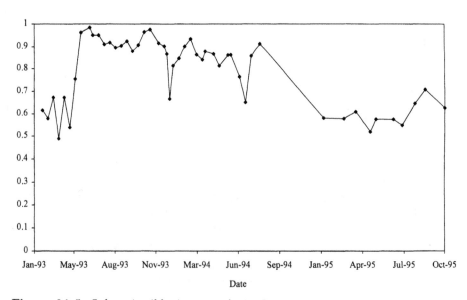

Figure 24-5. Sulawesi wild pigs on sale in Langowan market, January 1993 to October 1995, as a proportion of the total number of wild pigs on sale at that market. The remainder of the wild pigs on sale are babirusas.

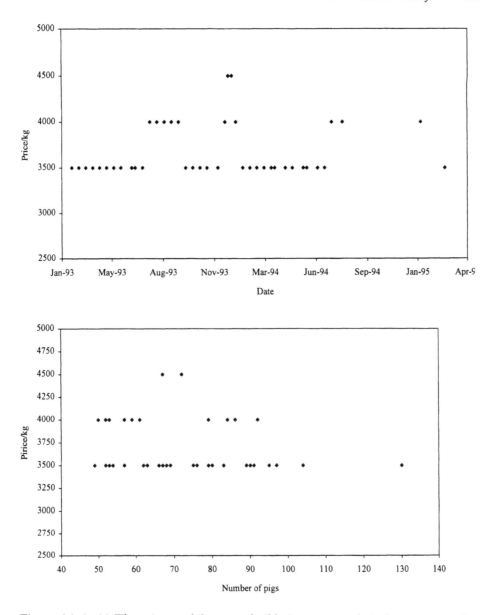

Figure 24-6. **(a)** The price per kilogram of wild pig meat on sale in Langowan market as a function of time, January 1993 to October 1995. **(b)** The price per kilogram of wild pig meat in Langowan market, plotted against the total number of pigs observed for sale that day. Both graphs show actual prices (not corrected for inflation). The level of inflation in Minahasa was low over the period, so that the use of actual prices does not have a significant effect on the results.

of government officials that babirusa meat was being openly sold in Langowan market. Subsequent law enforcement included government support for the prosecution of a dealer and visits to the markets by government officials. This had a dramatic effect on the number of babirusas sold openly, which continued for at least a year after the law enforcement. The action may have simply led to babirusa meat being discarded, or it might have been sold secretly in the market or directly from dealers' houses. Secret sales are unlikely to have occurred because the market monitor was a local woman who posed as a customer.

The price charged per kg of wild pig meat does not show a trend with time (see figure 24-6a), nor is it affected by the quantity of meat on sale at the market (figure 24-6b). Higher prices coincide with Christmas and New Year and with the mid-summer clove-harvesting season. At Christmas, wild pig is a popular feast food, whereas in the clove-harvesting season consumers have additional income. The fact that the price increases at these times, but quantity sold shows no increase, suggests that over the monitoring period the wild pig dealers were not increasing supply in response to periods of increased demand. A relationship between price and quantity sold is likely to develop over the course of a day; dealers bargain with customers to ensure the sale of all their meat by the end of the day, when the quality of the meat is poorer. However, the data suggest that, at the beginning of the day, there is no attempt by dealers to relate prices to the supply of wild pig meat in the market.

The market monitoring thus demonstrates the following:

1. The two wild pig species are treated as identical goods by consumers, so there is no preference shown between the legally obtained Sulawesi wild pig and the illegally obtained babirusa. Considering that snaring is indiscriminate, it is clear that, from the dealers' point of view, the two species are indistinguishable in every way except for their legal status.

2. The differing legal status of the two species did have an effect on the dealers' behavior, on the one occasion on which the market was visited by law enforcement officials. This effect was to increase the proportion of total sales that were Sulawesi wild pig, without any change in price or the total quantity of wild pig meat on sale in the market. The effect lasted into the medium term and suggests that it is possible for the dealers to continue their business and sell only legally obtained wild pig meat. What happened to the unsold babirusas is unclear.

3. The wild pig dealers increase their prices but not the quantity of meat they sell at festival times. The reason for this can be explored using the records of dealer X.

An Individual Dealer

Data from the accounts of dealer X allow us to examine the incentives for the dealer and deduce the level of profits that he is making from his trade. Because he lists the monies paid to individual hunters, we can deduce the amount that the hunters are earning, and whether they make a good wage in relation to other jobs. This information is extremely valuable for predicting what will happen to wild pig populations in the near future. If dealer X and his hunters are making good profits, then it is likely that more dealers will continue to enter the market, and wild pig hunting will continue to increase.

Over the period for which records are available, there are no clear trends in the total number of pigs bought by dealer X, even at Christmas when the market price of wild pigs increases (figure 24-7). This confirms the results of the market monitoring (see figure 24-6). If it were easy for dealer X to increase production at these times, a clear seasonal trend in his pig purchases would be expected, to make the most of periods of increased demand. A possible explanation is that dealer X's vehicle already fills up each trip, and he already makes as many journeys as he can. It also suggests that it is not economically worthwhile for him to reduce the number of other animals that he carries (such as domestic dogs) in favor of more pigs.

The mean price that dealer X paid per pig increased over the period, from Rp7944 (standard deviation [σ] = 787.6) in December 1991 to April 1992 to Rp12,320 (σ = 858.9) in March to June 1994, a 55% increase in 2 years. This increase cannot be explained by inflation or increases in wages elsewhere in the economy because no other relevant parameters show changes of this order of magnitude. It is therefore likely that the price increase was generated by processes within the wild pig market. The regression of mean price on purchase date is not significant (figure 24-8), due to the large variability in the prices paid by dealer X in 1993. Neither the mean amount paid to each hunter nor the number of hunters from whom dealer X bought meat show any clear trend with time, although both are highly variable. The price paid per pig by dealer X to individual hunters seems to be unrelated either to the number of pigs an individual hunter supplies on a particular occasion, or to the rate at which a hunter supplies pigs to the dealer (defined as the total number supplied divided by the time between the first and last recorded transaction).

From 1994 to 1997 there has been a change in the behavior of individual

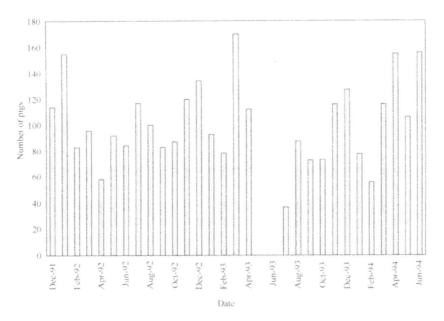

Figure 24-7. The number of wild pigs purchased each week by the case study dealer, December 1991 to June 1994.

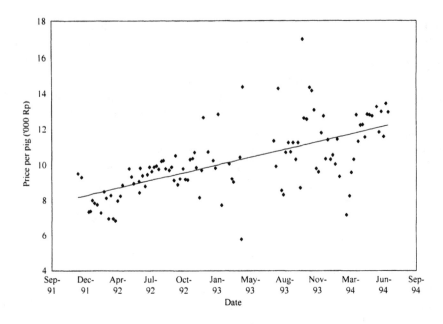

Figure 24-8. The price paid per pig by the case study dealer, December 1991 to June 1994, plotted with a linear regression line of the trend in price with time. The r^2 of the regression is 0.39. Actual prices are shown, rather than real prices.

hunters. Many dealers now buy meat per kilogram rather than per pig, which has led to an effective increase in the price paid to hunters. Hunters with a regular arrangement with one dealer now hold back meat to sell to these dealers. Together with the increasing number of dealers and hunters working in the wild meat trade, this suggests that the wild meat trade is becoming more competitive.

The Structure of the Wild Pig Meat Market

Usually, when the sustainability of hunting is considered, the assumption is made that hunters and dealers are free to enter and leave the wildlife trade market. This means that the market is competitive and that the hunters and dealers only consider short-term profits when they decide whether or not to operate in an area. The opposite extreme is a monopolistic market, when the resource is owned, so that access to it is controlled. The dealer can then manipulate the hunting rate with a view to long-term profits. It is important for conservation to know which market structure applies in a particular case, because in a competitive market the resource will usually be hunted to much lower population sizes than in a monopolistic market. In a monopolistic market, the earnings of the traders (figure 24-9) are maintained at a high level (super-normal profits) because there is some barrier to entry into the market, such as that the resource is privately owned or that a large investment is required to start trading. The fact that some of the activities in the market (selling protected species) are illegal might also constitute a barrier to

Figure 24-9. Traders or dealers bring the wild meat to markets. *Photo courtesy of Lynn Clayton.*

entry, although how significant that barrier is depends on attitudes to law enforcement. It is not likely to be a significant barrier in North Sulawesi at present. In reality, although a market may tend toward being either competitive or monopolistic, it is usually intermediate in structure. Moreover, resources can either be sustainably used or hunted to extinction in both market structures, depending on the economic conditions (Clark 1990). The evidence that the wild pig meat market in North Sulawesi is competitive so far is that large numbers of dealers and hunters are entering the trade and that the price paid by dealer X to his hunters is rising.

The other aspect of market structure crucial to understanding if trade is sustainable is whether it is close to equilibrium or not. At equilibrium, traders in a competitive market earn "normal profits." Normal profits, in the economic sense, mean that the traders are earning the same as they would in the nearest comparable employment. For example, hunters have the choice between being employed by dealers to hunt wild pigs or working as agricultural laborers. In a competitive market, people will continue to enter the profession of wild pig hunter until earnings fall to the point where they are equivalent to agricultural wages, at which point the number of hunters remains constant. The market is at competitive equilibrium.

To see how close to equilibrium the market is, we can compare the earnings of hunters and dealers with those in alternative employment. If they are very different, then other forms of evidence are needed to decide whether the market is in competitive disequilibrium or whether it is likely to be monopolistic.

Hunter-Dealer Transactions. Data from dealer X show that the hunters are currently earning about the same as they would in a comparable profession such as agricultural laboring (see appendix 24-1). The fact that they are not making excess profits could be because the hunters are effectively employees of the dealers or because they are competing among themselves to supply the dealers. The hunters are tied to particular dealers because the dealer provides them with string for making snares, but they are still able to hold back meat to sell to other dealers. Thus, it seems that the situation is intermediate between competition and employment. The hunters are only able to sell to dealers who are in the market. Thus, the recent expansion of the number of dealers in the market is likely to have led to an increase in the price paid to hunters, as the choice available to hunters increases. It also has led to an increase in the number of hunters, from about 400 in 1995 to about 1000 in 1997. The observed price increase to dealer X's hunters (figure 24-8) and the hunters' new practice of holding meat back for other dealers are both consistent with an expansion of the market over the past few years.

The Case Study Dealer. Dealer X's profits per trip can be calculated from his records (see appendix 24-2). The calculations suggest that he is making supernormal profits of Rp462,120 per trip. This is a large amount; it implies that the dealer's annual wage is 7.6 times the wage for his nearest alternative employment as a small businessman, and 38 times the average wage in Minahasa district. It seems to be strong evidence for the wild pig meat market not being in competitive equilibrium, evidence that concurs with the observation that many new dealers are entering the trade. The fact that new people are entering means that there is no significant monopolistic element to the market. There are major barriers to entry into the market, which might affect the rate at which people enter:

1. The need for capital. This is both to buy and maintain a truck, as well as to pay the hunters in cash for their meat, before the meat is sold in the market.
2. The illegality of some of the dealer's activities. Buying protected species is, in theory, a serious offense.

In summary, the detailed market data show that wild meat dealers are making large profits and that the market is expanding rapidly. The fact that hunters are not making major profits is less relevant because it is the dealers who decide the level of resource use; the hunters are involved as employees rather than decision makers. This means that hunting rates are likely to continue to increase rapidly in the near future without some enforcement. This control will inevitably lead to declines in populations of wild pigs and other traded species (such as macaques and anoas). In order to find the most effective way to tackle this decline, we need to look at the current state of law enforcement in the province and to draw lessons from previous experience of trying to control illegal hunting of protected species.

LAW ENFORCEMENT AND KNOWN PROSECUTIONS

Indonesian Wildlife Law

Current Indonesian wildlife law is centered on the Conservation of Natural Resources and Ecosystems Act (number 5) of 1990 (Republik Indonesia 1990). This includes laws on the protection of plant and animal species, the use of conservation areas, and penalties for offenders. A total of 545 species are protected under this act, including 94 mammals. This law specifies that it is illegal to catch, trade (alive or dead), or kill a protected species. The trade in or possession of parts of any protected species is also forbidden. The penalty for violation of this regulation is a maximum prison sentence of 5 years and a maximum fine of Rp100,000,000 (US$45,454). The law authorizes designated civil servants and police to investigate offenses against it. Additions to this law have been Government Regulation numbers 13 and 18 of 1994, on Hunting and Government Regulation and Utilisation of Conservation Areas, respectively.

Prior to this Act, Indonesian wildlife legislation was based on the Government Regulation of Wildlife Protection 1931 (Natuur bescherming 1931). Under this law, 36 species of animals were protected, including the anoa, babirusa, tarsier (*Tarsius spectrum*), and rusa deer (*Cervus timorensis*). It remains in force today as supporting legislation to the 1990 act. This law forbade the killing, trapping, and trade in these species, although seven species, including the babirusa, anoa, and rusa deer, could be hunted with a permit from the Chief Provincial Official. Penalties for offenses against this act were 3 months imprisonment or a fine of Rp500 (Departemen Pertanian Republik Indonesia 1953). During the 1930s the purchase price of one wild pig in the forest was Rp2.5, so this fine was 200 times the value of the pig to the hunter. By contrast, today's fine constitutes 8000 times the value of the pig to the hunter. Thus, the 1990 legislation represents a considerable toughening of the law with respect to wildlife protection. There has been a general tendency for countries to increase penalties for illegal wildlife trade in recent years, usually without a corresponding increase in law enforcement measures (Leader-Williams and Milner-Gulland 1993).

Babirusa. Enforcement actions taken during L.C.'s 5-year field study of the babirusa included overnight checkpoints at which dealers' vehicles were examined, and removal and confiscation of hunters' snares set in babirusa habitat. No action against babirusa poachers is known to have taken place prior to the start of the field study in 1989. In September 1989, local police detained two Minahasan dealers' vehicles near the study site in the Gorontalo district, which were being loaded with 17 babirusa and two anoa. Fifteen local hunters present were informed of the wildlife laws by local police officials. The meat was confiscated but no other action was taken. With continued vigilance, this one study site remained safe from hunters until November 1992, when local police acting on local information detained a single dealer carrying approximately 20 babirusas. Two hundred snares were removed from the forest by village police or surrendered by hunters. Ten babirusas and one anoa were discovered trapped in snares in the forest and were released; several had injured legs because they had been ensnared for several days. A prosecution was prepared against the dealer and the ringleader of the hunters by local police, with support from national authorities. Both dealer and ringleader were repeat offenders who were aware of the Indonesian wildlife laws. After 14 months the case was settled out of court, the dealer incurring a fine of US$500. This case had a considerable deterrent effect on other dealers, substantially increasing their perception of the probability of capture and conviction.

Dealers tended to be more responsive, and more susceptible, to enforcement measures than hunters. Hunters tended to reoffend more often, probably because they correctly perceived the chance of prosecution to be small, and could move to another district to reduce the probability of the authorities prosecuting them. Thus, of the 15 hunters warned in 1989, eight definitely reoffended, four definitely did not reoffend, and the subsequent activities of three are unknown. Twice-yearly checkpoints along the Trans-Sulawesi Highway (the only road linking the west and eastern ends of North Sulawesi province, along which all dealers must travel), targeting dealers rather than hunters, were effective in protecting L.C.'s babirusa study site from hunting. Enforcement measures were also effective in the end market: government officials were notified that babirusa were being sold in Langowan market, as a result of which the number of babirusas sold openly in the market declined. However, this enforcement was not sustained, and the number of babirusas sold increased again (see figure 24-5). Thus, the effect of law enforcement in the market was strong and lasted more than a year before fading. This was the first such law enforcement exercise conducted, so dealers' perceptions of the risk of receiving a high penalty for the possession of a protected species probably rose dramatically as a result. The frequency of subsequent law enforcement actions will allow dealers to judge more accurately the true probability of capture and adjust their responses accordingly.

Macaques. One prosecution for trade in *M. nigra* is known, at Tangkoko-Batuangus Nature Reserve in Minahasa (Y.Y. Muskita, personal communication). Two local villagers were discovered carrying 11 dead macaques (two adult males, four adult females, five immature animals) on December 30, 1990. The villagers were caught by an Indonesian researcher and staff of the Indonesian Wildlife Department, and the incident was reported to the local police. After 2 months no action had been taken, whereupon publicity for the case was raised by the researcher, through articles in local and national newspapers. Police action began after a staff member in the Wildlife Department reported the matter to the district magistrate's office. A further 4 months later the case was brought to court and the two hunters were each imprisoned for 6 months. The villagers were reportedly unaware that *M. nigra* is a protected species, and neither was known to hunt again following their release.

Maleo. In Sulawesi one prosecution is known (M. Argeloo, personal communication). This involved the arrest of three collectors by Park staff in Dumoga-Bone National Park (now renamed Bogani Nani Wartabone National Park). One of these was sentenced to 1 week's imprisonment. Most egg collectors are apparently aware that the maleo is protected by law, and local police often needed further information on existing Indonesian wildlife laws (M. Argeloo, personal communication). Traditional controls on harvesting of maleo eggs have been applied in the past (Argeloo and Dekker 1996), and where they are implemented today, the current National wildlife protection regulations apparently play no part in trade regulation and are ignored.

A MODEL OF THE TRADE IN WILD PIG MEAT

Clayton (1997) developed a spatially structured model to explore the trade in babirusas and Sulawesi wild pigs and predict the likely long-term equilibrium population distribution of these two species in North Sulawesi. This model calculated the costs to the dealers and hunters of getting to each part of North Sulawesi and the amount of meat that they are likely to obtain in each area. In the long term, hunting will only continue in an area if the revenues from hunting at least cover the costs of hunting there. In the wild pig model, we assume a competitive market. Our detailed study of the trade in wild meat in this chapter confirms that this is the correct assumption.

The model showed that the likely outcome of trade in wild pig meat is a dramatic decline in the babirusa population, with the population finally stabilizing at around 4% of the unexploited population size. The model predicts a less dramatic decline in the Sulawesi wild pig population, to around 37% of the unexploited population size. The only areas where the babirusa and Sulawesi wild pig are predicted to survive into the long run are those that are distant from roads, and/or in mountainous regions, where the costs of getting to the area are too high for it to be worthwhile hunting there (figure 24-10). The babirusa is predicted to become extinct due to hunting in most parts of the northern arm of Sulawesi, whereas the Sulawesi wild pig would survive in more places. This is because the Sulawesi wild pig, with its higher reproductive rate and higher density, is more resilient to hunting, and so can withstand a higher hunting rate than the babirusa.

These results might be taken to mean that, because wild pig populations are not completely exterminated by hunting, wild pig hunting will stabilize at levels that are sustainable. However, although it is likely that, under current conditions, there will be some wild pigs left in northern Sulawesi when the trade reaches equilibrium, these pigs will be confined to remote areas and will not be generally present in the forests of northern Sulawesi. Sustainability also depends on the remaining population being viable; it is unlikely that a babirusa population of 200 animals (4% of carrying capacity) would be viable in the long term. Sulawesi wild pig hunting is more likely to be sustainable. The key assumption of the model is that conditions remain as they are today, but other processes might affect the conclusions. For example, if logging started in all the current concessions, the habitat available to babirusas would be reduced considerably. Economic changes, such as changes in consumer income or in travel costs for the dealers, would also alter the results.

One interesting aspect of the model is the effect that the unusual situation in North Sulawesi has on the results. If the wild pigs were hunted by local people, rather than by Minahasans, then the costs of travel would be so much lower that the babirusa population would be almost entirely wiped out (0.2% of carrying capacity), and the Sulawesi wild pig population would be only 13% of carrying capacity. The effect of the recent road improvements can be demonstrated by rerunning the model for the roads in their 1950s state. This leads to babirusa and Sulawesi wild pig populations of 31% and 66% of carrying capacity, respectively, which is probably a viable size.

Another feature of the system affecting the outcome is that the babirusa and Sulawesi wild pig are hunted together, as if they were one species. This means that the joint population size in an area determines whether it is worthwhile hunting there. Consequently, because the Sulawesi wild pig has a higher carrying capacity

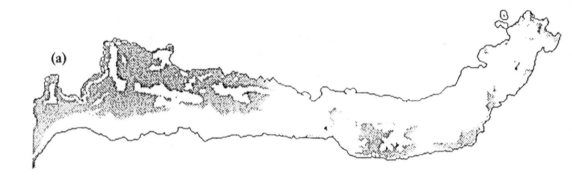

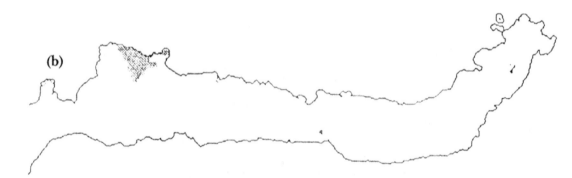

Figure 24-10. The predicted population sizes and distributions of wild pigs in northern Sulawesi at the long-term equilibrium, from the model of Clayton et al. (1997). The shade of grey shows the size of the population; white indicates an absence of pigs; and black shows that the pigs are at a carrying capacity. **(a)** Sulawesi wild pig and **(b)** Babirusa.

and reproductive rate, the Sulawesi wild pig population size dominates the dealers' decision making. This can be seen by assuming that the dealers trade separately in each species. If dealers were only interested in babirusas, then it would not be worthwhile hunting at all, whereas if they were interested only in Sulawesi wild pigs, the situation would hardly change from the current one. Thus, the model suggests that the babirusa is being hunted to local extinction in many areas because of the presence of Sulawesi wild pigs there. This is similar to the situation observed for rhinos and elephants in the Luangwa Valley, Zambia, in the 1980s, when the rhinos were only worth hunting because of the large elephant population (Milner-Gulland and Leader-Williams 1992). As was discussed in the previous section, the wild pig meat dealers carry several other species as well as wild pigs.

Because the wild pigs are the major source of income for the dealers, these other species are unlikely to make a major difference to the results of the model. However, the results of the model will be important for the future of endangered species, such as the anoas and macaques, that could be hunted to local extinction on the back of the wild pig meat trade.

Despite not making any a priori assumptions about range decline, the model's predictions are in agreement with observations. The model predicts that babirusa range size will continue to decline, as has been documented over the period when dealer X has been operating. It also predicts that the major range decline should coincide with the improvement in the road system, as has been observed.

POLICIES FOR CONTROLLING ILLEGAL TRADE

The model of wild pig hunting was used to investigate the effects of various law enforcement policies on babirusa hunting rates (Clayton et al. 1997). The analysis showed that the introduction of fines on dealers selling babirusa in the market would be effective in reducing hunting pressure on the babirusa. This method requires law enforcement officers to patrol markets regularly and to be able to distinguish babirusa meat from Sulawesi wild pig meat. It also allows the hunters to continue to trade in wild pig meat while encouraging them to alter their behavior so as to trap the unprotected Sulawesi wild pigs rather than babirusas. Even if nonselective snaring continued, so that babirusas were killed but discarded, the model shows that the policy could substantially reduce hunting pressure on both species by making some areas unprofitable for hunting. However, hunters could select for Sulawesi wild pigs, and against the protected babirusa and anoa, by setting snares in disturbed secondary forest rather than primary forest. An expected fine of Rp45,000 (the average price of a babirusa in the market) would make it unprofitable for dealers to sell any babirusas at all.

The expected fine is composed of two separate components: the actual fine received by the dealer if caught, multiplied by the dealer's perceived probability of being captured and receiving that fine. Thus, law enforcement officials have two main options to reduce the illegal wildlife trade: either impose stiffer penalties or increase the perceived probability of capture (or a combination of the two). If hunters and dealers were risk neutral, these two methods would have equivalent deterrent effect, and cost implications might suggest that stiffer penalties (particularly fines) were a better option for Wildlife Departments with limited budgets. The approach chosen by Indonesia, in common with many other countries, has been to increase the maximum penalty imposed. However, theory and empirical evidence suggest that stiffer penalties are unlikely to be an effective deterrent to illegal wildlife use when the perceived probability of detection is low. Because of the way people assess risks, punishment in the future, or punishment with a very low probability of happening, is perceived as less serious than punishment that is immediate or has a high probability of occurring. Thus, a prison sentence of 10 years with a probability of 0.1 will be perceived as less severe than a prison sentence of 5 years with a probability of 0.2. This means that the perceived probability of capture and conviction is a much more relevant factor in peoples' decision making (Eide 1994; Ehrlich 1973; Avio and Clark 1978). Thus, despite the continuing expense, increasing the perceived probability of being prosecuted is probably the most effective method of deterring the illegal wildlife trade.

Prosecutions for illegal trade in wildlife are still rare in North Sulawesi. The structure of the majority of the trade is unusual, in that all dealers pass along a single road to get from the forest to their end markets, of which there are only a few. Thus, law enforcement should be much easier in North Sulawesi than in many areas where wildlife is illegally traded. Random checkpoints on the road or inspections of the markets could be effective if performed only once or twice a year. So why are fines not imposed on dealers openly selling legally protected species? In the past there has been a general lack of concern to preserve endemic species in the Province and a shortage of information on current laws; also, officials might not have been readily available to distinguish protected species. In L.C.'s experience there is a willingness to act among the people of North Sulawesi, but economic pressures are strong, annual per capita income being just Rp720,766 (US\$330) in Minahasa, and Rp462,850 (US\$210) in Gorontalo (Kantor Statistik, Manado 1993). Recent evidence suggests that Gorontalonese people are starting to become hunters, rather than just Minahasans, because it provides more immediate cash than small-scale agriculture.

If law enforcement is to be effective, there is a need for swift prosecutions of offenders and clear guidelines for village police to follow when they catch dealers. Pre-Christmas enforcement measures would be particularly useful in reducing dealer profits because the demand for wild pigs increases considerably just before Christmas. Guidelines must also be drawn up for disposing of confiscated meat without encouraging demand. Economic analysis of the market is an important component of devising an effective law enforcement policy. In a competitive market, law enforcement measures such as fines, which increase the costs of dealing, would reduce the number of dealers and the harvest rate. In a market with barriers to entry, however, increasing costs often have little effect until the supernormal profits are reduced to zero. In this case, raising the entry barrier is a more effective way of reducing harvest rates. For example, if buying a truck is the major barrier to entry, confiscating the truck of a dealer caught with a protected species might be an effective policy.

We have shown here that, with increasing political will, a rapid, considerable, and relatively long-term reduction in the illegal trade in species carried along the Trans-Sulawesi Highway to the markets of Minahasa could be achieved relatively easily. This would benefit the babirusa particularly, but the reduction in dealer profits would be expected to reduce hunting pressure on all the species traded in this way: Sulawesi wild pigs, macaques, bats, and to some extent anoas. However, species that are consumed locally are much less tractable to law enforcement. This category includes maleo and anoas, both listed as endangered on the 1996 IUCN Red List of Threatened Animals, and thus urgently in need of conservation attention.

Wild meat hunting in North Sulawesi is predominantly for commercial gain rather than subsistence. This means that for effective management, it is essential to look not only at the ecology of the species involved, and the effects of hunting on them, but also at the market in wild meat. Until the incentives for people to hunt are fully understood, any policies aimed at controlling the trade will be hit-and-miss in their effects. Our long-term study of the trade in wild meat has led to some insights into the behavior of hunters and dealers. These can be translated into actions which, by controlling the trade, might help to ensure the sustainability of the wild meat harvest in the future.

ACKNOWLEDGMENTS

We gratefully acknowledge the support and cooperation of the following Institutions in Indonesia: the Centre for Research and Development in Biology of the Indonesian Institute of Sciences (L.I.P.I.), the Directorate General for Nature Conservation and Forest Protection (P.H.P.A.), the Indonesian Forestry Society (M.P.I.), Sam Ratulangi University, and regional and local police offices of North Sulawesi. This work was funded by the Darwin Initiative of the UK Department of the Environment, the Wildlife Conservation Society, the Leverhulme Trust, and the People's Trust for Endangered Species. We gratefully acknowledge the support of Fauna and Flora International to L.C. while writing this chapter. Additional support was provided by the British Airways Environment Programme and the British and International Federations of University Women. We thank David Macdonald, Kathy MacKinnon, Widodo S. Ramono, James Richardson, Effendy Sumardja, Tony Whitten, Soetikno Wirjoatmodjo, Bart Wowor, and many Minahasan dealers for their help and advice.

Appendix 24-1. Hunter Wages

Data from 1991 to 1995 are used to make a rough calculation of hunter profits. Dealer X pays Rp11,200 per pig to the hunters. Each dealer has approximately 33 hunters on his books (in 1995, when there were 12 dealers and around 400 hunters operating in North Sulawesi). Dealer X buys an average of 24 pigs per trip, paying Rp268,800. This represents an average weekly wage to a hunter of Rp8,145, and a yearly wage of Rp423,564. The hunters work in Gorontalo district, where the average wage is Rp462,850, similar to the wage calculated for hunters. Thus, the hunters are making normal profits and are likely to be working in a competitive market.

Appendix 24-2. Case Study: Dealer's Profits

Animal	Market Price	Purchase Price	Net Revenue
Wild pigs (market)	45,000	11,200	33,800
Wild pigs (bulk)	22,500	11,200	11,300
Dogs	20,000	5000	15,000
Macaques	15,000	3000	12,000
Bats	1500	500	1000

All prices are in Rp (Rp2200 = US$1) and are means per individual animal. The wild pig prices are for the period during which the dealer's records and market data overlap: January 1993 to July 1994. The other prices are for 1995. The wild pigs are sold partly on the market and partly in bulk to restaurants. There are no data available on the prices of pigs sold in bulk, but we assume a 50% discount because the restaurants may buy half the dealer's cargo each week, thus representing a substantial and reliable purchaser. Note that the dealer achieves a substantial mark-up when selling any species in the market; the prices in the market are three to five times those paid to hunters.

Mean Revenues per Trip from Each Type of Animal

Animal	Number Carried	Value/ Individual	Revenues	% of Total
Wild pigs (market)	12	33,800	405,600	57
Wild pigs (bulk)	12	11,300	135,600	19
Dogs	10	15,000	150,000	21
Macaques	1	12,000	12,000	2
Bats	5	1000	5000	1
Total			708,200	

The number of individuals carried is calculated as the mean number per week over the entire period of the dealer's records. This includes both weeks in which trips were made and weeks without trips. Note that wild pigs form the bulk of the dealer's revenues (76%).

Costs

Vehicle Costs: ach trip costs the dealer approximately Rp89,428 in fuel, parts, maintenance, and amortised vehicle purchase costs (Clayton et al. 1997).

Opportunity Cost of Time: The average wage per week for a person in the nearest alternative employment (owner of a medium-sized shop) is approximately Rp70,000. Each trip involves a week's full-time work, as the dealer leaves home on Monday evening and returns home at 6 AM on Friday. He then sells from his house for a few hours on Friday and at the market all day on Saturday.

Food on Trip and Table in the Market: Rp10,500/trip.

String: The dealer buys string to give to hunters for snares. The hunters receive the string as payment, together with the money received per pig. Each hunter receives approximately Rp100,000 worth of string a year. Assuming 33 hunters per dealer, this gives a string cost per trip of Rp63,460.

Opportunity Cost of Capital: The major capital that the dealer requires is a truck. The cost of a truck is Rp3,300,000. This money could have been invested elsewhere. The rate of return on capital is assumed to be 20% per annum in Minahasa (the bank interest rate). The investment in the truck thus represents an approximate opportunity cost of Rp12,690 per trip.

Super-normal Profits: Mean profits per trip for the case study dealer are calculated as follows:
708200 − 89,428 − 70,000 − 10,500 − 63,460 − 12,690 = 462,122

PART V

Synthesis

25

Hunting for Sustainability:
The Start of a Synthesis

ELIZABETH L. BENNETT AND JOHN G. ROBINSON

This book contains 24 other chapters, written by people from a range of disciplines, and based on work in a considerable diversity of countries, and under different biological, social, and economic conditions. Nevertheless, some clear patterns and themes emerge, and there are surprisingly few contradictions in findings of the different studies. In this chapter, we identify generalizations that arise from the book, and synthesize these with information from the literature to answer our initial questions: Is hunting as practiced in tropical forests today sustainable? Which factors affect sustainability? Which general conditions promote sustainability and damage sustainability? And what can be done to promote sustainable use of tropical forest wildlife in the future?

I. WILDLIFE AND PEOPLE

Investigating the relationship between wildlife and people is not a trivial academic exercise. The use of wildlife in human culture is ubiquitous. And the impact of humans on wildlife is so pervasive that the very survival of many species depends on our understanding and managing that use.

A. Importance of Wildlife to People

1. Wildlife in Tropical Forests Remains an Important Resource for Inhabitants of Tropical Forests. Wild species are hunted (summarized in Bennett and Robinson this volume) (a) for their nutritional value; (b) so that live animals, meat, hides, fur, and other body parts can be sold for cash; (c) for cultural reasons, either to obtain artefacts, or for other cultural purposes such as proving manhood; and (d) for multiple combinations of these.

2. Hunting Remains an Important Social and Cultural Tradition for Most Inhabitants of Tropical Forests. People will hunt even when they have alternative sources of income or nutrition (Bennett et al. this volume; Eves and Ruggiero this volume; Hill and Padwe this volume; Jorgenson this volume; Lee this volume; Madhusudan and Karanth this volume).

3. Wildlife Can Be Important Pests on Agricultural Crops and Livestock. Hunting of wildlife can reduce pest populations and minimize agricultural damage (Jorgenson this volume; Lee this volume). Frequently, such hunted wildlife is subsequently consumed or otherwise used.

B. Impact of Hunting on Wildlife

Although early formulations of sustainability (IUCN/UNEP/WWF 1980) assumed that wildlife could be harvested without significantly affecting the wild population, these were clearly erroneous: Hunting has a significant impact on wildlife populations.

1. Hunting Lowers Population Densities of Hunted Species. For tropical forest game species, hunting is largely additive to natural mortality and reduces population densities (Alvard this volume; Bennett et al. this volume; Eves and Ruggiero this volume; FitzGibbon et al. this volume; Hart this volume; Peres this volume). Redford (1992) estimated that mammal populations are reduced by 70% under light hunting and 95% under heavy hunting.

2. Hunting Can Reduce Average Body Size of Hunted Species. By selecting for large-bodied animals, hunting can decrease average body mass of a population (Bennett 1998).

3. Hunting Can Lower the Average Age of First Reproduction in a Population. Possibly mediated through a decrease in resource or reproductive competition among animals, young animals might mature sooner (Hart this volume).

4. Hunting Can Lead to an Increase in Average Female Fecundity. This is expected theoretically (Caughley 1977; Eltringham 1984) but has not been demonstrated for any tropical forest species.

5. Hunting Reduces the Proportion of Animals in Older Age Classes. Although the mechanisms generating this shift in age structure vary and are poorly understood (Leeuwenberg and Robinson this volume), a general observation is that the proportion of older animals is reduced in hunted populations (Bodmer this volume; Hart this volume; Leeuwenberg and Robinson this volume; Peres this volume).

6. Hunting Can Lead to a Decrease in Future Production of Hunted Populations. Despite possible lowering of the age of sexual maturity and increases in fecundity, tropical forest species densities often decrease to a fraction of the original following hunting, and the proportion of breeding adults is often greatly reduced. This generally contributes to lower future production (Robinson and Redford 1991b). A low production limits the potential for harvest.

7. Hunting Can Locally Extirpate Certain Vulnerable Species. In general, species are remarkably resistant to local extirpation through hunting. However, sustained heavy hunting will extirpate vulnerable species, often large-bodied species with low intrinsic rates of natural increase (Alvard this volume; Bennett et al. this volume; Clayton and Milner-Gulland this volume).

8. Hunting Changes the Size Structure of the Biological Community, Decreasing the Representation of Large-Bodied Species. Selection by hunters of large-bodied animals changes the structure of the biological community, with species in which animals are large bodied becoming relatively less abundant (Hart this volume; Peres this volume).

9. Hunting Changes the Composition of the Biological Community. The representation of different guilds and trophic levels in a community changes through the preference of hunters for certain classes of animals, and these changes can have wide repercussions throughout the forest. The preference for hunting frugivorous and granivorous species (Robinson 1996) can affect patterns of pollination, seed dispersal, and seed predation (Redford 1992). The human hunting of ungulates can reduce the populations of predators that depend on them (Madhusudan and Karanth this volume).

10. Hunting Leads to a Significant Decrease in the Production of the Biological Community. Reductions in the representation of large-bodied species in the community, and in the production of individual species, contributes to the lower production of the community as a whole (Puri 1992; Hart this volume; Stearman this volume).

II. THE SUSTAINABILITY OF HUNTING IN TROPICAL FORESTS

That hunting reduces specific wildlife populations and affects the biological community does not by itself indicate that hunting is not sustainable (Bennett and Robinson this volume). Sustainability is merely "the ability to maintain something undiminished over some period of time" (Lélé and Norgaard 1996). To assess whether a resource harvest is sustainable, one must also define what needs to be maintained, at what scale and in what form, and why one wants to maintain it.

A. Defining Sustainable Hunting

Within the context of hunting in tropical forests, there is a general consensus among authors in this volume on what are the minimum criteria necessary to define sustainability (Bennett and Robinson this volume). That consensus derives from the agreement that the core goal of management ("why one wants to maintain the entity") is the conservation of natural systems, whether for biocentric or anthropocentric reasons. Under these conditions, the following criteria for sustainability can be defined.

1. Harvested Populations Cannot Show a Consistent Decline in Numbers. Following the onset of harvesting, population densities almost always decline. In addition, many species have populations that fluctuate in density over time. However, these short-term declines are distinct from consistent declines over time in a harvested population, which often indicate that the annual harvest of animals from a population is greater than the annual production, after taking into account natural mortality (modeled in Robinson and Redford 1991b). The absence of a consistent decline in animal densities is a sine qua non of the ecological sustainability of harvest (Robinson 1993).

2. Harvested Populations Cannot Be Reduced to Densities Where They Are Vulnerable to Local Extinction. There is a broad consensus that if populations are reduced to low densities, or to low population numbers, they might be unable to recover and are in danger of extirpation. The mechanisms of small population extinction are well understood, and there is a general consensus of when "small is too small" (although perhaps not absolute agreement; see Caughley 1994; Hedrik et al. 1996).

3. Harvested Populations Cannot Be Reduced to Densities Where the Ecological Role of the Species in the Ecosystem Is Impaired. Species occur as part of biological communities. Interactions with other species affect the probabilities of survival of those other species, and the biological integrity of the community as a whole (Robinson and Redford 1991c). Broadly speaking, therefore, species have a role within a biological community. However, even though this criterion is widely used in discussions of sustainability, and indeed is codified in Article IV of the text of the Convention on International Trade in Endangered Species of Wild Flora and Fauna (CITES), until it can be measured unambiguously, it can only be used in a general way (see Bennett and Robinson this volume).

Taken together, these three criteria establish the lower limits on the densities of harvested populations and define the ecologically sustainable levels of harvest, but they do not establish socioeconomically sustainable levels of harvests (Robinson 1993). Recognizing the difficulty of defining human needs and aspirations, the fourth criterion remains rather general, but needs to be included with the other three when hunting sustainability is assessed.

4. Harvested Populations Cannot Be Reduced to Densities Where They Cease to Be a Significant Resource to Human Users. The term *significant* is not defined precisely, but its meaning can be assessed by considering the importance of wildlife to people as specified in the preceding section. The population density of a harvested species required to maintain socioeconomic sustainability will sometimes be higher than that required to maintain ecological sustainability.

B. Is Hunting in Tropical Forests Sustainable Today?

Many of the authors in this book used one or more of the criteria listed above to assess sustainability of hunting, and it is clear that overall, the hunting of many species important to local inhabitants of tropical forests is not presently sustainable. Biodiversity in the forest is declining in many areas, species are being extirpated, and the wildlife resource base on which people depend is declining. Given that people have hunted in tropical forests for many thousands of years, it is clear that at least in historical times, hunting of those species still surviving today must have been sustainable. What has changed? Many of the chapters in this book have identified physical, biological, social, cultural, institutional, and economic conditions that affect the sustainability of hunting, and these conditions are summarized below.

III. PHYSICAL FACTORS AFFECTING SUSTAINABILITY

The physical configuration of human land uses across the landscape influences the extent to which wildlife populations are buffered from hunting pressure.

A. Proximity to "Source" Areas

Proximity of a hunted area to a protected or other "source" area for wildlife increases hunting sustainability (Bodmer and Puertas this volume; Fimbel et al. this volume; Hart this volume; Hill and Padwe this volume; Robinson and Bennett this volume). The hunting area has to be close enough to a source population so that wildlife, through the course of their movements or dispersal, can repopulate the areas hunted.

B. Accessibility

Ease of access by people to a hunted area influences sustainability of hunting. Accessibility is influenced directly by physical factors (distance, relief, physical barriers) and indirectly through human mediation (roads and other transport mechanisms).

　　1. Easy Access of Outsiders to an Area Decreases Local Sustainability. If an area is easily accessible, outsiders can enter the area to hunt, thereby increasing the hunting pressure on the wildlife resources (Auzel and Wilkie this volume; Fa this volume; Fimbel et al. this volume; Madhusudan and Karanth this volume; Noss this volume).

　　2. Proximity of Hunted Areas to Market and Other Commercial Centers Decreases Sustainability. As proximity increases, market hunting tends to increase (Auzel and Wilkie this volume; Bennett et al. this volume; Clayton and Milner-Gulland this volume), communities can easily buy hunting technologies such as cartridges, snare wire, and batteries for night hunting (Bennett et al. this volume; Stearman this volume), and they become more involved in a cash economy so their tendency to sell wild meat to buy commodities increases (Stearman this volume).

IV. BIOLOGICAL FACTORS AFFECTING HUNTING SUSTAINABILITY

The number of tropical forest animals that can be harvested is limited by their supply. Supply is limited by the productivity of each species and the vulnerability or resilience of each species to harvest, in particular, and human activities, in general.

A. Biological Production

Hunting in forests with low production is less likely to be sustainable. Production is the addition to the wildlife population, whether or not it accumulates and survives to the end of the period under consideration (Banse and Mosher 1980). If a population is stable through time, that production is harvested or animals die naturally. In comparison with savanna and grassland ecosystems, forests have a very low production (Robinson and Bennett this volume).

1. Variation in Wildlife Densities and Wildlife Production Across Tropical Forests Is Considerable. In comparisons across habitat types, evergreen forests with closed canopies in general have much lower biomasses of harvestable species than more open, often deciduous forests, and thus invariably lower production (Robinson and Bennett this volume). Regionally, there is considerable variation in wildlife production across forest types: monodominant mbau forests in central Africa have much lower production than biologically diverse forests (Hart this volume), and upland *terra firme* forests in Amazonia have lower production than *varzea* forests (Peres this volume).

2. Hunting Usually Lowers Population Densities to Levels of Less Than Maximum Productivity. Population densities of 65% to 90% of carrying capacity (K) have been suggested as maximal for productivity (Robinson and Redford 1991b), yet in tropical forests even "light" hunting reduces populations, on average, to about 30% of carrying capacity (Redford 1992).

B. Vulnerability or Resilience to Harvest

The harvest of highly vulnerable species is less likely to be sustainable.

1. Species with Low Intrinsic Rates of Population Increase (r_{max}) Are Less Resilient to Harvest. As a group, primates and carnivores tend to have low rates for their body mass, whereas ungulates and rodents tend to have high rates (Bodmer 1995a; Bodmer and Puertas this volume; Clayton and Milner-Gulland this volume; Lee this volume; Peres this volume). However, r_{max} is not the only predictor of vulnerability (O'Brien and Kinnaird this volume).

2. Species Whose Nesting, Predator Avoidance, or Social Behavior Allow Easy Harvest Are Especially Vulnerable to Harvest (Clayton and Milner-Gulland this volume; FitzGibbon et al. this volume; O'Brien and Kinnaird this volume). Many species change their behavior in response to hunting, which lowers the risks (Mitchell and Tilson 1986; Ojasti 1991; Lahm 1993b; Sompud 1996).

3. Species That Are Intrinsically Rare Are Less Resilient to Harvest. Such species are susceptible to demographic instabilities (Hill and Padwe this volume).

4. Species That Have the Ability to Recolonize from Other Source Areas Are More Resilient to Harvest. Species with wide-ranging movements or those characterized by long-distance dispersal are more likely to recolonize hunted areas (Fimbel et al. this volume; Hart this volume; Robinson and Bennett this volume).

5. Species That Favor Disturbed Habitats, Such as Gardens and Secondary Forests, Are Often More Resilient to Hunting. Although being in closer proximity to humans increases mortality, this seems frequently to be offset by the greater productivity of the species in these habitats (Linnares 1976; Wilkie 1989; Alvard this volume; Clayton and Milner-Gulland this volume; Jorgenson this volume; Lee this volume).

V. SOCIAL FACTORS AFFECTING HUNTING SUSTAINABILITY

Social factors act on sustainability primarily by their influence on the rates of harvest of wild species.

A. Human Population Density

Increases in effective human population density decrease the probability that hunting will be sustainable. Human population densities affect sustainability through their impact on harvest needs. Higher human populations increase harvest needs and increase the difficulty of sustaining hunting (Fa this volume; FitzGibbon et al. this volume; Lee this volume). If a goal is biodiversity conservation, theory predicts that the carrying capacity for people depending exclusively on game meat will not greatly exceed one person per km^2, even under the most productive circumstances (Robinson and Bennett this volume). Actual densities of people in tropical forests where hunting is both ecologically and socioeconomically sustainable are usually much lower (Hill and Padwe this volume; Mena et al. this volume; Robinson and Bennett this volume).

1. Increased Growth of Human Population Increases Effective Human Population Density. Populations of people indigenous to tropical forests frequently increase as they come into contact with the broader society. This increase is brought about through a constellation of influences, including the increased production of children through loss of traditional taboos and the need for labor in agriculture, as well as decreased mortality, especially of children because of access to modern medicines (Stearman this volume).

2. Decreased Forest Area Increases Effective Human Population Density. Effective population density increases if people's access to forest is curtailed, either through forest conversion and fragmentation or directly through loss of tenure, usufruct rights, or other social dislocations (FitzGibbon et al. this volume; Hart this volume; Leeuwenberg and Robinson this volume; O'Brien and Kinnaird this volume; Stearman this volume).

3. Increased Immigration of Human Populations Increases Effective Human Population Density. Immigration into tropical forest areas generally increases the number of people that depend on local wildlife resources (Auzel and Wilkie this volume; Eves and Ruggiero this volume; Leeuwenberg and Robinson this volume; Townsend this volume).

4. Increased Sedentarism Increases the Effective Human Population Density Around Settlements. Hunting near settlements results in these locations becoming depleted, and hunters must progressively exploit areas further away (Bennett et al. this volume; FitzGibbon et al. this volume; Stearman this volume). This creates concentric rings of game depletion.

B. Immigration

Immigration into tropical forests areas decreases the probability of hunting sustainability. Increased immigration into tropical forests to obtain land (Hart this volume) or jobs (Auzel and Wilkie this volume; Eves and Ruggiero this volume; Fimbel et al. this volume) is well documented around the world. In addition simply to increasing effective population density, immigration also tends to change the pattern of wildlife harvests.

1. Immigrants Often Hunt a Narrower Range of Species. This focus can deplete populations of these favored species (Stearman this volume; Redford and Robinson 1987).

2. Immigration Brings in People Who Convert the Forest to Other Land Uses. Not only does this decrease forest area, but new immigrants often consider some wildlife species as "agricultural pests" to be extirpated (Lee this volume; Stearman this volume).

C. Sedentarism

Increased sedentarism is frequently associated with the loss of hunting sustainability. Many indigenous people living in tropical forests move across the landscape in response to local wildlife depletion. The increasing sedentarism of such people is frequently the policy of missionaries and government agencies (Eves and Ruggiero this volume; Leeuwenberg and Robinson this volume; Mena et al. this volume; Stearman this volume). Sedentarism allows people greater access to schools, markets, and jobs, but where people hunt, it also leads to local wildlife depletion due to the effective increase in local human population densities. Increased sedentarism is also associated with an increased reliance on swidden agriculture (Stearman this volume). Involvement with agriculture in turn requires greater market involvement, which is associated with wildlife depletion (Robinson and Bennett this volume).

D. Social Differentiation

Increased social differentiation, especially in societies indigenous to tropical forests, decreases the likelihood of hunting sustainability. Because many traditional societies are integrated into market economies, people's desire for material possessions increases, and the society increasingly evinces social and economic distinctions. Where wildlife is an important commodity, this results in increased harvest rates, and frequently this results in loss of sustainability (Stearman this volume).

VI. CULTURAL FACTORS AFFECTING HUNTING SUSTAINABILITY

Cultural factors act on sustainability primarily by influencing the diversity of species harvested and, within a species, the number of animals and age-sex classes harvested. An increased tolerance to harvest a wide array of species or an increased ability to do so might lessen the hunting pressure on other species, but only does so if overall hunting pressure does not increase because of other factors. Increased harvest of a given species, or harvesting a wider range of age-sex classes, will generally tend to detract from sustainability.

A. Hunting Practices

Changes in traditional hunting practices generally decrease the probability that hunting will be sustainable. Traditional hunting practices are more likely to be sustainable, especially when part of the repertoire of human groups who have lived for millennia in tropical forests. Many of these practices were adopted by people moving into the forest in historical times. But hunting practices are changing. Generally, these changes decrease the probability that hunting will be sustainable, although not always (Ayres et al. 1991).

1. Social Taboos Against Hunting Certain Species Are Breaking Down. Social taboos often proscribe the harvest of certain species (Bennett et al. this volume; Mena et al. this volume) or prescribe harvest only under certain conditions (Mitchell and Tilson 1986; Hill and Padwe this volume). These taboos can be widespread across a region or limited to a certain ethnic group, clan, or family. Taboos can be formal, as in the case of many indigenous societies, or informal, as found in many other forest-dwelling peoples (Redford and Robinson 1987).

2. The System of Traditional Hunting Territories in Many Areas Is Disappearing. Hunting territories belonging to specific hunters or families are characteristic of many traditional groups and serve to disperse hunting pressure and assign stewardship to a specific person or group of people. The loss of this system allows more open access to wildlife resources, with concomitant declines in the probability of hunting sustainability (Bodmer this volume; Eves and Ruggiero this volume; FitzGibbon et al. this volume; Hart this volume).

3. Traditional Hunting Methods Such as Hunting Zone Rotation, Outlier Camps, and Trekking Are Decreasing. These practices serve to disperse the impact of hunting across the landscape (Leeuwenberg and Robinson this volume; Stearman this volume). Their abandonment tends to concentrate hunting and to lead to local extirpation and depletion of game species.

4. Many Recent Immigrants into Tropical Forests Have Hunting Practices Derived from Different Social and Natural Environments. This lack of experience and knowledge frequently leads to unsustainable hunting practices (Eves and Ruggiero this volume; FitzGibbon et al. this volume). Recent immigrants also often do not have the local experience and knowledge, or the long-term commitment to the area for future generations.

B. Hunting Technology

Advances in hunting technology generally increase harvests and decrease the sustainability of those harvests. The advent of new hunting technology in general serves to increase the diversity of species harvested and, within a species, the number and age-sex class of animals harvested.

1. Use of Wire Snares Increases Indiscriminate Harvest. In comparison with many traditional hunting methods (Hart this volume), wire snares are not as selective, taking a wider variety of species and sizes of animals (Lee this volume; Noss this volume; but see Madhusudan and Karanth this volume). If snares are not regularly checked, there is a great deal of waste, and killed animals must be discarded (Noss this volume). Animals that escape from snares are frequently wounded, and presumably have a higher probability of subsequent death. All of these factors lead to overall higher mortality rates and a decreased probability of sustainability.

2. *Use of Firearms and Other Hunting Technologies Increases Hunting Efficiency.* Firearms are the tool of choice for hunting in tropical forests and have greatly improved the ease of wildlife harvest, especially of canopy-dwelling species (Yost and Kelley 1983; Mena et al. this volume). Access to flashlights, shotgun cartridges, batteries, outboard motors, gasoline, and so forth all increase efficiency. Whether this increased efficiency results in increased harvests, decreased time spent hunting, or a mixture of both is debated (Chagnon and Hames 1979; Hames 1979; Yost and Kelley 1983; Mena et al. this volume), but it is clear that the use of firearms increases the range of species taken (Eves and Ruggiero this volume; Mena et al. this volume; Stearman this volume), increases the injury rate of hunted animals (Leeuwenberg and Robinson this volume; Townsend this volume), and allows additional hunting for commercial markets (Bennett et al. this volume; Bodmer et al. this volume).

3. *Use of New Hunting Technology Requires Money for Purchase, and This Can Increase Harvest Rates.* The need for money to buy new hunting technologies encourages forest-dwelling people to participate in the market economy. Where the money for these purchases comes from the sale of wildlife (Bodmer and Puertas this volume; Stearman this volume), the use of new hunting technology increases wildlife harvests.

C. Hunting Proficiency

Loss of hunting proficiency by rural inhabitants favors lower levels of hunting. The decline in hunting can lead to increases in the sustainability of hunting. The taking of money-earning jobs by hunters can lead to loss of hunting skills (Jorgenson this volume). This is turn can lead to decreasing levels of hunting.

D. Use of Domestic Animals

The use of domestic animals can decrease the harvest of wild animals and increase the probability that a parallel wild harvest will be sustainable. Consumption of canned meat and meat from domestic animals and aquaculture can decrease reliance on meat from wild species (Fa et al. this volume; Jorgenson this volume; Lee this volume). However, the acceptability of domestic animals as a substitute for animal protein of wild origin varies with region. In Asia and Africa (King 1994; McRae 1997), a broad generalization is that wild meat is preferred over domestic meat by rural communities. Indeed, in certain groups, domestic meat is considered unclean

(Wildlife Conservation Society and Sarawak Forest Department 1996). In general, the reverse is true of rural communities in Latin America (Castro et al. 1976; Jorgenson this volume). In addition, people do not only hunt to secure animal protein. Hunting is frequently culturally and socially important, and it continues even if people have access to domestic animals (Bennett et al. this volume; Fimbel et al. this volume; Hill and Padwe this volume; Stearman this volume).

E. Organized Religions

Religious traditions, prescriptions, and proscriptions on the eating of wildlife affect harvest rates. Islam, Buddhism, and Hinduism, but generally not Christianity, have religious laws governing the killing and eating of wildlife. These laws constrain consumption rates of wildlife in certain regions, affecting harvest rates and increasing the probability that hunting is sustainable (Alvard this volume; FitzGibbon et al. this volume; Lee this volume).

VII. INSTITUTIONAL FACTORS AFFECTING HUNTING SUSTAINABILITY

Institutions, whether international, national, regional, or local, act to establish rules of conduct and to regulate relationships among other institutional actors. Here we are concerned with those institutions that designate the uses of the landscape, establish guidelines for the management of those areas, and regulate the management of these areas and the use of wildlife resources.

In most countries, national governments act as proprietors of wildlife, with wildlife property rights being ceded to varying extents to more local authorities, landowners, and other selected users. In general, management will be most effective where the scale of the management goals, the capacity of the institution to act at that scale, and the resource regimen are concordant (Naughton-Treves and Sanderson 1995).

A. National and Regional Government

National governments are often effective at managing wildlife resources where the aim is to minimize harvest of wildlife and control the commercial trade of wildlife products (Clayton and Milner-Gulland this volume; Madhusudan and Karanth this volume).

1. The Establishment of Protected Areas by National Governments Serves to Create Protected Sources for Wildlife Populations. Protected areas, parks, and reserves act as reservoirs or refuges for wildlife. If adequately protected or if they are inaccessible, protected areas can be a *source* of wildlife, replenishing stocks overharvested in adjacent lands (Bodmer and Puertas this volume; Fimbel et al. this volume; Hart this volume; Hill and Padwe this volume; Robinson and Bennett this volume).

2. The Imposition of Blanket Bans on the Sale of Wildlife, or Hunting for Sale, Through National Laws and Enforcement Can Decrease Overall Wildlife Harvests. Banning the commercial sale of wildlife can ensure that the hunting that does exist is sustainable, but it is only effective if there is adequate enforcement (Clayton and Milner-Gulland this volume). If there is not, banning just makes hunting or sale technically illegal and renders further management difficult.

3. The Regulation of the Commercial Trade in Wildlife Through National Laws and Enforcement Can Decrease Overall Wildlife Harvests. To have this effect, there must be adequate enforcement (Eves and Ruggiero this volume; FitzGibbon et al. this volume; Madhusudan and Karanth this volume). But even sporadic enforcement can have a lasting regulatory effect (Clayton and Milner-Gulland this volume).

B. Local Government

Local authorities are often effective at managing wildlife resources where the aim is making hunting more sustainable (Leeuwenberg and Robinson this volume). Managing wildlife in tropical forests so as to achieve sustainable hunting occurs at a scale where national agencies are often impotent and where local institutions are often more cost effective and efficient. They can also draw on local and traditional knowledge and management structures (Bodmer and Puertas this volume; Leeuwenberg and Robinson this volume; Townsend this volume):

1. *The establishment of protected areas by local government serves to create protected sources for wildlife populations* (Bodmer and Puertas this volume).
2. *The management of location and intensity of hunting can serve to disperse hunting pressure* (Leeuwenberg and Robinson this volume, Townsend this volume).

C. Scientific Institutions

Scientifically monitoring the results of management contributes to hunting sustainability (Lee 1993). Information on the impact of different management decisions can inform management strategies, a process known as adaptive management. Scientific institutions, and scholarship in general have been very active in informing and furthering wildlife management by local governments and communities (Bodmer and Puertas this volume; Leeuwenberg and Robinson this volume).

D. Religious Institutions

The pacification and settlement of indigenous peoples by religious institutions can decrease the sustainability of hunting by indigenous communities. Religious institutions involved in proselytizing to indigenous people foster increased sedentarism, social differentiation, loss of traditional taboos, and involvement in local market economies, all of which contribute to a decline in the sustainability of hunting (Stearman 1994; Stearman and Redford 1992; Stearman this volume).

VIII. ECONOMIC FACTORS AFFECTING HUNTING SUSTAINABILITY

A. Commercialization

Commercialization of wildlife trade decreases sustainability of hunting. Increased commercialization of the wild meat trade from tropical forests increases the number of consumers, increases the amount of hunting, increases the entrance of nonresident (often commercial) hunters into a region, and increases the incentive to hunt. Hunting effort or intensity increases, although harvests often fall off as wildlife resources are exhausted. Wildlife populations are decreased to levels where harvests are not economically sustainable (Clayton and Milner-Gulland this volume; Eves and Ruggiero this volume; Fimbel et al. this volume; Hart this volume; O'Brien and Kinnaird this volume; Stearman this volume).

1. Commercialization Increases Hunting Intensity by Local People. Involvement of local people in commercial wildlife hunting can dramatically increase wildlife harvest rates (Bodmer and Puertas this volume; Griffin and Griffin this volume; Hart this volume).

2. Commercialization Increases Entrance of Nonresident, Commercial Hunters into a Region. Involvement of outsiders in commercial wildlife harvests can increase harvests within an area (Marsh and Gait 1988; Bodmer and Puertas this volume; Hart this volume).

3. Commercialization Can Increase Hunting Intensity on Rare Species. If a commercial harvest is maintained economically by more common species, then hunting for the trade can simultaneously extirpate rare species (Clayton and Milner-Gulland this volume).

4. Commercialization Can Increase Hunting of Noncommercial Species. If local people are involved in the commerce of marketable species, they often turn to other noncommercial species for their own subsistence (Bodmer and Puertas this volume). Hunting of these species might not be sustainable (e.g., primates; see Bodmer and Puertas this volume). If the large species in an area are reduced through commercial trade, subsistence hunters increasingly hunt smaller, less resilient species (Bennett and Gumal in press).

5. Commercialization, Where Wild Meat Is Preferred, Results in an Increased Skew of Consumption by the Wealthy and an Increasing Demand for Wild Meat. As a commercial market for wild meat develops, hunters frequently sell the preferred species or pieces, and themselves consume the less preferred species or pieces (Bodmer and Puertas this volume; Eves and Ruggiero this volume). Demand for these preferred items, frequently in urban centers, can lead to increased harvest rates.

6. Commercialization, Where the Hunting Communities Are Newly Entering the Market Economy, Can Result in Debt Peonage. Access to capital allows traders to supply new hunting technology (e.g., guns, wire snares, flashlights, etc.) to hunters, who frequently remain in continuous debt to the traders (Clayton and Milner-Guland this volume; Hart this volume; Stearman this volume). This debt peonage serves to increase hunting intensity.

B. Market Value of Wildlife

Increased economic value of wildlife decreases the sustainability of hunting. Animals of species with very high market value are more likely to be overhunted. These tend to be species that are important in the international trade (Eves and Ruggiero, this volume; Fimbel et al., this volume;) or have a high cultural significance (such as hornbills; see Bennett et al. 1997). In contrast there is little variation in the value of animals hunted and sold for their meat. Although there is some variation in per kg prices of meat in local markets reflecting local gustatory preferences (Fa et al. this volume), meat is meat (Auzel and Wilkie this volume; Clayton and Milner-Gulland this volume; Lee this volume), and differences in market prices probably have little direct effect on hunting pressure.

C. Income of Hunters

Increases in hunter income, if reinvested in improving hunting technology, can decrease the sustainability of hunting. Buying firearms, cartridges, batteries, outboard motors, and fuel all increase hunting efficiency, and can thus increase harvests.

1. Income of Hunters Can Be Generated by the Commercial Sale of Wildlife to Local Markets. Increased involvement in the commercial sale of wildlife can allow hunters to purchase new hunting technologies (Bennett et al. this volume; Bodmer and Puertas this volume; Stearman this volume), which allow higher harvest rates.

2. Income of Hunters Can Be Generated by Men Temporarily Leaving Communities to Take Jobs. In many rural communities, men leave their communities for periods of time to take jobs, often in extractive industries like logging (Auzel and Wilkie this volume; Bennett et al. this volume), and oil fields (Mena et al. this volume). They return with the means to improve their hunting technology.

D. Income of Consumers

Changes in consumer income and buying power have different effects on the consumption of wild meat, depending on the region. Because commercialization of wildlife tends to decrease the sustainability of hunting, any change in consumption of wild meat brought about by changes in consumer income has a direct impact on sustainability.

1. Increased Income of Consumers Increases Consumption of Wild Meat in Asia and Africa. In Asia and Africa (Balinga 1977; Srikosamatara et al. 1992; King 1994; McRae 1997), a broad generalization is that wild meat is preferred by consumers because of its taste over domestic meat. Accordingly, increased income of consumers at the local level leads to increased commercial demand for wild meat (Auzel and Wilkie this volume; Lee this volume), which often leads to a decrease in the sustainability of hunting.

2. Increased Income and Buying Power of Consumers Decreases Consumption of Wild Meat in Latin America. In Latin America (Castro et al. 1975–1976; Bendayán 1991; TCA 1995), a broad generalization is that domestic meat is preferred by consumers in rural areas, although there is a speciality market for wild meat in larger towns. Increased income of consumers at the local level leads to a decreased demand for wild meat (Ayres et al. 1991), which can increase the sustainability of local hunting.

IX. APPROACHES THAT ENHANCE HUNTING SUSTAINABILITY

The preceding analysis was essentially reductionist—isolating specific factors that have a definable impact on the sustainability of hunting in tropical forests. In reality, the broader biological, social, cultural, economic, political, and institutional context means that many of these factors combine and act simultaneously. Accordingly, we consider here general approaches, the aim of which is generally perceived to be that of achieving hunting sustainability. These general approaches are defined by constellations of the factors identified above. Each factor identified here cross-refers to the discussion above by section number (e.g., I.A.1).

A. *National Parks and Wildlife Sanctuaries*

The establishment of totally protected areas (national parks, wildlife sanctuaries) and multiple use areas (indigenous reserves, forest reserves, wildlife reserves, etc.) allows a national government to regulate land uses. All serve either to prohibit hunting, or to regulate or control access to wildlife (Madhusudan and Karanth this volume). Many national parks in tropical forests contain significant human populations, but the existence of the protected area can enhance the sustainability of hunting for the following reasons:

- It acts as a "source" of wildlife for nearby hunting communities (III.A). If contiguous to lands that are hunted, the protected area serves as a refuge for wildlife where populations can maintain high densities and can replenish nearby areas that have been more heavily hunted (VII.A.1).
- Protected area regulations limit effective human population density and immigration into the area (V.A.3, V.B.1).
- Protected area regulations often limit conversion of forest to other land uses (V.B.2).

All of these factors serve to increase the sustainability of hunting within and around the protected area. For this approach to be effective, however, national agencies must be able to enforce land use regulations. "Paper parks" with little government authority often are worse than no protected area at all. If national governmental presence is absent, and if regional and local institutions do not take their place, then there is little regulation or control of access to the protected areas (FitzGibbon et al. this volume). National governments also frequently lack the training and expertise or do not allocate the financial resources to establish their presence in protected areas in tropical forests.

B. Wildlife Management at a National Level

National governments also can seek to manage the hunting of wildlife at a national level. National agencies have been most successful when managing sport hunting (Geist 1988), in which harvest rates can be kept low. They have been less successful at managing subsistence or commercial hunting, and the general approach in most countries has been to prohibit hunting, either in general or for selected species. If managed well, national governments can enhance the sustainability of hunting through

- controlling access of outsiders to key wildlife areas (III.B.1);
- controlling or banning the movement and trade of wildlife (III.B.2; VII.A.3);
- establishing and enforcing laws against hunting the species most vulnerable to hunting (IV.B);
- establishing and enforcing laws on use of different modern hunting technologies (VI.B);
- establishing programs to provide domestic animals for rural protein needs (V.D);
- establishing and maintaining protected areas (VII.A.1; IX.A);
- imposing blanket bans on hunting (VII.A2).

To be effective, however, all such regulations must be enforced. This means that the institutional capacity, staffing levels, training, motivation, and back-up from the judiciary or other local mechanisms must be present. At present, any or all of these are lacking in many tropical forest countries.

C. Community-Based Wildlife Conservation

Devolution and decentralization of authority over wildlife resources from national governments to local communities is presently in vogue (Western and Wright 1994; IIED 1994; Lutz and Caldecott 1996). Although some of the impetus for this derives from concerns of social justice and equity, some of it also is a recognition that local communities are often more effective managers of wildlife resources. Much of this devolution and decentralization of authority to local governments over wildlife tends to be externally initiated and imposed by national governments, conservation organizations, and scientific institutions (Bodmer and Puertas this volume), but often the local community co-opts the process (Leeuwenberg and Robinson this volume). Where successful, community-based management can enhance sustainability of hunting through:

- restricting the access of outsiders to the resources of the community, thereby lowering the effective human population density (V.A.);

- decreasing immigration into the region (V.A.3);
- strengthening traditional systems of resource exploitation (V.B.1; VI.A);
- resisting increased commercialization and the resulting social stratification (V.D; VIII.A);
- establishing local protected areas (VII.B.1);
- dispersing the location and intensity of hunting (VII.B.2).

Community-based management is less successful where the wildlife resources are not adequate to support hunting (Stearman 1994), a situation common in low-productivity tropical forests (Robinson and Bennett this volume) and where community access to adequate land is limited (Townsend this volume). Communities also must have the capacity to manage their land or wildlife resources, and this depends on appropriate cultural traditions; adequate political, legal and economic power; and strong community institutions. In today's complex social, political, and economic landscapes, few communities by themselves are likely to have this concatenation of circumstances (Rettig et al. 1989).

D. *Integrated Conservation and Development Projects*

The intent of integrated conservation and development projects (ICDPs), in the words of Brandon and Wells (1992), is to promote "protected area conservation," and "achieve this by promoting socioeconomic development and providing local people with alternative income sources which do not threaten plants and animals within the protected area." ICDPs should promote sustainable hunting by

- restricting the access of outsiders to the resources of the community, thereby lowering the effective human population density (V.A);
- decreasing immigration into the region (V.A.3);
- strengthening traditional systems of resource exploitation (V.B.1; VI.A);
- resisting increased commercialization and the resulting social differentiation (V.D; VIII.A);
- establishing local protected areas (VII.B.1);
- dispersing the location and intensity of hunting (VII.B.2);
- requiring traditional hunting practices and techniques in managed areas (VI.A; VI.B);
- establishing livestock schemes to meet the animal protein needs of people (VI.D);
- establishing protected areas (VII.A.1; VII.B.2);
- regulating commercial trade in wildlife (VII.A.3);
- drawing up scientific and technical expertise (VII.C);
- providing people with alternative sources of livelihood so that their dependence on wild meat for protein and income is reduced (VI.D).

Many ICDPs in tropical forests allocate rights to harvest wildlife to local communities as a means to help achieve local socioeconomic development (Wells and

Brandon 1992). Unfortunately, examples of sustainable socioeconomic development through increased wildlife harvest in tropical forests are nonexistent. The increased involvement of people in the market economy, and specifically the commercial sale of wildlife, can deplete wildlife populations (VIII.A). Changes in hunting practices and technology can exacerbate the situation (VI.A; VI.B). Immigration of people, drawn by the increased economic opportunities leads to increases in effective human population density (V.A.3). All of these factors decrease the probability that hunting will be sustainable.

X. CONDITIONS THAT DETRACT FROM SUSTAINABILITY

There are also general socioeconomic conditions and approaches that make it difficult to achieve hunting sustainability.

A. Tropical Forest Logging Operations

Although logging tropical forests changes the relative abundance of individual species within the biological community, logged forests can continue to support significant wildlife populations (Johns 1997). Populations of some species preferred by hunters (deer, pigs, and peccaries) actually increase following logging (Bennett and Dahaban 1995; Frumhoff 1995), although generally this is a function of logging intensity. Nevertheless, activities associated with logging significantly increase harvest rates and decrease the probability that hunting is sustainable. This is because

- logging roads increase access to the forest (III.B);
- logging operators themselves frequently subsist on wild meat, so their presence increases the human population density in the area (V.A);
- generally being outsiders to the area, the loggers often do not follow traditional hunting practices (VI.A);
- they receive a cash income, so they have the money to buy modern technologies for hunting (VI.B)
- logging opens up the forest to human exploitation, so it frequently spawns an associated commercial wild meat trade, which greatly reduces sustainability (VIII.A).

All of these factors combine to cause a decline in densities of wildlife (Bennett and Dahaban 1995) and declining hunting success of local people (Marsh and Gait 1988).

B. Road Construction

Road building or road improvement generally decreases sustainability of hunting by

- increasing access to forest areas to outsiders (III.A.1);
- increasing effective proximity of the forest to market areas (III.A.2);
- frequently leading to increased immigration to the forest area (V.A; V.B);
- allowing local peoples the opportunity to commercialize the wildlife harvest (VIII.A);
- allowing access to hunting technologies that facilitate indiscriminate and excessive hunting (VI.B).

Nevertheless, road building does not invariably have these effects, as has been demonstrated by Ayres et al. (1991) in their study of the impact of a road to a community in the Brazilian Amazon. Roads also can contribute to effective enforcement of wildlife management (Madhusudan and Karanth this volume).

C. Government Translocation

Government efforts to move people into tropical forests so as to relieve "surplus" populations elsewhere (projects such as the Transmigration in Indonesia and the POLAMAZONIA in Brazil) decrease the probability of any hunting being sustainable in local areas as follows

- by greatly increasing human population densities in the area (V.A);
- being outsiders, transmigrants are unlikely to follow traditional hunting practices local to the area (VI.A).

D. "Pacification" and Sedentarization of Indigenous People

Programs, either sponsored by national government or evangelical organizations, for material or spiritual reasons, to pacify and sedentarize indigenous people can destroy the sustainability of hunting in local contexts by

- providing increased access between the community and outside (III.B);
- increasing the effective local human population density (V.A.4);
- increasing the reliance on swidden agriculture (V.C.1);
- increasing social stratification (V.D);
- often breaking down taboos against hunting certain species (VI.A.1);
- breaking down the system of traditional hunting territories (VI.A.2);
- decreasing hunting zone rotation, outlier camps and trekking (VI.A.3);
- often providing improved technologies which can be used for hunting (VI.B).

The only way in which the effect of these factors will be reduced would be if the use of domestic animals is encouraged, to decrease the reliance on wild meat. Usually this does not occur, at least until wild meat supplies have been reduced to the point that they no longer provide sufficient protein, at which point other alternatives are sought (Bennett et al. this volume) or people become increasingly malnourished (Stearman 1994).

E. Commercial Trade in Wild Meat

The market in fresh and dried meat is much more extensive in Africa and Asia than in Latin America, possibly largely because of a cultural preference for wild meat over domestic meat in the former two continents. This greatly reduces sustainability as follows:

- by greatly increasing the number of consumers eating meat from an area of forest, the effective human population density is greatly increased (VI.A), even if they do not live in situ;
- commercialization of the trade increases social stratification among local hunting communities (V.D);
- the trend toward commercialization and cash economies causes breakdowns in traditional hunting practices (VI.A);
- the cash income from the trade permits the purchase of modern hunting technologies (VI.B.1; VI.B.2). In turn, this requires more wildlife to be sold to replenish supplies (VI.B.3; VIII.C.1);
- commercialization increases hunting levels, entry of outside commercial hunters, and intensity of hunting on rare species even if they are not sold commercially (VIII.A).

F. Commercial Trade in Wildlife Skins

The commercial trade in wildlife skins, leather, fur, and feathers has a more extensive history in Latin America than in Africa and Asia. The reasons why the trade decreases sustainability are similar to those for the commercial trade in wild meat: increasing social stratification amongst local hunting communities, breakdowns in traditional hunting practices, the cash income from the trade permits the purchase of modern hunting technologies, and increasing hunting levels on target species. The trade is characterized by "boom-and-bust" cycles—increases in the commercial value of a product are followed by overhunting; wildlife populations then decline until the commerce is no longer economically viable (Iriarte and Jaksíc 1986; Bodmer et al. 1990a; Redford and Robinson 1991). The history of the commercial trade in wildlife products is a history of nonsustainability; arguably there is not a single example of a commercial trade in wildlife products from tropical forests that has been sustainable over a significant period of time (but see Bodmer et al. 1988b for a description of a 40-year commercial harvest of peccaries).

XI. TOWARD SUSTAINABILITY OF HUNTING IN TROPICAL FORESTS

We posed two questions at the beginning of this book:

- Are hunting rates as practiced by rural peoples sustainable?
- If not, what are the biological, social, and cultural implications of this?

The content of the ensuing chapters show that in tropical forests throughout the world today, hunting rates for many species generally are clearly not sustainable. Unless governments, aid agencies, rural planners, and conservation organizations working in tropical forest areas recognize this fact, the situation will continue to deteriorate. The implications are manifold, both for biodiversity conservation and the well-being of rural communities and economies. This volume has identified many of the factors that affect whether hunting is sustainable or not, and specified what approaches should be followed to enhance sustainability. Only if the fundamental issue of hunting becomes central to all planning for tropical forest areas will the forests and their wildlife be conserved, and will rural populations in many tropical areas be sustained into the next century.

Appendix

Calculating Maximum Sustainable Harvests and Percentage Offtakes

JOHN G. ROBINSON

A number of researchers in this volume have used Robinson and Redford's (1991b) production model to calculate the sustainability of harvests. This model can provide a first assessment of sustainability in the absence of detailed information about the demographic structure of hunted populations and the impact of hunting on that structure (data that are required for most sustainability models). The approach depends on estimating population production, which is then compared with actual harvests to obtain a measure of sustainability (production is defined as the addition to the population through births and immigrations during a specified time period, whether the animals survive, emigrate, or die during the period (Banse and Mosher 1980).

The approach adopted by Robinson and Redford (1991b) was to calculate the maximum possible production of a population, and then compare this with the actual harvest. This model allows managers to evaluate whether an actual harvest is not sustainable (whenever harvest exceeds maximum possible production) but not whether an actual harvest *is* sustainable (because maximum possible production might not have been attained in the specific case).

If in a specific case, the density of a hunted population is known, then this approach estimates maximum annual production *at this density* ($P_{max(D)}$) by multiplying the observed density (D) by the maximum multiplication rate of the population (λ_{max}) and holding the population stable through time. Thus:

$$P_{max(D)} = (D \times \lambda_{max}) - D = (\lambda_{max} - 1)D$$

If the observed harvest exceeds the maximum annual production at this density, then it is not sustainable. λ_{max} is the maximum finite rate of increase and is the exponential of the intrinsic rate of natural increase (e^r), which in turn can be estimated using Cole's equation:

$$1 = e^{-r_{max}} + be^{-r_{max}(a)} - be^{-r_{max}(w+1)}$$

where a is the species-specific age of first reproduction, w is the age of last reproduction, and b is the annual birth rate of female offspring. These reproductive parameters are available for many forest species that are commonly hunted (Robinson and Redford 1986b), although they are often derived from captive animals.

This calculation makes the assumption that hunted populations in tropical forests have the capacity to exhibit maximum rates of increase (λ_{max}). This is not unreasonable if hunted populations are reduced to densities at which intrapopulation competition does not influence population growth rates. This might often be the case. Redford (1992), based on comparisons with unhunted sites, estimated that mammal populations in tropical forests are significantly reduced even under light hunting.

If density of a hunted species is not known in a specified area, then the Robinson and Redford model can estimate maximum annual production by assuming that maximum production would be achieved when the population density was at 0.6K (thus assuming density dependence), where K was assumed to be the density of a general unhunted, undisturbed population (Robinson and Redford 1986a). Maximum production could be calculated by multiplying the density at the point of maximum production by the maximum multiplication rate of the population, and holding the population stable through time. Thus:

$$P_{max} = (0.6K \times \lambda_{max}) - 0.6K = (\lambda_{max} - 1)\, 0.6K$$

This formulation allows an estimate of maximum production even in the absence of a reliable measure of population density or demography in a given area. If the observed harvest exceeds the maximum annual production, then it is not sustainable.

This method of calculating population growth rate assumes no prereproductive or adult mortality (Hayssen 1984), so to estimate the proportion of that production that can be harvested by human hunters, Robinson and Redford (1991b) suggested using the average life span of a species as an index of the number of animals that would have died in the absence of human hunting. They suggested that with very short lived species, natural mortality is high and human harvest can take a higher proportion of the production without reducing the standing population. They proposed that human harvest might be able to take 0.6 of the production in very short lived species (those whose age of last reproduction is less than five years), 0.4 of the production in short-lived species (those whose age of last reproduction is between 5 and 10 years), and 0.2 of the production in long-lived species (those whose age of last reproduction is over 10 years). This modification of λ_{max} by a factor f_{RR} of 0.6, 0.4, or 0.2 (Slade et al. 1998) means that the effective rate of population growth λ_{RR} is calculated as follows:

$$\lambda_{RR} = 1 + (\lambda_{max} - 1) f_{RR}$$

The maximum possible production (and maximum possible sustainable harvest) is calculated as follows:

$$P_{RR} = (\lambda_{RR} - 1) D$$

where D is either a site-specific estimate or equal to $0.6K$. These multiplicative factors, f_{RR}, are ad hoc, and Slade et al. (1998) suggest that they tend generally to overestimate growth rates and maximal production. These authors suggest some alternative computations of λ that depend on more realistic estimates of prereproductive and adult survival.

Nevertheless, use of λ_{RR} has some currency in the literature, and the approach has been used previously to propose maximum possible sustainable harvests for a number of tropical forest mammal species (Alvard 1993a, 1993b; Fa et al. 1995; FitzGibbon et al. 1995b, 1996; Noss 1995; Alvard et al. 1997; Wilkie et al. 1998; Muchaal and Ngandjui 1999). Predictions based on this approach are generally consistent with other indicators of sustainability (Robinson and Redford 1994b).

This approach also can be used to calculate the maximum percentage of sustainable offtakes from wildlife populations. The percentage offtake from a population is the percentage of the standing population or biomass that is harvested annually. The magnitude of the sustainable offtake depends in a specific case largely on the interaction between nonhunting and hunting mortality rates, as well as on whether hunting mortality is additive or compensatory (Caughley 1985). Calculation of the maximum percentage sustainable offtake rates has been attempted in well-studied populations of temperate wildlife species, using models that depend on considerable knowledge of the numbers and demographic structure of the population. For species with significant nonhunting annual mortality and high rates of population increase, suggested offtake rates are a significant proportion of the population, being in the 30% to 50% range for many temperate ungulates (Crete et al. 1981; Gore et al. 1985; Adams 1985). Much lower offtake rates are suggested for more "K-selected" species, for example, less than 1.6% of the population of adult female polar bears (Taylor et al. 1987).

In the absence of detailed demographic information in tropical forest wildlife species, a first approximation of the maximum percentage of sustainable offtakes can be derived from estimates of population growth rates:

$$\text{maximum \% sustainable offtake} = (\lambda_{RR} - 1) \times 100.$$

Maximum percentage sustainable offtakes for a variety of African and Latin American species, calculated by a number of different investigators are presented in table A-1. For species with very short life spans, predicted maximum percentage sustainable offtakes can be a high proportion of the standing population. For ungulates, with short life spans, they fall within the range of many temperate species, with much of the variation arising from a lack of concensus on the value of λ for many of the duiker species. However, for species with long life spans and low rates of population increase, such as primates, maximum percentage sustainable offtakes are generally very low.

Table A-1. Maximum Percentage Sustainable Offtakes, Derived Theoretically from Effective Rates of Population Increase (λ_{RR}), for Some Tropical Forest Mammal Populations

Species	% Offtake	Reference
Insectivores		
Petrodomus tetradactylus	70	FitzGibbon et al. 1996
Rhynchocyon chrysopygus	34	FitzGibbon et al. 1996
Edentates		
Dasypus novemcinctus	40	Robinson and Redford 1991b
Rodents		
Heliosciurius and *Funisciurus* spp.	55	FitzGibbon et al. 1996
Atherurus africanus	13	Noss this volume
Atherurus africanus	16	Fa et al. 1995
Cricetomys emini	40	Fa et al. 1995
Dasyprocta spp.	80	Robinson and Redford 1991b
Agouti paca	19	Robinson and Redford 1991b
Primates		
Mandrillus leucophaeus	4	Fa et al. 1995
Piliocolobus pennanti	3	Fa et al. 1995
Colobus satanus	4	Fa et al. 1995
Cercopithecus erythrotis	3	Fa et al. 1995
Cercopithecus mitis	2	FitzGibbon et al. 1996
Cercopithecus nictitans	2	Fa et al. 1995
Cercopithecus pogonias	>1	Fa et al. 1995
Cercopithecus preussi	2	Fa et al. 1995
Papio cynocephalus	3	FitzGibbon et al. 1996
Cebus apella	3	Robinson and Redford 1991b
Alouatta spp.	3	Robinson and Redford 1991b
Ateles spp.	3	Robinson and Redford 1991b
Lagothrix lagothricha	3	Robinson and Redford 1991b
Pholidota		
Phataginus tricuspis	20	Fa et al. 1995
Ungulates		
Cephalophus monticola	25	Fimbel et al. this volume.
Cephalophus monticola	25	Fa et al. 1995
Cephalophus ogilbyi	10	Fa et al. 1995
Red duikers[a]	10	Fimbel et al. this volume
Duikers[b]	7	Noss this volume.
Duikers[c]	5	FitzGibbon et al. 1996
Tapirus terrestris	4	Robinson and Redford 1991b
Tayassu pecari	26	Robinson and Redford 1991b
Tayassu tajacu	50	Robinson and Redford 1991b
Mazama americana	20	Robinson and Redford 1991b
Mazama gouazoubira	25	Robinson and Redford 1991b

[a]Includes *Cephalophus nigrifrons, C. leucogaster, C. dorsalis,* and *C. callipygus.*

[b]Includes *C. monticola, C. callipygus,* and *C. dorsalis.*

[c]Includes *C. monticola, C. natalensis, Neotragus moschatus,* and *Sylvicapra grimmia.*

References

AASANA (Administración de Aereopuerto y Servicios Auxiliares a la Navigación Aerea). 1992. *Datos meterológicos de la estación de ASAANA en Trinidad.* Trinidad, Bolivia: Administración de Aereopuerto y Servicios Auxiliares a la Navigación Aerea.

ACCT/CERDOTOLA. 1987. *Langues et Dialectes.* Congo: author.

Adams, N. E. Jr. 1985. "Deer Harvest Management, Welder and McCan Ranch, Texas." In S. L. Beasom and S. F. Roberson, eds., *Game Harvest Management,* pp. 165–74. Kingsville, TX: Caesar Kleberg Wildlife Research Institute.

Adams, R. E. W. 1991. *Prehistoric Mesoamerica* (revised edition). Norman, OK: University of Oklahoma Press.

Adams, R. E. W. ed. 1977. *The Origins of Maya Civilization.* Albuquerque: University of New Mexico Press.

Addo, F., E. O. A. Asibey, K. B. Quist, and M. B. Dyson. 1994. "The Economic Contribution of Women and Protected Areas: Ghana and the Bushmeat Trade." In M. Munasinghe and J. McNeely, eds., *Protected Area Economics and Policy: Linking Conservation and Sustainable Development,* pp. 99–115. Washington, DC: World Bank and World Conservation Union (International Union for Conservation of Nature and Natural Resources).

Adeola, M. O. and E. Decker. 1987. Utilisation de la faune sauvage en milieu rural au Nigeria. *Nature Faune* 3: 15–21.

Adouki, D. 1996. *Guide Pratique Juridique: Guide à la Législation Congolaise Concernant la Gestion des Aires Protegées au Congo.* New York: Projet PARCS Congo, Wildlife Conservation Society.

Ajayi, S. S. 1971. Wildlife as a source of protein in Nigeria: Some priorities for development. *Nigerian Field* 36: 115–27.

Ajayi, S. S. 1974. Giant rats for meat and some taboos. *Oryx* 12: 379–80.

Alcorn, J. 1993. Indigenous people and conservation. *Conserv Biol* 7: 424–6.

Ali, S. and S. D. Ripley. 1983. *Handbook of the Birds of India and Pakistan* (Compact edition). Bombay: Oxford University Press.

Allen, C. M. and S. R. Edwards. 1995. The sustainable-use debate: Observations from the IUCN. *Oryx* 29: 92–8.

Allen, M. S. 1985. "The Rain Forest of Northeastern Luzon and Agta Foragers." In P. B. Griffin and A. Estioko-Griffin, eds., *The Agta of Northeastern Luzon: Recent Studies,* pp. 45–68. Cebu City, The Philippines: University of San Carlos Publications.

Almquist, A. 1992. Horticulture and hunting in the Congo Basin: A case from Northcentral Zaire [unpublished report]. New York: Wildlife Conservation Society.

Alvard, M. 1993a. Testing the ecologically noble savage hypothesis: Conservation and subsistence hunting in Amazonian Peru [Ph.D. dissertation]. Albuquerque, NM: University of New Mexico.

Alvard, M. 1993b. A test of the "ecologically noble savage hypothesis": Interspecific prey choice by neotropical hunters. *Hum Ecol* 21:355–87.

Alvard, M. 1994. Conservation by native peoples: Prey choice in a depleted habitat. *Hum Nature* 5: 127–54.

Alvard, M. 1995a. Shotguns and sustainable hunting in the neotropics. *Oryx* 29: 58–66.

Alvard, M. 1995b. Intraspecific prey choice by Amazonian hunters. *Curr Anthropol* 36: 789–818.

Alvard, M. 1995c. Conservation by native peoples: Prey choice in a depleted habitat. *Hum Nature* 5: 127–54.

Alvard, M. 1996. The sustainability of Wana hunting in Morowali Nature Reserve, Sulawesi Tengah [unpublished report to the Indonesian Institute of Science (LIPI)]. Bogor, Indonesia.

Alvard, M. 1997. Home range size of tropical horticulturalists: The Posangke Wana of Central Sulawesi, Indonesia. Presented at the 22nd Annual meeting of the Human Biology Association, St. Louis, MO.

Alvard, M. and H. Kaplan. 1991. "Procurement Technology and Prey Mortality Among Indigenous Neotropical Hunters." In M. C. Stiner, ed., *Human Predators and Prey Mortality*, pp. 79–104. Boulder, CO: Westview Press.

Alvard, M., J. G. Robinson, K. H. Redford, and H. Kaplan. 1997. The sustainability of subsistence hunting in the neotropics: Data from two native communities. *Conserv Biol* 11: 977–82.

Anadu, P. A., P. O. Elamah, and J. F. Oates. 1988. The bushmeat trade in southwestern Nigeria: A case study. *Hum Ecol* 16: 199–208.

Anderson, D. R., J. L. Laake, B. R Crain, and K. P. Burnham. 1979. Guidelines for line transect sampling of biological populations. *J Wildlife Management* 43: 70–8.

Anonymous. 1928. El potrero de Moka. *La Guinea Española*. Número extraordinario.

Anonymous. 1954. *Resumen Estadistico del Africa Española*. Madrid: Dirección de Marruecos y Colonias. Dirección de Agricultura de los Territorios Españoles del Golfo de Guinea.

Anonymous. 1955. *Memoria (Años 1949–1955)*. Madrid: Gobierno General de los Territorios Españoles del Golfo de Guinea.

Anonymous. 1983. *Little-known Asian Animals with a Promising Economic Future*. Washington, DC: National Academy of Sciences Press.

Anonymous. 1988. The monster that stalked the Gulf of Mexico. *U.S. News & World Report* 105(12): 9–10.

Anonymous. 1996. Factors influencing sustainability. A report of the activities of the IUCN Sustainable Use Initiative Program, First World Conservation Congress, Montreal, Canada, October 14–23, 1996.

Anstey, S. 1991. *Wildlife Utilisation in Liberia*. Gland, Switzerland: World Wide Fund for Nature and FDA Wildlife Survey.

Argeloo, M. 1994. The maleo *Macrocephalon maleo*: New information on the distribution and status of Sulawesi's endemic megapode. *Bird Conserv Int* 4: 383–93.

Argeloo, M. and R. W. R. J. Dekker. 1996. Exploitation of megapode eggs in Indonesia: The role of traditional methods in the conservation of megapodes. *Oryx* 30: 59–64.

Arita, H. T., J. G. Robinson, and K. H. Redford. 1990. Rarity in neotropical forest mammals and its ecological correlates. *Conserv Biol* 4: 181–92.

Ascher, W. and R. Healy. 1990. "Natural Resource Policymaking in Developing Countries." In *Environment, Economic Growth and Income Distribution.* Durham, NC: Duke University Press.

Asibey, E. O. A. 1974. Wildlife as a source of protein south of the Sahara. *Biol Conserv* 6:32–9.

Asibey, E. O. A. and G. Child. 1991. Wildlife management for rural development in sub-Saharan Africa. *Nature Faune* 7:36–47.

Atkinson, J. 1989. *The Art and Politics of Wana Shamanism.* Berkeley: University of California Press.

August, P. V. 1983. The role of habitat complexity and heterogeneity in structuring tropical mammal communities. *Ecology* 64: 1495–507.

Auzel, P. 1996a. "Agriculture/Extractivisme et Exploitation Forestière." In *Etude de la dynamique des modes d'exploitation du milieu dans le nord de IUFA de Pokala, nord Congo.* Bomassa, Republic of Congo: Wildlife Conservation Society.

Auzel, P. 1996b. "Evaluation de l'Impact de la Chasse sur la Faune des Forêts d'Afrique Centrale, Nord Congo." In *Mise au point de méthodes basées sur l'analyse des pratiques et les résultats de chasseurs locaux.* Bomass, Republic of Congo: Wildlife Conservation Society.

Avio, K. L. and Clark, C. S. 1978. The supply of property offences in Ontario: Evidence on the deterrent effect of punishment. *Can J Economics* 10: 1–19.

Ayres, J. M. 1986. Uakaris and Amazonian flooded forest [Ph.D. dissertation]. Cambridge, England: University of Cambridge.

Ayres, J. M., D. de M. Lima, E. de S. Martins, and J. L. Barreiros. 1991. "On the Track of the Road: Changes in Subsistence Hunting in a Brazilian Amazonian Village." In J. G. Robinson and K. H. Redford, eds., *Neotropical Wildlife Use and Conservation*, pp. 82–92. Chicago: Chicago University Press.

Ayres, J. M., A. R. Alves, H. L. Queiroz, M. Marmontel, E. Moura, D M. Lima, A. Azevedo, M. Reis, P. Santos, R. Silveira, and D. Masterson 1999. "Mamirauá: The Conservation of Biodiversity in an Amazonian Flooded Forest." In C. Padoch, J. M. Ayres, M. Pinedo-Vasquez, and A. Henderson, eds., *Várzea: Diversity, Development and Conservation of Amazonia's Whitewater Floodplains*, pp. 203–216. Bronx, NY: The New York Botanical Garden Press.

Bahuchet, S. 1972. Etude écologique d'un campement de Pygmées BaBinga (Région de la Lobaye, République Centrafricaine). *J Agriculture Trop Botan Appl* 19:509–59.

Bahuchet, S., ed. 1979. *Pygmées de Centrafrique: Etudes Ethnologiques, Historiques, et Linguistiques sur les "Ba.Mgenga" (aka/baka) du Nord-ouest du Bassin Congolais.* Paris: Société d'Etudes Linguistiques et Anthropologiques de France.

Bahuchet, S. 1985. *Les Pygmées Aka et la Forêt Centrafricaine: Ethnologie Écologique.* Paris: Société d'Etudes Linguistiques et Anthropologiques de France.

Bahuchet, S. 1989. "Les Pygmées Aka et la Forêt Centrafricaine." In C. M. Hladik, S. Bahuchet, and I. De Garine, eds. *Se Nourrir en Forêt Equatoriale: Anthropologie Alimentaire des Populations des Régions Forestiéres Humides d'Afrique*, pp. 19–23. Paris: United Nations Educational, Scientific, and Cultural Organization/Man and the Biosphere Program.

Bahuchet, S. 1991. "Ethnoécologie du Pays Aka." In J. M. C. Thomas and S. Bahuchet, eds., *Encyclopédie des Pygmées Aka.* Paris: Peeters Press.

Bahuchet, S. 1992. *Dans la Forêt d'Afrique Centrale, les Pygmées Aka et Baka.* Paris: Peeters-Selaf.

Bahuchet, S. 1993a. "History of the Inhabitants of the Central African Rain Forest: Perspectives from Comparative Linguistics." In C. M. Hladik, A. Hladik, O. F. Linares, H. Pagezy, A. Semple, and M. Hadley, eds., *Tropical Forests, People and Food*, pp. 37–54. Paris: UNESCO.

Bahuchet, S. 1993b. *Situation des Populations Indigènes de Forêts Denses et Humides.* Brussels: Commision Européenne.

Bailey, R. C. 1985. The socioecology of Efe Pygmy men in the Ituri Forest, Zaire [Ph.D. dissertation]. Cambridge, MA: Harvard University.

Bailey, R. C. and R. Aunger, Jr. 1989. Net hunters vs. archers: Variation in women's subsistence strategies in the Ituri forest. *Hum Ecol* 17: 273–97.

Bailey, R. C., S. Bahuchet, and B. Hewlett. 1992. "Development in the Central African Rainforest: Concern for Forest Peoples." In K. Cleaver, M. Munasinghe, M. Dyson, N. Egli, A. Peuker, and F. Wencélius, eds. *Conservation of West and Central African Rainforests*, pp. 202–211. World Bank Environment Paper Number 1. Washington, DC: The World Bank.

Bailey, R. C. and N. R. Peacock. 1988. "Efe Pygmies of Northeast Zaire: Subsistence Strategies in the Ituri Forest." In I. de Garine and G. A. Harrison, eds., *Coping with Uncertainty in Food Supply*, pp. 88–117. Oxford, England: Clarendon.

Balée, W. 1987. Cultural forests of the Amazon. *Garden* 6: 12–4.

Balée, W. 1988. Indigenous adaptations to Amazonian palm forests. *Principes* 32: 47–54.

Balée, W. 1989. "The Culture of Amazonian Forests." In D. A. Posey and W. Balée, eds., *Resource Management in Amazonia: Indigenous and Folk Strategies*, pp. 1–21. Bronx, NY: New York Botanical Garden.

Balée, W. 1992. "People of the Fallow: A Historical Ecology of Foraging in Lowland South America." In K. H. Redford and C. Padoch, eds., *Conservation of Neotropical Forests: Building on Traditional Resource Use*, pp. 35–57. New York: Columbia University Press.

Balée, W. 1994. *Footprints of the Forest: Kaapor Ethnobotany—The Historical Ecology of Plant Utilization by an Amazonian People.* New York: Columbia University Press.

Balinga, V. S. 1977. Usos competidores de la fauna. *Unasylva* 29: 22–5.

Balmford, A., N. Leader-Williams, and M. J. B. Green. 1992. "Protected Areas of Afrotropical Forest: History, Status and Prospects." In M. Collins and J. A. Sayer, eds., *Tropical Rain Forests—An Atlas for Conservation*, Vol. II. Africa, pp. 69–80. London: Macmillan.

Banse, K. and S. Mosher. 1980. Adult body mass and annual production/biomass relationships of field populations. *Ecol Monogr* 50: 355–79.

Barnes, R. F. W. and W. K. L. Jensen. 1987. How to count elephants in forests. *IUCN Afr Elephant Rhino Specialist Group Tech Bull* 1: 1–6.

Barrera de Jorgenson, A. 1993. Chicle extraction and conservation in Quintana Roo, Mexico [M.S. thesis]. Gainesville: University of Florida.

Barrera de Jorgenson, A. and J. P. Jorgenson. 1995. "Use of Forest Resources and Conservation in Quintana Roo, Mexico." In J. A. Bissonette and P. R. Krausman, eds., *Integrating People and Wildlife for a Sustainable Future*, pp. 16–20. Bethesda, MD: The Wildlife Society.

Barrett, C. B. and P. Arcese. 1995. Are integrated conservation-development projects (ICDPs) sustainable? On the conservation of large mammals in sub-Saharan Africa. *World Dev* 23: 1073–84.

Basilio, A. 1962. *La Vida Animal en la Guinea Española.* Madrid: Instituto de Estudios Africanos, CSIC.

Bauchot, R. and H. Stephan. 1966. Encéphalisation et niveau évolutif chez les simiens. *Mammalia* 33: 225–75.

Beasom, S. L. and J. F. Roberson, eds. 1985. *Game Harvest Management.* Kingsville, TX: Caesar Kleberg Wildlife Research Institute.

Becker, C. D. and E. Ostrom. 1995. Human ecology and resource sustainability: The importance of institutional diversity. *Annu Rev Ecol Systematics* 26: 113–33.

Beckerman, S. 1978. Comment on Ross. *Curr Anthropol* 19: 17–8.

Beckerman, S. and T. Sussenbach. 1983. "A Quantitative Assessment of the Dietary Contribution of Game Species to the Subsistence of South America Tropical Forest Tribal Peoples." In J. Clutton-Brock and C. Grigson, eds., *Animals and Archaeology: Hunters and Their Prey*, pp. 337–50. Oxford, England: British Archaeological Reports, International Series.

Bell, R, H. V. 1987. "Conservation with a Human Face: Conflict and Reconciliation in African Land Use Planning." In D. Anderson and R. Grove, eds., *Conservation in Africa: People, Policies and Practice*, pp. 79–101. Cambridge, England: Cambridge University Press.

Bellwood, P. S. 1980. The peopling of the Pacific. *Sci Am* 243: 174–85.

Bellwood, P. S. 1985. *Prehistory of the Indo-Malaysian Archipelago*. Sydney: Academic Press.

Bendayán A., N. Y. 1991. Influencia socioeconómica de la fauna silvestre como recurso alimentario en Iquitos [thesis of licensure]. Iquitos, Peru: Universidad Nacional de la Amazonía Peruana.

Bennett, C. F. 1962. The Bayano Cuna Indians, Panama: An ecological study of diet and livelihood. *Ann Assoc Am Geographers* 52: 32–50.

Bennett, E. L. 1992. *A Wildlife Survey of Sarawak*. Kuching, Sarawak: Forest Department. New York: Wildlife Conservation International.

Bennett, E. L. 1998. *The Natural History of Orang-Utan*. Kota Kinabalu, Sabah, Malaysia: Natural History Publications (Borneo).

Bennett, E. L., Caldecott, J. O., Kavanagh, M., and Sebastian, A. C. 1985. Conservation status of Sarawak's primates. *Sarawak Gazette* 61: 30–4.

Bennett, E. L. and Z. Dahaban. 1995. "Wildlife Responses to Disturbances in Sarawak and their Implications for Forest Management." In R. B. Primack and T. E. Lovejoy, eds., *Ecology, Conservation, and Management of Southeast Asian Rainforests*, pp. 66–85. New Haven and London: Yale University Press.

Bennett, E. L. and M. T. Gumal. In press. "The inter-relationships of commercial logging, hunting and wildlife in Sarawak, and recommendations for forest management." In R. Fimbel, A. Grajal and J. G. Robinson, eds., *Wildlife-Logging Interactions in Tropical Forests*. New York: Columbia University Press.

Bennett, E. L., A. J Nyaoi., and J. Sompud. 1997. Hornbills and culture in northern Borneo: Can they continue to co-exist? *Biol Conserv* 82: 41–6.

Bennett Hennessey, A. 1995. A study of the meat trade in Ouesso, Republic of Congo. Report to Wildlife Conservation Society, Bronx, New York.

Bernard, H. R. 1988. *Research Methods in Cultural Anthropology*. Thousand Oaks, CA: SAGE Publications.

Berry, J. W., J. M. H. van de Koppel, C. Sénéchal, R. C. Annis, S. Bahuchet, L. L. Cavalli-Sforza, and H. A. Witkin. 1986. *On the Edge of the Forest: Cultural Adaptation and Cognitive Development in Central Africa*. Berwyn: Swets North America.

Birdsell, J. 1958. On population structure in generalized hunting and collecting populations. *Evolution* 12: 189–205.

Birdsell, J. 1968. "Some Predictions for the Pleistocene Based on Equilibrium Systems Among Recent Hunter-Gatherers." In R. Lee and I. De Vore, eds., *Man the Hunter*, pp. 229–40. Chicago: Aldine Press.

Blackburn, T. M. and J. H. Lawton. 1994. Population abundance and body size in animal assemblages. *Phil Trans R Soc [B]* 343: 33–9.

Blake, S. 1993. A reconnaissance survey in the Likouala Swamps of Northern Congo and its implications for conservation [M.S. thesis]. Edinburgh, Scotland: University of Edinburgh.

Blake, S. 1994. A reconnaissance survey in the Kabo Logging Concession South of the Nouabalé-Ndoki National Park, Northern Congo. Unpublished report, Bomassa, Republic of Congo: Wildlife Conservation Society.

Bodmer, R. E. 1989. Ungulate biomass in relation to feeding strategy within Amazonian forests. *Oecologia* 81: 547–50.

Bodmer, R. E. 1993. Managing wildlife with local communities: the case of the Reserva Comunal Tamshiyacu-Tahuayo. Paper prepared for the Liz Claiborne Art Ortenberg Foundation Community Based Conservation Workshop, Airlie, Virginia, pp. 1–31.

Bodmer, R. E. 1994. "Managing Wildlife with Local Communities in the Peruvian Amazon: The Case of the Reserva Comunal Tamshiyacu-Tahuayo." In D. Western, R. M. Wright and S. Strum, eds., *Natural Connections*, pp. 113–34. Washington, DC: Island Press.

Bodmer, R. E. 1995a. Managing Amazonian wildlife: Biological correlates of game choice by detribalized hunters. *Ecol Applications* 5: 872–7.

Bodmer, R. E. 1995b. "Susceptibility of Mammals to Overhunting in Amazonia." In J. Bissonette and P. Krausmanm. eds., *Integrating People and Wildlife for a Sustainable Future*, pp. 292–5. Bethesda, MD: The Wildlife Society.

Bodmer, R. E., N. Y. Bendayán A., L. Moya I., and T. G. Fang. 1990a. Manejo de ungulados en la Amazonia Peruana: Analisis de su caza y commercialización. *Boletín de Lima* 70: 49–56.

Bodmer, R. E., J. F. Eisenberg, and K. H. Redford. 1997a. Hunting and the likelihood of extinction of Amazonian mammals. *Conserv Biol* 11: 460–6.

Bodmer, R. E., T. G. Fang, and L. Moya. 1988a. Primates and ungulates: A comparison of susceptibility to hunting. *Primate Conservation* 9: 79–83.

Bodmer, R. E., T. G. Fang, and L. Moya. 1988b. Ungulate management and conservation in the Peruvian Amazon. *Biol Conserv* 45: 303–10.

Bodmer, R. E., T. G. Fang, and L. Moya. 1990b. Fruits of the forest. *Nature* 343: 109.

Bodmer, R. E., T. G. Fang, L. Moya I., and R. Gill. 1994. Managing wildlife to conserve Amazonian forests: Population biology and economic considerations of game hunting. *Biol Conserv* 67: 29–35.

Bodmer, R. E., J. W. Penn, P. E. Puertas, L. Moya I., and T. G. Fang. 1997b. "Conserving Amazonian Forests Through Sustainable Use of Natural Resources: Interdisciplinary, Community Based Management of the Reserva Comunal Tamshiyacu-Tahuayo." In C. Freese, ed., *Harvesting Wild Species*. Baltimore and London: Johns Hopkins University Press.

Booth, A. 1988. *Agricultural Development in Indonesia*. Sydney: Allen & Unwin.

Bowland, A. E. 1990. The response of red duikers *Cephalophus natalensis* to drive counts. *Koedoe* 33: 47–54.

BPS (Biro Pusat Statistik), Government of Indonesia. 1980. *Analisa Ringkas Hasil Sensus Penduduk*. No. 03320.8411, Jakarta, Indonesia.

BPS, Government of Indonesia. 1984a. *Food Balance Sheet in Indonesia in 1982*. Jakarta, Indonesia.

BPS, Government of Indonesia. 1984b. *Statistik Energi 1984*. No. 04140.8601, Jakarta, Indonesia.

BPS, Government of Indonesia. 1984c. *Statistik potensa desa, sensus Pertanian 1983*. No. 04340.8602, Seri E2. Jakarta, Indonesia.

BPS, Government of Indonesia. 1986a. *Penduduk Indonesia 1985 Menurut Propinsi*. Jakarta, Indonesia.

BPS, Government of Indonesia. 1986b. *Statistik Kenderaan Bermotor dan Panjang Jalan.* Jakarta, Indonesia.

BPS, Government of Indonesia. 1986c. *Indikator Kesejahteraan Rakyat 1984.* Jakarta, Indonesia.

Braak, C.J.F. ter. 1988. CANOCO. An extension of DECORANA to analyse species-environment relationships. *Vegetatio* 75: 159–60.

Brandon, K. E. and M. Wells. 1992. Planning for people and parks: design dilemmas. *World Dev* 20: 557–70.

Brockelman, W. Y. and R. Ali. 1987. "Methods of Surveying and Sampling Forest Primate Populations." In C. W. Marsh and R. A. Mittermeier, eds., *Primate Conservation in Tropical Rain Forest*, pp. 23–62. New York: Alan R. Liss.

Bromley, D. W. 1994. "Economic Dimensions of Community-Based Conservation." In D. Western and R. M. Wright, eds., *Natural Connections: Perspectives in Community-Based Conservation*, pp. 428–47. Washington, DC: Island Press.

Brookfield, H. 1988. "The New Great Age of Clearance and Beyond." In J. S. Denslow and C. Padoch, eds., *People of the Tropical Rain Forest*, pp. 209–24. Los Angeles: University of California Press.

Brosius, J. P. 1986. River, forest and mountain: The Penan Gang landscape. *Sarawak Museum J* 36, No. 57 (New Series): 173–84.

Brosius, J. P. 1990. Penan hunter-gatherers of Sarawak. *AnthroQuest* 42: 1–7.

Browder, J. O. 1992. The limits to extractivism. *BioScience* 42: 174–82.

BSP (Biodeversity Support Program). 1990. *Sangha-Ngoko Rainforest Area. Global Inventory Modeling and Mapping Studies.* Washington, DC: Department of the Interior.

Buckland, S., D. Anderson, K. Burnham, and J. Laake. 1993. *Distance Sampling: Estimating Abundance of Biological Populations.* New York and London: Chapman and Hall.

Bugo, H. 1995. "The Significance of the Timber Industry in the Economic and Social Development of Sarawak." In R. B. Primack and T. E. Lovejoy, eds., *Ecology, Conservation, and Management of Southeast Asian Rainforests*, pp. 221–40. New Haven and London: Yale University Press.

Burgess, N., C. D. FitzGibbon, and P. Clarke. 1996. "Coastal Forests." In T. McClanahan, ed., *Ecosystems and Their Conservation in East Africa*, pp. 329–59. New York: Oxford University Press.

Burgman, M. A., S. Ferson, and H. R. Akaçaya. 1993. *Risk Assessment in Conservation Biology.* London: Chapman and Hall.

Burnham, K. P., D. R. Anderson, and J. L. Laake. 1980. Estimation of density from line transect sampling of biological populations. *Wildlife Monogr* 72: 1–202.

Butynski, T. and S. Koster. 1994. Distribution and conservation status of primates in Bioko Island, Equatorial Guinea. *Biodiversity Conserv* 3: 893–909.

Butynski, T., C. D. Schaaf, and G. W. Hearn. 1995a. Status and conservation of ungulates on Bioko Island (Fernando Poo), Equatorial Guinea [unpublished report]. Atlanta, GA: Zoo Atlanta.

Butynski, T., C. D. Schaaf, and G. W. Hearn. 1995b. Status and conservation of duikers on Bioko Island. *Gnusletter* 14: 23–28.

Bynum, E. 1995. Hybridization between *Macaca tonkeana* and *Macaca hecki* in Central Sulawesi, Indonesia [Ph.D. dissertation]. New Haven, CT: Yale University.

Caldecott, J. O. 1986a. *Hunting and Wildlife Management in Sarawak: Final Report of Conservation Management Study for Hunted Wildlife in Sarawak.* Kuala Lumpur, Malaysia: World Wildlife Fund.

Caldecott, J. O. 1986b. *An Ecological and Behavioural Study of the Pig-Tailed Macaque*. Basel, Switzerland: Karger.

Caldecott, J. O. 1988. *Hunting and Wildlife Management in Sarawak*. Gland, Switzerland and Cambridge, England: International Union for Conservation of Nature and Natural Resources.

Calouro, A. M. 1995. Caça de subsistência: sustentabilidade e padrões de uso entre seringueiros ribeirinhos e não-ribeirinhos Estado do Acre [M.Sc. thesis]. Brasília, Brazil: Universidade de Brasília.

Campbell, K. L. I. and H. Hofer. 1995. "People and Wildlife: Spatial Dynamics and Zones of Interaction." In A. R. E. Sinclair and P. Arcese, eds., *Serengeti II: Research, Management and Conservation of an Ecosystem*, pp. 535–74. Chicago: University of Chicago Press.

Cana, F. R., ed. 1911. French Congo. *Encyclopedia Britannica*, 11th ed., pp. 99–102.

Cañadas, L. 1983. *El Mapa Bioclimático y Ecológico del Ecuador*. Quito, Ecuador: Ministerio de Agricultura y Ganaderia.

Cant, J. 1977. A census of the agouti *Dasyprocta punctata* in seasonally dry forest at Tikal, Guatemala, with some comments on strip censusing. *J Mammalogy* 58: 688–90.

Carneiro, R. 1968. "Slash and Burn Cultivation among the Kuikuru and Its Implications for Cultural Development in the Amazon Basin." In Y. Cohen, ed. *Man and Adaptation: The Cultural Present*, pp. 132–45. Chicago: Aldine.

Carroll, R. W. 1986. *The Creation, Development, Protection, and Management of the Dzanga-Sangha Dense Forest Sanctuary and the Dzanga-Ndoki National Park in Southwestern Central African Republic*. New Haven, CT: Yale University School of Forestry and Environmental Studies.

Carroll, R. W. 1988. Relative density, range extension, and conservation potential of the lowland gorilla in the Dzanga-Sangha region of southwestern Central African Republic. *Mammalia* 52: 309–23.

Cashdan, E. A. 1985. Coping with risk: Reciprocity among the Basarwa of northern Botswana. *Man* 20: 454–74.

Cartmill, M. 1993. *A View to a Death in the Morning*. Cambridge and London: Harvard University Press.

Castro, N., J. Revilla, and M. Neville. 1975–1976. Carne de monte como una fuente de proteinas en Iquitos, con referencia especial a monos. *Rev Forestal Peru* 6: 19–32.

Castroviejo, J. 1995. *Estudio Sobre la Zonificación y Uso Multiple de las Areas Protegidas de la Isla de Bioco*. Seville, Spain: Amigos de Doñana.

Castroviejo, J., J. Juste, and R. Castelo. 1986. *Proyecto de Investigación y Conservación de la Naturaleza en Guinea Ecuatorial*. Madrid: Secretaria de Estado para la Cooperación Internacional y para Iberoamérica, Oficina de Cooperación con Guinea Ecuatorial.

Castroviejo, J., J. Juste, B. J. Perez del Val, R. Castelo, and R. Gil. 1994a. Diversity and status of sea turtle species in the Gulf of Guinea islands. *Biodiversity Conserv* 3: 828–36.

Castroviejo, J., J. Juste, B. J. Perez del Val, and R. Castelo. 1994b. The Spanish co-operation programme in Equatorial Guinea: A ten year review of research and nature conservation in Bioko. *Biodiversity Conserv* 3: 951–61.

Caughley, G. 1966. Mortality patterns in mammals. *Ecology* 48: 834–9.

Caughley, G. 1977. *Analysis of Vertebrate Populations*. New York: John Wiley & Sons.

Caughley, G. 1985. "Harvesting of Wildlife: Past, Present, and Future." In S. L. Beasom and S. F. Roberson, eds., *Game Harvest Management*, pp. 3–14. Kingsville, TX: Caesar Kleberg Wildlife Research Institute.

Caughley, G. 1994. Directions in conservation biology. *J Animal Ecol* 63: 215–44.

Caughley, G. and A. Gunn. 1996. *Conservation Biology in Theory and Practice*. Oxford, England: Blackwell Science.

Caughley, G. and A. R. E. Sinclair. 1994. *Wildlife Ecology and Management*. Boston: Blackwell Scientific Publications.

Cavalli-Sforza, L. L., ed. 1986. *African Pygmies*. New York: Academic Press.

CEDI/PETI and Museo Nacional (Centro Ecumênico de Documentaçãoe informação, Projeto de Estudo sobre Terras Indígenas no Brasil, and Museo Nacional). 1990. *Terras Indigenas no Brasil*. Rio de Janiero: Universidade Federal de Rio de Janiero.

Chagnon, N. A. 1968. *Yanomamo: The Fierce People*. New York: Holt, Rinehart and Winston.

Chagnon, N. A. and R. B. Hames 1979. Protein deficiency and tribal warfare in Amazonia: New data. *Science* 203: 910–3.

Chambers, R. 1994. The origins and practice of participatory rural appraisal. *World Development* 22: 953–69.

Chapman, J. G. and G. A. Feldhamer, eds. 1982. *Wild Mammals of North America*. Baltimore and London: The Johns Hopkins University Press.

Chardonnet, P., H. Fritz, N. Zorzi, and E. Feron. 1995. "Current Importance of Traditional Hunting and Major Contrasts in Wild Meat Consumption in Sub-Saharan Africa." In J. A. Bissonette and P. R. Krausman, eds., *Integrating People and Wildlife for a Sustainable Future*, pp. 304–7. Bethesda, MD: The Wildlife Society.

Charnov, E. L. 1976. Optimal foraging: The marginal value theorem. *Theor Pop Biol* 9: 129–36.

Chicchón, A. 1992. Chimane resource use and market involvement in the Beni Biosphere Reserve, Bolivia [Doctoral Dissertation]. Gainesville, FL: University of Florida.

Child, G. and V. Wilson. 1964. Delayed effects of tsetse control hunting on a duiker population. *J Wildlife Management* 28: 866–8.

Chin, S. C. 1985. Agriculture and resource utilization in a lowland rainforest Kenyah community. *Sarawak Museum J* 25, No. 56 (New Series): Special Monograph Number 4.

Chopra, K. 1993. The value of non-timber forest products: An estimation for tropical deciduous forests in India. *Economic Botany* 47: 251–7.

Choudhury, A. 1995. Further observations on the Phayre's leaf monkey (*Trachypithecus phayrei*) in Cachar, Assam. *J Bombay Natural History Soc* 91: 203–10.

Christensen, N. L. et al. 1996. The report of the Ecological Society of America Committee on the scientific basis for ecosystem management. *Ecol Applications* 6: 665–91.

Chua, T. K. 1996. Editorial: hunting and wildlife conservation. *AZAM News* 5(4): 1–2.

CIEH. 1990. *République Populaire du Congo: Précipitations Journalières de 1966 à 1980*. Ouagadougou, Congo: CIEH.

CITES (Convention on International Trade in Endangered Species of Wild Flora and Fauna). 1994. Include in Appendix II of the Convention all species of *Entandrophragma*. Proposal to the Convention on International Trade in Endangered Species of Wild Fauna and Flora submitted by Germany.

Clark, C. D. 1990. The trading networks of the northeastern Cagayan Agta Negritos [M.A. thesis]. Manoa, HI: Department of Anthropology, University of Hawaii.

Clark, C. W. 1990. *Mathematical Bioeconomics*, 2nd ed. New York: Wiley & Son.

Clayton, L. M. 1996. Conservation biology of the babirusa, *Babyrousa babyrussa*, in Sulawesi, Indonesia [D.Phil. dissertation]. Oxford, England: University of Oxford.

Clayton, L. M., M. Keeling, and E. J. Milner-Gulland. 1997. Bringing home the bacon: A spatial model of wild pig harvesting in Sulawesi, Indonesia. *Ecol Applications* 7: 642–52.

Cohen, M. and G. Armelegos, eds. 1984. *Paleopathology and the Origins of Agriculture*. Orlando, FL: Academic Press.

Cohen, M. N. 1989. *Health and the Rise of Civilization*. New Haven and London: Yale University Press.

Colchester, M. 1993. "Slave and Enclave Towards a Political Ecology of Equatorial Africa." In Veber, Dahl, Wilson, and Waelhe, eds. *Never Drink From the Same Cup: Proceedings of the Conference on Indigenous Peoples in Africa*. IWGIA Document No. 74. Copenhagen: IWGIA and Centre for Development Research.

Cole, L. 1954. The population consequences of life history phenomena. *Q Rev Biol* 29: 103–37.

Colell, M., C. Maté, and J. E. Fa. 1994. Hunting among Moka Bubis in Bioko: Dynamics of faunal exploitation at the village level. *Biodiversity Conserv* 3: 939–50.

Collar, N. J., M. J. Crosby, and A. J. Stattersfield. 1994. *Birds to watch 2. The world list of threatened birds*. BirdLife Conservation Series No 4. Cambridge, England: BirdLife International.

Collett, S. F. 1981. Population characteristics of *Agouti paca* (Rodentia) in Colombia. Publications of the Museum, Michigan State University, Biological Series 5, pp. 489–602.

Colyn, M., A. Dudu, M. Mankoto, and M. A. Mbaelele. 1987a. Exploitation du petit et moyen gibier des forêts ombrophiles du Zaire. *Nature Faune* 3: 22–39.

Colyn, M., A. Dudu, M. Mankoto, and M. A. Mbaelele 1987b. "Données sur l'Exploitation du 'Petit et Moyen Gibier' des Forêts Ombrophiles du Zaire." In M. Colyn, M. A. Dudu, M. Mankoto, and M. A. Mbaelele, eds., *International Symposium and Conference on Wildlife Management in Sub-Saharan Africa*, pp. 110–45. Paris: UNESCO.

Coomes, O. T. 1992. Making a living in the Amazon rain forest: Peasants, land, and economy in the Tahuayo River Basin of northeastern Peru [Ph.D. dissertation]. Madison, WI: University of Wisconsin.

Coquery-Vidrovitch, C. 1972. *Le Congo au Temps des Grandes Compagnies Concessionaires 1898–1930*. Paris: Mouton and Co.

Cornelius, D. and J. Creed. 1996. Conflict in Alaska's rainforest. *Defenders* 71: 27–36.

Cotgreave, P. 1993. The relationship between body size and abundance in animals. *Trends Ecol Evolution* 8: 244–8.

Cramer, J. S. 1991. *The Logit Model: An Introduction for Economists*. New York: Arnold.

Crete, M., R. J. Taylor, and P. A. Jordan. 1981. Optimization of moose harvest in southwestern Quebec. *J Wildlife Management* 45: 598–611.

CTFT (Centre Technique Forestier Tropical). 1985. *Inventaire des Ressources Forestières du Sud Cameroun*. Nogent-sur-Marne, France: Département Forestièr du CIRAD, Centre Technique Forestier Tropical.

Dachary, A. C. and S. M. Arnaiz Burne. 1983. *Estudios Socioeconomicos Preliminares de Quintana Roo: Sector Agropecuario y Forestal (1902–1980)*. Puerto Morelos, Quintana Roo, Mexico: Centro de Investigaciones de Quintana Roo.

Dale, A. P. and M. B. Lane. 1994. Strategic perspectives analysis: a procedure for participatory and political social impact assessment. *Society Natural Resources* 7: 253–67.

Damuth, J. 1987. Interspecific allometry of population density in mammals and other animals: The independence of body mass and species energy use. *Biol J Linnaean Soc* 31: 193–246.

Daniel, J. C. 1983. *The Book of Indian Reptiles*. Bombay: Bombay Natural History Society.

DaSilva, N. J. and J. W. Sites. 1995. Patterns of diversity of neotropical Squamate reptile species with emphasis on the Brazilian Amazon and the conservation potential of Indigenous Reserves. *Conserv Biol* 9: 873–901.

Dawes, R. 1980. Social dilemmas. *Annu Rev Psychol* 31: 169–93.

De Steven, D. and F. E. Putz. 1984. Impact of mammals on early recruitment of a tropical canopy tree, *Dipteryx panamensis*, in Panama. *Oikos* 43: 207–16.

Dekker, R. W. R. J. and P. J. K. McGowan. 1995. *Megapodes: An Action Plan for Their Conservation 1995–1999*. Gland, Switzerland: International Union for Conservation of Nature and Natural Resources and World Pheasant Association.

Dekker, R. W. R. J. and J. Wattel. 1987. "Egg and Image: New and Traditional Uses for the Maleo (*Macrocephalon maleo*)." In A. W. Diamond and F. Filion, eds., *The Value of Birds: Proceedings of a Workshop Held During the 19th World Conference of the I.C.B.P. in June 1986 at Queen's University, Kingston, Ontario*. Technical publication no. 6. East Anglia, U.K.: International Council for Bird Preservation.

Delobeau, J. M. 1989. *Yamonzombo et Yandenga: Les Relations Entre Les Villages Monzombo et les Campements Pygmées Aka dans la Sous-préfecture de Mongoumba (Centrafrique)*. Paris: Société d'Etudes Linguistiques et Anthropologiques de France.

Demesse, L. 1978. *Changements Techno-économiques et Sociaux chez les Pygmées BaBinga (Nord Congo et Sud Centrafrique)*. Paris: Société d'Etudes Linguistiques et Anthropologiques de France.

Denevan, W. 1966. The Aboriginal Cultural Geography of the Llanos De Mojos of Bolivia. Berkeley: University of California Press.

Denevan, W. 1992. "Native American Populations in 1492: Recent Research and a Revised Hemispheric Estimate." In W. Denevan, ed., *The Native Population of the Americas in 1492*, 2nd ed., pp. xvii–xxxviii. Madison, WI: University of Wisconsin Press.

DENR (Department of Energy and National Resources). 1992. *Management Plan for the Northern Sierra Madre Natural Park*. Quezon City, Philippines: Department of Energy and Natural Resources.

Departemen Pertanian Republik Indonesia. 1953. *Semua Undang-undang dan Peraturan-peraturan Mengenai Perlindungan Alam dan Pemburuan di Indonesia*. Vorkink-Van Hoeve, Bandung: Djawatan Penjelidikan alam (Kebun Raya Indonesia), Bagian Perlindungan Alam dan Pemburuan, Bogor.

Dewey, K. G. 1981. Nutritional consequences of the transformation from subsistence to commercial agriculture in Tabasco, Mexico. *Hum Ecol* 9: 157–81.

Diamond, J. M. 1984. "Historic Extinctions: A Rosetta Stone for Understanding Prehistoric Extinctions." In P. S. Martin and R. G. Klein, eds., *Quaternary Extinctions: A Prehistoric Revolution*, pp. 824–62. Tucson, AZ: University of Arizona Press.

Diamond, J. M. 1989. "Overview of Recent Extinctions." In D. Western and M. Pearl, eds., *Conservation for the Twenty-first Century*, pp. 37–41. New York: Oxford University Press.

Diong, C. H. 1982. Population biology and management of the feral pig (*Sus scrofa* L.) in Kipahulu Valley, Maui [Ph.D. dissertation]. Manoa, HI: Department of Zoology, University of Hawaii.

Dirzo, R. and A. Miranda. 1990. Contemporary neotropical defaunation and forest structure, function, and diversity: A sequel to John Terborgh. *Conserv Biol* 4: 444–7.

Doungoubé, G. 1990. "Central African Republic: The Dzanga-Sangha Dense Forest Reserve." In A. Kiss, ed., *Living with Wildlife: Wildlife Resource Management with Local Participation in Africa*, pp. 75–9. Washington, DC: World Bank.

Dounias, E., A. Hladik, and C. M. Hladik. 1996. "De la Ressource Disponible à la Ressource Exploitée: Méthode de Quantification des Ressources Alimentaires dans les Régions Forestières et les Savanes du Cameroun." In A. Froment, I. de Garine, C. Binam Bikoi, and J. F. Loung, eds., *Bien Manger et Bien Vivre: Anthropologie Alimentaire et Développement en Afrique Intertropicale: du Biologique au Social*, pp. 55–66. Paris: Orstom.

Dounias, E. In press. The art of trapping by the Mvae Farmers of the southern rain forest of Cameroon. *Hum Ecol.*

Downing, R. 1981. Deer harvest sex ratios: A symptom, a prescription, or what? *Wildlife Soc Bull* 9: 8–13.

Downing, R., W. Moore, and J. Knight. 1965. Comparison of deer census techniques applied to a known population in a Georgia enclosure. *Proceedings of the Annual Conference of the Southeast Association of Game and Fish Comm.*, pp. 26–30. Frankfort, KY: Southeastern Association of Game and Fish Commissioners.

Dryer, P. 1985. The contribution of non-domesticted animals to the diet of Etolo, southern highlands province, Papua New Guinea. *Ecol Food Nutrition* 17: 101–15.

Dubost, G. 1980. L'écologie et la vie sociale du céphalophe bleu (*Cephalophus monticola* Thunberg), petit ruminant forestier africain. *Z Tierpsychol* 54: 205–66.

Dubost, G. 1983. Le comportement de *Cephalophus monticola* Thunberg et *C. dorsalis* Gray, et la place des céphalophes au sein des ruminants. *Mammalia* 47: 141–77, 281–310.

Dubost, G. 1984. Comparison of the diets of frugivorous forest ruminants of Gabon. *J Mammalogy* 65: 298–316.

Dubost, G. and F. Feer. 1992. Saisons de reproduction des petits ruminants dans le nord-est du Gabon en fonction des variations des ressources alimentaires. *Mammalia* 56: 25–43.

Dufour, D. 1983. "Nutrition in the Northwest Amazon: Household Dietary Intake and Time-energy Expenditure." In R. B. Hames and W. T. Vickers, eds., *Adaptive Responses of Native Amazonians*, pp. 329–55. New York: Academic Press.

Duivenvoorden, J. F. and J. M. Lips. 1995. A land-ecological study of soils, vegetation, and plant diversity in Colombia Amazonia. *Tropenbos Series* 12: 1–438.

Dunbar-Brander, A. A. 1923. *Wild Animals of Central India.* London: Edward Arnold.

Edwards, A. E. 1992. The diurnal primates of Korup National Park, Cameroon: Abundance, productivity, and polyspecific associations [M.S. thesis]. Gainesville, FL: University of Florida.

Edwards, C. R. 1986. The human impact on the forest in Quintana Roo, Mexico. *J Forest History* 30: 120–7.

Eide, E. 1994. *Economics of Crime: Deterrence and the Rational Offender.* Amsterdam: North-Holland.

Eilers, H. 1985. Protected areas and indigenous peoples. *Cultural Survival Q* 9: 6–9.

Eisenberg, J. F. 1980. "The Density and Biomass of Tropical Mammals." In M. E. Soulé and B. A. Wilcox, eds., *Conservation Biology: An Evolutionary Ecological Perspective*, pp. 34–55. Sunderland, MA: Sinauer Associates.

Eisenberg, J. F. 1981. *The Mammalian Radiations.* Chicago: University of Chicago.

Eisenberg, J. F. and M. C. Lockhart. 1972. An ecological reconnaissance of Wilpattu National Park, Ceylon. *Smiths Contrib Zool* 101: 1–118.

Eisenberg, J. F. and J. Seidensticker. 1976. Ungulates in southern Asia: A consideration of biomass estimates for selected habitats. *Biol Conserv* 10: 293–308.

Eisenberg, J. F. and R. W. Thorington, Jr. 1973. A preliminary analysis of a neotropical mammal fauna. *Biotropica* 5: 150–61.

Eisentraut, M. 1973. *Die wirbeltierfauna von Fernando Poo und Westkamerun.* Bonn Zoological Monograph 3. Bonn, Germany: Zool Forschung Museum Alexander Koenig.

Eiten, G. 1972. The cerrado vegatation of Brasil. *Botan Rev* 38: 201–341.

Eiten, G. 1990. "Vegetação." In M. Novaes et al., eds., *Cerrado: Caracterização, ocupação e perspectivas*, pp. 9–67. Brasília, Brazil: Secretaria do Meio Ambiente Ciência e Tecnologia (SEMATEC).

Ellen, R. 1975. Non-domesticated resources in Nuaulu ecological relations. *Social Sci Information* 14: 129–50.

Ellison, P. T. 1990. Human ovarian function and reproductive ecology: New hypotheses. *Am Anthropol* 92: 933–52.

Elmendorf, M. 1976. *Nine Mayan Women*. New York: Schenkman Publishing Company.

Eltringham, S. K. 1984. *Wildlife Resources and Economic Development*. Chichester: Wiley and Sons.

Eltringham, S. K. 1994. Can wildlife pay its way? *Oryx* 28: 163–8.

Emmons, L. H. 1984. Geographic variation of densities and diversities of non-flying mammals in Amazonia. *Biotropica* 163: 210–22.

Emmons, L. H. 1989. Tropical rain forests: why they have so many species and how we may lose this biodiversity without cutting a single tree. *Orion* 8: 8–14.

Emmons, L. H. 1992. Ecological considerations on the farming of game animals: Capybaras yes, pacas no. *Vida Silvestre Neotropical* 1: 54–5.

Emmons, L. H. and F. Feer. 1990. *Neotropical Rainforest Mammals. A Field Guide*. Chicago: University of Chicago Press.

Erhlich, I. 1973. Participation in illegitimate activities—A theoretical and empirical investigation. *J Political Economy* 81: 521–65.

Erhlich, P. R. 1988. "The Loss of Diversity: Causes and Consequences." In E. O. Wilson and F. M. Peter, eds., *Biodiversity*, pp. 21–7. Washington, DC: National Academy Press.

Erhlich, P. R. and Wilson, E. O. 1991. Biodiversity studies: Science and policy. *Science* 253: 758–62.

Estioko, A. A. and P. B. Griffin. 1975. The Ebukid Agta of Northeastern Luzon. *Philippine Q Culture Soc* 3: 237–44.

Estioko-Griffin, A. A. 1984. The ethnography of southeastern Cagayan Agta hunting [M.A. thesis]. Diliman, Quezon City: Department of Anthropology, University of the Philippines.

Eussen, J. 1980. Biological and ecological aspects of alang-alang [*Imperata cylindrica* (L.) Beauv.] Proceedings of BIOTROP Workshop on Alang-Alang, Bogor, 27–29 July 1976. *BIOTROP* 5 (Special Publication): 15–22.

Eves, H. E. 1995. Pilot study investigation of the socioeconomics of natural resource utilization in the Kabo logging concession, northern Congo. Unpublished report to the Wildlife Conservation Society, the World Bank, and the Government of the Congo.

Eves, H. E. 1996. The socioeconomics of natural resource utilization in northern Congo. Unpublished report to the Wildlife Conservation Society, the World Bank, and the Government of the Congo.

Fa, J. E. 1992a. *Conservación de los Ecosistemas Forestales en Guinea Ecuatorial*. Gland, Switzerland and Cambridge, England: International Union for Conservation of Nature and Natural Resources.

Fa, J. E. 1992b. Conservation in Equatorial Guinea. *Oryx* 26: 87–94.

Fa, J. E., J. Juste, J. Perez del Val and J. Castroviejo. 1995. Impact of market hunting on mammal species in Equatorial Guinea. *Conserv Biol* 9: 1107–15.

Fa, J. E. and A. Purvis 1997. Body size, diet and population density in Africotropical forest mammals: A comparison with neotropical species. *J Anim Ecol* 66: 98–112.

Falconer, J. 1991. The significance of forest resources in rural economies of southern Ghana: Summary of a study on non-timber forest projects. Unpublished report.

London: Forestry Department, Government of Ghana and Overseas Development Administration, UK.

Falconer, J. 1992. Non-timber forest products in southern Ghana. *ODA Forestry Series 2*.

FAO (Food and Agricultural Organization). 1982. *National Conservation Plan for Indonesia*. Vol. VI: Sulawesi. Field Report of UNDP/FAO National Parks Development Project. Jakarta, Indonesia.

FAO. 1991. Deuxième rapport intermédiaire sur l'état des forêts tropicales. *Nature Faune* 7: 10–2.

FAO/PNUD. 1984. *Guinea Ecuatorial: Desarrollo agrícola*. Preparación de proyectos y control de progreso de realización (Proyecto FAO/PNUD–EQG/81/007). Malabo: Ministerio de Agricultura, Ganadería y Desarrollo Rural.

FAO/WHO (Food and Agriculture Organization/World Health Organization). 1973. *Energy and Protein Requirements: Report of a Joint FAO/WHO ad Hoc Expert Committee*. WHO Technical Report, series no. 522. Geneva, Switzerland: World Health Organization.

Fay, J. M. 1993. *Ecological and Conservation Implications of Development Options for the Dzanga-Sangha Special Reserve and the Dzanga-Ndoki National Park*. Report to Deutsche Gesellschaft für Technische Zusammenarbeit (GTZ).

Fay, J. M. and M. Agnagna. 1992. Census of gorillas in the northern Republic of the Congo. *Am J Primatol* 27: 275–84.

Fay, J. M., M. Agnagna, and J. M. Moutsambote. 1990. *A Survey of the Proposed Nouabale Conservation Area in Northern Congo*. New York: Wildlife Conservation Society.

Fay, J. M., R. G. Ruggiero, and S. Blake. 1994. Use of a traditional hunting method to estimate prey availability in forests of northern Republic of Congo. Internal research report. Nouabalé-Ndoki National Park, Bomassa, Congo.

Fearnside, P. M. 1992. Reservas extrativistas: Uma estratégia de uso sustentado. *Ciência Hoje* 14: 15–7.

FED/DHV. 1989. *Proyecto de Rehabilitación de Cacao: Isla de Bioko. Uso Actual y Potencial de las Tierras*. Malabo, República de Guinea Ecuatorial: Ministerio de Agricultura, Ganadería, Pesca y Forestal.

Feer, F. 1988. Stratégies ecologiques de deux espèces de bovidés sympatriques de la forêt sempervirente africaine (*Cephalophus callipygus* et *C. dorsalis*): Influence du rhythme d'activité [Ph.D. thesis]. Paris: Université Pierre et Marie Curie.

Feer, F. 1989a. Occupation de l'espace par deux bovidés sympatriques de la forêt dense africaine (*Cephalophus callipygus* et *C. dorsalis*): Influence du rhythme d'activité. *La Terre et la Vie/Revue d'Ecologie* 44: 225–48.

Feer, F. 1989b. Comparison des régimes alimentaires des *Cephalophus callipygus* et *C. dorsalis*, Bovidés sympatriques de la forêt sempervirente africaine. *Mammalia* 53: 563–604.

Feer, F. 1993. "The Potential for Sustainable Hunting and Rearing of Game in Tropical Forests." In C. M. Hladik, A. Hladik, O. F. Linares, H. Pagezy, A. Semple, and M. Hadley, eds., *Tropical Forests, People and Food: Biocultural Interactions and Applications to Development*, pp. 691–708. Paris: UNESCO.

Feldmann, F. 1994. "Community Environmental Action: The National Policy Context." In D. Western and R. M. Wright, eds., *Natural Connections: Perspectives in Community-based Conservation*, pp. 393–402. Washington, DC.: Island Press.

Fenchel, T. 1974. Intrinsic rate of natural increase: The relationship with body size. *Oecologia* 14: 317–26.

Feron, E. M. 1995. New food sources, conservation of biodiversity and sustainable development: Can unconventional animal species contribute to feeding the world? *Biodiversity Conserv* 4: 233–40.

Ferraro, P. J. and R. A. Kramer. 1995. *A Framework for Affecting Household Behavior to Promote Biodiversity Conservation. Environment and Natural Resource Policy and Training (EPAT) Project*. Arlington, VA: Winrock International Environmental Alliance.

Fimbel, C. 1994. The relative use of abandoned farm clearings and old forest habitats by primates and a forest antelope at Tiwai, Sierra Leone, West Africa. *Biol Conserv* 70: 277–86.

FitzGibbon, C. D. 1994. The distribution and abundance of the Golden-rumped Elephant-shrew *Rhynchocyon chrysopygus* in Kenyan coastal forests. *Biol Conserv* 67: 153–60.

FitzGibbon, C. D., H. Leirs, and W. Verheyen. 1995a. Population dynamics, distribution and habitat choice of the lesser pouched rat, *Beamys hindei*. *J Zool* 236: 499–512

FitzGibbon, C. D., H. Mogaka, and J. H. Fanshawe. 1995b. Subsistence hunting in Arabuko-Sokoke Forest, Kenya and its effects on mammal populations. *Conserv Biol* 9: 1116–26.

FitzGibbon, C. D., H. Mogaka, and J. H. Fanshawe. 1996. "Subsistence Hunting and Mammal Conservation in a Kenyan Coastal Forest: Resolving a Conflict." In V. J. Taylor and N. Dunstone, eds., *The Exploitation of Mammal Populations*, pp. 147–59. London: Chapman and Hall.

FitzGibbon, C. D. and G. B. Rathbun. 1994. Surveying *Rhynchocyon* elephant-shrews in tropical forest. *Afr J Ecol* 32: 50–7.

Fleuret, P. and A. Fleuret. 1980. Nutrition, consumption and agricultural change. *Hum Organization* 39: 250–60.

Flowers, N. M. 1983. "Seasonal Factors in Subsistence, Nutrition, and Child Growth in a Central Brazilian Indian Community." In R. Hames and W. Vickers, eds., *Adaptive Responses of Native Amazonians*, pp. 357–90. New York: Academic Press.

Flowers, N. M., D. R. Gross, M. L. Ritter, and D. W. Werner. 1982. Variation in swidden practices in four central Brazilian Indian societies. *Hum Ecol* 10: 203–17.

FMB (Fundacion Moises Bertoni). 1992. *Reserva Natural Mbaracayú. Plan operativa*. Asuncion, Paraguay: FMB.

Fonseca, G. A. B. and J. G. Robinson. 1990. Forest size and structure: Competitive and predatory effects on small mammal communities. *Biol Conserv* 53: 265–94.

Fragoso, J. M. 1991. "The Effect of Hunting on Tapirs in Belize." In J. G. Robinson and K. H. Redford, eds., *Neotropical Wildlife Use and Conservation*, pp. 154–62. Chicago: University of Chicago Press.

Franta, M., E. Ndongo, F. Salomon, I. Segorb, and R. Mba. 1984. Encuesta nutricional en la isla de Bioko a los niños menores de 5 años. *Dia Mundial de la Alimentación, Malabo*.

Freeman, D. 1970. *Report on the Iban*. London: Althone Press.

Freese, C. H., P. G. Heltne, N. Castro, and G. Whitesides. 1982. Patterns and determinants of monkey densities in Peru and Bolivia, with notes on distributions. *Int J Primatol* 3: 53–90.

Friedl, E. 1975. *Women and Men: An Anthropologist's View*. New York: Holt, Rinehart and Winston.

Froment, M. E. 1887. Trois affluents français du Congo Rivières Alim, Likouala, Sanga. *Bull Soc Geogr Lille* 7: 458–74.

Frumhoff, P. C. 1995. Conserving wildlife in tropical forests managed for timber. *BioScience* 45: 456–64.

Fujita, M. S. and M. D. Tuttle. 1991. Flying foxes (Chiroptera: Pteropodidae): Threatened animals of key ecological and economic importance. *Conserv Biol* 5: 455–63.

Gade, D. 1976. Naturalization of plant aliens: The volunteer orange in Paraguay. *J Biogeogr* 3: 269–79.

Gadgil, M. 1990. India's deforestation: Patterns and processes. *Soc Natural Resources* 3: 131–43.

Gadgil, M. 1991. Conserving India's biodiversity: The societal context. *Evolutionary Trends Plants* 51: 3–8.

Gadgil, M. 1992. Conserving biodiversity as if people matter: A case study from India. *Ambio* 21: 266–70.

Gadgil, M. and K. C. Malhotra. 1985. Ecology is for the people. *South Asian Anthropologist* 6: 1–14.

Gadgil, M., S. N. Prasad, and R. Ali. 1983. Forest management and forest policy in India: A critical review. *Social Action* 33: 127–55.

Gadsby, E. L. 1990. The status and distribution of the drill, *Mandrillus leucophaeus*, in Nigeria. Unpublished report to Wildlife Conservation International, New York.

Gadsby, E. L. and P. D. Jenkins. 1992. Report on wildlife and hunting in the proposed Etinde Forest Reserve. Unpublished report to Overseas Development Administration, Cameroon.

Gartlan, J. S. 1989. *La conservation des ecosystèmes forestiers du Cameroun*. Gland, Switzerland: International Union for Conservation of Nature and Natural Resources.

Gautier-Hion, A., L. H. Emmons, and G. DuBost. 1980. A comparison of the diets of three major groups of primary consumers of Gabon (primates, squirrels and ruminants). *Oecologia* 45: 182–9.

Geddes, W. R. 1976. *Migrants of the Mountains*. Oxford, England: Clarendon Press.

Geist, V. 1988. How markets in wildlife meat and parts, and the sale of hunting privileges, jeopardize wildlife conservation. *Conserv Biol* 2: 15–26.

Geist, V. 1994. Wildlife conservation as wealth. *Nature* 368: 491–2.

Getz, W. M. and R. G. Haight. 1989. *Population Harvesting: Demographic Models of Fish, Forest and Animal Resources*. Princeton, NJ: Princeton University Press.

Giaccaria, B. and A. Heide 1984. *Xavante, povo autêntico*. São Paulo, Brasil: Editoria Salesiana Dom Bosco.

Gibson, C. and S. Marks. 1995. Transforming rural hunters into conservationists: An assessment of community-based wildlife management programs in Africa. *World Dev* 23: 941–57.

Giles-Vernick, T. 1996. A dead people? Migrants, land and history in the rainforests of the Central African Republic [Ph.D. Dissertation]. Baltimore: Johns Hopkins University.

Glander, K. E., J. Tapia R., and A. Fachin T. 1984. The impact of cropping on wild populations of *Saguinus mystax* and *Saguinus fuscicollis* in Peru. *Am J Primatol* 7: 89–97.

Glanz, W. 1982. "The Terrestrial Mammal Fauna of Barro Colorado Island: Censuses and Long-term Changes." In E. G. Leigh, Jr., A. S. Rand and D. M. Windsor, eds., *The Ecology of a Tropical Forest*, pp. 455–68. Washington, DC: Smithsonian Institution Press.

Glanz, W. 1990. "Neotropical Mammal Densities: How Unusual Is the Community on Barro Colorado Island." In A. Gentry, ed., *Four Neotropical Rainforests*, pp. 287–313. New Haven, CT: Yale University Press.

Glanz, W. 1991. "Mammalian Densities at Protected vs. Hunted Sites in Central Panama."

In J. G. Robinson and K. H. Redford, eds., *Neotropical Wildlife Use and Conservation.* pp. 163–73. Chicago: University of Chicago Press.

Glave, E. J. 1889–1890. The Congo River of To-day. *Century Illustrated Monthly Magazine* 39: 618–20.

Godoy, R., R. Lubowski, and A. Markandya. 1993. A method for the economic valuation of non-timber tropical forest products. *Economic Botany* 47: 220–33.

Gómez-Pompa, A. and A. Kaus. 1992. Taming the wilderness myth. *BioScience* 42: 271–9.

Góngora-Arones, E. 1987. *Etnozoología Lacandona: La Herpetofauna de Lacanja-Chansayab. Cuadernos de Divulgación.* No. 31. Veracruz, México: Instituto Nacional de Investigaciones sobre Recursos Bióticos.

Gonzalez-Kirchner, J. P. 1994. *Ecología y Conservación de los Primates de Guinea Ecuatorial.* Cantabria: Ceiba Ediciones.

Good, K. R. 1987. "Limiting Factors in Amazonian Ecology." In M. Harris and E. B. Ross, eds,. *Food and Evolution. Toward a Theory of Human Food Habits.* pp. 407–21. Philadelphia: Temple University Press.

Goodman, M. J., P. B. Griffin, A. Estioko-Griffin, and J. S. Grove. 1985. The compatibility of mothering and hunting among the Agta hunter-gatherers of the Philippines. *Sex Roles* 12: 1199–209.

Gordillo, G. 1988. *Campesinos al Asalto del Cielo: De la Expropiación Estatal a la Apropiación Campesina.* México: Siglo XXI Editores.

Gore, H. G., W. F. Harwell, M. D. Hobson, and W. J. Williams. 1985. "Buck Permits as a Management Tool in South Texas." In S. L. Beasom and S. F. Roberson, eds. *Game Harvest Management*, pp. 149–63. Kingsville, TX: Caesar Kleberg Wildlife Research Institute.

Gragson, T. L. 1989. Allocation of time to subsistence and settlement in a Ciri Khonome Pume village of the Llanos of Apure, Venezuela [Ph.D. dissertation]. University Park, PA: Pennsylvania State University.

Greenberg, L. 1992. Garden hunting among the Yucatec Maya: A coevolutionary history of wildlife and culture. *Ethnoecologica* 1: 23–33.

Gregor, T. 1985. *Anxious Pleasures: The Sexual Lives of an Amazonian People.* Chicago: The University of Chicago Press.

Griffin, M. B. 1996. Change and Stability: Agta Kinship in a History of Uncertainty [Ph.D. dissertation]. Urbana, IL: Department of Anthropology, University of Illinois.

Griffin, P. B. 1984. "Forager Resource and Land Use in the Humid Tropics." In C. Schrire, ed., *The Agta of Northeastern Luzon, the Philippines*, pp. 95–121. Orlando, FL: Academic Press.

Griffin, P. B. 1985. "Population Movements and Socio-economic Change in the Sierra Madre." In P. B. Griffin and A. Estioko-Griffin, eds., *The Agta of Northeastern Luzon: Recent Studies*, pp. 85–101. Cebu City: University of San Carlos Publications.

Griffin, P. B. 1989. "Hunting, Farming and Sedentism in a Rain Forest Foraging Society." In S. Kent, ed., *Farmers as Hunters: The Implications of Our Sedentarism*, pp. 70–80. Cambridge: Cambridge University Press.

Griffin, P. B. 1991. Philippine Agta forager-serfs: Commodities and exploitation. *Senri Ethnological Papers* 30: 199–222.

Griffin, P. B. and A. Estioko-Griffin, eds. 1985. *The Agta of Northeastern Luzon.* Cebu City: San Carlos Publications, University of San Carlos.

Groombridge, B., ed. 1994. *1994 IUCN Red List of Threatened Animals.* Gland, Switzerland and Cambridge, England: International Union for Conservation of Nature and Natural Resources.

Gross, D. 1983. "Village Movement in Relation to Resources in Amazonia." In R. Hames and W. Vickers, eds., *Adaptive Responses of Native Amazonians*, pp. 429–49. New York: Academic Press.

Gross, D., G. Eiten, N. M. Flowers, F. M. Leoi, M. L. Ritter, and D. Werner. 1979. Ecology and acculturation among native peoples of central Brazil. *Science* 206: 1043–50.

Grzimek, B. 1975. *Grzimek's Animal Life Encyclopedia. Mammals I–IV*. Vol. 10–13. New York: Van Nostrand Reinhold.

Guillemard, F. H. H. 1886. *The Cruise of the Marchesa to Kamschatka and New Guinea with Notices of Formosa, Liu-Liu, and Various Islands of the Malay Archipelago*. London: Murray.

Guinea, E. 1949. *En el País de los Bubis. Relato Ilustrado de Mi Primer Viaje a Fernando Poo*. Madrid: Instituto de Estudios Africanos.

Gumal, M. T. and Bennett, E. L. 1995. Report on a field trip to Penan communities of Belaga District, 11th–14th December 1995. Unpublished report. National Parks and Wildlife Office, Sarawak Forest Department.

Gunatilleke, I. A. U. N., C. V. S. Gunatilleke, and P. Abeygunawardena. 1993. Interdisciplinary research towards management of non-timber forest resources in lowland rain forests of Sri Lanka. *Economic Botany* 47: 282–90.

Haaga, J., J. Mason, F. Z. Omoro, V. Quinn, A. Raferty, K. Test, and L. Wasonga. 1986. Child malnutrition in rural Kenya: A geographical and agricultural classification. *Ecol Food Nutrition* 18: 297–307.

Haila, Y. 1988. Calculating and miscalculating density: The role of habitat geometry. *Ornis Scand* 19: 88–92.

Hall, J. 1993. Report on the strategic planning mission for the creation of a protected area in the Lobéké Region of Southeastern Cameroon: Assessment of timber exploitation, safari hunting, and preliminary vegetation analsysis. Unpublished report to the Wildlife Conservation Society, New York.

Hamblin, N. L. 1984. *Animal Use by the Cozumel Maya*. Tucson, AZ: University of Arizona Press.

Hamblin, N. L. 1985. "The Role of Marine Resources in the Maya Economy: A Case Study from Cozumel, Mexico." In: M. Pohl, ed. *Prehistoric Maya Environment and Subsistence Economy*, pp 159–73. Papers of the Peabody Museum of Archaeology and Ethnology, Vol. 77. Cambridge, MA: Harvard University.

Hames, R. 1979. A comparison of the efficiencies of the shotgun and the bow in neotropical forest hunting. *Hum Ecol* 7: 219–52.

Hames, R. 1983. "The Settlement Pattern of a Yanomamo Population Bloc: A Behavioral Ecological Interpretation." In R. Hames and W. Vickers, eds., *Adaptive Responses of Native Amazonians*, pp. 393–427. New York: Academic Press.

Hames, R. 1987. Game conservation or efficient hunting? In B. McCay and J. Acheson, eds., *The Question of the Commons*, pp. 97–102. Tucson, AZ: University of Arizona Press.

Hames, R. 1991. "Wildlife Conservation in Tribal Societies." In M. L. Oldfield and J. Alcorn, eds., *Culture, Conservation and Ecodevelopment*, pp. 172–99. Boulder, CO: Westview Press.

Hamilton, A. C. 1981. The quaternary history of African forests: Its relevance to conservation. *Afr J Ecol* 19: 1–6.

Hannah, L. 1992. African people, African parks. Report to USAID, Biodiversity Support Program, and Conservation International.

Harako, R. 1976. The Mbuti as hunters: a study of ecological anthropology of the Mbuti Pygmies. *Kyoto Univ Afr Stud* 10: 37–99.

Harako, R. 1981. "The Cultural Ecology of Hunting Behavior Among Mbuti Pygmies in the Ituri Forest, Zaire." In R. S. O. Harding and G. Teleki, eds., *Omnivorous Primates*. New York: Columbia University Press.

Hardin, G. 1968. The tragedy of the commons. *Science* 162: 1243–8.

Hardin, R. 1997. Culture Contact and Control of Resources in the Southwestern Central African Republic [Ph.D. dissertation]. New Haven, CT: Yale University.

Hardouin, J. 1995. Minilivestock: From gathering to controlled production. *Biodiversity Conserv* 4: 220–32.

Harris, M. and E. B. Ross, eds. 1987. *Food and Evolution. Toward a Theory of Human Food Habits*. Philadelphia: Temple University Press.

Harris, S., W. J. Cresswell, P. G. Forde, W. J. Trewhella, T. Woollard, and S. Wray. 1990. Home-range analysis using radio-tracking data—A review of problems and techniques particularly as applied to the study of mammals. *Mammal Rev* 20: 97–123.

Hart, J. 1979. Nomadic Hunters and Village Cultivators: A Study of Subsistence Interdependence in the Ituri Forest of Zaire [M.S. thesis]. East Lansing, MI: Michigan State University.

Hart, J. A. 1978. From subsistence to market: A case study of the Mbuti net hunters. *Hum Ecol* 6: 325–53.

Hart, J. A. 1986. Comparative dietary ecology of a community of frugivorous forest ungulates in Zaire [Ph.D. dissertation]. East Lansing, MI: Michigan State University.

Hart, J. A. and J. Hall. 1996. Status of eastern Zaire's forest parks and reserves. *Conserv Biol* 10: 316–27.

Hart, J. A. and T. B. Hart. 1984. The Mbuti of Zaire: Political change and the opening of the Ituri Forest. *Cultural Survival Q* 8: 18–20.

Hart, J. A., M. Katembo, and K. Punga. 1996. Diet, prey selection and ecological relations of leopard and golden cat in the Ituri Forest, Zaire. *Afr J Ecol* 34: 364–79.

Hart, J. A. and G. Petrides. 1987. "A Study of Relationships Between Mbuti Hunting Systems and Faunal Resources in the Ituri Forest, Zaire." In A. Lugo, J. Ewel, S. Hecht, P. Murphy, C. Padoch, M. Schmink, and D. Stone, eds., *People and the Tropical Forest*. pp. 12–5. Washington, DC: U.S. Man and the Biosphere Program, U.S. State Department.

Hart, T. B. 1985. The ecology of a single-species-dominant forest and of a mixed forest in Zaire, Africa [Ph.D. dissertation]. East Lansing, MI: Michigan State University.

Hart, T. B., J. A. Hart, and P. Murphy. 1989. Monodominant and species-rich forests of the humid tropics: Causes for their co-occurrence. *Am Naturalist* 133: 613–33.

Hart, T. B. and J. A. Hart. 1986. The ecological basis for hunter gatherer subsistence in the African rain forest: The Mbuti of eastern Zaire. *Hum Ecol* 14: 29–55.

Hawkes, K. 1993. Why hunter-gatherers work. *Curr Anthropol* 34: 341–61.

Hawkes, K., K. Hill, and J. O'Connell. 1982. Why hunters gather: Optimal foraging and the Aché of eastern Paraguay. *Am Ethnologist* 2: 379–98.

Hayssen, V. 1984. Basal metabolic rate and the intrinsic rate of increase: An empirical and theoretical examination. *Oecologia (Berlin)* 64: 419–24.

Hayssen, V. 1993. *Asdell's Patterns of Mammalian Reproduction: A Compendium of Species-Specific Data*. Ithaca, NY: Comstock Publishing Associates.

Headland, T. N. 1986. Why foragers do not become farmers: A historical study of a changing ecosystem and its effect on a Negrito hunter-gatherer group in the Philippines [Ph.D. dissertation]. Manoa, HI: Department of Anthropology, University of Hawaii.

Hearn, G. W. and R. W. Berghaier. 1996. Census of diurnal primate groups in the Gran Caldera Volcanica de Luba, Bioko Island, Equatorial Guinea, January, 1996. Unpublished report to the Government of Equatorial Guinea, Beaver College, Glenside.

Hecketsweiler, P. 1990. *La Conservation des Ecosystèmes Forestiers du Congo*. Gland, Switzerland: IUCN.

Hedges, S. 1996. "The Anoas, Lowland Anoa (*Bubalus depressicornis*) and Mountain Anoa (*Bubalus quarlesi*): Review of their Biology, Distribution and Status." In S. Hedges, ed., *Asian Wild Cattle and Buffaloes: Status Report and Conservation Action Plan*. Gland, Switzerland: International Union for Conservation of Nature and Natural Resources/ Species Survival Commission Asian Cattle Specialist Group.

Hedrik, P. W., R. C. Lacy, F. W. Allendorf, and M. E. Soulé. 1996. Directions in conservation biology: Comments on Caughley. *Conserv Biol* 10: 1312–20.

Heinen, J. T. 1996. Human behavior, incentives, and protected area management. *Conserv Biol* 10: 681–4.

Henley, P. 1982. *The Panare. Tradition and Change on the Amazonian Frontier*. New Haven, CT: Yale University Press.

Hennemann, W. W. 1983. Relationship among body mass, metabolic rate and the intrinsic rate of natural increase in mammals. *Oecologia* 56: 104–8.

Hennessey, A. B. 1995. A Study of the Meat Trade in Ouesso, Republic of the Congo. Report to Deutsche Gesellschaft für Technische Zusammenarbeit (GTZ) and Wildlife Conservation Society (WCS).

Hernandez-Camacho, J. and R. W. Cooper. 1976. "The Nonhuman Primates of Colombia." In R. W. Thorington, Jr., and P. G. Heltne, eds., *Neotropical Primates: Field Studies and Conservation*, pp. 35–69. Washington, DC: National Academy of Science.

Hewlett, B. 1977. Notes on the Mbuti and Aka pygmies of Central Africa [Master's thesis]. California: California State University.

Hickson, S. J. 1889. *A Naturalist in North Celebes*. London: Murray.

Hilbourn, R., C. J. Walters, and D. Ludwig. 1995. Sustainable exploitation of renewable resources. *Annu Rev Ecol Systematics* 26: 45–67.

Hill, K. 1996. "The Mbracayu Reserve and the Aché of Paraguay." In K. H. Redford and J. A. Mansour, eds., *Traditional Peoples and Biodiversity Conservation in Large Tropical Landscapes*, pp. 159–96. Arlington, VA: The Nature Conservancy.

Hill, K. and K. Hawkes. 1983. "Neotropical hunting among the Aché of eastern Paraguay. Shotguns, blowguns, and spears: The analysis of technological efficiency." In R. Hames and W. Vickers, eds., *Adaptive Responses of Native Amazonians*, pp. 139–88. New York: Academic Press.

Hill, K., K. Hawkes, H. Kaplan, and A. Hurtado 1987. Foraging decisions among Aché hunter-gatherers: New data and implications for optimal foraging models. *Ethol Sociobiol* 8: 1–36.

Hill, K. and A. Hurtado. 1989. Hunter-Gatherers of the New World. *Am Scientist* 77: 434–43.

Hill, K. and A. Hurtado. 1996. *Aché Life History: The Ecology and Demography of a Foraging People*. New York: Aldine de Gruyter.

Hill, K., H. Kaplan, K. Hawkes, and A. Hurtado. 1984. Seasonal variance in the diet of Aché hunter-gatherers of eastern Paraguay. *Hum Ecol* 12: 145–80.

Hill, K., H. Kaplan, K. Hawkes, and A. Hurtado. 1985. Men's time allocation to subsistence work among the Aché of eastern Paraquay. *Hum Ecol* 13: 29–47.

Hill, K., J. Padwe, C. Bejyvagi, A. Bepurangi, F. Jakugi, R. Tykuarangi, and T. Tykuarangi. 1997. Impact of hunting on large vertebrates in the Mbaracayú Reserve, Paraguay. *Conserv Biol* 11: 1339–53.

Hill, K. and T. Tykuarangi. 1996. "Case Study: The Mbaracayú Reserve and the Aché of Paraguay." In K. H. Redford and J. A. Mansour, eds., *Traditional Peoples and Biodiversity Conservation in Large Tropical Landscapes*, pp. 159–96. Washington, DC: The Nature Conservancy.

Hilty, S. L. and W. Brown. 1986. *A Guide to Birds of Colombia*. Princeton, NJ: Princeton University Press.

Hjerpe, K. No date. Like food for nourishment, nutrition and socioeconomic change: The case of the Yuquí Indians of Bolivia. Unpublished manuscript.

Hladik, C. M., S. Bahuchet, and I. de Garine, 1990. *Food and Nutrition in the African Rain Forest*. Paris, France: UNESCO-MAB.

Hodson, T., F. Enhlander, and H. O' Keefee 1995. Rain forest preservation, markets and medicinal plants: Issues of property rights and present value. *Conserv Biol* 9: 1319–21.

Hofer, H., K. L. I. Campbell, M. East, and S. A. Huish. 1996. "The Impact of Game Meat Hunting on Target and Non-target Species in the Serengeti." In V. J. Taylor and N. Dunstone, eds., *The Exploitation of Mammal Populations*, pp. 117–46. London: Chapman and Hall.

Holmberg, A. R. 1969. *Nomads of the Long Bow: The Sirionó of Eastern Bolivia*. American Museum of Natural History. Garden City, NJ: The Natural History Press.

Holmes, D. and K. Phillipps. 1996. *The Birds of Sulawesi*. New York: Oxford University Press.

Holmes, R. 1995. "Small Is Adaptive: Nutritional Anthropometry of Native Amazonians." In L. Sponsel, ed., *Indigenous Peoples and the Future of Amazonia: An Ecological Anthropology of an Endangered World*, pp. 121–48. Tucson: The University of Arizona Press.

Hooijer, D. A. 1960. The orang-utan in Niah prehistory. *Sarawak Museum J* 9: 408–21.

House, W. J. 1991. The nature and determinants of socioeconomic inequality among peasant households in southern Sudan. *World Dev* 19: 867–84.

Howell, K. M. 1981. Pugu Forest Reserve: Biological values and development. *Afr J Ecol* 19: 73–81.

HPI (Heifer Project International). 1996. *Boyo Rural Integrated Farmer's Alliance, Cameroon: Project Summary*. Little Rock, AR: Heifer Project International.

Hudson, J. L. 1990. Advancing methods in zooarchaeology: an ethnoarchaeological study among the Aka [Ph.D. Dissertation]. Santa Barbara, CA: University of California.

Hutterer, K. L. 1988. "The Prehistory of the Asian Rain Forests." In J. S. Denslow and C. Padoch, eds., *Peoples of the Tropical Rain Forest*, pp. 63–72. Los Angeles: University of California Press.

ICFRE (Indian Council for Forestry Research and Education). 1995. *Forestry Statistics India*. Dehra Dun, India: ICFRE.

Ichikawa, M. 1978. The residential groups of the Mbuti Pygmies. *Senri Ethnological Studies* 1: 131–88.

Ichikawa, M. 1983. An examination of the hunting-dependent life of the Mbuti Pygmies, eastern Zaire. *Afr Stud Monogr* 4: 55–76.

Ichikawa, M. 1986. Ecological basis of symbiosis, territoriality and intra-based cooperation of the Mbuti Pygmies. *Sprache Gesch Afr* 7: 161–88.

IIED (International Institute for Environment and Development). 1994. *Whose Eden? An Overview of Community Approaches to Wildlife Management*. Nottingham, England: Russell Press.

Infield, M. 1988. Hunting, trapping and fishing in villages within and on the periphery of the Korup National Park. Report to World Wide Fund for Nature, Gland Switzerland.

Irby, L. R., J. E. Swenson, and S. T. Stewart. 1989. Two views of the impacts of poaching on bighorn sheep in the upper Yellowstone Valley, Montana, USA. *Biol Conserv* 47: 259–72.

Iriarte, J. A. and F. M. Jaksíc. 1986. The fur trade in Chile: an overview of seventy five years of export data (1920–1984). *Biol Conserv* 38: 243–53.

Irvine, D. 1987. Resource management by the Runa Indians of the Ecuadorian Amazon [Ph.D. dissertation]. Palo Alto, CA: Stanford University.

ITTO (International Tropical Timber Organization). 1988. *The Case for Multiple Use Management of Tropical Hardwoods.* Cambridge, MA: Harvard Institute for International Development.

IUCN (International Union for Conservation of Nature and Natural Resources). 1991. *Atlas of Tropical Rainforests.* Gland, Switzerland: IUCN Special Publication.

IUCN. 1992. *Protected Areas of the World: A Review of National Systems.* Vol. 4. Nearctic and Neotropics. Cambridge, England: World Conservation Monitoring Centre.

IUCN. 1994. *IUCN Red List Categories.* Gland, Switzerland: IUCN Species Survival Commission.

IUCN. 1996. *IUCN List of Threatened Species.* Gland, Switzerland: IUCN.

IUCN/UNEP (United Nations Environment Programme)/WWF (World Wildlife Fund). 1980. *World Conservation Strategy. Living Resource Conservation for Sustainable Development.* Gland, Switzerland: IUCN/UNEP/WWF.

IUCN/UNEP/WWF. 1991. *Caring for the Earth: A Strategy for Sustainable Living.* Gland, Switzerland: IUCN/UNEP/WWF.

Ivlev, V. S. 1961. *Experimental Ecology of the Feeding of Fishes.* New Haven, CT: Yale University Press.

Janson, C. H. and L. Emmons. 1990. "Ecological Structure of the Non-flying Mammal Community at Cocha Cashu Biological Station, Manu National Park, Peru." In A. Gentry, ed., *Four Neotropical Rainforests*, pp. 339–57. New Haven, CT: Yale University Press.

Janzen, D. H. 1978. A bat-generated fig seed shadow in rain forest. *Biotropica* 10: 121.

Jarman, P. J. 1974. The social organisation of antelope in relation to their ecology. *Behaviour* 48: 215–67.

Jenkins, C. and K. Milton 1993. "Food Resources and Survival Among the Hagahai of Papua New Guinea." In C. M. Hladik, A. Hladik, O. F. Linares, H. Pagezy, A. Semple, and M. Hadley, eds., *Tropical Forests, People and Food*, pp. 281–93. Paris: UNESCO.

Johansson, P. 1995. A bird in the pot or two in the bush? *New Scientist* 2006: 52.

Johns, A. G. 1997. *Timber Production and Biodiversity Conservation in Tropical Rain Forests.* Cambridge, England: Cambridge University Press.

Johnsingh, A. J. T. 1986. Diversity and conservation of carnivorous mammals in India. Proceedings of the Indian Academy of Sciences Supplement 73–86.

Johnson, A. and M. Baksh. 1987. "Ecological and Structural Influences on the Proportions of Wild Foods in the Diets of Two Machiguenga Communities." In M. Harris and E. Ross, eds., *Food and Evolution. Toward a Theory of Human Food Habits*, pp. 387–406. Philadelphia: Temple University Press.

Jones, D. N., R. W. R. J. Dekker, and C. S. Roselaar. 1995. *The Megapodes: Bird Families of the World.* Vol. 3. Oxford, England: Oxford University Press.

Jones, J. 1990. A native movement and march in Eastern Bolivia: Rationale and response. Development Anthropology Network. *Bull Inst Dev Anthropol* 8: 1–8.

Jones, P. J. 1994. Biodiversity in the Gulf of Guinea: An overview. *Biodiversity Conserv* 3: 772–784.

Jorgenson, J. P. 1993. Gardens, wildlife densities, and subsistence hunting by Maya Indians in Quintana Roo, Mexico [Ph.D. dissertation]. Gainesville: University of Florida.

Jorgenson, J. P. 1995a. Maya subsistence hunters in Quintana Roo, Mexico. *Oryx* 29: 49–57.

Jorgenson, J. P. 1995b. "A Profile of Maya Subsistence Hunters in Southeastern Mexico." In J. A. Bissonette and P. R. Krausman, eds., *Integrating People and Wildlife for a Sustainable Future*. Bethesda, MD: The Wildlife Society.

Jorgenson, J. P. No date. The impact of hunting on wildlife in the Maya forest of Mexico and suggestions for its management. Unpublished manuscript.

Jori, F. 1996. *Etude sur la Faisabilité de l'Elevage Commercial d'Espèces Sauvages au Gabon*. Gabon: World Wide Fund for Nature.

Jori, F., G. A. Mensah, and E. Adjanohoun. 1995. Grasscutter production: An example of rational exploitation of wildlife. *Biodiversity Conserv* 4: 257–65.

Joshi, N. V. and M. Gadgil 1991. On the role of refugia in promoting prudent use of biological resources. *Theor Pop Biol* 40: 211–29.

Juste, B. J. 1989. *Zonación Ecologica y Evaluación del Impacto Ambiental de Usos Actuales en la Isla de Bioco*. Malabo: Ministerio de Agricultura y Forestal, Organización de las Naciones Unidas para la Agricultura y la Alimentación.

Juste, B. J. 1992. *Zonación Ecologica y Evaluación del Impacto Ambiental de Usos Actuales en la Isla de Bioco*. Malabo: Ministerio de Agricultura y Forestal. Organización de las Naciones Unidas para la Agricultura y la Alimentación.

Juste, B. J., and J. E. Fa. 1994. Biodiversity conservation in the Gulf of Guinea islands: Taking stock and preparing action. *Biodiversity Conserv* 3: 759–71.

Juste, B. J. and J. Perez del Val. 1995. Altitudinal variation in the subcanopy fruit bat guild in Bioko island, Equatorial Guinea, Central Africa. *J Trop Ecol* 11: 141–6.

Juste, J. and J. Cantero. 1991. *Informe Nacional Sobre Medio Ambiente y Desarrollo: Guinea Ecuatorial*. Malabo: Conferencia de las Naciones Unidas Sobre Medio Ambiente y Desarrollo (UNCED).

Juste, J., J. Fa, J. Perez del Val, and J. Castroviejo. 1995. Market dynamics of bushmeat species in Equatorial Guinea. *J Appl Ecol* 32: 454–67.

Kahurananga, J. 1981. Population estimates, densities, and biomass of large herbivores in Simanjiro plains, northern Tanzania. *Afr J Ecology* 19: 225–38.

Kalivesse, A. 1991. L'approvisionnement des marchés de Bangui en viande de chasse. *Nature Faune* 17: 14–20.

Kancewick, M. and E. Smith. 1991. Subsistence in Alaska: Towards a native priority. *UMKC Law Rev* 59: 645–77.

Kantor Statistik, Manado. 1993. *Sulawesi Utara dalam Angka*. Manado, Indonesia: Kantor Statistik, Propinsi Sulawesi Utara.

Kaplan, H. and K. Hill 1985. Food sharing among Aché foragers: Tests of explanatory hypotheses. *Curr Anthropol* 26: 223–45.

Kaplan, H., K. Hill, K. Hawkes, and A. Hurtado. 1984. Food sharing among the Aché hunter-gatherers of eastern Paraguay. *Curr Anthropol* 25: 113–5.

Kaplan, J. E., J. W. Larrick, and J. A. Yost. 1984. Workup on the Waorani. *Natural History* 93: 69–75.

Kaplan, R. 1997. Was democracy just a moment? *Atlantic Monthly* 280: 55–80.

Karanth, K. U. 1987. Status of wildlife and habitat conservation in Karnataka. *J Bombay Natural History Soc* 83(suppl): 166–79.

Karanth, K. U. 1988. Population structure, density and biomass of large herbivores in a south Indian tropical forest [M.S. thesis]. Gainesville: University of Florida.

Karanth, K. U. 1991. "Ecology and Management of the Tiger in Asia." In N. Maruyama, B. Bobek, Y. Ono, W. Regelin, L. Bartos, and P. R. Ratcliffe, eds., *Wildlife Conservation: Present Trends and Perspectives for the 21st Century*, pp. 156–9. Tokyo: Japan Wildlife Research Centre.

Karanth, K. U. 1995. Estimating tiger *Panthera tigris* populations from camera-trap data using capture-recapture models. *Biol Conserv* 71: 331–8.

Karanth, K. U. and M. E. Sunquist. 1992. Population structure, density and biomass of large herbivores in the tropical forests of Nagarahole, *India J Trop Ecol* 8: 21–35.

Karanth, K. U. and M. E. Sunquist. 1995. Prey selection by tiger, leopard and dhole in tropical forests. *J Animal Ecol* 64: 439–50.

Karsenty, A. and H. F. Maître. 1994. L'exploitation et la gestion durable des forêts tropicales: Pour des nouveaux outils de régulation. *Bois Forêts Trop* 240: 37–51.

Kay, C. 1994. Aboriginal overkill: The role of Native Americans in structuring western ecosystems. *Hum Nature* 5: 359–79.

Keel, S. 1987. *Informe de Viaje al Terreno del Banco Mundial*. Asunción, Paraguay: Centro de Datos para la Conservación.

Keel, S., A. Gentry, and L. Spinzi. 1993. Using vegetation analysis to facilitate the selection of conservation sites in eastern Paraguay. *Conserv Biol* 7: 66–75.

Kelsey, M. G. and T. E. S. Langton. 1984. *The Conservation of the Arabuko-Sokoke Forest, Kenya*. Study Report no. 4. East Anglia, England: International Council for Bird Preservation.

Kensinger, K. M. 1995. *How Real People Ought to Live. The Cashinahua of Eastern Peru*. Prospect, IL: Waveland Press.

King, S. 1994. Utilization of wildlife in Bakossiland, West Cameroon, with particular reference to primates. *TRAFFIC Bull* 14: 63–73.

King, V. T. 1978. "Introduction." In V. T. King, ed., *Essays on Borneo Societies*, pp. 1–36. Oxford, England: Oxford University.

Kingdon, J. 1997. *The Kingdon Field Guide to African Mammals*. London: Academic Press.

Kinnaird, M. F. and T. G. O'Brien. 1995a. Variation in fruit resources and the effects on vertebrate frugivores: The role of disturbance regimes in Sulawesi rainforests. Unpublished report to Wildlife Conservation Society and PHPA, Bogor, Indonesia.

Kinnaird, M. F. and T. G. O'Brien. 1995b. *Tangkoko-DuaSudara Nature Reserve, North Sulawesi: Draft Management Plan 1996–2000*. J. L. Gatot Subroto, Jakarta, Indonesia: PHPA, Bina Program. Gedung Manggala Wanabakti.

Kinniard, M. F. and T. G. O'Brien. 1996. Ecotourism in the Tangkoko-DuaSudara Nature Reserve: Opening Pandora's box? *Oryx* 30: 65–73.

Kinnaird, M. F., T. G. O' Brien, and S. Suriyadi. 1996. Population fluctuation in Sulawesi red-knobbed hornbills: Tracking figs in space and time. *Auk* 113: 431–40.

Kitanishi, K. 1994. The exchange of forest products (*Irvingia* nuts) between the Aka hunter-gatherers and the cultivators in northeastern Congo. *Tropics* 4: 79–92.

Kitanishi, K. 1995a. Seasonal changes in the subsistence activities and food intake of the Aka hunter-gathers in northeastern Congo. *Afr Stud Monogr* 16: 73–118.

Kitanishi, K. 1995b. *Difference in the Subsistence Activities and Interdependence Between the Old and the Young Men Among the Aka Hunter-Gatherers in Northeastern Congo*. African Studies Monographs. Kyoto, Japan: Kyoto University.

Kleiman, D. G. 1970. Reproduction in the female green acouchi, *Myoprocta pratti* Pocock. *J Reprod Fertil* 23: 55–65.

Kleiman, D. G., M. E. Allen, K. V. Thompson, and S. Lumpkin, eds. 1997. *Wild Mammals in Captivity*. Chicago: University of Chicago Press.

Kleymeyer, C. D. 1994. "Cultural Traditions and Community-based Conservation." In D. Western and R. M. Wright, eds., *Natural Connections: Perspectives in Community-based Conservation*, pp. 323–46. Washington, DC: Island Press.

Knowlton, F. F. 1972. Preliminary interpretation of coyote population mechanics with some management implications. *J Wildlife Management* 36: 369–82.

Kock, R. 1995. Wildife utilization: Use it or lost it—A Kenyan perspective. *Biodiversity Conserv* 4: 241–56.

Kohlhaas, A. K. 1993. Behavior and ecology of *Macaca nigrescens*: Behavioral and social responses to the environment and fruit availability [Ph.D. dissertation]. Boulder: University of Colorado.

Koppert, G. 1996. In A. Froment, I. de Garine, C. Binam Bikoi, and J. F. Loung, eds., *Bien Manger et Bien Vivre: Anthropologie Alimentaire dans les Régions Forestières et les Savanes du Cameroun*, pp. 89–98. Paris: Orstom.

Koppert, G. J. A., E. Dounias, A. Froment, and P. Pasquet. 1993. "Food Consumption in Three Forest Populations of the Southern Coastal Area of Cameroon: Yassa-Mvae-Bakola." In C. M. Hladik, A. Hladik, O. F. Linares, H. Pagezy, A. Semple, and M. Hadley, eds., *Tropical Forests, People and Food*, pp. 295–310. Paris: UNESCO.

Koppert, G. and C. M. Hladik. 1990. "Measuring Food Consumption." In C. M. Hladik, S. Bahuchet, and I. de Garine, eds., *Food and Nutrition in the African Rain Forest*, pp. 59–61. Paris, France: UNESCO-MAB.

Koster, S. and J. Hart. 1988 Methods of estimating ungulate populations in tropical forest. *Afr J Ecol* 26: 117–26.

Kothari, A., S. Suri, and N. Singh. 1995. People and protected areas: Rethinking conservation in India. *Ecologist* 25: 188–94.

Kramer, R. A., N. Sharma, P. Shyamsundar, and M. Munasinghe. 1994. *Cost and Compensation Issues in Protecting Tropical Rainforests: Case Study of Madagascar*. Environment Working Paper No. 62. Washington, DC: The World Bank, Environment Department, African Technical Department.

Kranz, K. 1986. *Cephalophus sylvicultor*. *Mammalian Species* 225: 1–7.

Krebs, C. J. 1985. *Ecología: Estudio de la Distribución y la Abundancia*, 2nd ed. Mexico City: Editorial Harla.

Krebs, J. R. 1979. "Foraging Strategies and Their Social Significance." In P. Maileo and J. G. Vandenbergh, eds. *Handbook of Behavioral Neurobiology*, Vol. 3, pp. 225–70. New York: Plenum.

Kruyt, A. 1930. De To Wana op Ost-Celebes. *Tijdschrift Voor Indische Taal-, Land-, en Volkenkunde* 70: 398–625.

KSPSU (Kantor Statistik Propinsi Sulawesi Utara). 1983. *Sulawesi Utara Dalam Angka*.

KSPSU. 1988. *Sulawesi Utara Dalam Angka*.

KSPSU. 1990. *Sulawesi Utara Dalam Angka*.

KSPSU. 1992. *Sulawesi Utara Dalam Angka*.

KSPSU. 1993. *Sulawesi Utara Dalam Angka*.

KSPSU. 1994. *Sulawesi Utara Dalam Angka*.

Kumar, A., V. Menon, and S. Gupta. 1993. Sustainable use of wildlife: Views and perspectives. Unpublished compilation, TRAFFIC/World Wildlife Fund, New Delhi.

Kwapena, N. 1984. Traditional conservation and utilization of wildlife in Papua New Guinea. *Environmentalist* 4(suppl 7): 22–6.

Laake, J., S. Buckland, D. Anderson, and K. Burnham. 1993. *Distance Sampling: Abundance Estimation of Biological Populations—Distance Users Guide version 2.0*. Fort Collins, CO: Colorado Co-operative Fish and Wildlife Research Unit, Colorado State University.

Lahm, S. A. 1993a. "Utilization of Forest Resources and Local Variation of Wildlife Populations in Northeastern Gabon." In C. M. Hladik, A. Hladik, O. F. Linares, H. Pagezy, A. Semple, and M. Hadley, eds. *Tropical Forests, People and Food: Biocultural Interactions and Applications to Development*, pp. 213–26. Paris: The Parthenon Publishing Group.

Lahm, S. A. 1993b. Ecology and economics of human/wildlife interaction in northeastern Gabon [Ph.D. dissertation]. New York: New York University.

Lande, R. 1988. Genetics and demography in biological conservation. *Science* 241: 1455–60.

Langub, J. 1988. "The Penan Strategy." In J. S. Denslow and C. Padoch, eds., *People of the Tropical Rain Forest*, pp. 207–8. Berkeley: University of California Press.

Laurent, E. 1992. Wildlife utilization survey of villages surrounding the Rumpi Hills Forest Reserve. Mundemba, Cameroun. Deutche Gesellschaft Technische Zusammenarbeit-Korup Project.

Lavigne, D. M., C. J. Callaghan, and R. J. Smith. 1996. "Sustainable Utilization: The Lessons of History." In V. J. Taylor and N. Dunstone, eds., *The Exploitation of Mammal Populations*, pp. 250–265. London: Chapman and Hall.

Lazcano-Barrero, M. A., O. A. Flores-Villela, M. Benabid-Nisenbaum, J. A. Hernández-Gómez, M. P. Chávez-Peón, and A. Cabrera-Aldave. 1988. *Estudio y Conservación de los Anfibios y Reptiles de México: Una Propuesta*. Cuadernos de Divulgación Inireb No. 25. Xalapa, Veracruz, México: Instituto Nacional de Investigaciones sobre Recursos Bióticos.

Leader-Williams, N., and E. J. Milner-Gulland. 1993. Policies for the enforcement of wildlife laws: The balance between detection and penalties in Luangwa Valley, Zambia. *Conserv Biol* 7: 611–7.

Lee, J. C. 1980. *An Ecogeographic Analysis of the Herpetofauna of the Yucatán Peninsula*. Miscellaneous publication. Lawrence, KS: Museum of Natural History, University of Kansas.

Lee, K. N. 1993. *Compass and Gyroscope: Integrating Science and Politics for the Environment*. Washington, DC: Island Press.

Lee, P. C., J. Thornback, and E. L. Bennett. 1988. *Threatened Primates of Africa*. The IUCN Red Data Book. Gland, Switzerland, and Cambridge, United Kingdom: International Union for Conservation of Nature and Natural Resources.

Lee, R. and I. DeVore. 1968. *Man the Hunter*. Chicago: Aldine.

Lee, R. J. 1995. Population survey of the crested black macaque (*Macaca nigra*) at Manembonembo Nature Reserve in North Sulawesi, Indonesia. *Primate Conserv* 16: 63–5.

Lee, R. J. 1996. Population surveys and conservation status of the crested black macaque (*Macaca nigra*) in North Sulawesi, Indonesia. *Primate Conserv* 16: 63–5.

Lee, R. J. 1997. Impact of hunting and habitat disturbance on the population dynamics and behavioral ecology of the crested black macaque (*Macaca nigra*) [Ph.D. dissertation]. Eugene, OR: University of Oregon.

Leeuwenberg, F. 1994. Análise etno-zoológica e manejo da fauna cinegética na Reserve Indigena Xavante Pimentel Barbosa, Mato Grosso, Brazil. Unpublished report to Centro de Pesquisa Indígena, Wildlife Conservation Society, World Wildlife Fund, and Gaia Foundation.

Leeuwenberg, F. 1995. Relatório viagem Reserva Indígena Rio Breu, Alto Juruá, Acre. Unpublished report to Centro de Pesquisa Indígena, Brasil.

Leeuwenberg, F., L. Pinder, and S. Lara Resende. 1997. "Manejo de Populações Silvestres." In J. M. Barbanti Duarte, ed., *Biologia e Conservação de Cervídeos Sul-Americanos: Blastoceros, Ozotoceros, e Mazama*, pp. 110–23. Brasil: UNESP (Universidade Estadual de Sao Paulo), FUNEP (Fundação de Estudos e Pesquisas em Agronomia, Medicina Veterinaria e Zootecnica), FAPESP (Fundação de Amparo a Pesquisa do Estado de Sao Paulo).

Lefkovitch, L. P. 1967. A theoretical evaluation of population growth after removing individuals from some age groups. *Bull Entomol Res* 57: 437–45.

Lehm, Z. 1991. La demanda territorial del Pueblo Sirionó. Unpublished report to CIDDEBENI, Trinidad, Bolivia.

Lélé, S. and R. B. Norgaard. 1996. Sustainability and the scientist's burden. *Conserv Biol* 10: 354–65.

Leopold, A. 1933. *Game Management*. Madison, WI: University of Wisconsin Press.

Leopold, A. 1949. *A Sand County Almanac*. New York: Oxford University Press.

Leopold, A. S. 1977. *Fauna Silvestre de México: Aves y Mamíferos de Caza*. Instituto Mexicano de Recursos Naturales Renovables, México.

Letouzy, R. 1985. *Notice de la Carte Phytogéographique du Cameroun au 1: 500,000*. Toulouse, France: Institute de la Carte Internationale de la Végétation.

Leung, W. T. W. and M. Flores. 1961. Tabla de Composición de Alimentos Para Uso en América Latina. Instituto de Nutrición de Centro América y Panama, Guatemala.

Lewin, R. 1983. What killed the giant mammals? *Science* 221: 1036–7.

Lewontin, R. C. 1965. "Selection for Colonizing Ability." In H. G. Baker and G. L. Stebbins, eds., *The Genetics of Colonizing Species*. pp 77–94. New York: Academic Press.

Lima, D. 1991. Kin saints and the forest: The study of Amazonian caboclos in the middle Solimões region [Ph.D. dissertation]. Cambridge, England: University of Cambridge.

Linares, O. 1976. Garden Hunting in the American Tropics. *Hum Ecol* 4: 331–349.

Little, P. D. 1994. "The Link Between Local Participation and Improved Conservation: A Review of Issues and Experiences." In D. Western and R. M. Wright, eds., *Natural Connections: Perspectives in Community-based Conservation*, pp. 347–72. Washington, DC: Island Press.

Lopez Escartin, N. 1991. *Données de Base sur la Population du Congo*. Brazzaville, Congo: Centre Français sur la Population et le Développement, Ministère de la Coopération et du Développement.

Loudiyi, D. 1995. Population dynamics: A monitoring tool for priority conservation areas. A methodology for field data collection. Dzanga-Sangha Dense Forest Reserve, Dzanga-Ndoki National Park, CAR. Report to the World Wildlife Fund, Washington, DC.

Ludwig, D., R. Hilborn, and C. Walters. 1993. Uncertainty, resource exploitation and conservation: Lessons from history. *Science* 260: 17–36.

Lutz, E. and J. Caldecott. 1996. *Decentralization and Biodiversity Conservation*. Washington, DC: The World Bank.

Lynch, O. J. and J. B. Alcorn 1994. "Tenurial Rights and Community-Based Conservation." In D. Western and R. M. Wright, eds., *Natural Connections: Perspectives in Community-based Conservation*, pp. 373–92. Washington, DC: Island Press.

MacAndrews, C. and L. Saunders. 1997. Conservation and national park financing in Indonesia. USAID/NRMP Occasional Papers No. 6.

MacArthur, R. H. and E. R. Pianka 1966. On the optimal use of a patchy environment. *Am Nat* 100: 603–9.

Macdonald, A. A. 1993a. "The Babirusa." In Oliver, W. L. R., ed., *Status Survey and Conservation Action Plan: Pigs, Peccaries and Hippos*, pp. 161–71. Gland, Switzerland: International Union for Conservation of Nature and Natural Resources.

Macdonald, A. A. 1993b. "The Sulawesi Warty Pig." In Oliver, W. L. R., ed., *Status Survey and Conservation Action Plan: Pigs, Peccaries and Hippos*, pp. 155–60. Gland, Switzerland: International Union for Conservation of Nature and Natural Resources.

MacKinnon, J. and K. MacKinnon. 1981. *Cagar Alam Gn. Tangkoko- DuaSaudara, Sulawesi Utara Management Plan 1981–1986.* Bogor, Indonesia: PHPA, Jl. Ir. H. Juanda, No. 100.

MacKinnon, J. R. 1981. The distribution and status of wild pigs in Indonesia. Unpublished report to International Union for Conservation of Nature and Natural Resources/ SSC Pigs and Peccaries Specialist Group.

MacKinnon, J. R. 1992. "Species Survival Plan for the Orangutan." In G. Ismail, M. Mohammed, and S. Omar, eds., *Forest Biology and Conservation in Borneo*, pp. 209–19. Publication No.2. Kota Kinabalu, Sabah: Centre for Borneo Studies.

MacKinnon, K. 1983. Report of a World Health Organization (WHO) consultancy to Indonesia to determine population estimates of the Cynomolgus or long-tailed macaque, *Macaca fascicularis* (and other primates) and the feasibility of semi-wild breeding projects of this species. WHO Primate Resources Programme Feasibility Study: Phase II, Mimeo.

Madroño, A. and E. Z. Esquivel. 1995. Reserva Natural del Bosque Mbaracayu: Su importancia en la conservación de aves amenazadas, cuasi-amenazadas y endemicas del Bosque Atlantico del Interior. *Cotinga* 4: x–xx.

Magurran, A. E. 1988. *Ecological Diversity and Its Measurement.* London: Croom Helm.

Maisels, F., A. Gautier-Hion, and J-P. Gautier. 1994. Diets of two sympatric colobines in Zaire: More evidence of seed-eating in forests on poor soils. *Int J Primatol* 15: 681–701.

Makana, J. R., T. Hart and J. Hart. In press. "Forest Structure and Diversity of Lianas and Understory Treelets in Monodominant and Mixed Forest in the Ituri, Zaire." In Delmaire, F., ed., *Measuring and Monitoring Forest Biological Diversity: The International Network of Biodiversity Plots.* Washington, DC: Smithsonian Institution Press.

Malhotra, K. C. and M. Gadgil. 1988. "Coping with Uncertainty in Food Supply: Case Studies Among Pastoral and Non-pastoral Nomads of Western India. In I. de Garine and G. A. Harrison, eds., *Coping with Uncertainty in Food Supply*, pp. 379–404. Oxford, England: Clarendon Press.

Malhotra, K. C., S. B. Khomne, and M. Gadgil. 1983. Hunting strategies among three non-pastoral nomadic groups of Maharashtra. *Man in India* 63: 21–39.

Malonga, R. 1995. Circuit commercial de viande de chasse à Brazzaville. Eléments préliminaires de la première phase d'étude. Unpublished report to Wildlife Conservation Society.

Maltby, E., P. J. Dugan, and J. C. Lefeuvre. 1992. *Conservation and Development: The Sustainable Use of Wetland Resources.* Gland, Switzerland: International Union for Conservation of Nature and Natural Resources.

Mani, M. S. 1974. *Ecology and Biogeography of India.* The Hague: W. Junk.

March M. I. J. 1987. Los Lacandones de México y su relación con los mamíferos silvestres: Un estudio etnozoológico. *Biótica* 12: 43–56.

Marks, S. A. 1973. Prey selection and annual harvest of game in a rural Zambian community. *East Afr Wildlife J* 11: 113–28.

Marks, S. A. 1994. Local hunters and wildlife surveys: a design to enhance participation. *Afr J Ecol* 32: 233–54.

Marks, S. A. 1996. Local hunters and wildlife surveys: An assessment and comparison of counts for 1989, 1990, and 1993. *Afr J Ecol* 34: 237–57.

Marsh, C. W. and B. Gait. 1988. Effects of logging in rural communities: A comparative study of two villages in Ulu Kinabatangan. *Sabah Soc J* 8: 394–434.

Marsh, C. W. and W. L. Wilson. 1981. *A Survey of Primates in Peninsular Malaysian Forests.* Kuala Lumpur, Malaysia: Universiti Kebangsaan.

Marshall, A. G. 1985. Old World phytophagous bats (Megachiroptera) and their food plants: A survey. *Zool J Linn Soc* 83: 351–69.

Marshall, G. 1991. FAO and tropical forestry. *Ecologist.* 21: 66–72.

Martin, F. W. and R. M. Ruberte. 1978. *Survival and Subsistence in the Tropics.* Mayaguez, Puerto Rico: Antillian College.

Martin, G. H. G. 1983. Bushmeat in Nigeria as a natural resource with environmental implications. *Environ Conserv* 10: 125–32.

Martin, G. H. G. 1985. Carcass composition and palatability of some wild animals commonly used as food. *World Animal Rev* 53: 40–4.

Martins, E. S. 1992. A caça de subsistência de extrativistas na Amazônia: Sustentabilidade, biodiversidade e extinção de espécies [M.Sc. thesis]. Brasília: Universidade de Brasília.

Mas, J., A. Yumbe, N. Solé, R. Capote, and T. Cremades. 1995. Prevalence, geographical distribution and clinical manifestations of onchocerciasis on the Island of Bioko (Equatorial Guinea). *Trop Med Parasitol* 46: 13–18.

Maté, C. and M. Colell 1995. Relative abundance of forest cercopithecines in Arihá, Bioko Island, Republic of Equatorial Guinea. *Folia Primatol* 64: 49–54.

May, R. M. 1981. "Models for Single Populations." In R. M. May, ed., *Theoretical Ecology*, pp. 5–29. Oxford: Blackwell.

Maybury-Lewis. 1984. *A Sociedade Xavante.* Rio de Janiero: Livraria Francisco Alves, and Oxford: Oxford University Press.

Mazurek, R. R. de S. No date. Subsistence hunting among the Waimiri-Atroari indians in Central Amazon, Brazil. Unpublished manuscript.

M'bokolo, E. 1992. "From the Cameroon Grasslands to the Upper Nile." In B. A. Ogot, ed. *General History of Africa V: Africa from the Sixteenth to the Eighteenth Century*, pp. 515–45. UNESCO. Berkeley, CA: University of California Press.

Mbuyi, M. 1978. "La Population." In *L'Atlas du Zaire.* Paris: Edts Jeune Afrique.

McCullough, D. R. 1974. Status of larger mammals in Taiwan. A report to World Wildlife Fund. Tourism Bureau, Taipei, Taiwan.

McCullough, D. R. 1977. "Essential Data Required on Population Structure and Dynamics in Field Studies of Threatened Herbivores." In: *Threatened Deer: Proceedings of a Working Meeting of the Deer Specialist Group of the IUCN Survival Service Commission*, pp. 302–17. Morges, Switzerland: International Union for the Conservation of Nature and Natural Resources.

McCullough, D. R. 1979. *The George Reserve Deer Herd: The Population Ecology of a K-selected Species.* Ann Arbor: University of Michigan Press.

McCullough, D. R. 1982. "The Theory and Management of Odoicoleus Populations." In C. M. Wemmer, ed., *Biology and Management of the Cervidae*, pp. 535–49. Washington, DC: Smithsonian Institution.

McCullough, D. R. 1991. "Refuges and Zoning to Control Exploitation of Hunted Wildlife." In Y-S. Lin and K.-H. Chang, eds., *Proceedings of the First International Symposium on Wildlife Conservation, Republic of China*, pp. 9–21, Taipei, Taiwan: Council of Agriculture.

McCullough, D. R. 1994. What do herd compositions counts tell us? *Wildlife Soc Bull* 22: 295–300.

McCullough, D. R. 1996a. Spatially structured populations and harvest theory. *J Wildlife Management* 60: 1–9.

McCullough, D. R. 1996b. "Metapopulation Management: What Patch Are We in and Which Corridors Should We Take?" In D. R. McCullough, ed., *Metapopulations and Wildlife Conservation*, pp. 405–10. Washington, DC: Island Press.

McKay, G. M. 1973. Behavior and ecology of the Asiatic elephant in southeastern Ceylon. *Smiths Contrib Zool* 125: 1–113.

McNeely, J. A. 1988. *Economics and Biological Diversity: Developing and Using Economic Incentives to Conserve Biological Resources.* Gland, Switzerland: IUCN.

McNeely, J. A. 1994. Lessons from the past: Forests and biodiversity. *Biodiversity Conserv* 3: 3–20.

McRae, M. 1997. Road kill in Cameroon. *Natural History*, 2/97: 36–47; 74–5.

Mech, L. 1977. Wolf-pack buffer zones as prey reservoirs. *Science* 198: 320–1.

Medway, Lord. 1959. Food bones in Niah Cave excavations. *Sarawak Museum J* 13: 627–36.

Medway, Lord. 1977. Mammals of Borneo. Monographs of the Malaysian Branch of the Royal Asiatic Society, No. 7.

MEFP (Ministére des Eaux et Forêts). 1991. Rapport annuel d'activités de la direction régionale de l'economie forestière de la pêche et de l'environnement de la Sangha. Report to the Ministry, Government of Congo, Brazzaville.

MEFP. 1993. Rapport annuel d'activités de la direction régionale des eaux et forêts de la Likouala. Report to the Ministry, Government of Congo, Brazzaville.

MEFP. 1994. Rapport annuel d'activités de la direction régionale des eaux et forêts de la Likouala. Report to the Ministry, Government of Congo, Brazzaville.

Meggars, B. 1982. "Archeological and Ethnographic Evidence Compatible with the Model of Forest Fragmentation." In G. Prance, ed., *Biological Diversification in the Tropics*, pp. 483–96. New York: Columbia University Press.

Meijer, W. and G. H. S. Wood. 1964. *Dipterocarps of Sabah, North Borneo.* Sabah Forest Record No. 5. Sandakan, Sabah, Malaysia: Forest Department.

Mellink, E., J. Aguirre, and E. Garcia Moya. 1986. *Utilización de la Fauna Silvestre en el Altiplano Potosino-Zacatecano.* México: Colegio de Postgrados (Institución de Enseñanza e Investigación en Ciencias Agrícolas).

Melnyk, M. and N. Bell. 1996. The direct-use values of tropical moist forest foods: The Huottuja (Piaroa) Amerindians of Venezuela. *Ambio* 25: 468–72.

Mercader, J. In press. Changing perceptions of tropical forest history: Archaeological discoveries in Ituri Forest. D. R. Congo. *Afr Archaeol Rev.*

Milberg, P. and T. Tyrberg. 1993. Naive birds and noble savages—A review of man-caused prehistoric extinctions of island birds. *Ecography* 16: 229–50.

Milner-Gulland, E. J. 1994. A population model for the management of the Saiga Antelope. *J Appl Ecol* 31: 25–39.

Milner-Gulland, E. J., and N. Leader-Williams. 1992. A model of incentives for the illegal exploitation of black rhinos and elephants: Poaching pays in Luangwa Valley, Zambia. *J Appl Ecol* 29: 388–401.

Mitchell, A. H. and R. L. Tilson, 1986. "Restoring the Balance: Traditional Hunting and Primate Conservation in the Mentawai Islands, Indonesia." In: J. G. Else and P. C. Lee, eds., *Primate Ecology and Conservation*, pp. 249–60. Cambridge: Cambridge University Press.

Mitchell, C. L. and E. Raez L. 1991. The impact of human hunting on primate and game bird populations in the Manu biosphere reserve in southeastern Peru. Report to Wildlife Conservation International, New York.

Mitchell, J. and H. Slim. 1991. Listening to rural people in Africa: The semi-structured interview in rapid rural appraisal. *Disasters* 15: 68–72.

Mittermeier, R. A. 1991. "Hunting and Its Effects on Wild Primate Populations in Suriname." In J. G. Robinson and K. H. Redford, eds., *Neotropical Wildlife Use and Conservation*, pp. 93–107. Chicago: University of Chicago Press.

Mittermeier, R. A. and D. Cheney. 1987. "Conservation of Primates and Their Habitats." In B. Smuts, D. Cheney, R. Seyfarth, R. Wrangham, and T. Struhsaker, eds., *Primate Societies*. pp. 477–90. Chicago: Chicago University Press.

Moffat, A. S. 1996. Ecologists look at the big picture. *Science* 273: 1490.

Mogaka, H. 1992. A report on a study of hunting in Arabuko-Sokoke Forest Reserve. Unpublished report to Kenya Indigenous Forest Conservation Project. Overseas Development Administration, Nairobi, Kenya.

Montes de Oca, I. 1989. *Geografía y Recursos Naturales de Bolivia*. La Paz, Bolivia: Academia Nacional de Ciencias de Bolivia.

Mordi, R. A. 1991. *Attitudes Toward Wildlife in Botswana*. New York: Garland Publishing.

Moreno. 1947. Orígen y vicisitudes del antiguo reino de Moka. *Arch Inst East Afr* 27: 7–30.

Morris, H. S. 1978. "The Coastal Melanau." In V. T. King, ed., *Essays on Borneo Societies*, pp. 37–58. Oxford: Oxford University.

Mossman, A. S. 1989. "Appropriate Technology for Rural Development." In R. J. Hudson, K. R. Drew, and L. M. Basin, eds., *Wildlife Production Systems: Economic Utilisation of Wild Ungulates*, pp. 446–57. Cambridge, England: Cambridge University Press.

Moussa, M. 1992. "Le Grand Braconnage en Centrafrique." In K. Cleaver, et al. eds., *Conservation of West and Central African Rainforests*, pp. 182–8. Washington, DC: World Bank.

Moutsambote, J. M., T. Yumoto, M. Mitani, T. Nishihara, S. Suzuki, and S. Kuroda. 1994. Vegetation and list of plant species identified in the Nouabalé-Ndoki Forest, Congo. *Tropics* 3: 277–93.

Muchaal, P. and G. Ngandjui. 1995. Wildlife Populations in the Western Dja Reserve (Cameroon): An assessment of the impact of village hunting and alternatives for sustainable utilisation. Unpublished Report, ECOFAC-Cameroon, ECOFAC, Brazzaville, Congo.

Muchaal, P. K. and G. Ngandjui. 1999. Wildlife populations in the western Dja Reserve (Cameroon): an assessment of the impact of village hunting. *Conserv Biol* 13:385–396

Mudar, K. M. 1985. "Bearded Pigs and Beardless Men: Predator-prey Relationships Between Pigs and Agta in Northeastern Luzon, Philippines." In P. B. Griffin and A. E. Estioko, eds., *The Agta of Northeastern Luzon: Recent Studies*, pp. 69–84. Cebu City: San Carlos Publications.

Mudar, K. M. 1986. A morphometeric analysis of the five subspecies of *Sus barbatus*, the bearded pig [M.S. thesis]. East Lansing, MI: Department of Zoology, Michigan State University.

Munasinghe, M. and M. Wells. 1992. "Protection of Natural Habitats and Sustainable Development of Local Communities." In K. Clearer, M. Munaisinghe, M. Dyson, N. Egli, A. Peuker, and F. Wencélius, eds. *Conservation of West and Central African Rainforests*, pp. 161–8. World Bank Environment Paper Number 1. Washington, DC: The World Bank.

Murphree, M. W. 1994. "The Role of Institutions in Community-based Conservation." In D. Western and R. M. Wright, eds., *Natural Connections: Perspectives in Community-based Conservation*, pp. 403–27. Washington, DC: Island Press.

Murphy, J. 1990. Indigenous forest use and development in the Maya Zone of Quintana Roo, Mexico [M.E.S. thesis]. Ontario: York University.

Murphy, P. G. and A. E. Lugo. 1986. Ecology of tropical dry forest. *Annu Rev Ecol Systematics* 17: 67–88.

Murphy, Y. and R. F. Murphy. 1972. *Women of the Forest*. New York: Columbia University Press.

Murtadza, A., A. Madroño, and E. Z. Esquivel, eds. 1995. Reserva Natural del Bosque Mbaracayú: Su importancia en la conservación de aves amenazadas, cuasi-amenazadas y endemicas del Bosque Atlantico del Interior. *Cotinga* 4: x–xx.

Musser, G. 1987. "The Mammals of Sulawesi." In T. C. Whitmore, ed., *Biogeographical Evolution of the Malay Archipelago*, pp. 73–93. Oxford, England: Oxford University Press.

Myers, N. 1981. The hamburger connection: how Central America's forests become North America's hamburgers. *Ambio* 10: 3–8.

Myers, N. 1988. Threatened biotas: "Hot spots" in tropical forests. *Environmentalist* 8: 187–208.

Myers, P., A. Taber, and I. Gamarra de Fox. 1995. "Mammalogy in Paraguay." In G. Caballos ed., *The Conservation of South American Mammals*.

Nair, S. S., P. V. K. Nair, H. C. Sharathchandra, and M. Gadgil. 1977. An ecological reconnaissance of the proposed Jawahar National Park. *J Bombay Natural History Soc* 74: 401–35.

Napier, J. R. and P. H. Napier. 1967. *A Handbook of Living Primates*. London: Academic Press.

Nations, J. D. and R. B. Nigh. 1980. The evolutionary potential of Lacandon Maya sustained-yield tropical forest agriculture. *J Anthropol Res* 36: 1–30.

Naughton-Treves, L. and S. Sanderson 1995. Property, politics and wildlife conservation. *World Dev* 23: 1265–75.

Ndinga, A. 1996. Governance, civic society and participation. In *Environment and Development in Africa Participatory Processes and New Partnerships*, pp. 50–9. Denmark: Scandinavia Seminar College.

Negreros, P. 1991. Ecology and management of mahogany (*Swietenia macrophylla* King) regeneration in Quintana Roo, Mexico [Ph.D. dissertation]. Ames, IA: Iowa State University.

Newmark, W. D., D. N. Manyanza, D. G. M. Gamassa, and H. I. Sariko. 1994. The conflict between wildlife and local people adjacent to protected areas in Tanzania: Human density as a predictor. *Conserv Biol* 8: 245–55.

Ngoko-Sangha. 1911. Annual Report from Compagnie Ngoko-Sangha. C.O.A.M. A.E.F. 8Q20.

Nicoll, M. and G. Rathbun. 1990. African insectivora and elephant-shrews: an action plan for their conservation. International Union for the Conservation of Nature and Natural Resources, Gland.

Nietschmann, B. 1973. *Between Land and Water*. New York: Seminar Press.

Nitecki, M. H., ed. 1984. *Extinctions*. Chicago: University of Chicago Press.

Njiforti H. 1996. Preferences and present demand for bushmeat in North Cameroon: some implications for wildlife conservation. *Environ Conserv* 23: 149.

Noguez-Galvez, A. M. 1991. Changes in soil properties following shifting cultivation in Quintana Roo, Mexico [M.S. thesis]. Gainesville: University of Florida.

Noss, A. J. 1995. Duikers, cables and nets: A cultural ecology of hunting in a central African forest [Ph.D. dissertation]. Gainesville: University of Florida.

Noss, A. J. 1997a. Challenges to integrated conservation and development or community-based conservation in Central Africa. *Oryx* 31: 180–8.

Noss, A. J. 1997b. The economic importance of communal net hunting among the BaAka of the Central African Republic. *Hum Ecol* 25: 71–89.

Noss, A. J. 1998a. Cable snares and bushmeat markets in a Central African Forest. *Environ Conserv* 25: 228–33.

Noss, A. J. 1998b. The impacts of BaAka net hunting on rainforest wildlife. *Biol Conserv* 86: 161–7.

Noss, A. J. 1998c. The impacts of cable snare hunting on wildlife populations in the forests of the Central African Republic. *Conserv Biol* 12: 390–8.

Noss, A. J. In press a. Censusing rainforest game species with communal net hunts. *Afr J Ecol*.

Noss, A. J. In press b. "The Aka of the Central African Republic." In W. Weber, A. Vedder, H. Simons Morland, L. White, and T. Hart, eds., *African Rain Forest Ecology and Conservation*. New Haven: Yale University Press.

Nosti, J. 1947. *Notas Geográficas, Físicas y Económicas Sobre los Territorios Españoles del Golfo de Guinea*. Madrid: Instituto de Estudios Africanos (CSIC).

Novaro, A. J. 1995. Sustainability of harvest of culpeo foxes in Patagonia. *Oryx* 29: 18–22.

Nowak, R. and J. Paradiso. 1983. *Walker's Mammals of the World*, 4th ed. Baltimore and London: Johns Hopkins University Press.

Nowak, R. M. 1991. *Walker's Mammals of the World*, 5th ed. Baltimore and London: Johns Hopkins University Press.

O'Brien, C. J. 1994. Seasonal strategies of carcass acquisition by Plio-Pleistocene hominids in East Africa [Ph.D. dissertation]. Madison, WI: University of Wisconsin.

O'Brien, T. G. and M. F. Kinnaird. 1994. Notes on the density and distribution of the endemic Sulawesi Tarictic Hornbill (*Penelopides exarhatus*) in Tangkoko-Dua Saudara Nature Reserve, North Sulawesi. *Trop Biodiversity* 2: 252–60.

O'Brien, T. G. and M. F. Kinnaird. 1996. Changing population of birds and mammals in North Sulawesi. *Oryx* 30: 150–6.

Oates, J. F. 1996a. Habitat alteration, hunting and the conservation of folivorous primates in African forests. *Aust J Zool* 21: 1–9.

Oates, J. F. 1996b. *Status Survey and Conservation Action Plan*, revised ed. *African Primates*. Gland, Switzerland: International Union for Conservation of Nature and Natural Resources/World Wildlife Fund.

Oates, J. F., G. H. Whitesides, A. G. Davies, P. G. Waterman, S. M. Green, G. A. Dasilva, and S. Mole. 1990. Determinants of tropical forest primate biomass: New evidence from West Africa. *Ecology* 71: 328–43

Ojasti, J. 1984. Hunting and conservation of mammals in Latin America. *Acta Zool Fenn* 172: 177–81.

Ojasti, J. 1991. "Human Exploitation of the Capybara." In J. G. Robinson and K. H. Redford, eds., *Neotropical Wildlife Use and Conservation*, pp. 236–52. Chicago and London: University of Chicago Press.

Ojasti, J. 1993. *Utilización de la Fauna Silvestre en America Latina. Situación y Perspectivas para un Manejo Sostenible*. Guia Fao Conservacion 25. Rome, Italy: Food and Agricultural Organization.

Olson, M. 1965. *The Logic of Collective Action: Public Goods and the Theory of Groups*. Cambridge, MA: Harvard University Press.

Olson, S. L. 1989. "Extinction on Islands: Man as a Catastrophe." In D. Western and M.

Pearl, eds., *Conservation for the Twenty-first Century*, pp. 50–3. Oxford, England: Oxford University Press.

Olson, S. L. and H. F. James 1982. Fossil birds from the Hawaiian Islands: Evidence for wholesale extinction by man before Western contact. *Science* 217: 633–5.

Otañez Toxqui, G., and B. Equihua Enriquez. 1981. Comercialización del chicle en México. Boletín Técnico del Instituto Nacional de Investigaciones Forestales, Number 70, 74, México.

Pachecho, V., B. D. Patterson, J. L. Patton, L. H. Emmons, S. Solari, and C. F. Ascorra. 1993. *List of Mammal Species Known to Occur in Manu Biosphere Reserve*. Publicaciones del Museo de Historia Natural. Lima, Peru: Universidad de San Marcos.

Pacheco, T. 1987. Impacto de un inventario forestal sobre la fauna silvestre en la zona del Río Lobo (Requena-Peru). *Matero* 1: 9–17.

Padoch, C. and W. de Jong 1992. "Diversity and Change in Ribereño Agriculture." In K. H. Redford and C. Padoch, eds., *Conservation of Neotropical Forests: Working from Traditional Resource Use*, pp. 158–74. New York: Columbia University Press.

Padoch, C. and N. L. Peluso 1996. "Borneo People and Forests in Transition: An Introduction." In C. Padoch and N. L. Peluso, eds., *Borneo in Transition. People, Forests, Conservation, and Development*, pp. 1–9. Kuala Lumpur: Oxford University Press.

PAGASA (Philippine Atmospheric, Geophysical, and Astronomical Administration) 1975. *Extremes of Temperature, Rainfall, Winds, and Sea Level Pressure of the Philippines as of 1972*. Quezon City: Philippine Atmospheric, Geophysical, and Astronomical Administration.

Panwar, H. S. 1987. "Project Tiger: The Reserves, the Tigers and Their Future." In R. L. Tilson and U. S. Seal, eds., *Tigers of the World*, pp. 110–7. Park Ridge, NJ: Noyes Publications.

Parra Lara, A. 1986. Uso y manejo tradicional de la fauna silvestre y su relación con otras actividades productivas en San Pedro Jicayan, Oaxaca. Cuadernos de Divulgación INIREB (Instituto Nacional de Investigaciones sobre Recursos Bióticos), No. 27, Xalapa, Veracruz Mexico.

Pascal, J. P., S. Shyamsunder, and V. M. Meher-Homji. 1982. Forest map of south India. Sheet: Mercara-Mysore. Karnataka Forest Department, and Institut Français de Pondichery, Pondicherry.

Pasquet, P. and G. J. A. Koppert. 1993. "Activity Patterns and Energy Expenditure in Cameroonian Tropical Forest Populations." In C. M. Hladik, A. Hladik, O. F. Linares, H. Pagezy, A. Semple, and M. Hadley, eds., *Tropical Forests, People and Food*, pp. 311–20. Paris: UNESCO.

Payne, J. C. 1992. A field study of techniques for estimating densities of duikers in Korup National Park, Cameroon [M.S. thesis]. Gainesville: University of Florida.

Pearce, J., and Ammann, K. 1995. *Slaughter of the Apes: How the Tropical Timber Industry Is Devouring Africa's Great Apes*. London: World Society for the Protection of Animals.

Peres, C. A. 1990. Effects of hunting on Western Amazonian primate communities. *Biol Conserv* 54: 47–59.

Peres, C. A. 1991. Humboldt's woolly monkeys decimated by hunting in Amazonia. *Oryx* 25: 89–95.

Peres, C. A. 1993a. Structure and spatial organization of an Amazonian terra firme forest primate community. *J Trop Ecol* 9: 259–76.

Peres, C. A. 1993b. Notes on the primates of the Juruá River, western Brazilian Amazonia. *Folia Primatol* 61: 97–103.

Peres, C. A. 1993c. Biodiversity conservation by native Amazonians: A pilot study in the

Kaxinawá Indigenous Reserve of Rio Jordão, Acre, Brazil. Unpublished Report to the World Wildlife Fund, Washington, DC.

Peres, C. A. 1994. Indigenous reserves and nature conservation in Amazonian forests. *Conserv Biol* 8: 586–8.

Peres, C. A. 1996. Population status of white-lipped *Tayassu pecari* and collared peccaries *T. tajacu* in hunted and unhunted Amazonian forests. *Biol Conserv* 77: 115–23.

Peres, C. A. 1997a. Primate community structure at twenty western Amazonian flooded and unflooded forests. *J Trop Ecol* 13: 381–405.

Peres, C. A. 1997b. Effects of habitat quality and hunting pressure on arboreal folivore densities in Neotropical forests: A case study of howler monkeys (*Alouatta* spp.). *Folia Primatol* 68: 199–222.

Peres, C. A. 1999. "The Structure of Nonvolant Mammal Communities in Different Amazonian Forest Types." In J. F. Eisenberg and K. H. Redford, eds., *Mammals of the Neotropics*. Vol 3, pp. 564–81. Chicago: University of Chicago Press.

Peres, C. A., L. C. Schiesari, and C. L. Dias-Leme. 1997. Vertebrate predation of Brazil-nuts, an agouti-dispersed Amazonian seed crop (*Bertholletia excelsa*, Lecythidaceae): A test of the escape hypothesis. *J Trop Ecol* 13: 69–79.

Peres, C. A. and J. Terborgh. 1995. Amazonian nature reserves: An analysis of the defensibility status of existing conservation units and design criteria for the future. *Conserv Biol* 9: 34–46.

Perez del Val, J. 1996. *Las Aves de Bioko, Guinea Ecuatorial. Guía de campo*. Leon: Edilesa.

Perez del Val, J., J. E. Fa, J. Castroviejo, and F. J. Purroy. 1994. Species richness and endemism of birds in Bioko. *Biodiversity Conserv* 3: 868–92.

Peters, R. H. and K. Wassenberg. 1983. The effect of body size on animal abundance. *Oecologia* 60: 89–96.

Peterson, J. T. 1978. *The Ecology of Social Boundaries: Agta foragers in the Philippines*. Urbana, IL: University of Illinois Press.

Peterson, R. 1991. "To Search for Life": A study of spontaneous immigration, settlement and land use on Zaire's Ituri Forest frontier [M.S. thesis]. Madison: University of Wisconsin.

Peterson, R. T., and E. L. Chalif. 1973. *A Field Guide to Mexican Birds: Mexico, Guatemala, Belize (British Honduras), El Salvador*. Boston: Houghton Mifflin Company.

Pfeffer, P. and J. O. Caldecott, 1986. The bearded pig (*Sus barbatus*) in East Kalimantan and Sarawak. *J Malaysian Branch R Asiatic Soc* 59: 81–100.

Pinkerton, E. 1989. Introduction: Attaining better fisheries management through co-management—prospects, problems, and propositions. In E. Pinkerton, ed., *Co-operative Management of Local Fisheries: New Directions for Improved Management and Community Development*, pp. 3–33. Vancouver: University of British Columbia Press.

Plotkin, M., and L. Famolare, eds., 1992. *Sustainable Harvest and Marketing of Rain Forest Products*. Washington, DC: Island Press.

Plumptre, A. J. and S. Harris 1995. Estimating the biomass of large mammalian herbivores in a tropical montane forest: A method of faecal counting that avoids assuming a "steady state" system. *J Appl Ecol* 32: 11–120.

Poffenberger, M. 1994. "The Resurgence of Community Forest Management in Eastern India." In D. Western and R. M. Wright eds., *Natural Connections: Perspectives in Community-Based Conservation*, pp. 53–79. Washington, DC: Island Press.

Pohl, M. E. D. 1976. Ethnozoology of the Maya: An analysis of fauna from five sites in Peten, Guatemala [Ph.D. dissertation]. Cambridge, MA: Harvard University.

Ponce, C. F. 1987. *Manejo de Fauna Silvestre y Desarollo Rural. Informacion Sobre Siete Especies de America Latina y el Caribe*. Lima, Peru: Food and Agricultural Organization/ Programma Naciones Unidas del MedioAmbíente.

Posey, D. 1988. "Kayapó Indian Natural Resource Management." In J. S. Denslow and C. Padoch, *Peoples of the Tropical Rain Forest*, pp. 89–91. Los Angeles: University of California Press.

Posey, D. A. 1982. Keepers of the forest. *Garden* 8: 8–12, 32.

Prater, S. H. 1980. *The Book of Indian Animals*, 3rd ed. Bombay: Oxford University Press.

Prescott-Allen, R. and C. Prescott-Allen, 1982. *What's Wildlife Worth?* Washington, DC: International Institute for Environment and Development.

Prins, H. H. T. and J. M. Reitsma 1989. Mammalian biomass in an African equatorial rain forest. *J Animal Ecol* 58: 851–61.

Pulliam, H. R. 1988. Sources, sinks, and population regulation. *Am Naturalist* 132: 652–61.

Pulliam, H. R. and B. J. Danielson 1991. Sources, sinks, and habitat selection: A landscape perspective on population dynamics. *Am Naturalist* 137(suppl): 50–6.

Puri, G. S., V. M. Meher-Homji, R. K. Gupta, and S. Puri. 1983. *Forest Ecology*. Vol. I. New Delhi and Oxford, England: IBH Publishers.

Puri, R. 1992. Mammals and hunting on the Lurak River: recommendations for management of faunal resources in the Cagar Alam Kayan-Mentarang. Paper presented at the Borneo Research Council Second Biennial International Conference, 13–17 July 1992, Kota Kinabalu, Sabah, Malaysia.

Putzer, H. 1984. "The Geological Evolution of the Amazon Basin and Its Mineral Resources." In H. Sioli, ed., *The Amazon: Limnology and Landscape Ecology of a Mighty Tropical River and Its Basin*, pp. 15–46. Dordrecht: W. Junk.

Pyke, G., H. R. Pulliam, and E. L. Charnov. 1977. Optimal foraging: A selective review of theory and test. *Q Rev Biol* 52: 137–54.

Rabinowitz, A. 1991. *Chasing the Dragon's Tail: The Struggle to Save Thailand's Wild Cats*. New York: Doubleday.

Rabinowitz, A. 1995. Helping a species go extinct: The Sumatran rhino in Borneo. *Conserv Biol* 9: 482–8.

Rabinowitz, A. 1998. Killed for a cure. *Natural History* 98: 22–4.

Radam. 1973–1981. *Projeto Radam Brasil: Levantamento de Recursos Naturais*. Vol. 1–18, Rio de Janeiro: Departamento Nacional de Produção Mineral, Ministério das Minas e Energia.

Rai, N. K. 1990. *Living in a Lean-to: Philippine Negrito Foragers in Transition*. No. 80 Anthropological Papers. Ann Arbor: University of Michigan, Museum of Anthropology.

Rajanathan, R. 1992. Differential habitat use by primates in Samunsam Wildlife Sanctuary, Sarawak, and its application to conservation management [M.S. thesis]. Gainesville: University of Florida.

Ralls, K. 1973. *Cephalophus maxwelli*. *Mammalian Species* 111: 1–4.

Ramirez, M. 1984. Population recovery in the moustached tamarin *(Saguinus mystax)*: Management strategies and mechanisms of recovery. *Am J Primatol* 7: 245–59.

Rangarajan, M. 1996. *Fencing the Forest: Conservation and Ecological Change in India's Central Provinces 1860–1914*. New Delhi: Oxford University Press.

Rappaport, R. 1968. *Pigs for the Ancestors: Ritual in the Ecology of a New Guinea People*. New Haven, CT: Yale University Press.

Rasker, R., M. V. Martin, and R. L. Johnson. 1992. Economics: Theory versus practice in wildlife management. *Conserv Biol* 6: 338–49.

Raval, S. R. 1994. Wheel of life: Perceptions and concerns of the resident peoples for Gir National Park in India. *Soc Natural Resources* 7: 305–20.

Redfield, R., and A. Villa Rojas. 1962. *Chan Kom, a Maya village*. No. 448, Carnegie Institution of Washington [first published in 1934]. Chicago: University of Chicago Press.

Redford, K. H. 1989. Monte Pascoal—Indigenous rights and conservation in conflict. *Oryx* 23: 33–6.

Redford, K. H. 1990. The ecologically noble savage. *Orion* 9: 24–9.

Redford, K. H. 1991. The ecologically noble savage. *Cultural Survival Q* 15: 46–8.

Redford, K. H. 1992. The empty forest. *BioScience* 42: 412–22.

Redford, K. H. 1993. "Hunting in Neotropical Forests: A Subsidy from Nature." In C. M. Hladik, A. Hladik, O. F. Linares, H. Pagezy, A. Semple, and M. Hadley, eds., *Tropical Forests, People and Food*, pp. 227–46. Paris: UNESCO.

Redford, K. H. No date. Natural Histories. Hunting by Humans, Tropical Forest Composition and the Definition of Natural. Unpublished manuscript.

Redford, K. H. and J. F. Eisenberg. 1992. *Mammals of the Neotropics*. Vol. 2. *The Southern Cone*. Chicago: University of Chicago Press.

Redford, K. H. and J. A. Mansour, eds. 1996. *Traditional Peoples and Biodiversity Conservation in Large Tropical Landscapes*. Washington, DC: America Verde Publications, The Nature Conservancy.

Redford, K. H., and J. G. Robinson. 1987. The game of choice: patterns of Indian and colonist hunting in the neotropics. *Am Anthropol* 89: 650–67.

Redford, K. H., and J. G. Robinson. 1990. A research agenda for studies of subsistence hunting in the Neotropics. *Florida J Anthropol* (Special Publication), No. 6, pp. 117–20.

Redford, K. H. and J. G. Robinson. 1991. "Subsistence and Commercial Uses of Wildlife in Latin America." In J. G. Robinson and K. H. Redford, eds., *Neotropical Wildlife Use and Conservation*, pp. 6–23. Chicago: University of Chicago Press.

Redford, K. H., and A. Stearman. 1993. Forest-dwelling native Amazonians and the conservation of biodiversity: Interests in common or in collision? *Conserv Biol* 7: 248–55.

Reed, R. 1995. *Prophets of Agroforestry. Guaraní Communities and Commercial Gathering*. Austin, TX: University of Texas Press.

Reed, R. 1997. *Forest Dwellers, Forest Protectors. Indigenous Models of International Development*. Boston: Allyn and Bacon.

Republik Indonesia. 1990. Undang-undang Republik Indonesia. Nomor 5, Tahun 1990. Government of Indonesia, Jakarta.

République Centrafricaine. 1992a. Portant règlement intérieur du Parc National Dzanga-Ndoki, Arreté No. 008. Bangui: Ministère des Eaux, Forêts, Chasse, Pêche, Tourisme et de l'Environnement.

République Centrafricaine. 1992b. Portant règlement intérieur de la Reserve Spéciale de Forêt Dense Dzanga-Sangha, Arreté No. 007. Bangui: Ministère des Eaux, Chasse, Pêche, Tourisme et de l'Environnement.

Rettig, R. B., F. Berkes, and E. Pinkerton. 1989. "The Future of Fisheries Co-management: A Multi-disciplinary Assessment." In E. Pinkerton. ed., *Co-operative Management of Local Fisheries: New Directions for Improved Management and Community Development*, pp. 273–89. Vancouver: University of British Columbia Press.

Rewald, M. 1989. Stump sprouts in the milpa cycle and their role in forest regeneration [M.E.S. thesis]. Ontario: York University.

Reyes Castillo, P. 1981. La fauna silvestre en el plan Balancán-Tenosique. Cuadernos de Divulgación. No. 4, INIREB (Instituto Nacional de Investigaciones sobre Recursos Bióticos), Xalapa, Veracruz, México.

Robinette, W., C. Loveless, and D. Jones. 1974. Field tests of strip census methods. *J Wildlife Management* 38: 91–6.

Robinson, J. G. 1993. Limits to caring: Sustainable living and the loss of biodiversity. *Conserv Biol* 7: 20–8.

Robinson, J. G. 1994. Carving up tomorrow's planet. *Int Wildlife* 24: 30–7.

Robinson, J. G. 1996. "Hunting Wildlife in Forest Patches: An Ephemeral Resource." In J. Schelhas and R. Greenberg, eds., *Forest Patches in Tropical Landscapes*, pp. 111–30. Washington, DC: Island Press.

Robinson, J. G. and K. H. Redford. 1986a. Body size, diet, and population density of neotropical forest mammals. *Am Naturalist* 128: 665–80.

Robinson, J. G. and K. H. Redford, eds. 1986b. Intrinsic rate of natural increase in neotropical forest mammals: Relationship to phylogeny and diet. *Oecologia* 68: 516–20.

Robinson, J. G. and K. H. Redford, eds. 1991a. *Neotropical Wildlife Use and Conservation*. Chicago: Chicago University Press.

Robinson, J. G. and K. H. Redford. 1991b. "Sustainable Harvest of Neotropical Forest Mammals." In J. G. Robinson and K. H. Redford. eds., *Neotropical Wildlife Use and Conservation*, pp. 415–29. Chicago: University of Chicago Press.

Robinson, J. G. and K. H. Redford. 1991c. "Preface." In J. G. Robinson and K. H. Redford, eds., *Neotropical Wildlife Use and Conservation*, pp. XV–XVII. Chicago: University of Chicago Press.

Robinson, J. G. and K. H. Redford. 1994a. "Community-based Approaches to Wildlife Conservation in Neotropical Forests." In D. Western and R. M. Wright, eds., *Natural Connections: Perspectives in Community-based Conservation*, pp. 300–22. Washington, DC: Island Press.

Robinson, J. G. and K. H. Redford. 1994b. Measuring the sustainability of hunting in tropical forests. *Oryx* 28: 249–56.

Robinson, N. A. 1993. *Agenda 21: Earth's Action Plan*. New York: Oceana Publication.

Robinson, W. L. and E. G. Bolen 1984. *Wildlife Ecology and Management*. New York: Macmillan.

Rodgers, W. A., and H. S. Panwar. 1988. Planning a protected area network in India. Vol. I and II. Unpublished report. Wildlife Institute of India, Dehra Dun.

Roesler, M. 1997. "Shifting Cultivation in the Ituri Forest (Haut-Zaire): Colonial Intervention, Present Situation, Economic and Ecological Prospects." In *Civilisations: Les Peuples des Forets Tropicales*, pp. 44–61. Special issue. Brussels: Institut de Sosiologie, Université Libre.

Rogers, A. 1991. Conserving resources for children. *Hum Nature* 1: 73–82.

Roosevelt, A. 1994. "Strategy for a New Synthesis." In A. Roosevelt, ed., *Amazonian Indians from Prehistory to the Present: Anthropological Perspectives*, pp. 1–20. Tucson: University of Arizona Press.

Roosevelt, A. C., M. Lima da Costa, C. Lopez Machado, M. Michab, N. Mercier, H. Valladas, J. Feathers, W. Barnett, M. Imazio da Silveira, A. Henderson, J. Sliva, B. Chernoff, D. S. Reese, J. A. Holman, N. Toth, and K. Schick 1996. Paleoindian cave dwellers in the Amazon: The peopling of the Americas. *Science* 272: 373–84.

Rosenbaum, B., T. G. O'Brien, M. Kinnaird, and J. Supriatna. 1998. Population densities of Sulawesi crested black macaques (*Macaca nigra*) on Bacan and Sulawesi, Indonesia: Effects of habitat disturbance and hunting. *Am J Primatol* 44: 89–106.

Ross, C. 1988. The intrinsic rate of natural increase and reproductive efforts in primates. *Oecologia* 214: 199–219.

Ross, C. 1992. Environmental correlates of the intrinsic rate of natural increase in primates. *Oecologia* 90: 383–90.

Ross, E. 1978. Food taboos, diet, and hunting strategy: The adaptation to animals in Amazon cultural ecology. *Curr Anthropol* 19: 1–19.

Rubio-Torgler, H. 1995. Demanda de fauna por parte de comunidades indígenas Emberá y estratégias para el manejo de especies de caza en el área de influencia del Parque Nacional Natural Utría (Chocó, Colombia). II Congreso Internacional sobre Manejo de Fauna Silvestre en la Amazonia, 7–12 de Mayo de 1995, Iquitos, Perú.

Ruddle, K. 1970. The hunting technology of the Maracá Indians. *Antropologica* 25: 21–63.

Ruggiero, R. G. and H. E. Eves. In press. Bird-mammal associations in forest openings in northern Congo (Brazzaville). *Afr J Ecol*.

Rylands, A. 1991. *The Status of Conservation Areas in the Brazilian Amazon*. Washington, DC: World Wildlife Fund.

Saffirio, J. and R. Hames. 1983. *The Forest and the Highway*. Cultural Survival Report 11. Cambridge, MA: Cultural Survival.

Saffirio, J. and R. Scaglion. 1982. Hunting efficiency in acculturated and unacculturated Yanomama villages. *J Anthropol Res* 38: 315–27.

Saharia, V. B. 1982. *The Wildlife of India*. Dehra Dun: Natraj.

Sahlins, M. 1968. "Notes on the Original Affluent Society." In R. B. Lee and I. DeVore, eds., *Man the Hunter*, pp. 85–9. Chicago: Aldine Press.

Sahlins, M. 1972. *Stone Age Economics*. Chicago: Aldine Press.

Salazar, D. J. and A. K. Lenard. 1994. Conservation and the nature of goods. *Society Natural Resources* 7: 331–48.

Salzano, F., J. Neel and D. Mayberry-Lewis. 1967. Further studies on the Xavante Indians. I. Demographic data on two additional villages: Genetic structure of the tribe. *Am J Hum Genet* 19: 463–89.

Sanchez, P. A. 1981. "Soils of the Humid Tropics." In P. A. Sanchez, ed., *Blowing in the Wind: Deforestation and Long-term Implications*, pp. 347–410. Williamsburg, VA: College of William and Mary.

Sanchez, T. F. 1973. "The Climate of Paraguay." In J. R. Gorham, ed., *Paraguay: Ecological Essays*, pp. 33–8. Miami: Academy of the Arts and Sciences of the Americas.

Sanderson, S. E. and K. H. Redford. 1997. "Biodiversity politics and the contest of ownership of the world's biota." In R. Kramer, C. van Schaik, and J. Johnson, eds., *Last Stand: Protected Areas and the Defense of Tropical Biodiversity*, pp. 115–32. Oxford and New York: Oxford University Press.

Sankhala, K. 1978. *Tiger: The Story of the Indian Tiger*. London: Collins.

Santana, E., L. I. Iñiguez Dávalos, and S. Navarro. 1990. Utilización de la fauna silvestre por las comunidades rurales de la Reserva de la Biósfera Sierra de Manantlán. *Tiempos de Ciencia/Revista de Difusión Científica* 18: 36–43.

Sanvicente, M. 1996. "Conservatión y Aprovechamiento de Amortiguamiento de la Reserva de la Biósfera Calakmul." In C. Campos Roso, A. Ulloa, and H. Rubio Torgler, eds. *Manejo de Fauna con Comunidades Rurales*, pp. 72–85. Santafé de Bogotá, Colombia: Fundación Natura.

Sato, H. 1983. Hunting of the Boyela, slash-and-burn agriculturalists, in the central Zaire forest. *Afr Stud Monogr* 4: 1–54.

Sato, H. 1992. Notes on the distribution and settlement pattern of hunter-gatherers in northwestern Congo. *Afr Stud Monogr* 13:203–16.

Schaaf, C. D., T. M. Butynski, and G. W. Hearn. 1990. The drill (*Mandrillus leucophaeus*) and other primates in the Gran Caldera Volcanica de Luba: Results of a survey conducted 7–22, 1990. Unpublished report to the Government of Equatorial Guinea. Atlanta: Zoo Atlanta.

Schaaf, C. D., T. T. Struhsaker, and G. W. Hearn 1990. Recommendations for biological conservation areas on the island of Bioko, Equatorial Guinea. Unpublished report to the Government of Equatorial Guinea. Atlanta: Zoo Atlanta.

Schaller, G. B. 1967. *The Deer and the Tiger: A Study of Wildlife in India*. Chicago: University of Chicago Press.

Schaller, G. B. 1983. Mammals and their biomass on a Brazilian ranch. *Arquivos Zool* 31: 1–36.

Schwartzman, S. 1989. "Extractive Reserves: The Rubber Tappers' Strategy for Sustainable Use of the Amazon Rain Forest." In J. O. Browder, ed., *Fragile Lands of Latin America: Strategies for Sustainable Development*, pp. 150–65. Boulder, CO: Westview Press.

Schweithelm, J., N. Wirawan, J. Elliot, and A. Khan. 1992. Sulawesi Parks Program, land use and socio-economic survey Lore Lindu National Park and Morowali Nature Reserve. Report to the Nature Conservancy, Jakarta, Indonesia.

Scott, W. H. 1979. Semper's Kalingas 120 years later. *Philippine Sociol Rev* 27: 93–101.

Seber, G. A. F. 1982. *The Estimation of Animal Abundance and Related Parameters*. New York: Macmillan.

Shaw, J. 1991. The outlook for sustainable harvests of wildlife in Latin America. In: J. G. Robinson and K. H. Redford, eds., *Neotropical Wildlife Use and Conservation*, pp. 24–34. Chicago: Chicago University Press.

Shaw, J. H. 1985. *Introduction to Wildlife Management*. New York: McGraw-Hill.

Sih, A. 1987. Prey refuges and predator-prey stability. *Theor Pop Biol* 31: 1–12.

Silva, J. L. and S. D. Strahl. 1991. "Human impact on populations of Chachalacas, Guans, and Curassows (Galliformes: Cricidae) in Venezuela." In J. G. Robinson and K. H. Redford, eds., *Neotropical Wildlife Use and Conservation*, pp. 37–52. Chicago: University of Chicago Press.

Silva, M., and J. Downing. 1995. *CRC Handbook of Mammalian Body Masses*. Boca Raton, FL: CRC Press.

Silva, N. J. da, and J. W. Sites. 1995. Patterns of diversity of neotropical squamate reptile species with emphasis on the Brazilian Amazon and the conservation potential of indigenous reserves. *Conserv Biol* 9: 873–901.

Simons, J. L. 1980. Resources, population, environment: An oversupply of false bad news. *Science* 208: 1431–7.

Sinclair, A. R. E. 1989. "Population Regulation in Animals." In J. M. Cherrett, ed., *Ecological Concepts*, pp. 197–241. Oxford, England: Blackwell.

Skorupa, J. P. 1986. "Responses of Rainforest Primates to Selective Logging in Kibale Forest, Uganda: A Summary Report." In Benirschke, K., ed., *Population Primates: The Road to Self-sustaining Populations*, pp. 57–70. New York: Springer-Verlag.

Slade, N. A., R. Gomulkiewicz, and H. M. Alexander. 1998. Alternatives to Robinson and Redford's method of assessing overharvest from incomplete demographic data. *Conserv Biol* 12: 148–55.

Smith, E. 1991. *Inujjuamiut Foraging Strategies: Evolutionary Ecology of an Arctic Hunting Economy*. New York: Aldine.

Smith, F. D. M., G. C. Daily, and P. R. Erhlich. 1995. "Human Population Dynamics and Biodiversity Loss." In T. M. Swans, ed., *The Economics of Biodiversity: The Forces Driving Global Change*, pp. 125–41. Cambridge, England: Cambridge University Press.

Smith, N. J. H. 1976. Utilization of game along Brazil's Transamazon highway. *Acta Amazonica* 6: 455–66.

Smith, R. 1993. *Drama Bajo el Manto Amazonico. El Turismo y Otros Problemas de los Huao-rani en la Actualidad.* Quito, Ecuador: Abya-Yala.

Smole, W. 1976. *The Yanoama Indians: A Cultural Geography.* Austin: University of Texas Press.

Snook, L. 1993. Stand dynamics of mahogany (*Swietenia macrophylla* King) and associated species after fire and hurricane in the tropical forests of the Yucatán Peninsula [Ph.D. dissertation]. New Haven, CT, Yale University.

Soemantri, S. 1983. "Pola Perkembangan dan Perbandingan antara Daerah Angka Kema-tian Bayi [Patterns of Development and Regional Comparisons of Infant Death Rates]." In *Seminar Tingkat kematian Bayi di Indonesia*, pp. 173–192. Jakarta: Biro Pusat Statistik.

Sokal, R. R. and F. J. Rolf. 1981. *Biometry,* 2nd ed. New York: W. H. Freeman & Co.

Somaiah, K. K. 1953. Working plan for the Eastern deciduous forests of Coorg. Unpub-lished report, Coorg Forest Department, Madikeri, India.

Sompud, J. 1996. Calling adaptations by primates, hornbills, and argus pheasants to hunting in Sabah and Sarawak [M.S. Thesis]. Bangi, Malaysia: Universiti Keban-gasaan Malaysia.

Sowls, L. K. 1984. *The Peccaries.* Tucson, AZ: University of Arizona Press.

Speth, J. and K. Spielmann. 1983. Energy source, protein metabolism and hunter-gatherer subsistence strategies. *J Anthropol Archeol* 2: 1–31.

Srikosamatara, S., B. Siripholdej, and V. Suteethorn 1992. Wildlife trade in Lao P.D.R. and between Lao P.D.R. and Thailand. *Natl History Bull Siam Soc* 40: 1–47.

Stearman, A. M. 1984. The Yuquí Connection: Another look at Sirionó deculturation. *Am Anthropologist* 86: 630–50.

Stearman, A. M. 1987. *No Longer Nomads: The Sirionó Revisited.* Lanham, Maryland: Hamilton Press.

Stearman, A. M. 1989a. *Yuquí: Forest Nomads in a Changing World.* New York: Holt, Rine-hart and Winston.

Stearman, A. M. 1989b. Yuquí foragers in the Bolivian Amazon: Subsistence strategies, prestige, and leadership in an acculturating society. *J Anthropol Res* 45: 219–44.

Stearman, A. M. 1989c. "Yuqui': Forest Nomads in a Changing World." In G. Spindler and L. Holt, eds. *Case Studies in Cultural Anthropology Series.* New York: Holt, Rinehart and Winston.

Stearman, A. M. 1990. The effects of settler incursion on fish and game resources of the Yuquí, a native Amazonian society of eastern Bolivia. *Hum Organization* 49: 373–85.

Stearman, A. M. 1991. Making a living in the tropical forest: Yuquí foragers in the Bolivian Amazon. *Hum Ecol* 19: 245–60.

Stearman, A. M. 1994. Losing game. *Natural History* 1/94: 6–10.

Stearman, A. M. 1995. "Neotropical Foraging Adaptations and the Effects of Accultura-tion on Sustainable Resource Use: The Yuquí of Lowland Bolivia." In L. E. Sponsel, ed., *Indigenous Peoples and the Future of Amazonia. An Ecological Anthropology of an Endangered World*, pp. 207–24. Tucson: University of Arizona Press.

Stearman, A. M. and K. H. Redford. 1992. Commercial hunting by subsistence hunters: Sirionó Indians and Paraguayan caiman in lowland Bolivia. *Hum Organization* 51: 235–44.

Stearman, A. M. and K. H. Redford. 1995. Game management and cultural survival: The Yuquí Ethnodevelopment Project in lowland Bolivia. *Oryx* 29: 29–34.

Stebbing, E. P. 1929. *The Forests of India.* Vol. 1. John Lane. London: The Bodley Head Limited.

Steel, E. A. 1994. *Study of the Value and Volume of Bushmeat Commerce in Gabon*. Libreville: World Wildlife Fund.

Stelfox, J. G., D. G. Peden, H. Epp, R. J. Hudson, S. W. Mbugua, J. L. Agatsiva and C. I. Amuyunzu. 1986. Herbivore dynamics in southern Narok, Kenya. *J Wildlife Management* 50: 339–47.

Steward J., ed. 1944–49. *Handbook of South American Indians*, Vols. 1–7. Bureau of American Ethology Bulletin. Washington, DC: US Government Printing Office.

Stiles, D. 1981. Hunters of the northern East African coast: Origins and historical processes. *Africa* 51: 848–61.

Stoll, H. L. 1992. Boycott des bois tropicaux. *Nature Faune*. 8.

Strickland, S. S. 1986. Long-term development of Kajamaan subsistence: An ecological study. *Sarawak Museum J* 36(New Series): 117–72.

Stromayer, K. A. K. and A. Ekobo. 1991. Biological surveys of southeastern Cameroon. Unpublished report to the Wildlife Conservation Society and European Community.

Struhan, R. 1983. *The Australian Museum Complete Book of Australian Mammals*. London: Angus & Robertson.

Struhsaker, T. T. 1997. *Ecology of an African Rain Forest: Logging in Kibale*. Gainesville: University of Florida Press.

Struhsaker, T. T. and L. Leland 1979. Socioecology of five sympatric monkey species in the Kibale forest, Uganda. *Adv Study Behavior* 9: 159–228.

Strum, S. C. 1994. "Lessons Learned." In D. Western and R. M. Wright, eds., *Natural Connections, Perspectives in Community-based Conservation*. pp. 512–23. Washington, DC: Island Press.

Suárez, E., J. Stallings, and L. Suárez. 1995. Small-mammal hunting by two ethnic groups in north-eastern Ecuador. *Oryx* 29: 35–42.

Sugardjito, J., C. H. Southwick, J. Supriatna, A. K. Kohlhaas, S. Baker, J. Erwin, K. Froehlich, and N. Lerche, 1989. Population survey of macaques in North Sulawesi. *Am J Primatol* 18: 285–301.

Sukumar, R. 1989. *The Asian Elephant: Ecology and management*. Cambridge, England: Cambridge University Press.

Supriatna, J. 1991. Hybridization between *Macaca maura* and *M. tonkeana*: A test of species status using behavioral and morphogenetic analyses [Ph.D. dissertation]. New Mexico: University of New Mexico.

Tanaka, J. 1978. A study of the comparative ecology of African gatherer-hunters with special reference to San (Bushman-speaking people) and Pygmies. *Senri Ethnol Stud* 1: 189–212.

Tanno, T. 1976. The Mbuti net hunters in the Ituri forest, eastern Zaāre—their hunting activities and band composition. *Kyoto Univ Afr Stud* 10: 101–35.

Tarwotjo, I., I. J. Susanto, and A. Sommer. 1978. Characterization of vitamin A deficiency and design of intervention program. Paper presented at the XI International Congress of Nutrition, Rio de Janeiro, Brazil.

Taylor, M. K., D. P. DeMaster, F. L. Bunnell, and R. E. Schweinsburg. 1987. Modeling the sustainable harvest of female polar bears. *J Wildlife Management* 51: 811–20.

TCA (Tratado de Cooperacion Amazonica.) 1995. *Uso y conservación de la fauna silvestre en la Amazonia*. Lima, Peru: Tratado de Cooperación Amazonica.

TELESIS. 1991. *Sustainable Economic Development Options for the Dzanga-Sangha Reserve Central African Republic*. Providence, RI: TELESIS (USA), Inc.

Terborgh, J. 1983. *Five New World Primates: A Study in Comparative Ecology*. Princeton, NJ: Princeton University Press.

Terborgh, J. 1988. The big things that run the world: A sequel to E. O. Wilson. *Conserv Biol* 2: 402–3.

Terborgh, J. 1990. "An Overview of Research at Cocha Cashu Biological Station." In A. Gentry, ed., pp. 48–59. *Four Neotropical Rainforests*. New Haven, CT: Yale University Press.

Terborgh, J. 1992. Maintenance of diversity in tropical forests. *Biotropica* 24: 283–92.

Terborgh, J., L. H. Emmons, and C. Freese. 1986. La fauna silvestre de la Amazonia: El despilfarro de un recurso renovable. *Bol Lima* 46: 77–85.

Terborgh, J., S. K. Robinson, T. A. Parker III, C. A. Munn, and N. Pierpoint. 1990. Structure and organization of an Amazonian forest bird community. *Ecol Monogr* 6: 213–38.

TFAP (Tropical Forestry Action Program). 1994. *Tropical Forestry Action Program Update*. Rome, Italy: TFAP, Food and Agricultural Organization.

Thiollay, J. M. 1986. Structure comparée du peuplement avien dans trois sites de forêt primaire en Guyane. *Rev Ecol* 41: 59–105.

Thiollay, J. M. 1992. Influence of selective logging on bird species diversity in a Guianan rain forest. *Conserv Biol* 6: 47–63.

Thompson, S. 1987. Body size, duration of parental care, and the intrinsic rate of natural increase in eutherian and metatherian mammals. *Oecologia* 71: 201–9.

Thomson, J. A., D. L. Hess, K. D. Dahl, S. A. Iliff-Sizemore, R. L. Stouffer, and D. P. Wolf. 1992. The Sulawesi crested black macaque (*Macaca nigra*) menstrual cycle: Changes in perineal tumescence and serum estradiol, progesterone, follicle-stimulating hormone, and leuteinizing hormone levels. *Biol Reprod* 46: 879–84.

Thorbecke, E. and T. van der Pluijm, 1993. *Rural Indonesia: Socio-economic Development in a Changing Environment*. New York: New York University Press.

Torres C. B. and A. Smith B. 1992. Cuando los nativos hablan de sus animales. Reserva de Biosfera del Manu. *Bol Lima* 77: 75–90.

Townsend, W. 1995. Living on the edge: Sirionó hunting and fishing in lowland Bolivia [Ph.D. dissertation]. Gainesville: University of Florida.

Townsend, W. 1996. *Nyao Itô: Caza y Pesca de los Sirionó*. La Paz, Bolivia: Instituto de Ecología, Universidad Mayor de San Andrés, FUND-ECO.

Townsend, W. 1997. "La Participación Comunal en el Manejo de Fauna Silvestre en el Oriente de Bolivia." In T. G. Fang, R. E. Bodmer, R. Aquino, and M. H Valqui, eds., *Manejo de Fauna Silvestre en la Amazonia*, pp. 105–9. La Paz, Bolivia: OFAVIM.

Tshombe, R. 1994. Intégration des populations locales dans l'aménagement et la gestion de la RFO. Unpublished Report. Wildlife Conservation Society, Bronx, NY.

Turnbull, C. M. 1961. *The Forest People*. London: Chatto and Windus.

Turnbull, C. M. 1965. *Wayward Servants: The Two Worlds of the African Pygmies*. Garden City, NY: Natural History Press.

Tyndale-Biscoe, C. H. and R. B. Makenzie 1976. Reproduction in *Didelphis marsupialis* and *D. albiventris* in Colombia. *J Mammalogy* 57: 249–65.

Ulijaszek, S. J. and S. P. Poraituk 1993. "Making Sago in Papua, New Guinea: Is It Worth the Effort?" In C. M. Hladik, A. Hladik, O. F. Linares, H. Pagezy, A. Semple, and M. Hadley, eds., *Tropical Forests, People and Food*, pp. 271–80. Paris: UNESCO.

UNEP (United Nations Environment Programme). 1995. *Global Biodiversity Assessment*. Cambridge, England: Cambridge University Press.

UNESCO. 1984. Action plan for biosphere reserves. *Natural Resources* 20: 1–12.

UNESCO. 1985. Action plan for biosphere reserves. *Environ Conserv* 12: 17–27.

Usher, M. B. 1976. Extensions to modes, used in renewable resource management, which incorporate an arbitrary structure. *J Environmental Management* 4: 123–40.

Valdes, L. 1928. Memoria redactada por el General Luis Valdes Cavanilles para el estudio de todas las posibilidades de explotaciones agrícolas y forestales en gran escala. Madrid.

Van Ketelhodt, H. F. 1977. The lambing interval of the blue duiker, *C. monticola* in captivity, with observations on its breeeding and care. *S Afr J Wildlife Res* 7: 41–4.

Vanoverberg, M. 1937. Negritos of Eastern Luzon. *Anthropos* 32: 905–28.

Vanoverberg, M. 1938. Negritos of Eastern Luzon. *Anthropos* 33: 119–64.

Verschuren, J. 1989. Habitats mammals and conservation in the Congo. *Bull Inst R Sci Naturelles Belg Biol* 59: 169–80.

Vickers, W. T. 1979. Cultural adapation to Amazonian habitats: The Siona-Secoya of eastern Ecuador [doctoral dissertation]. Gainesville: Department of Anthropology, University of Florida.

Vickers, W. T. 1980. "An Analysis of Amazonian Hunting Yields as a Function of Settlement Age." In W. T. Vickers and K. M. Kensinger, eds., *Working Papers on South American Indians*, pp. 7–29. Bennington, VT: Bennington College.

Vickers, W. T. 1983. "The Territorial Dimensions of Siona-Secoya and Encabellado Adaptation." In R. Hames and W. Vickers, eds., *Adaptive Responses of Native Amazonians*, pp. 451–78. New York: Academic Press.

Vickers, W. T. 1984. The faunal components of lowland South American hunting kills. *Interciencia* 9: 366–76.

Vickers, W. T. 1988. Game depletion hypothesis of Amazonian adaptation: Data from a native community. *Science* 239: 1521–2.

Vickers, W. T. 1991. "Hunting Yields and Game Composition Over Ten Years in an Amazonian Village." In J. G. Robinson and K. H. Redford, eds., *Neotropical Wildlife Use and Conservation*, pp. 53–81. Chicago: University of Chicago Press.

Vickers, W. T. 1994. From opportunism to nascent conservation: The case of the Siona-Secoya. *Hum Nature* 5: 307–37.

Villa Rojas, A. 1987. *Los Elegidos de Dios: Etnografía de los Mayas de Quintana Roo.* México: Instituto Nacional Indigenista.

Vitug, M. D. 1993. *The Politics of Logging: Power from the Forest.* Manila: Philippine Center for Investigative Journalism.

Wagley, C. 1951 [1985]. "Cultural Influences on Population: A Comparison of Two Tupí Tribes." In P. J. Lyon, ed., *Native South Americans. Ethnology of the Least Known Continent*, pp. 377–84. Prospect Heights, IL: Waveland Press [reprint].

Wagley, C. 1977 [1983]. *Welcome of Tears.* Prospect Heights, IL: Waveland Press [reprint].

Wallace, A. R. 1869. *The Malay Archipelago.* London: Macmillan & Co.

Walters, C. J. 1986. *Adaptive Management of Renewable Resources.* New York: Macmillan.

Waterman, P. G., J. A. M. Ross, E. L. Bennett, and A. G. Davies. 1988. A comparison of the floristics and leaf chemistry of the tree flora in two Malaysian rain forests, and the influence of leaf chemistry on populations of colobine monkeys in the Old World. *Biol J Linn Soc* 34: 1–32.

WCMC (World Conservation Monitoring Centre). 1996. *Checklist of CITES Species.* Cambridge, England: World Conservation Monitoring Centre.

WCS (Wildlife Conservation Society). 1995a. *Saving the Tiger: A Conservation Strategy.* WCS Policy Report No. 3. New York: Wildlife Conservation Society.

WCS. 1995b. Tangkoko-Duasudara Nature Reserve, North Sulawesi Draft Management Plan 1996–2000. Unpublished report to the Directorate of Nature Conservation Directorate-General of Forestry (PHPA), Republic of Indonesia.

WCS. 1996. The Lobéké Forest, Southeastern Cameroon, Summary of Activities. Unpublished report. Wildlife Conservation Society, New York.

WCS/SFD (Wildlife Conservation Society/Sarawak Forest Department). 1996. *A Master Plan for Wildlife in Sarawak*. New York: Wildlife Conservation Society, and Kuching, Sarawak: Sarawak Forest Department.

Webb, L. J. 1982. "The Human Face in Forest Management." In E. G. Hallsworth, ed., *Socio-economic Effects and Constraints in Tropical Forest Management*, pp. 159–75. London: Wiley & Sons.

Webber, I. L. 1980. "Social Organization and Change in Modern Yucatán." In E. H. Mosely and E. D. Terry, eds., *Yucatán: A World Apart*, pp. 172–201. Tuscaloosa: University of Alabama Press.

Wells, M. and K. Brandon. 1992. *People and Parks: Linking Protected Area Management with Local Communities*. Washington, DC: The World Bank.

Wells, M. P. 1994. "A Profile and Interim Assessment of the Annapurna Conservation Project Area, Nepal." In D. Western and R. M. Wright. eds., *Natural Connections: Perspectives in Community-based Conservation*, pp. 261–81. Washington, DC: Island Press.

Wemmer, C. M., J. West, D. Watling, L. Collins, and L. Kang. 1983. External characters of the Sulawesi palm civet, *Macrogalidia musschenbroeckii* Schlegel, 1879. *J Mammalogy* 64: 133–6.

Werner, D. 1983. "Why Do the Mekranoti Trek?" In R. Hames and W. Vickers, eds., *Adaptive Responses of Native Amazonians*, pp. 225–38. New York: Academic Press.

Werner, D. 1990. *Amazon Journey. An Anthropologist's Year Among Brazil's Mekranoti Indians*. Englewood Cliffs, NJ: Prentice-Hall.

Western, D. 1979. Size, life history and ecology in mammals. *Afr J Ecol* 17: 185–204.

Western, D., and R. M. Wright. 1994. *Natural Connections: Perspectives in Community-based Conservation*. Washington, DC: Island Press.

Wetterberg, G. B., M. Ferreira, W. L. S. Brito, V. G. Araújo. 1976. Fauna Amazonica preferida como alimento. Technical series #4, p. 17, Food and Agricultural Organization Brazil.

White, L. J. T. 1992. Vegetation history and logging disturbance: Effects on rain forest mammals in the Lope Reserve, Gabon (with special emphasis on elephants and apes) [Ph.D. dissertation]. Edinburgh, Scotland: University of Edinburgh.

White, L. J. T. 1994. Biomass of rain forest mammals in the Lopé reserve, Gabon. *J Animal Ecol* 63: 499–512.

White, L. J. T., M. E. Rogers, C. E. G. Tutin, E. A. Williamson, and M. Fernandez. 1995. Herbaceous vegetation in different forest types in the Lopé Reserve, Gabon: Implications for keystone food availability. *Afr J Ecol* 33: 124–41.

White, L. J. T., C. E. G. Tutin, and M. Fernandez. 1993. Group composition and diet of forest elephants (*Loxodonta africana cyclotis* Matschie 1900) in Lopé reserve, Gabon. *Afr J Ecol* 31: 181–99.

Whitesides, G. H., J. F. Oates, S. M. Green, and R. P. Kluberdanz. 1988. Estimating primate densities from transects in a west African rain forest: A comparison of techniques. *J Animal Ecol* 57: 345–67.

Whitmore, T. C. 1984a. A vegetation map of Malesia at scale 1: 5 million. *J Biogeogr* 11: 461–71.

Whitmore, T. C. 1984b. *Tropical Rain Forests of the Far East*, 2nd ed. Oxford: Clarendon Press.

Whitmore, T. C. 1995. "Comparing Southeast Asian and Other Tropical Rainforests." In R. B. Primack and T. E. Lovejoy, eds., *Ecology, Conservation, and Management of Southeast Asian Rainforests*, pp. 5–15. New Haven and London: Yale University Press.

Whitten, A. J., S. J. Damanik, J. Anwar, and N. Hisyam. 1987b. *The Ecology of Sumatra*. Yogyakarta, Indonesia: Gadjah Mada University Press.

Whitten, A. J., M. Mustafa, and G. S. Henderson. 1987a. *The Ecology of Sulawesi*. Yogyakarta, Indonesia: Gadjah Mada University Press.

Wilder, R. 1988. Damage report from a capricious hurricane: unequal treatment at Gilbert's hands. *U.S. News & World Report* 105(13): 68.

Wilkie, D. S. 1989. Impact of roadside agriculture on subsistence in the Ituri Forest of Northeastern Zaire. *Am J Phys Anthopol* 78: 485–94.

Wilkie, D. S. 1996. "Logging in the Congo: Implications for indigenous foragers and farmers." In L. Sponsel, T. Headland, and R. C. Bailey, eds., *Tropical Deforestation: The Human Dimension*. New York: Columbia University Press.

Wilkie, D. S. and B. Curran. 1991. Why do Mbuti hunters use nets? Ungulate hunting efficiency of bows and nets in the Ituri rain forest. *Am Anthropol* 93: 680–9.

Wilkie, D. S. and B. Curran. 1993. Historical trends in forager and farmer exchange in the Ituri Rainforest of northeastern Zaire. *Hum Ecol* 21: 389–417.

Wilkie, D. S., B. Curran, and G. A. Morelli. 1998a. Deforestation or defaunation: The impact of natural resource exploitation in the Ituri forest of northeastern Zaire. *Conserv Biol* 12: 137–47.

Wilkie, D. S., B. Curran, R. Tshombe, and G. A. Morelli. 1998b. Modeling the sustainability of subsistence farming and hunting in the Ituri Forest of Zaire. *Conserv Biol* 12: 137–47.

Wilkie, D. S and J. T. Finn. 1990. Slash-burn cultivation and mammal abundance in the Ituri Forest, Zaire. *Biotropica* 22: 90–9.

Wilkie, D. S., J. Sidle, and G. C. Boundzanga. 1992. Mechanized logging, market hunting, and a bank loan in Congo. *Conserv Biol* 6: 570–80.

Wilkie, D. S. and J. G. Sidle. 1990. Social and environmental assessment of the timber production capacity extension project of the Société Forestière Algéro-Congolaise in the People's Republic of the Congo. Unpublished report to USDA/OICD, Washington, DC.

Wilkie, D. S., J. G. Sidle, G. C. Boundzanga, S. Blake, and P. Auzel. In press. "Defaunation or Deforestation: Commercial Logging and Market Hunting in Northern Congo." In R. Fimbel, A. Grajal, and J. G. Robinson. eds., *Wildlife Logging Interactions in Tropical Forests*. New York: Columbia University Press.

Williams, A., A. Mwinyi, and J. Said. 1996. A Population Survey of the Mini-antelope Ader's duiker (*Cephalophus adersi*), Zanzibar blue duiker (*Cephalophus monticola sudevalli*), and Suni (*Neotragus moschatus moschatus*) of Unguja, Zanzibar. Unpublished report to the Commission for Natural Resources, Zanzibar.

Williams, G. 1966. *Adaptation and Natural Selection*. Princeton, NJ: Princeton University Press.

Wilson, V. J. 1987. *Action Plan for Duiker Conservation*. Bulawayo: Chipangali Wildlife Trust.

Wilson, V. J. 1990. *News Sheet No. 6: Congo Report*. Bulawayo: Chipangali Wildlife Trust.

Wilson, V. J. and H. H. Roth. 1967. The effects of tsetse control operations on common duiker in Eastern Zambia. *East Afr Wildlife J* 5: 53–64.

Wilson, V., J. Schmidt, and J. Hanks. 1984. Age determination and body growth for the common duiker *Sylvicapra grimmia* (Mammalia). *J Zool (London)* 202: 283–97.

Wilson, V. J. and B. L. P. Wilson. 1991. La chasse traditionelle et commerciales dans le sud-ouest du Congo. *Tauraco Res Rep* 4: 279–88.

Wirawan, N. 1981. *Ecological Survey of the Proposed Lore Lindu National Park, Central Sulawesi.* Ujung Pandang, Sulawesi: Universitas Hasanuddin.

Wuerthrner, G. 1987. Alaskan natives—the first ecologists? *Earth First* May: 22–4.

WWF (World Wildlife Fund.) 1980a. Morowali Nature Reserve: A plan for Conservation. Bogor, Indonesia. Unpublished report to the Directorate of Nature Conservation, Republic of Indonesia.

WWF. 1980b. Cagar-Alam Gunong Tangkoko-Dua Saudara Management Plan 1981–1986. Bogor, Indonesia. Unpublished report to the Directorate of Nature Conservation, Republic of Indonesia.

WWF. 1985. *Proposals for a Conservation Strategy for Sarawak.* Compiled by L. Chan, M. Kavanagh, Earl of Cranbrook, J. Langub and D. R. Wells. Kuching, Malaysia: WWF Malaysia.

Yalden, D. W. 1996. "Historical Dichotomies in the Exploitation of Mammals." In V. J. Taylor and N. Dunstone, eds., *The Exploitation of Mammal Populations*, pp. 16–27. London: Chapman and Hall.

Yost, J. A. 1981. "Los Waorani: Un Pueblo de la Selva." In *Ecuador: A la Sombra de los Volcanes*, pp. 97–115. Quito: LibriMundo.

Yost, J. A. and P. Kelley. 1983. "Shotguns, Blowguns and Spears: The Analysis of Technological Efficiency." In R. B. Hames and W. T. Vickers, eds., *Adaptive Responses of Native Amazonians*, pp. 198–224. New York: Academic Press.

Zuraina, M. 1982. The west mouth, Niah, in the prehistory of South-east Asia. *Sarawak Museum J* (Special Monograph no. 3) 31: 1–200.

Index